80 Chapter 5 • Sprinkler Systems

The ...
mal pressure on the system into which it is installed is ... psi (... bar) ... pressure in the system can vary from 116 psi to 134 psi (8 bar to 9.2 bar). This range is acceptable.

5.3.3 Alarm Devices.

5.3.3.1 Mechanical waterflow devices including, but not limited to, water motor gongs, shall be tested quarterly.

5.3.3.2* Vane-type and pressure switch–type waterflow devices shall be tested semiannually.

A.5.3.3.2 Data concerning reliability of electrical waterflow switches indicate no appreciable change in failure rates for those tested quarterly and those tested semiannually. Mechanical motor gongs, however, have additional mechanical and environmental failure modes and need to be tested more often.

The testing frequency for waterflow alarms was changed from quarterly to semi-annually for the 2002 edition of NFPA 25 but included vane-type devices only. For the 2008 edition of NFPA 25, the technical committee changed the test frequency for pressure switch–type waterflow devices to semi-annual to match that of vane-type devices. The initial change to a semi-annual frequency was made to correlate with the testing frequency for waterflow switches with that of *NFPA 72®, National Fire Alarm Code®*. In 1996, NFPA 72 changed the test frequency for waterflow devices from quarterly to semi-annually based on a survey of test results for these devices. Commentary Table 5.1 contains the results of this survey and clearly indicates that semi-annual testing does not illustrate a significant difference in failure rates from that of quarterly testing. Note, however, that testing on an annual versus semi-annual basis results in an 85 percent increase in failures of waterflow devices. Based on these results, semi-annual testing is appropriate.

EXHIBIT 5.11 System Pressure Gauge.

COMMENTARY TABLE 5.1 Testing Frequency Survey

Frequency of Testing	Devices Tested	Devices Failed	Percent of Failure
Quarterly	13,455	161	1.20
Semiannual	3,355	51	1.52
Annual	1,772	50	2.82
Total	18,582	262	1.41

The waterflow alarm is a key part of the sprinkler system and must be tested. Paragraphs 5.3.3.1 and 5.3.3.2 take into account that each type of alarm has its own failure modes and that some are more susceptible than others to environmental conditions. For example, water motor gongs (see Exhibit 5.12) may be affected more by environmental conditions (bird or wasp nests or freezing) than an electric alarm initiating device such as a pressure switch or vane-type flow switch.

NFPA 25 does not specify who may perform the required tests other than to require a ... general, and the licensing ... mine who is permitted to ... n 13.2.6.
... accomplished by opening ...
... ction Systems Handbook

---

112 Chapter 7 • Private Fire Service Mains

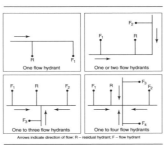

Arrows indicate direction of flow: R – residual hydrant; F – flow hydrant

EXHIBIT 7.8 Suggested Flow Test Arrangements. (Source: NFPA 291, 2007, Figure 4.3.4)

EXHIBIT 7.9 Verifying Hydrant Outlet Coefficient.

The tester proceeds to the flow hydrant and measures the opening (assuming no flow aid is attached, such as stream straighteners or playpipes). At this point, the tester also should determine the hydrant coefficient of discharge (see Exhibit 7.9). That determination can be made by inserting one hand into the hydrant and comparing the shape of the outlet to those shown in Exhibit 7.10.

After obtaining the size of the hydrant opening and the hydrant coefficient of discharge, the tester should open up the flow hydrant. When a clean steady stream of water is present, the tester should measure the flowing pressure by inserting the Pitot tube and gauge assembly (see Exhibit 7.6) into the middle of the stream, one-half the opening diameter away from the edge of the opening.

The tester should record the Pitot gauge reading while someone else simultaneously records the gauge pressure on the test hydrant. The pressure recorded by the Pitot tube assembly is used to calculate flow. Flow can be found listed in the flow tables in NFPA 291, *Recommended Practice for Fire Flow Testing and Marking of Hydrants*, or can be calculated by inserting the outlet coefficient (c), outlet diameter (d), and velocity pressure (p) into the following formula:

$$Q = 29.83 \, cd^2 \sqrt{p}$$

EXHIBIT 7.10 Three General Types of Hydrant Outlets and Their Coefficients of Discharge. (Source: NFPA 291, 2007, Figure 4.7.1)

Outlet smooth and rounded (coef. 0.90) | Outlet square and sharp (coef. 0.80) | Outlet square and projecting into barrel (coef. 0.70)

2008 Water-Based Fire Protection Systems Handbook

Water-Based Fire Protection Systems Handbook

Water-Based Fire Protection Systems Handbook

SECOND EDITION

Edited by

David R. Hague, P.E., CFPS
Principal Fire Safety and Systems Engineer, NFPA

With the complete text of the 2008 edition of NFPA® 25, *Standard for the Inspection, Testing, and Maintenance of Water-Based Fire Protection Systems*

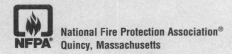

National Fire Protection Association®
Quincy, Massachusetts

Product Manager: Debra Rose
Project Editor: Irene Herlihy
Permissions Editor: Josiane Domenici
Copy Editor: Marla Marek
Composition: Modern Graphics, Inc.

Art Coordinator: Cheryl Langway
Cover Designer: Cameron, Inc.
Manufacturing Manager: Ellen Glisker
Printer: Courier/Westford

Copyright 2007
National Fire Protection Association®
One Batterymarch Park
Quincy, Massachusetts 02169-7471

All rights reserved.

Notice Concerning Liability: Publication of this handbook is for the purpose of circulating information and opinion among those concerned for fire and electrical safety and related subjects. While every effort has been made to achieve a work of high quality, neither the NFPA® nor the contributors to this handbook guarantee the accuracy or completeness of or assume any liability in connection with the information and opinions contained in this handbook. The NFPA and the contributors shall in no event be liable for any personal injury, property, or other damages of any nature whatsoever, whether special, indirect, consequential, or compensatory, directly or indirectly resulting from the publication, use of, or reliance upon this handbook.

This handbook is published with the understanding that the NFPA and the contributors to this handbook are supplying information and opinion but are not attempting to render engineering or other professional services. If such services are required, the assistance of an appropriate professional should be sought.

NFPA codes and standards are made available for use subject to Important Notices and Legal Disclaimers, which appear at the end of this handbook and can also be viewed at *www.nfpa.org/disclaimers.*

Notice Concerning Code Interpretations: This second edition of *Water-Based Fire Protection Systems Handbook* is based on the 2008 edition of NFPA 25, *Standard for the Inspection, Testing, and Maintenance of Water-Based Fire Protection Systems.* All NFPA codes, standards, recommended practices, and guides are developed in accordance with the published procedures of the NFPA by technical committees comprised of volunteers drawn from a broad array of relevant interests. The handbook contains the complete text of NFPA 25 and any applicable Formal Interpretations issued by the Association. These documents are accompanied by explanatory commentary and other supplementary materials.

The commentary and supplementary materials in this handbook are not a part of the Standard and do not constitute Formal Interpretations of the NFPA (which can be obtained only through requests processed by the responsible technical committees in accordance with the published procedures of the NFPA). The commentary and supplementary materials, therefore, solely reflect the personal opinions of the editor or other contributors and do not necessarily represent the official position of the NFPA or its technical committees.

The following are registered trademarks of the National Fire Protection Association:

NFPA®
National Fire Protection Association®
Building Construction and Safety Code® and NFPA 5000®
National Fire Alarm Code® and NFPA 72®
Life Safety Code® and NFPA 101®
NEC® and National Electrical Code®

NFPA No.: 25HB08
ISBN-10: 0-87765-751-3
ISBN-13: 978-0-87765-751-4

Library of Congress Card Control No.: 2007931937

Printed in the United States of America
08 09 10 11   5 4 3 2

Contents

List of Figures, Tables, and Exhibits   vii

Preface   xiii

About the Contributors   xv

About the Editor   xvii

PART ONE
NFPA 25, *Standard for the Inspection, Testing, and Maintenance of Water-Based Fire Protection Systems*, 2008 edition, with Commentary   1

1 Administration   3
- 1.1 Scope   3
- 1.2 Purpose   5
- 1.3 Application   7
- 1.4 Units   7

2 Referenced Publications   9
- 2.1 General   9
- 2.2 NFPA Publications   9
- 2.3 Other Publications   10
- 2.4 References for Extracts in Mandatory Sections   10

3 Definitions   11
- 3.1 General   11
- 3.2 NFPA Official Definitions   11
- 3.3 General Definitions   13
- 3.4 Deluge Foam-Water Sprinkler and Foam-Water Spray Systems Definitions   32
- 3.5 Valve Definitions   32
- 3.6 Water-Based Fire Protection System Definitions   33

4 General Requirements   43
- 4.1 Responsibility of the Property Owner or Occupant   43
- 4.2 Impairments   48
- 4.3 Corrective Action   49
- 4.4 Records   49
- 4.5 Inspection   49
- 4.6 Testing   50
- 4.7 Maintenance   53
- 4.8 Safety   54
- 4.9 Electrical Safety   56

5 Sprinkler Systems   63
- 5.1 General   63
- 5.2 Inspection   64
- 5.3 Testing   75
- 5.4 Maintenance   85
- 5.5 Component Action Requirements   89

6 Standpipe and Hose Systems   93
- 6.1 General   93
- 6.2 Inspection   95
- 6.3 Testing   95
- 6.4 Maintenance   101
- 6.5 Component Action Requirements   101

7 Private Fire Service Mains   103
- 7.1 General   103
- 7.2 Inspection   105
- 7.3 Testing   109
- 7.4 Maintenance   115
- 7.5 Component Action Requirements   115

8 Fire Pumps   119
- 8.1 General   119
- 8.2 Inspection   129
- 8.3 Testing   134
- 8.4 Reports   150
- 8.5 Maintenance   150
- 8.6 Component Replacement Testing Requirements   151

9 Water Storage Tanks   157
- 9.1 General   157
- 9.2 Inspection   163

| vi | Contents |

9.3 Testing 169
9.4 Maintenance 171
9.5 Automatic Tank Fill Valves 173
9.6 Component Action Requirements 174

10 Water Spray Fixed Systems 177

10.1 General 177
10.2 Inspection and Maintenance Procedures 184
10.3 Operational Tests 192
10.4 Ultra-High-Speed Water Spray System (UHSWSS) Operational Tests 195
10.5 Component Action Requirements 195

11 Foam-Water Sprinkler Systems 199

11.1 General 199
11.2 Inspection 203
11.3 Operational Tests 208
11.4 Maintenance 217
11.5 Component Action Requirements 219

12 Water Mist Systems 223

12.1 Responsibility of the Owner or Occupant 223
12.2 Inspection and Testing 226
12.3 Maintenance 241
12.4 Training 242

13 Valves, Valve Components, and Trim 245

13.1 General 245
13.2 General Provisions 251
13.3 Control Valves in Water-Based Fire Protection Systems 259
13.4 System Valves 263
13.5 Pressure Reducing Valves and Relief Valves 275
13.6 Backflow Prevention Assemblies 281
13.7 Fire Department Connections 283

14 Obstruction Investigation 287

14.1 General 287
14.2 Obstruction Investigation and Prevention 287
14.3 Ice Obstruction 293

15 Impairments 295

15.1 General 295
15.2 Impairment Coordinator 295
15.3 Tag Impairment System 296
15.4 Impaired Equipment 297
15.5 Preplanned Impairment Programs 298
15.6 Emergency Impairments 301
15.7 Restoring Systems to Service 302

Annexes

A Explanatory Material 305
B Forms for Inspection, Testing, and Maintenance 307
C Possible Causes of Pump Troubles 309
D Obstruction Investigation 315
E Informational References 333

PART TWO
Supplements 335

1 Microbiologically Influenced Corrosion in Fire Sprinkler Systems 337
2 Foam Environmental Issues 347
3 The Role of the Inspector 355
4 Sample Inspection, Testing, and Maintenance Program 369
5 Technical/Substantive Changes from the 2002 Edition to the 2008 Edition of NFPA 25 383

Answer Key 395

NFPA 25 Index 399

Commentary Index 407

Important Notices and Legal Disclaimers 419

List of Figures, Tables, and Exhibits

NFPA 25 Figures and Tables

Number	Caption
Table A.1.4	Metric Conversions
Figure A.3.3.9(a)	Typical Fire Hydrant Connection.
Figure A.3.3.9(b)	Flush-Type Hydrant.
Figure A.3.3.9.1	Dry Barrel Hydrant.
Figure A.3.3.9.2	Hydrant with Monitor Nozzle.
Figure A.3.3.9.3	Wall Hydrant.
Figure A.3.3.9.4	Wet Barrel Hydrant. *(Courtesy of the Los Angeles Department of Water and Power.)*
Figure A.3.3.13(a)	Hose House of Five-Sided Design for Installation over a Private Hydrant.
Figure A.3.3.13(b)	Steel Hose House of Compact Dimensions for Installation over a Private Hydrant. House is shown closed; top lifts up, and doors on front side open for complete accessibility.
Figure A.3.3.13(c)	Hose House That Can Be Installed on Legs, As Pictured, or on a Wall Near, but Not Directly over, a Private Hydrant.
Figure A.3.3.16.1	Conventional Pin Rack.
Figure A.3.3.16.2	Horizontal Rack.
Figure A.3.3.16.3	Constant Flow Hose Reel.
Figure A.3.3.16.4	Semiautomatic Hose Rack Assembly.
Figure A.3.3.22.1(a)	Standard Monitor Nozzles; Gear Control Nozzles Also Are Permitted.
Figure A.3.3.22.1(b)	Alternative Arrangement of Standard Monitor Nozzles.
Figure A.3.3.26	Pressure Vacuum Vent.
Figure A.3.3.27	Proportioner.
Figure A.3.3.27.1	Bladder Tank Proportioner.
Figure A.3.3.27.2	In-Line Balanced Pressure Proportioner.
Figure A.3.3.27.3	Line Proportioner.
Figure A.3.3.27.4	Standard Balanced Pressure Proportioner.
Figure A.3.3.27.5	Standard Pressure Proportioner.
Figure A.3.6.3	Typical Private Fire Service Main. [24:Figure A.3.3.11]
Table 5.1	Summary of Sprinkler System Inspection, Testing, and Maintenance
Figure A.5.2.4.4	Refrigerator Area Sprinkler System Used to Minimize the Chances of Developing Ice Plugs.
Figure A.5.2.7	Sample Hydraulic Nameplate.
Figure A.5.3.1.1	Sprinkler Operating Element Identification.
Figure A.5.3.4.1	Isothermal Lines — Lowest One-Day Mean Temperature (°F). [24:Figure A.10.5.1]
Table 5.3.4.1(a)	Antifreeze Solutions To Be Used If Nonpotable Water Is Connected to Sprinklers
Table 5.3.4.1(b)	Antifreeze Solutions To Be Used If Potable Water Is Connected to Sprinklers
Table 5.5.1	Summary of Component Replacement Action Requirements
Table 6.1	Summary of Standpipe and Hose Systems Inspection, Testing, and Maintenance
Table 6.2.2	Standpipe and Hose Systems
Table 6.5.1	Summary of Component Replacement Action Requirements
Table 7.1	Summary of Private Fire Service Main Inspection, Testing, and Maintenance
Table 7.2.2.1.2	Exposed Piping
Table 7.2.2.3	Mainline Strainers
Table 7.2.2.4	Dry Barrel and Wall Hydrants
Table 7.2.2.5	Wet Barrel Hydrants
Table 7.2.2.6	Monitor Nozzles
Table 7.2.2.7	Hose Houses
Table 7.5.1	Summary of Component Replacement Action Requirements
Table 8.1	Summary of Fire Pump Inspection, Testing, and Maintenance
Figure A.8.1.2(a)	Impeller Between Bearings, Separately Coupled, Single-Stage Axial (Horizontal) Split Case. *(Courtesy of Hydraulic Institute Standard for Centrifugal, Rotary and Reciprocating Pumps.)*
Figure A.8.1.2(b)	Overhung Impeller, Close-Coupled, Single-Stage, End Suction. *(Courtesy of Hydraulic Institute Standard for Centrifugal, Rotary and Reciprocating Pumps.)*
Figure A.8.1.2(c)	Turbine-Type, Vertical, Multistage, Deep Well. *(Courtesy of Hydraulic Institute Standard for Centrifugal, Rotary and Reciprocating Pumps.)*
Figure A.8.2.2	Wet Pit Suction Screen Installation.
Table A.8.2.2	Weekly Observations — Before Pumping
Table A.8.3.2.2	Weekly Observations — While Pumping

Number	Caption
Figure A.8.3.4.4(a)	Checking Angular Alignment. *(Courtesy of Hydraulic Institute Standard for Centrifugal, Rotary and Reciprocating Pumps.)*
Figure A.8.3.4.4(b)	Checking Parallel Alignment. *(Courtesy of Hydraulic Institute Standard for Centrifugal, Rotary and Reciprocating Pumps.)*
Figure A.8.3.5.3(1)	Fire Pump Retest.
Table 8.5.3	Summary of Fire Pump Inspection, Testing, and Maintenance
Table 8.6.1	Summary of Component Replacement Testing Requirements
Table 9.1	Summary of Water Storage Tank Inspection, Testing, and Maintenance
Figure A.9.3.1	Mercury Gauge.
Table 9.5.1.1	Summary of Automatic Tank Fill Valve Inspection and Testing
Table 9.6.1	Summary of Component Replacement Action Requirements
Table 10.1	Summary of Water Spray Fixed System Inspection, Testing, and Maintenance
Table 10.5.1	Summary of Component Replacement Action Requirements
Table 11.1	Summary of Foam-Water Sprinkler System Inspection, Testing, and Maintenance
Figure A.11.3.2	Foam System/Test Header Combination.
Table 11.5.1	Summary of Component Replacement Action Requirements
Table 12.2.2	Maintenance of Water Mist Systems
Table 12.3.4	Maintenance Frequencies
Table 13.1	Summary of Valves, Valve Components, and Trim Inspection, Testing, and Maintenance
Figure A.13.1(a)	Reduced Pressure Backflow Preventers (left) and Double Check Valve Assemblies (right).
Figure A.13.1(b)	Butterfly Post Indicator Valve. *(Courtesy of Henry Pratt Co.)*
Figure A.13.1(c)	Detector Check Valve.
Figure A.13.1(d)	Deluge Valve.
Figure A.13.1(e)	Dry Pipe Valve.
Figure A.13.1(f)	Dry Pipe System Accelerator. *(Courtesy of Reliable Automatic Sprinkler Co.)*
Figure A.13.1(g)	Vertical Indicator Post.
Figure A.13.1(h)	OS&Y Gate Valve.
Figure A.13.1(i)	Nonindicating-Type Gate Valve.
Figure A.15.3.1	Sample Impairment Tag.
Figure C.1	Possible Causes of Fire Pump Troubles.
Figure D.2.5(a)	Map of Hard Water Areas. *(Courtesy of Cast Iron Pipe Research Association.)*
Figure D.2.5(b)	Scale Deposition as a Function of the Alkalinity/pH Ratio.
Figure D.3.2(a)	Replacement of Elbow at End of Cross Main with a Flushing Connection Consisting of a 50 mm (2 in.) Nipple and Cap.
Figure D.3.2(b)	Connection of 65 mm (2½ in.) Hose Gate Valve with a 50 mm (2 in.) Bushing and Nipple and Elbow to 50 mm (2 in.) Cross Main.
Figure D.5.1	Arrangement for Flushing Branches from Underground Mains to Sprinkler Risers.
Table D.5.1	Flushing Rates to Accomplish Flow of 10 ft/sec (3 m/sec)
Figure D.5.3	Gridded Sprinkler System Piping.
Figure D.5.4(a)	Hydropneumatic Machine.
Figure D.5.4(b)	Schematic Diagram of Sprinkler System Showing Sequence To Be Followed Where Hydropneumatic Method Is To Be Utilized.

Commentary Exhibits and Tables

Number	Caption
C-Table 1.1	Leading Reasons for Unsatisfactory Sprinkler Performance
Exhibit 3.1	Double Check Valve Assembly.
Exhibit 3.2	Sectional Drain.
Exhibit 3.3	Air-Aspirating Foam Discharge Device. (Source: NFPA 16, 2007, Figure A.3.3.5.4)
Exhibit 3.4	Directional Non-Automatic Water Spray Nozzle.
Exhibit 3.5	Pressure Restricting Device. (Source: *Fire Protection Handbook,* 2003, Figure 10.18.8)
Exhibit 3.6	Bladder Tank Proportioning System.
Exhibit 3.7	Reduced-Pressure Principle Backflow Prevention Assembly.
Exhibit 3.8	Corrosion-Resistant Sprinkler. (Source: *Automatic Sprinkler Systems Handbook,* 2007, Exhibit 3.30)
Exhibit 3.9	Dry-Type Sprinkler.
Exhibit 3.10	ESFR Sprinkler.
Exhibit 3.11	Pendent Sprinkler.
Exhibit 3.12	Quick-Response Sprinkler.
Exhibit 3.13	Recessed Sprinkler.
Exhibit 3.14	Upright Sprinkler.
Exhibit 3.15	Combination Sprinkler/Standpipe System. [Source: NFPA 14, 2007, Figure A.7.1(a)]
Exhibit 3.16	Master Pressure Reducing Valve.
Exhibit 3.17	Horizontal Fire Pump Installation.
Exhibit 3.18	Antifreeze System. (Source: NFPA 13, 2007, Figure 7.6.3.1)
Exhibit 3.19	Deluge System. (Source: *Fire Protection Handbook,* 2003, Figure 10.11.13)
Exhibit 3.20	Dry Pipe System for Unheated Properties. (Source: *Fire Protection Handbook,* 2003, Figure 10.11.10)
Exhibit 3.21	Dry Pipe Valve. (Source: *Fire Protection Handbook,* 2003, Figure 10.11.9)
Exhibit 3.22	Typical Preaction Sprinkler System (*top*); Comparison of Operating Features for the Types of Preaction Systems (*bottom*). (Source: *Fire Protection Handbook,* 2003, Figures 10.11.11 and 10.11.12)
Exhibit 3.23	Wet Pipe Sprinkler System. (Source: *Fire Protection Handbook,* 2003, Figure 10.11.8)
Exhibit 4.1	Sample Information Sign.
Exhibit 4.2	Sample Confined Space Entry Sign. (Source: NFPA 22, 2003, Figure A.4.8.2)

Number	Caption
Exhibit 4.3	Sample Confined Space Entry Permit. (Source: OSHA, 29 CFR 1910.146, Appendix D)
Exhibit 4.4	OSHA Form 174 Material Safety Data Sheet. (Source: OSHA, 29 CFR 1910.1200, Subpart Z)
C-Table 4.1	Hazard/Risk Category Classifications
C-Table 4.2	Protective Clothing and Personal Protective Equipment (PPE) Matrix
Exhibit 4.5	Personal Protective Equipment.
Exhibit 5.1	Examples of Sprinklers Needing Replacement. *(Top row, left to right)*: (1) Recessed pendent sprinkler that has been painted. (2) Brass upright sprinkler that has been painted. (3) Pendent sprinkler installed in the upright position. *(Bottom row, left to right)*: (1) Upright sprinkler installed in the pendent position. (2) Loaded sprinkler from a candy factory coated with sugar. (3) Sprayed-on fireproofing covering an upright sprinkler.
Exhibit 5.2	Sprinkler Repaired and Returned to Service.
Exhibit 5.3	Glass Bulb Sprinkler Without Fluid.
Exhibit 5.4	Typical Obstruction.
Exhibit 5.5	Collection Device for a Leaking Sprinkler.
Exhibit 5.6	Spare Sprinkler Cabinet.
Exhibit 5.7	Pressure Gauge.
Exhibit 5.8	High-Temperature and Intermediate-Temperature Zones at Unit Heaters. (Source: NFPA 13, 2007, Figure 8.3.2.5)
Exhibit 5.9	Hydraulic Data Nameplate.
Exhibit 5.10	Plunge Test Apparatus and Test Sprinkler.
Exhibit 5.11	System Pressure Gauge.
C-Table 5.1	Testing Frequency Survey
Exhibit 5.12	Water Motor Alarm Gong *(top)* and Water Motor Gong with Wasp Nests *(bottom)*.
Exhibit 5.13	Antifreeze Testing. (1) Obtaining an antifreeze sample by draining a small quantity of antifreeze from the antifreeze system *(top)*. (2) Applying the antifreeze test sample to a refractometer *(middle)*. (3) Reading the refractometric index of the antifreeze sample by holding the refractometer up to a natural light source *(bottom)*.
Exhibit 5.14	Sprinkler Protected from Overspray with Plastic Bag.
Exhibit 5.15	Rust and Scale Removed from Dry Pipe System.
Exhibit 6.1	Hose Rack Assembly in Poor Condition.
Exhibit 6.2	Hose Rack Assembly in Proper Condition.
Exhibit 6.3	Severely Corroded Manual-Dry Standpipe.
Exhibit 6.4	Five-Year Flow Test. *(Top left)* Flow test from most remote standpipe. *(Top right)*. Flow test from second most remote standpipe. *(Bottom left)* Flow test from third most remote standpipe. *(Bottom right)* Fire department pump used for standpipe system flow test.
Exhibit 7.1	Public/Private Equipment Boundary. (Source: NFPA 24, 2007, Figure A.3.3.11)
Exhibit 7.2	Fire Protection Mainline Strainer.
Exhibit 7.3	Wet Barrel Hydrant.
Exhibit 7.4	Monitor Nozzle.

Number	Caption
Exhibit 7.5	Hydrant Flow Test. *(Top left)* A calibrated test gauge attached to the pressure hydrant. *(Top right)* Measuring outlet diameter on flow hydrant. *(Bottom left)* Flow hydrant at full flow. *(Bottom right)* Taking Pitot measurement at flow hydrant.
Exhibit 7.6	Typical Pitot Tube Assembly. (Source: *Fire Protection Handbook*, 2003, Figure 10.6.1)
Exhibit 7.7	Typical Hydrant Cap and Gauge. (Source: *Fire Protection Handbook*, 2003, Figure 10.6.2)
Exhibit 7.8	Suggested Flow Test Arrangements. (Source: NFPA 291, 2007, Figure 4.3.4)
Exhibit 7.10	Three General Types of Hydrant Outlets and Their Coefficients of Discharge. (Source: NFPA 291, 2007, Figure 4.7.1)
Exhibit 7.11	Hydrant with Indicating Pole.
Exhibit 8.1	Fire Pump Assembly.
Exhibit 8.2	Fire Pump Internal Components. (Courtesy of John Jensen)
Exhibit 8.3	Air Release Valve. (Courtesy of John Jensen)
Exhibit 8.4	Pressure Gauge.
Exhibit 8.5	Circulation Relief Valve.
Exhibit 8.6	Fire Pump Test Header.
Exhibit 8.7	Fire Pump Flowmeter.
Exhibit 8.8	Jockey Pump.
Exhibit 8.9	Pressure Switch.
Exhibit 8.10	Jockey Pump Installation with Fire Pump. (Source: NFPA 20, 2007, Figure A.5.24.4)
Exhibit 8.11	Piping Connection for Each Automatic Pressure Switch (for Electric Fire Pump and Jockey Pumps). [Source: NFPA 20, 2007, Figure A.10.5.2.1(a)]
Exhibit 8.12	Piping Connection for Pressure-Sensing Line (Electric Fire Pump). [Source: NFPA 20, 2007, Figure A.10.5.2.1(b)]
Exhibit 8.13	Ventilation Louvers.
Exhibit 8.14	Fire Pump Controller with Transfer Switch.
Exhibit 8.15	Fuel Tank for a Diesel Fire Pump.
Exhibit 8.16	Diesel Fire Pump Batteries.
Exhibit 8.17	Fire Pump Packing Gland and Drip Pocket.
Exhibit 8.18	Pressure-Sensing Line.
Exhibit 8.19	Stone in Impeller *(left)* and Size of Stone *(right)*. (Courtesy of John Jensen)
Exhibit 8.20	Handheld Tachometer Used to Measure Motor Speed in rpms.
Exhibit 8.21	Weekly Test.
Exhibit 8.22	Annual Test at Churn.
Exhibit 8.23	Discharge from Annual Fire Pump Test at Rated Capacity.
Exhibit 8.24	Discharge from Annual Test at Overload.
Exhibit 8.25	Coupling Alignment During Annual Test.
Exhibit 9.1	Steel Ground-Level Tank.
Exhibit 9.2	Elevated Storage Tank.
Exhibit 9.3	Wooden Gravity Tanks. (Courtesy of Hall-Woolford Tank Co., Inc.)
Exhibit 9.4	Wooden Gravity Tank Under Construction. (Courtesy of Hall-Woolford Tank Co., Inc.)

List of Figures, Tables, and Exhibits

Number	Caption
Exhibit 9.5	Fiberglass-Reinforced Plastic Tank.
Exhibit 9.6	Ground-Level Suction Tank. (Source: *Fire Protection Handbook,* 2003, Figure 10.2.1)
Exhibit 9.7	Steel Pressure Tank.
Exhibit 9.8	Embankment-Supported Fabric Suction Tank. [Source: NFPA 22, 2003, Figure B.1(e)]
Exhibit 9.9	Isothermal Lines — Lowest One-Day Mean Temperature.
Exhibit 9.10	Potential Obstruction Inside Steel Tank. (Courtesy of Conrady Consultant Services)
Exhibit 9.11	Anti-Vortex Plate. [Source: NFPA 22, 2003, Figure B.1(p)]
Exhibit 9.12	Electronic Water Level Indicator. (Courtesy of Potter Electric Signal Co.)
Exhibit 9.13	Electronic Water Temperature Sensor. (Courtesy of Potter Electric Signal Co.)
Exhibit 9.14	Wooden Tank with Space Below the Roof. [Source: NFPA 22, 2003, Figure B.1(i)]
Exhibit 9.15	Automatic Tank Fill Valve.
Exhibit 10.1	Discharging System.
Exhibit 10.2	Horizontal Chemical Tank Nozzle Arrangement. (Source: *Fire Protection Systems for Special Hazards,* 2004, Figure 6.3)
Exhibit 10.3	Pipe Rack Nozzle Arrangement. (Source: *Fire Protection Systems for Special Hazards*, 2004, Figure 6.16)
Exhibit 10.4	Typical Conveyor Belt Protection. [Source: NFPA 15, 2007, Figure A.7.2.3.3.1(a)].
Exhibit 10.5	Typical Transformer Water Spray System. (Source: NFPA 15, 2007, Figure A.7.4.4.1)
Exhibit 10.6	Strainer with Basket-Type Screen.
Exhibit 10.7	Testing a Heat Detector on a Water Spray System.
Exhibit 10.8	Dry Pilot Actuation Trim Options.
Exhibit 10.9	Automatic Spray Nozzles.
Exhibit 10.10	Low Velocity Non-Automatic Water Spray Nozzles *(top)* and High Velocity Non-Automatic Spray Nozzles *(bottom).*
Exhibit 10.11	Nozzle with Blow-Off Cap. (Source: *Fire Protection Handbook*, 2003, Figure 10.15.11)
Exhibit 11.1	Air Aspirating Discharge Device. (Courtesy of "Automatic" Sprinkler/Kidde Fire Fighting)
Exhibit 11.2	Hydraulically Operated Concentrate Valve.
Exhibit 11.3	Orifice Plate Proportioning Arrangement.
Exhibit 11.4	Typical Graph of Refractive Index Versus Foam Concentration. (Source: NFPA 11, 2005, Figure D.2.1.1.2)
Exhibit 11.5	Equipment Needed for Conductivity Method of Proportioning Measurement. (Source: NFPA 11, 2005, Figure D.2.1.2)
Exhibit 11.6	Typical Graph of Conductivity Versus Foam Concentration. (Source: NFPA 11, 2005, Figure 2.1.2.2)
Exhibit 12.1	Gas-Driven System with Water Stored in Tank with Air-Water Interface Greater than 10.8 ft^2 (1 m^2).
Exhibit 12.2	Pre-Engineered, Gas-Driven Single Fluid High Pressure Water Mist System with Water Stored in High Pressure Cylinders.
Exhibit 12.3	Engineered Water Mist System with Positive Displacement Pump Assembly That Has Unloader Valves on Individual Pumps and Sectional Control Valves for Separate Zones.
Exhibit 12.4	Composite Pump Curve for a Four-Motor, Eight-Pump Assembly, with All Eight Unloader Valves Set at 1595 psi (±72.5 psi) [110 bar (± 5 bar)].
Exhibit 12.5	Two Banks of Compressed Gas Cylinders for Use as Propellant for a Water Mist System.
Exhibit 13.1	Wet Pipe System Riser with Alarm Valve and Trim.
Exhibit 13.2	Butterfly Valve.
Exhibit 13.3	Detector Check Valve. (Courtesy of John Jensen)
Exhibit 13.4	OS&Y Indicating Valves.
Exhibit 13.5	Typical Valve Clearance Dimensions.
Exhibit 13.6	Discharge from a Main Drain Test.
Exhibit 13.7	Vane-Type Waterflow Switch with Retard *(left)* and Vane-Type Waterflow Alarm Switch — Small Pipe *(right).* (Courtesy of Potter Electric Signal Co.)
Exhibit 13.8	Pressure Switch.
Exhibit 13.9	Set Position for Alarm Valves *(top)* and Operating Position for Alarm Valves *(bottom).*
Exhibit 13.10	Set Position for Deluge Valves *(top)* and Operating Position for Deluge Valves *(bottom).*
Exhibit 13.11	Pressure Switch Waterflow Alarm.
Exhibit 13.12	Inoperable WPIV. (Courtesy of John Jensen)
Exhibit 13.13	Supervisory Switch.
Exhibit 13.14	Alarm Valve Inspection Points.
Exhibit 13.15	Alarm Valve Internal Inspection.
Exhibit 13.16	Preaction Valve with Trim Piping and Accessories *(left)* and Deluge Valve with Trim Piping and Accessories *(right).*
Exhibit 13.17	Externally Reset Deluge Valve.
Exhibit 13.18	Multiple System Manifold.
Exhibit 13.19	Typical Dry Pipe Valve.
C-Table 13.1	Dry System Water Delivery
Exhibit 13.20	Low Point Drain.
Exhibit 13.21	Corroded Pipe.
Exhibit 13.22	Pressure Reducing Valves for Both a Sprinkler and a Standpipe System.
Exhibit 13.23	Master Pressure Reducing Valve.
Exhibit 13.24	Pressure Relief Valve.
Exhibit 13.25	Pressure Reducing Valve Test Assembly. (Courtesy of Oliver Sprinkler Co., Inc.)
Exhibit 13.26	Typical PRV Installations. Pressure regulating device arrangement *(left)* and dual pressure regulating device arrangement *(right).* [Source: NFPA 14, 2007, Figures A.7.2.2(a) and A.7.2.2(b)]
Exhibit 13.27	Pressure Relief Valve with Sight Glass. (Courtesy of John Jensen)
Exhibit 13.28	FDC Inspection.
Exhibit 13.29	FDC Obstructed by Vegetation.
Exhibit 13.30	FDC Missing Caps.
Exhibit 13.31	FDC Showing Clear Access with Caps Intact.
Exhibit 14.1	Zebra Mussel.
Exhibit 14.2	Zebra Mussel Map. (Source: USGS)

Number	Caption
Exhibit 14.3	Work Gloves Removed from System Piping. (Source: Courtesy of John Jensen)
Exhibit 14.4	Weld Coupons Residing in System Piping.
Exhibit 14.5	Pipe Severely Corroded by MIC.
Exhibit 14.6	Localized Corrosion Caused by MIC.
Exhibit 15.1	Closed and Tagged Valve.
Exhibit 15.2	Impairment Notice. (Courtesy of Tyco/Fire & Security Simplex-Grinnell)
Exhibit D.1	Discs Attached to Piping from Which They Were Cut. (Source: *Automatic Sprinkler Systems Handbook,* 2007, Exhibit 6.15)
Exhibit D.2	Weld Coupons Found in ESFR Sprinkler Main During Obstruction Investigation.
Exhibit D.3	Work Gloves and Other Obstructing Material. (Courtesy of John Jensen)
Exhibit D.4	Pipe Infested with Zebra Mussels. (Source: Michigan Sea Grant)
Exhibit D.5	Relative Size of Adult Zebra Mussel. (Source: Michigan Sea Grant)
Exhibit D.6	Extent of Zebra Mussel Infestation.
Exhibit D.7	Pipe Affected by CaCO3 Deposits.
Exhibit D.8	Pipe Sample Showing Effect of Nodule Formation. (Courtesy of Bioindustrial Technologies, Inc.)
Exhibit D.9	Pipe Sample Showing Effect of Severe Nodule Formation and Corrosion. (Courtesy of Bioindustrial Technologies, Inc.)
Exhibit D.10	Close-Up View of Nodules. (Courtesy of Bioindustrial Technologies, Inc.)
Exhibit D.11	Pipe Sample with Pinhole Leak. (Courtesy of Bioindustrial Technologies, Inc.)
Exhibit D.12	Pipe Sample with Severe Pitting (Pits Within Pits). (Courtesy of Bioindustrial Technologies, Inc.)
Exhibit D.13	Copper Pipe with Localized Pitting. (Courtesy of Bioindustrial Technologies, Inc.)

Preface

The *Water-Based Fire Protection Systems Handbook* came about due to the interest expressed by the many individuals charged with maintaining the operational status of water-based fire protection systems. The questions presented in advisory service calls and the intriguing discussions encountered during the NFPA 25 Seminar made clear the many issues faced by those enforcing the standard or those charged with the hands-on work to keep systems functioning. These issues led to the development of the first edition and the update to the second edition. The commentary and supplements contained in this handbook are designed to help readers to understand better both the intent behind the requirements and the function of the system and components themselves.

Whereas the commentary is intended to clarify the requirements of NFPA 25, the handbook as a whole is designed to provide a training tool for those involved in performing the work and those involved in maintaining the operability of fire protection systems in their properties. Throughout the book, numerous references are made to activities that can be accomplished by the property owner. Supplements such as "The Role of the Inspector" are intended to outline the qualifications and responsibilities of the inspector who works for the fire protection contractor. Each chapter contains a review question section to permit use of this handbook as a textbook for formal or individual training. The handbook can also be used in preparation for certification or licensing testing.

History of NFPA 25

NFPA 25 is now 15 years old, but its origins date back much earlier. NFPA 25, *Standard for the Inspection, Testing, and Maintenance of Water-Based Fire Protection Systems,* now in its 5th edition, originated in 1939. A preliminary progress report of the sprinkler committee was presented during the 43rd Annual Meeting in Chicago during that year. In 1940, official adoption by the Association membership of the first edition of NFPA 13A, *Recommended Practice for the Inspection, Testing and Maintenance of Sprinkler Systems,* took place in Atlantic City during the 44th Annual Meeting. Since that time, 11 revisions have been adopted. The final edition, 1987, was officially withdrawn during the Fall 1993 Association Meeting because it was about to be replaced by the new NFPA 25.

During the Fall 1988 meeting, the NFPA membership adopted the first edition of NFPA 14A, *Recommended Practice for the Inspection, Testing, and Maintenance of Standpipe and Hose Systems,* presented by the Technical Committee on Standpipes. NFPA 14A was developed from existing industry standards and model building codes. It was short-lived, as it was withdrawn when NFPA 25 was first introduced.

Because both NFPA 13A and 14A were recommended practices, they contained no mandatory language, but offered suggestions for the proper care of sprinkler and standpipe systems. During the development of the final revision of NFPA 13A, several areas outside of the scope of the Technical Committee on Automatic Sprinklers needed to be addressed. Discussions on equipment such as underground mains, tanks, and pumps were necessary to provide guidance on the care of these systems. These issues surfaced once again during development of NFPA 14A. As a result, the Correlating Committee on Water Extinguishing

Systems recommended that a new document be developed as a standard (containing mandatory language) and that it include a chapter on each type of basic system under their committee scope. An ad hoc committee was formed in 1989 to develop a committee scope, document scope and purpose, and an outline draft of each chapter. The NFPA Standards Council approved the proposed project in 1990, a technical committee was appointed, and NFPA 25 was born. Although it met with some resistance initially (some felt that maintenance should be through a recommended practice and not mandated), the standard now has wide acceptance, with adoptions in over 35 states as of this writing. The first edition of NFPA 25 was issued in 1992, with revisions occurring in 1995, 1998, 2002, and the current 2008 edition.

Prior to the development of NFPA 25, maintenance of systems had been well documented as a significant contributing factor in the successful operation of a sprinkler system. First published in *Fire Journal* in 1965 and again in 1970, *Automatic Sprinkler Performance Tables, 1970 Edition*, indicated that sprinklers perform exceptionally well. However, when systems do fail, of the reasons reported for their failure, 65 percent of those failures can be directly attributed to lack of proper maintenance. The standard addresses each of the leading causes of failures from water control valves being shut off to inadequate water supplies (addressed in Chapter 7), obstruction to water distribution (addressed in Chapters 5 and 14), defective dry pipe valves (addressed in Chapter 13), and frozen systems (addressed in all systems chapters).

Acknowledgments

A book of this kind cannot be the product of a single individual's effort, and this handbook is no exception. I must begin by acknowledging the Technical Committee on the Inspection, Testing, and Maintenance of Water-Based Fire Protection Systems, with whom I've had the privilege of working these last twelve years. The men and women of this committee represent a wide variety of interests, from contractors and insurance representatives to building owners, fire marshals, manufacturers, and consumers, to name a few. They literally poured their hearts and souls into this effort. This work was enhanced by the work of Terry L. Victor, Tyco International and Technical Committee member and Jack Mawhinney for his contribution to the new chapter on water mist systems. I would be terribly remiss if I did not acknowledge the NFPA editorial staff who worked on this handbook: Debra Rose, product manager, who kept us all on track; Irene Herlihy, senior project editor, who along with Marla Marek, copyeditor, and David March, proofreader, put the finishing touches on this work; and Josiane Domenici, permissions editor, who acquired the permission needed to reprint numerous illustrations for this project. This book would not be a finished product without their efforts.

I also wish to thank the many contributors of artwork, which makes the handbook more useful. They are John Conrady of Conrady Consulting Services, Todd Havican of Argus Fire Control, Jack Hillman of Hall-Woolford Tank Co., Ken Linder of Swiss Re, Global Asset Protection Services and Committee Chairman, Richard Oliver of Oliver Sprinkler Co., Rebecca Pope of Bioindustrial Technologies Inc., and Ron Rispoli of Entergy Corp.

Finally, I thank my wife Cathy for her encouragement, sense of humor, and understanding for the time spent both away from home and at home while working on this project.

David R. Hague, P.E., CFPS

About the Contributors

Jack R. Mawhinney, P. Eng., FSFPE (Chapter 12)

Jack Mawhinney is a Senior Engineer with Hughes Associates, Inc., in Baltimore, Maryland, involved in research and development of water-based fire suppression systems. His work includes full-scale fire testing of suppression systems as well as failure analysis. Jack joined Hughes Associates in 1996, after 11 years as a Senior Research Officer at the Institute for Research in Construction (IRC) in Ottawa, Canada. Prior to joining IRC in 1984, Jack had over 15 years' experience in sprinkler contracting. He has authored many papers on water mist systems, including the chapters in NFPA's *Fire Protection Handbook*® and the *SFPE Handbook of Fire Protection Engineering*. Jack was chair of the committee for NFPA 750, *Standard on Water Mist Fire Protection Systems*, from 1991 to 2002. He is currently a member of NFPA 750, Water Mist, and NFPA 232, Records Storage, committees. Jack is a Fellow of the Society of Fire Protection Engineers.

Terry Victor (Supplement 3)

Terry Victor is the National Manager of Sprinkler Systems for SimplexGrinnell, a Tyco Fire & Security company. He has worked in the sprinkler industry since 1973 and has also held the positions of Designer, Design Manager, Senior Sales Representative, District Manager, and Total Service Manager. He is a certified Six Sigma Black Belt and is a NICET Level IV certified technician in two subfields. He actively serves the sprinkler industry as a member of the NFSA Engineering and Standards Committee and the NFSA Contractors Council, and is a member of six NFPA technical committees, including NFPA 13, Sprinkler System Installation Criteria; NFPA 13D & 13R, Residential Sprinkler Systems; NFPA 25, Inspection, Testing, and Maintenance of Water-Based Fire Protection Systems; NFPA 20, Fire Pumps; NFPA 16, Foam-Water Systems; NFPA 15, Water Spray Systems; and NFPA 214, Water Cooling Towers.

About the Editor

David R. Hague, P.E., CFPS, CET, is a Principal Fire Safety and Systems Engineer at the National Protection Association, where he is responsible for NFPA standards on foam extinguishing systems, gaseous extinguishing systems, standpipes, water spray fixed systems, foam-water sprinkler systems, wet and dry chemical systems, water tanks, inspection, testing, and maintenance of water-based systems, fire safety symbols, and protection of records. Dave writes and lectures about these subjects and serves as instructor for NFPA's sprinkler systems, fire pump, plan review, sprinkler hydraulics, and maintenance seminars. Dave has also written several books on the subject of water-based fire protection systems and commissioning of fire protection systems.

Dave is a Registered Professional Engineer and Certified Fire Protection Specialist and is NICET certified in Sprinkler and Special Hazards Systems Layout. Prior to joining NFPA, Dave worked as an engineering technician designing fire protection systems for the sprinkler industry.

PART ONE

NFPA® 25,
Standard for the Inspection, Testing, and Maintenance of Water-Based Fire Protection Systems,
2008 Edition, with Commentary

Part One of this handbook includes the complete text of the 2008 edition of NFPA 25, *Standard for the Inspection, Testing, and Maintenance of Water-Based Fire Protection Systems,* which consists of 15 mandatory chapters and 5 nonmandatory annexes. Working within the framework of NFPA's consensus-based codes- and standards-making process, the Technical Committee on Inspection, Testing, and Maintenance of Water-Based Fire Protection Systems prepared both the mandatory provisions found in Chapters 1 through 15 and the material found in the annexes.

The annex material is advisory or informational in nature and is provided to assist users in interpreting the mandatory standard provisions. It is not considered part of the requirements of the standard. An asterisk (*) following a standard paragraph number indicates that advisory material pertaining to that paragraph appears in Annex A. For the reader's convenience, in this handbook Annex A material appears immediately following the mandatory paragraph to which it relates.

The explanatory commentary in this handbook was prepared by the handbook editor, with the assistance of those persons mentioned in the acknowledgments, and is intended to provide the reader with an understanding of the provisions of the standard and to serve as a resource and reference for implementing the provisions of or enforcing the standard. It is not a substitute for the actual wording of the standard or the text of the many codes and standards that are incorporated by reference. The commentary immediately follows the standard text it discusses and is set in blue type for easy identification.

This edition of the handbook includes a frequently asked questions feature. The marginal FAQs are based on the questions most commonly asked of the NFPA 25 staff. The handbook also features a tool designed to help users easily identify important new or revised elements in the standard. New or revised material is identified with a "new" icon in the margin of the book next to that material. Case Studies appear throughout to illustrate "real-life" situations related to the subject matter of the standard. To assist the reader in assimilating the material, each chapter ends with Review Questions. An Answer Key is provided on p. 395.

Administration

CHAPTER 1

Chapter 1 of NFPA 25 covers the administrative requirements for the periodic inspection, testing, and maintenance of water-based fire protection systems. In addition to the scope and purpose of the standard, Chapter 1 provides guidance on the application of the standard and explains how units are expressed throughout. Note that the most important point when using NFPA 25 is that the purpose of the standard is to verify the operational status of a system and to ensure that the system will perform when needed.

1.1* Scope

This document establishes the minimum requirements for the periodic inspection, testing, and maintenance of water-based fire protection systems, including land-based and marine applications.

Section 1.1 defines the applications of NFPA 25 to help users determine if they are using the correct document. As noted, NFPA 25 contains the minimum requirements for the inspection, testing, and maintenance of a water-based system to ensure the system performs properly. The minimum requirements must be met in order to comply with this standard. Nothing in the standard is intended to prevent more frequent inspection, testing, and maintenance activities if the level of safety or performance of the system is at stake.

A.1.1 Generally accepted NFPA installation practices for water-based fire protection systems relevant to this standard are found in the following:

NFPA 13, *Standard for the Installation of Sprinkler Systems.*

NFPA 13R, *Standard for the Installation of Sprinkler Systems in Residential Occupancies up to and Including Four Stories in Height.*

NFPA 14, *Standard for the Installation of Standpipe and Hose Systems.*

NFPA 15, *Standard for Water Spray Fixed Systems for Fire Protection.*

NFPA 16, *Standard for the Installation of Foam-Water Sprinkler and Foam-Water Spray Systems.*

NFPA 20, *Standard for the Installation of Stationary Pumps for Fire Protection.*

NFPA 22, *Standard for Water Tanks for Private Fire Protection.*

NFPA 24, *Standard for the Installation of Private Fire Service Mains and Their Appurtenances.*

NFPA 750, *Standard on Water Mist Fire Protection Systems.*

For systems originally installed in accordance with one of these standards, the repair, replacement, alteration or extension of such systems should also be performed in accordance with that same standard. When original installations are based on other applicable codes or standards, repair, replacement, alteration, or extension practices should be conducted in accordance with those other applicable codes or standards.

1.1.1 This standard does not address all of the inspection, testing, and maintenance of the electrical components of the automatic fire detection equipment for preaction and deluge systems that are addressed by *NFPA 72, National Fire Alarm Code*. The inspection, testing, and maintenance required by this standard and *NFPA 72, National Fire Alarm Code*, shall be coordinated so that the system operates as intended.

Subsection 1.1.1 is new to the 2008 edition and clarifies that not all of the inspection, testing, and maintenance requirements for electrical devices that are used to actuate preaction and deluge systems are addressed by NFPA 25. For the complete requirements, reference to *NFPA 72®, National Fire Alarm Code®*, is necessary. The inspection, testing, and maintenance of electrical components, in some cases, may have to be coordinated with a licensed electrician or fire alarm technician.

1.1.2 The types of systems addressed by this standard include, but are not limited to, sprinkler, standpipe and hose, fixed water spray, and foam water. Included are the water supplies that are part of these systems, such as private fire service mains and appurtenances, fire pumps and water storage tanks, and valves that control system flow. The document also addresses impairment handling and reporting. This standard applies to fire protection systems that have been properly installed in accordance with generally accepted practices. Where a system has not been installed in accordance with generally accepted practices, the corrective action is beyond the scope of this standard. The corrective action to ensure that the system performs in a satisfactory manner shall be in accordance with the appropriate installation standard.

It is the intent of the standard to require inspection, testing, and maintenance of all water-based fire protection system regardless of the quality of the design and installation. The intent of the scope statement is to relieve the inspector of the burden of continually reverifying the design and installation of the system. The scope statement basically means that the function of the inspector is to look for signs of normal wear and tear or aging of the system and components, not to reverify acceptance criteria.

Inspections required by this standard are not intended to reveal installation flaws or code compliance violations. It is clearly not the intent of this standard to evaluate a system year after year for compliance with an installation standard. In many instances, the inspector performing the work is not necessarily trained to make this evaluation, nor is it cost effective for such an evaluation to take place each year. The inspections required by NFPA 25 are specifically intended to reveal damage or normal aging of the system and components with the goal to verify that the system will function as intended. Exceptions to this goal include modifications to the building layout, such as new or relocated walls or ceilings, changes in occupancy, or an addition is constructed. It is assumed that an existing system was reviewed and approved for compliance and commissioned properly when it was initially placed into service. Such issues as piping pitch should have been verified during commissioning of the system.

NFPA 25 is intended to address "normal wear and tear" of a system or system components. It is not the intent of NFPA 25 to place the burden of a complete system evaluation on the inspector. However, any deficiencies found during an inspection, whether the result of damage, aging, "wear and tear," or design or installation flaws, should be documented and reported to the building owner immediately.

It is not practical to expect an inspector to perform a detailed evaluation of the design and installation aspects of a fire protection system each year as the cost would be prohibitive. Such an evaluation should only be conducted by a registered or certified design professional and should only be conducted when required by the building code or insurance company. *NFPA 5000®, Building Construction and Safety Code®*, 2006 edition, as with most building codes, requires the code to apply to existing structures when specific situations occur:

> **Existing Buildings and Structures.** The provisions of this *Code* shall apply to existing buildings where any one of the following conditions applies:

(1) A change of use or occupancy classification occurs.
(2) A repair, renovation, modification, reconstruction, or addition is made.
(3) The building or structure is relocated.
(4) The building is considered damaged, unsafe, or a fire hazard.
(5) A property line that affects compliance with any provision of this *Code* is created or relocated.

[*5000:*1.3.4]

NFPA 1, *Uniform Fire Code*™, 2006 edition, has similar provisions for existing buildings:

Existing buildings that are occupied at the time of adoption of this *Code* shall remain in use provided that the following conditions are met:

(1) The occupancy classification remains the same.
(2) No condition deemed hazardous to life or property exists that would constitute an imminent danger.

[**1:**10.3.2]

A complete evaluation of the fire protection system should be necessary only where the items listed in 1.3.4 of *NFPA 5000*, 10.3.2 of NFPA 1, or your local building code apply. Notice that neither code requires an annual evaluation of the system or building, but such an evaluation is needed when changes occur.

For a description of the types of water-based fire protection systems and water supplies covered by NFPA 25, see the commentary in Chapter 3.

1.1.3 This standard shall not apply to sprinkler systems designed and installed in accordance with NFPA 13D, *Standard for the Installation of Sprinkler Systems in One- and Two-Family Dwellings and Manufactured Homes*.

Systems installed in accordance with NFPA 13D, *Standard for the Installation of Sprinkler Systems in One- and Two-Family Dwellings and Manufactured Homes*, are not covered by this standard due to their simplicity, although some of the procedures contained herein may be applied as good practice. For maintenance of these systems, see Annex A of NFPA 13D.

◀ FAQ
Does NFPA 25 apply to sprinkler systems in one- and two-family homes?

1.2* Purpose

The purpose of this document is to provide requirements that ensure a reasonable degree of protection for life and property from fire through minimum inspection, testing, and maintenance methods for water-based fire protection systems. In those cases where it is determined that an existing situation involves a distinct hazard to life or property, the authority having jurisdiction shall be permitted to require inspection, testing, and maintenance methods in excess of those required by the standard.

The purpose of NFPA 25, as explained in Section 1.2, is to make certain that the operational status of a system is maintained. This section provides the authority having jurisdiction (AHJ) with the flexibility to deal with extenuating circumstances such as product recalls or other situations specific to a particular area or project. The AHJ, who is often the fire marshal or building official, but can also be an insurance engineer or representative or the property owner, is usually the person reviewing plans or enforcing the standard.

A.1.2 History has shown that the performance reliability of a water-based fire protection system under fire-related conditions increases where comprehensive inspection, testing, and maintenance procedures are enforced. Diligence during an inspection is important. The inspection, testing, and maintenance of some items in the standard might not be practical or possible, depending on existing conditions. The inspector should use good judgment when making inspections.

> **CASE STUDY**
>
> **Sprinklers Control Fire at Power Plant**
>
> Sprinklers controlled a fire that broke out in a two-unit, coal-fired power plant when leaking hydraulic oil ignited. The sprinkler system and plant employees were so effective at extinguishing the blaze that the fire department wasn't called. An employee who brought this incident to NFPA's attention and documented it later said that the "wet pipe sprinkler system prevented major damage and lengthy production loss."
>
> The 10-story structure measured 588 by 132 ft and was of unprotected, noncombustible construction. A full-coverage, wet pipe sprinkler system protected the building, and the plant had an emergency team that was trained to the incipient level to respond to internal fire alarms.
>
> The plant and both of the 860-MV steam turbines were in full operation at 3:30 a.m., when operators noticed a large amount of smoke coming from an enclosed turbine deck. A supervisor investigated and discovered smoke and flames around a hydraulic coupling. He activated an alarm, which alerted the plant emergency team, and ordered a cutback in electrical production. Within three or four minutes, water from nine operating sprinklers brought the fire under control, and plant personnel used portable extinguishers to put out the remaining flames. The sprinklers were left on for about five minutes after extinguishment, and a fire watch was kept until repairs were made.
>
> The fire started when a pipe failed at an oil reservoir sight glass, allowing the hydraulic oil to leak onto a hot steam pipe under one of the two operating turbines. The heat from the pipe ignited the oil.
>
> One turbine was reduced in load for about 90 minutes, while the damage to pipe insulation and metal lagging was repaired. Damage to the plant, valued at $150 million, was estimated at less than $5,000. There were no injuries.
>
> Special thanks to Rich Schartel for his initiative and assistance in bringing us this report.

Sprinkler systems perform exceptionally well. Of those rare instances when they do fail, more than half (53.4 percent) of the system failures can be related to maintenance issues. Commentary Table 1.1 lists some types of failures related to maintaining the operational status of the fire protection system. NFPA 25 addresses all of the issues listed in Commentary Table 1.1.

COMMENTARY TABLE 1.1 Leading Reasons for Unsatisfactory Sprinkler Performance

Cause of Failure	Percentage of Cases
Water shut off	35.4
Inadequate maintenance	8.4
Obstruction to water distribution	8.2
System frozen	1.4

Note: Table statistics are based on 3134 fires reported to NFPA from 1925 to 1969 for which sprinkler performance was deemed unsatisfactory. Of these reported fires, 75 percent were in industrial facilities, 12 percent were in storage facilities, 5.6 percent were in stores, and 7.4 percent were in all other properties.

Source: Rohr, K. *U.S. Experience with Sprinklers: Who Has Them? How Well Do They Work?* National Fire Protection Association, Quincy, MA, 2001.

> **CASE STUDY**
>
> **Sprinkler System Not in Operation Due to Frozen Burst Pipes**
>
> Fire completely destroyed a former mill building in Rhode Island. The four-story heavy timber construction building, used for general storage, was partially vacant at the time of the fire. Dollar loss was approximately $7,000,000, and arson was determined to be the cause.
>
> Neighbors discovered the fire and notified the fire department. Approximately two minutes later, fire fighters arrived to find a fire in an empty storage area on the first floor. Flames spread horizontally and vertically as plywood-covered windows burned rapidly. Portions of the building started to collapse within 30 minutes, as Number 6 fuel oil stored in the building and a ruptured 2 in. natural gas main added fuel to the fire. It took fire fighters seven hours to contain the fire, and they spent several days wetting down hot spots.
>
> Although the building was fully covered by a wet pipe sprinkler system, the system was not in operation at the time of the fire because several pipes had frozen and burst. No automatic detection equipment was present.

1.3* Application

It is not the intent of this document to limit or restrict the use of other inspection, testing, or maintenance programs that provide an equivalent level of system integrity and performance to that detailed in this document. The authority having jurisdiction shall be consulted and approval obtained for such alternative programs.

A.1.3 An entire program of quality control includes, but is not limited to, maintenance of equipment, inspection frequency, testing of equipment, on-site fire brigades, loss control provisions, and personnel training. Personnel training can be used as an alternative even if a specific frequency differs from that specified in this standard.

The increasing availability of new products, materials, and technologies, such as ultrasound examination of piping systems (now recognized by the standard), makes the language in Sections 1.3 and A.1.3 necessary. Products or methods not specifically addressed in the standard are recognized in Section 1.3 provided that the level of system integrity and performance is not lowered.

A performance-based option for inspection, testing, and maintenance has been introduced in the 2008 edition of NFPA 25 (for details, see Chapter 4). This performance-based option is the first step in a detailed process for establishing alternate methods of frequencies for inspection, testing, and maintenance. Also, other inspection methods or procedures that may not be referenced in this standard can be used if they are demonstrated to be of equal value to the continued operation of the fire protection system. Such an alternate method must be approved by the AHJ.

Ultimately, it is the intent of this standard to verify the operational status of the fire protection system through either prescribed or performance-based methods.

◄ **FAQ**
Can the specified frequencies in NFPA 25 be altered?

1.4* Units

Metric units of measurement in this standard are in accordance with the modernized metric system known as the International System of Units (SI).

A.1.4 The liter and bar units, which are not part of but are recognized by SI, commonly are used in international fire protection. These units are provided in Table A.1.4 with their conversion factors.

TABLE A.1.4 Metric Conversions

Name of Unit	Unit Symbol	Conversion Factor
liter	L	1 gal = 3.785 L
liter per minute per square meter	L/min·m^2	1 gpm/ft^2 = 40.746 L/min·m^2
cubic decimeter	dm^3	1 gal = 3.785 dm^3
pascal	Pa	1 psi = 6894.757 Pa
bar	bar	1 psi = 0.0689 bar
bar	bar	1 bar = 10^5 Pa

Note: For additional conversions and information, see IEEE/ASTM-SI-10, *American National Standard for Use of the International System of Units (SI): The Modern Metric System.*

1.4.1 If a value for measurement as given in this standard is followed by an equivalent value in other units, the first stated shall be regarded as the requirement. A given equivalent value shall be considered to be approximate.

1.4.2 SI units have been converted by multiplying the quantity by the conversion factor and then rounding the result to the appropriate number of significant digits. Where nominal or trade sizes exist, the nominal dimension has been recognized in each unit.

All units of measure in this standard are direct mathematical conversions, except pipe diameters, in which case a "soft conversion" was used to recognize trade sizes for metric pipe.

SUMMARY

Chapter 1 of NFPA 25, as with other NFPA standards, establishes the ground rules for how and when the standard should be used. The chapter also notes particular situations to which the standard does not apply and explains how units are expressed and used throughout the document.

REVIEW QUESTIONS

1. Is it the intent of NFPA 25 to correct design or installation flaws?
2. Can the use of other inspection, test, and maintenance programs be in compliance with the requirements of NFPA 25?
3. What maintenance issues are most commonly responsible for sprinkler system failures?
4. What water-based fire protection systems are not required to be inspected, tested, and maintained under the requirements of NFPA 25?

REFERENCES CITED IN COMMENTARY

National Fire Protection Association, 1 Batterymarch Park, Quincy, MA 02169-7471.

NFPA 1, *Uniform Fire Code*™, 2006 edition.

NFPA 11, *Standard for Low-, Medium-, and High-Expansion Foam*, 2005 edition.

NFPA 13D, *Standard for the Installation of Sprinkler Systems in One- and Two-Family Dwellings and Manufactured Homes*, 2007 edition.

NFPA 72®, *National Fire Alarm Code*®, 2007 edition.

NFPA 5000®, *Building Construction and Safety Code*®, 2006 edition.

Rohr, K., "U.S. Experience with Sprinklers: Who Has Them? How Well Do They Work?," 2001.

Referenced Publications

CHAPTER 2

Chapter 2 of NFPA 25 lists the mandatory referenced publications. Annex E lists nonmandatory referenced publications. Because the information is located immediately after Chapter 1, Administration, the user is presented with the complete list of publications needed for effective use of the standard before reading the specific requirements. The provisions of the publications that are mandated by NFPA 25, *Standard for the Inspection, Testing, and Maintenance of Water-Based Fire Protection Systems*, are also requirements. Regardless of whether an actual requirement resides within NFPA 25 or is mandatorily referenced and appears only in the referenced publication, the requirement must be met to achieve compliance with NFPA 25.

2.1 General

The documents or portions thereof listed in this chapter are referenced within this standard and shall be considered part of the requirements of this document.

2.2 NFPA Publications

National Fire Protection Association, 1 Batterymarch Park, Quincy, MA 02169-7471.

NFPA 11, *Standard for Low-, Medium-, and High-Expansion Foam*, 2005 edition.
NFPA 13, *Standard for the Installation of Sprinkler Systems*, 2007 edition.
NFPA 13D, *Standard for the Installation of Sprinkler Systems in One- and Two-Family Dwellings and Manufactured Homes*, 2007 edition.
NFPA 14, *Standard for the Installation of Standpipe and Hose Systems*, 2007 edition.
NFPA 15, *Standard for Water Spray Fixed Systems for Fire Protection*, 2007 edition.
NFPA 16, *Standard for the Installation of Foam-Water Sprinkler and Foam-Water Spray Systems*, 2007 edition.
NFPA 20, *Standard for the Installation of Stationary Pumps for Fire Protection*, 2007 edition.
NFPA 22, *Standard for Water Tanks for Private Fire Protection*, 2003 edition.
NFPA 24, *Standard for the Installation of Private Fire Service Mains and Their Appurtenances*, 2007 edition.
NFPA 72®, National Fire Alarm Code®, 2007 edition.
NFPA 110, *Standard for Emergency and Standby Power Systems*, 2005 edition.
NFPA 307, *Standard for the Construction and Fire Protection of Marine Terminals, Piers, and Wharves*, 2006 edition.
NFPA 409, *Standard on Aircraft Hangars*, 2004 edition.
NFPA 750, *Standard on Water Mist Fire Protection Systems*, 2006 edition.
NFPA 1962, *Standard for the Inspection, Care, and Use of Fire Hose, Couplings, and Nozzles and the Service Testing of Fire Hose*, 2003 edition.

2.3 Other Publications

2.3.1 ASTM Publications.

ASTM International, 100 Barr Harbor Drive, P.O. Box C700, West Conshohocken, PA 19428-2959.

ASTM D 3359, *Standard Test Methods for Measuring Adhesion by Tape Test*, 1997.

2.3.2 Other Publications

Merriam-Webster's Collegiate Dictionary, 11th edition, Merriam-Webster, Inc., Springfield, MA, 2003.

2.4 References for Extracts in Mandatory Sections

NFPA 11, *Standard for Low-, Medium-, and High-Expansion Foam*, 2005 edition.
NFPA 13, *Standard for the Installation of Sprinkler Systems*, 2007 edition.
NFPA 14, *Standard for the Installation of Standpipe and Hose Systems*, 2007 edition.
NFPA 15, *Standard for Water Spray Fixed Systems for Fire Protection*, 2007 edition.
NFPA 16, *Standard for the Installation of Foam-Water Sprinkler and Foam-Water Spray Systems*, 2007 edition.
NFPA 20, *Standard for the Installation of Stationary Pumps for Fire Protection*, 2007 edition.
NFPA 24, *Standard for the Installation of Private Fire Service Mains and Their Appurtenances*, 2007 edition.
NFPA 96, *Standard for Ventilation Control and Fire Protection of Commercial Cooking Operations*, 2008 edition.
NFPA 750, *Standard on Water Mist Fire Protection Systems*, 2006 edition.
NFPA 820, *Standard for Fire Protection in Wastewater Treatment and Collection Facilities*, 2008 edition.
NFPA 1071, *Standard for Emergency Vehicle Technician Professional Qualifications*, 2006 edition.
NFPA 1141, *Standard for Fire Protection Infrastructure for Land Development in Suburban and Rural Areas*, 2008 edition.

Definitions

CHAPTER 3

Chapter 3 of NFPA 25 contains definitions of specialized terms used in the standard. This chapter also contains systems descriptions to assist the reader. The NFPA official definitions listed in Section 3.2 are used throughout NFPA codes and standards for consistency.

3.1 General

The definitions contained in this chapter shall apply to the terms used in this standard. Where terms are not defined in this chapter or within another chapter, they shall be defined using their ordinarily accepted meanings within the context in which they are used. *Merriam-Webster's Collegiate Dictionary*, 11th edition, shall be the source for the ordinarily accepted meaning.

3.2 NFPA Official Definitions

3.2.1* Approved. Acceptable to the authority having jurisdiction.

A.3.2.1 Approved. The National Fire Protection Association does not approve, inspect, or certify any installations, procedures, equipment, or materials; nor does it approve or evaluate testing laboratories. In determining the acceptability of installations, procedures, equipment, or materials, the authority having jurisdiction may base acceptance on compliance with NFPA or other appropriate standards. In the absence of such standards, said authority may require evidence of proper installation, procedure, or use. The authority having jurisdiction may also refer to the listings or labeling practices of an organization that is concerned with product evaluations and is thus in a position to determine compliance with appropriate standards for the current production of listed items.

The term *approved* does not come into play very often in this standard since during the course of an inspection, the inspector is not required to verify the use of approved components or installation drawings. Approved equipment or material should have been used for the initial installation and is only of concern during the repair or modification or replacement of an existing component.

3.2.2* Authority Having Jurisdiction (AHJ). An organization, office, or individual responsible for enforcing the requirements of a code or standard, or for approving equipment, materials, an installation, or a procedure.

A.3.2.2 Authority Having Jurisdiction (AHJ). The phrase "authority having jurisdiction," or its acronym AHJ, is used in NFPA documents in a broad manner, since jurisdictions and approval agencies vary, as do their responsibilities. Where public safety is primary, the authority having jurisdiction may be a federal, state, local, or other regional department or individual such as a fire chief; fire marshal; chief of a fire prevention bureau, labor department, or health department; building official; electrical inspector; or others having statutory authority. For insurance purposes, an insurance inspection department, rating bureau, or other insurance company representative may be the authority having jurisdiction. In many circumstances, the property owner or his or her designated agent assumes the role of the authority having jurisdiction; at government installations, the commanding officer or departmental official may be the authority having jurisdiction.

The authority having jurisdiction (AHJ) is that person or office enforcing the standard. In cases where the standard is to be legally enforced, the AHJ is usually a fire marshal or building official. The AHJ can also be an insurance company representative or the property owner. It is common for multiple AHJs to review the same project, enforcing this standard, other standards, or all.

3.2.3* Listed. Equipment, materials, or services included in a list published by an organization that is acceptable to the authority having jurisdiction and concerned with evaluation of products or services, that maintains periodic inspection of production of listed equipment or materials or periodic evaluation of services, and whose listing states that either the equipment, material, or service meets appropriate designated standards or has been tested and found suitable for a specified purpose.

A.3.2.3 Listed. The means for identifying listed equipment may vary for each organization concerned with product evaluation; some organizations do not recognize equipment as listed unless it is also labeled. The authority having jurisdiction should utilize the system employed by the listing organization to identify a listed product.

FAQ ▶
Do *classified* and *approved* mean the same as *listed*?

When system components or parts are replaced, the related installation standard, which should be followed, will require that all components be listed. As defined, a listing means that the component has been evaluated under the conditions in which it is expected to perform and that it meets certain product standards. Some testing laboratories may refer to "classified" or "approved" components, each of which meets the definition of *listed* in this standard.

3.2.4 Shall. Indicates a mandatory requirement.

The term *shall* indicates a requirement of this standard and mandates that a specific provision of NFPA 25 be followed. Mandatory requirements of NFPA 25 are found in the main body of the standard.

3.2.5 Should. Indicates a recommendation or that which is advised but not required.

The term *should* indicates a recommendation of the standard. It identifies a good idea or a best practice. When the term *should* is used with a provision of the standard, the provision is not meant to be a mandatory requirement. Recommended provisions are limited to Annexes A through E of NFPA 25. Some jurisdictions, however, consider the annex material to be enforceable. In such cases, the annex material should be treated as mandatory. It is not the intent of any NFPA standard to make annex material mandatory.
   The user of this handbook is reminded that any section number preceded by a letter (for example, A.3.4) is an annex item.

3.2.6 Standard. A document, the main text of which contains only mandatory provisions using the word "shall" to indicate requirements and which is in a form generally suitable for mandatory reference by another standard or code or for adoption into law. Nonmandatory provisions shall be located in an appendix or annex, footnote, or fine-print note and are not to be considered a part of the requirements of a standard.

3.3 General Definitions

3.3.1 Alarm Receiving Facility. The place where alarm or supervisory signals are received. This can include proprietary supervising stations, central supervising stations, remote supervising stations, or public fire service communications centers.

The need for this definition in NFPA 25 is to coordinate terminology with that of *NFPA 72®, National Fire Alarm Code®*. Alarm testing needs to be done in accordance with both documents.

3.3.2* Automatic Detection Equipment. Equipment that automatically detects heat, flame, products of combustion, flammable gases, or other conditions likely to produce fire or explosion and cause other automatic actuation of alarm and protection equipment.

A.3.3.2 Automatic Detection Equipment. Water spray systems can use fixed temperature, rate-of-rise, rate-compensation fixed temperature, optical devices, flammable gas detectors, or products of combustion detectors.

The term *automatic detection equipment* is normally used to describe an electronic detection system and should not be confused with a pilot sprinkler system where the sprinkler is used as a fixed temperature heat detector to hydraulically or pneumatically release the system actuation valve.

3.3.3 Automatic Operation. Operation without human intervention. This operation includes, but is not limited to, heat, rate of heat rise, smoke, or pressure change.

3.3.4 Deficiency. A condition in which the application of the component is not within its designed limits or specifications. [**1071,** 2006]

The term *deficiency* was added to the 2008 edition to indicate when a component must be repaired or replaced.

3.3.5 Discharge Device. A device designed to discharge water or foam-water solution in a predetermined, fixed, or adjustable pattern. Examples include, but are not limited to, sprinklers, spray nozzles, and hose nozzles. [**16,** 2003]

3.3.6 Double Check Valve Assembly (DCVA). This assembly consists of two internally loaded check valves, either spring-loaded or internally weighted, installed as a unit between two tightly closing resilient-seated shutoff valves as an assembly, and fittings with properly located resilient-seated test cocks.

Exhibit 3.1 shows a double check valve assembly.

3.3.7 Drain.

3.3.7.1 Main Drain. The primary drain connection located on the system riser and also utilized as a flow test connection.

Although flow can be estimated from a main drain connection, the main drain test is not intended to evaluate a water supply for hydraulic purposes. The intent of the main drain test is to verify that water supply valves are open or to reveal any changes in the condition of the water supply by a comparison of results to those of previous tests.

EXHIBIT 3.1 Double Check Valve Assembly.

EXHIBIT 3.2 Sectional Drain.

3.3.7.2 Sectional Drain. A drain located beyond a sectional control valve that drains only a portion of the system (e.g., a drain located beyond a floor control valve on a multistory building).

Referred to in NFPA 13, *Standard for the Installation of Sprinkler Systems*, as an "auxiliary drain," sectional drain connections can consist of a fitting with a plug or can be complete with a valve, nipple, and cap. The size and configuration of the auxiliary drain depend on the capacity of the trapped section of the system. Exhibit 3.2 illustrates a sectional drain for a dry pipe system.

3.3.8 Fire Department Connection. A connection through which the fire department can pump supplemental water into the sprinkler system, standpipe, or other system furnishing water for fire extinguishment to supplement existing water supplies.

 3.3.9* Fire Hydrant. A valved connection on a water supply system having one or more outlets and that is used to supply hose and fire department pumpers with water. [**1141,** 2008]

A.3.3.9 Fire Hydrant. See Figure A.3.3.9(a) and Figure A.3.3.9(b).

3.3.9.1* Dry Barrel Hydrant (Frostproof Hydrant). This is the most common type of hydrant; it has a control valve below the frost line between the footpiece and the barrel. A drain is located at the bottom of the barrel above the control valve seat for proper drainage after operation.

A.3.3.9.1 Dry Barrel Hydrant (Frostproof Hydrant). See Figure A.3.3.9.1.

3.3.9.2* Monitor Nozzle Hydrant. A hydrant equipped with a monitor nozzle capable of delivering more than 250 gpm (946 L/min).

A.3.3.9.2 Monitor Nozzle Hydrant. See Figure A.3.3.9.2.

3.3.9.3* Wall Hydrant. A hydrant mounted on the outside of a wall of a building, fed from interior piping, and equipped with control valves located inside the building that normally are key-operated from the building's exterior.

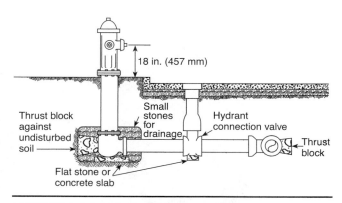

FIGURE A.3.3.9(a) Typical Fire Hydrant Connection.

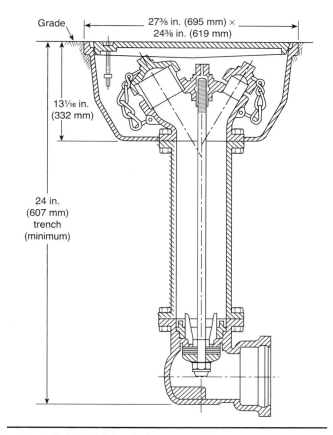

FIGURE A.3.3.9(b) Flush-Type Hydrant.

A.3.3.9.3 Wall Hydrant. See Figure A.3.3.9.3.

3.3.9.4* Wet Barrel Hydrant. A type of hydrant that sometimes is used where there is no danger of freezing weather. Each outlet on a wet barrel hydrant is provided with a valved outlet threaded for fire hose.

A.3.3.9.4 Wet Barrel Hydrant. See Figure A.3.3.9.4.

3.3.10* Foam Concentrate. A concentrated liquid foaming agent as received from the manufacturer. [**11,** 2005]

A.3.3.10 Foam Concentrate. For the purpose of this document, "foam concentrate" and "concentrate" are used interchangeably.

3.3.11 Foam Discharge Device. Any device that, when fed with a foam-water solution, produces foam. These devices shall be permitted to be non-air-aspirating (e.g., sprinklers, water nozzles) or air-aspirating (e.g., foam-water sprinklers, directional foam-water nozzles, foam nozzles). All discharge devices have a special pattern of distribution peculiar to the particular device.

Exhibit 3.3 illustrates an air-aspirating foam discharge device.

3.3.12 Hose Connection. A combination of equipment provided for connection of a hose to the standpipe system that includes a hose valve with a threaded outlet. [**14,** 2007]

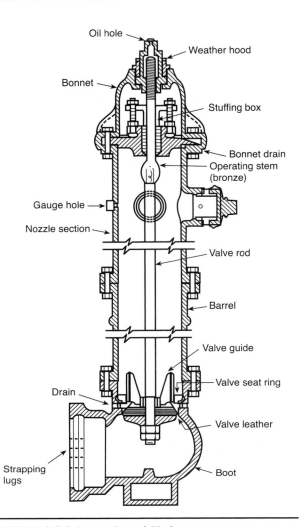

FIGURE A.3.3.9.1 Dry Barrel Hydrant.

FIGURE A.3.3.9.2 Hydrant with Monitor Nozzle.

EXHIBIT 3.3 Air-Aspirating Foam Discharge Device. (Source: NFPA 16, 2007, Figure A.3.3.5.4)

3.3.13* Hose House. An enclosure located over or adjacent to a hydrant or other water supply designed to contain the necessary hose nozzles, hose wrenches, gaskets, and spanners to be used in fire fighting in conjunction with and to provide aid to the local fire department.

A.3.3.13 Hose House. See Figure A.3.3.13(a) through Figure A.3.3.13(c).

3.3.14 Hose Nozzle. A device intended for discharging water for manual suppression or extinguishment of a fire.

3.3.15 Hose Station. A combination of a hose rack, hose nozzle, hose, and hose connection. [14, 2007]

3.3.16 Hose Storage Devices.

3.3.16.1* Conventional Pin Rack. A hose rack where the hose is folded vertically and attached over the pins.

A.3.3.16.1 Conventional Pin Rack. See Figure A.3.3.16.1.

3.3.16.2* Horizontal Rack. A hose rack where the hose is connected to the valve, then stack-folded horizontally to the top of the rack.

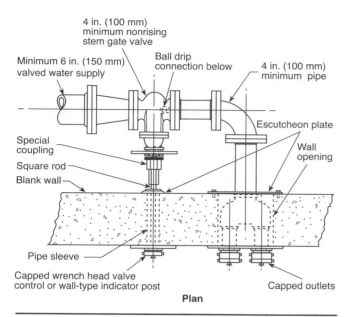

FIGURE A.3.3.9.3 Wall Hydrant.

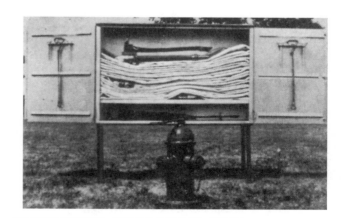

FIGURE A.3.3.9.4 Wet Barrel Hydrant. (Courtesy of the Los Angeles Department of Water and Power.)

FIGURE A.3.3.13(a) Hose House of Five-Sided Design for Installation over a Private Hydrant.

FIGURE A.3.3.13(b) Steel Hose House of Compact Dimensions for Installation over a Private Hydrant. House is shown closed; top lifts up, and doors on front side open for complete accessibility.

FIGURE A.3.3.13(c) Hose House That Can Be Installed on Legs, As Pictured, or on a Wall Near, but Not Directly over, a Private Hydrant.

FIGURE A.3.3.16.1 *Conventional Pin Rack.*

A.3.3.16.2 Horizontal Rack. See Figure A.3.3.16.2.

3.3.16.3* Hose Reel. A circular device used to store hose.

A.3.3.16.3 Hose Reel. See Figure A.3.3.16.3.

3.3.16.4* Semiautomatic Hose Rack Assembly. The same as a "conventional" pin rack or hose reel except that, after the valve is opened, a retaining device holds the hose and water until the last few feet are removed.

A.3.3.16.4 Semiautomatic Hose Rack Assembly. See Figure A.3.3.16.4.

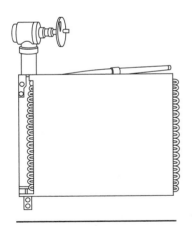

FIGURE A.3.3.16.2 *Horizontal Rack.*

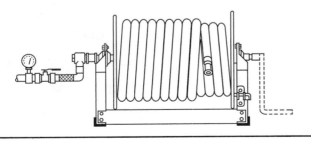

FIGURE A.3.3.16.3 *Constant Flow Hose Reel.*

FIGURE A.3.3.16.4 *Semiautomatic Hose Rack Assembly.*

3.3.17* Impairment. A condition where a fire protection system or unit or portion thereof is out of order, and the condition can result in the fire protection system or unit not functioning in a fire event.

A.3.3.17 Impairment. Temporarily shutting down a system as part of performing the routine inspection, testing, and maintenance on that system while under constant attendance by qualified personnel, and where the system can be restored to service quickly, should not be considered an impairment. Good judgment should be considered for the hazards presented.

> ***3.3.17.1 Emergency Impairment.*** A condition where a water-based fire protection system or portion thereof is out of order due to an unexpected occurrence, such as a ruptured pipe, an operated sprinkler, or an interruption of the water supply to the system.

An emergency impairment can be difficult to predict and plan for. However, some precautions can be taken to minimize system "downtime." These precautions should include the presence of spare sprinklers or nozzles, including pilot sprinklers, and repair kits for all valves, including alarm, dry pipe, deluge, preaction valves, and backflow prevention devices. (See the commentary in Chapter 15 for more information on impairments.) A reserve supply of foam concentrate is also essential to the timely recommissioning of a foam-water sprinkler or foam-water spray system.

◀ **FAQ**
What precautions can be taken to prepare for an emergency impairment?

> ***3.3.17.2 Preplanned Impairment.*** A condition where a water-based fire protection system or a portion thereof is out of service due to work that has been planned in advance, such as revisions to the water supply or sprinkler system piping.

Any system impairment can lead to the danger of a shut valve. In the 35.4 percent of system failures noted in Commentary Table 1.1 that were attributed to shut valves, it was not apparent whether the valve was shut due to maintenance (preplanned or emergency impairment) or from willful or unintentional neglect. Either way, during an impairment, it is essential to ensure that a fire prevention program is in place and that valves are opened completely during recommissioning of each system. An organized approach to monitoring system valves through a lock-out/tag-out program can ensure that no valve is left completely or partially closed. See the commentary in Chapter 15 for more information on impairments. (Also see NFPA 70E, *Standard for Electrical Safety in the Workplace*, 2004 edition, Article 120, for further information regarding lock-out/tag-out procedures.)

3.3.18 Inspection. A visual examination of a system or portion thereof to verify that it appears to be in operating condition and is free of physical damage. [**820**, 2008]

An inspection is a quick visual examination of a system or component to verify that the system or component is in apparent working order. The intent of the inspection, as described in NFPA 25, is to look for obvious signs of damage or wear and, most important, to verify that water control valves are open. Inspections are the most frequently required activity in this standard and provide a minimum assurance that the system and components are functioning. Testing and maintenance activities provide verification of system and component function.

◀ **FAQ**
Is a sprinkler system *inspection* different from a sprinkler system *evaluation* in the context of NFPA 25? If so, is *evaluation* defined by NFPA?

The term *evaluation* or *evaluate* is not defined in any NFPA code or standard, and thus the user should refer to the dictionary definition from *Merriam-Webster's Collegiate Dictionary*, 11th edition, as follows:

> **Evaluate.** To determine the significance, worth, or condition of usually by careful appraisal and study.

Evaluation of a system beyond a visual examination for operating condition and freedom from physical damage should be conducted by a registered design professional, not an inspector.

3.3.19 Inspection, Testing, and Maintenance Service. A service program provided by a qualified contractor or qualified property owner's representative in which all components unique to the property's systems are inspected and tested at the required times and necessary maintenance is provided. This program includes logging and retention of relevant records.

3.3.20 Maintenance. In water-based fire protection systems, work performed to keep equipment operable or to make repairs.

Maintenance, as defined in NFPA 25, includes not only the requirements of the standard but also practices and procedures recommended by the manufacturer.

3.3.21 Manual Operation. Operation of a system or its components through human action.

3.3.22 Nozzles.

3.3.22.1* Monitor Nozzle. A device specifically designed with large, clear waterways to provide a powerful, far-reaching stream for the protection of large amounts of combustible materials, aircraft, tank farms, and any other special hazard locations where large amounts of water need to be instantly available without the delay of laying hose lines. The nozzle is normally fitted with one of three interchangeable tips that measure $1\frac{1}{2}$ in., $1\frac{3}{4}$ in., and 2 in. (40 mm, 45 mm, and 50 mm) in diameter.

A.3.3.22.1 Monitor Nozzle. See Figure A.3.3.22.1(a) and Figure A.3.3.22.1(b).

3.3.22.2* Water Spray Nozzle. An open or automatic water discharge device that, when discharging water under pressure, will distribute the water in a specific, directional pattern.

A.3.3.22.2 Water Spray Nozzle. The selection of the type and size of spray nozzles should have been made with proper consideration given to such factors as physical character of the hazard involved, draft or wind conditions, material likely to be burning, and the general purpose of the system.

High-velocity spray nozzles, generally used in piped installations, discharge in the form of a spray-filled cone. Low-velocity spray nozzles usually deliver a much finer spray in the form of either a spray-filled spheroid or cone. Due to differences in the size of orifices or waterways in the various nozzles and the range of water particle sizes produced by each type, nozzles of one type cannot ordinarily be substituted for those of another type in an individual installation without seriously affecting fire extinguishment. In general, the higher the velocity and the coarser the size of the water droplets, the greater the effective "reach" or range of the spray.

Another type of water spray nozzle uses the deflector principle of the standard sprinkler. The angle of the spray discharge cones is governed by the design of the deflector.

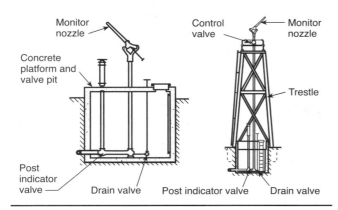

FIGURE A.3.3.22.1(a) *Standard Monitor Nozzles; Gear Control Nozzles Also Are Permitted.*

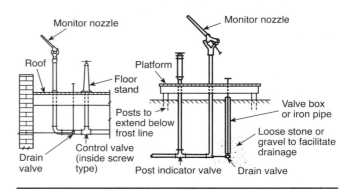

FIGURE A.3.3.22.1(b) *Alternative Arrangement of Standard Monitor Nozzles.*

Some manufacturers make spray nozzles of this type individually automatic by constructing them with heat-responsive elements as used in standard automatic sprinklers.

Exhibit 3.4 illustrates directional, non-automatic water spray nozzles. These nozzles are available with a variety of spray patterns ranging from 30 to 180 degrees. They can also be used in an automatic configuration with a fusible link with temperature ratings comparable to those of a standard spray sprinkler.

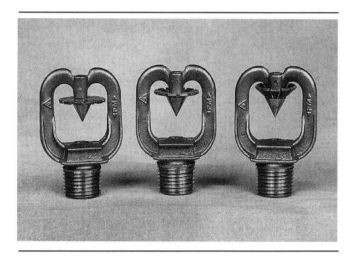

EXHIBIT 3.4 Directional Non-Automatic Water Spray Nozzles.

3.3.23 Orifice Plate Proportioning. This system utilizes an orifice plate(s) through which passes a specific amount of foam concentrate at a specific pressure drop across the orifice plate(s).

This type of system depends on a properly sized orifice plate to control the amount of foam concentrate from a foam concentrate pump.

3.3.24* Pressure Regulating Device. A device designed for the purpose of reducing, regulating, controlling, or restricting water pressure. [**14,** 2007]

A.3.3.24 Pressure Regulating Device. Examples include pressure reducing valves, pressure control valves, and pressure restricting devices.

3.3.25 Pressure Restricting Device. A valve or device designed for the purpose of reducing the downstream water pressure under flowing (residual) conditions only. [**14,** 2007]

A pressure restricting device (see Exhibit 3.5) is typically found on standpipe systems where flowing pressures exceed 100 psi for $1\frac{1}{2}$ in. (40 mm) hose connections or 175 psi for $2\frac{1}{2}$ in. (65 mm) hose connections.

3.3.26* Pressure Vacuum Vent. A venting device mounted on atmospheric foam concentrate storage vessels to allow for concentrate expansion and contraction and for tank breathing during concentrate discharge or filling. At rest (static condition), this device is closed to prevent free breathing of the foam concentrate storage tank.

A.3.3.26 Pressure Vacuum Vent. See Figure A.3.3.26.

3.3.27* Proportioners.

A.3.3.27 Proportioners. See Figure A.3.3.27.

3.3.27.1 Bladder Tank Proportioner.* This system is similar to a standard pressure proportioner, except the foam concentrate is contained inside a diaphragm bag that is

22 Chapter 3 • Definitions

EXHIBIT 3.5 *Pressure Restricting Device. (Source: Fire Protection Handbook, 2003, Figure 10.18.8)*

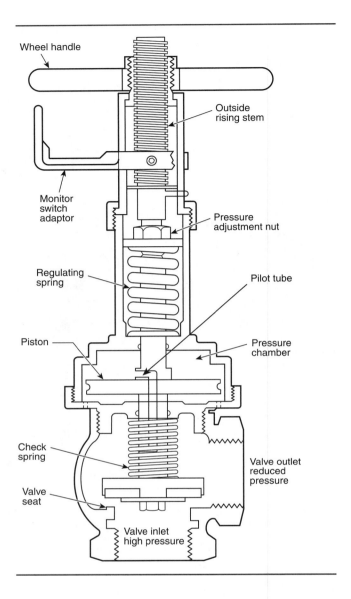

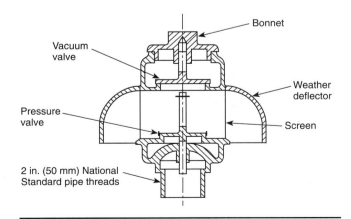

FIGURE A.3.3.26 *Pressure Vacuum Vent.*

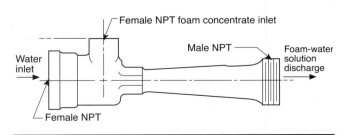

FIGURE A.3.3.27 *Proportioner.*

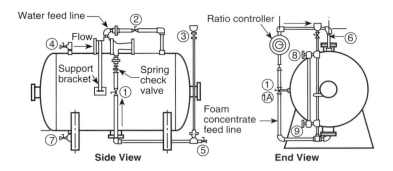

Valve no.	Valve description	Normal position	
	Description	Manual system	Auto system
1	Concentrate shutoff	Closed	Closed
1A	Auto. conc. shutoff	N/A	Closed
2	Water pres. shutoff	Open	Open
3	Fill cup shutoff	Closed	Closed
4	Tank water vent	Closed	Closed
5	Diaph. conc. vent	Closed	Closed
6	Water fill	Closed	Closed
7	Concentrate drain/fill	Closed	Closed
8	Upr. sight gauge (opt.)	Closed	Closed
9	Lwr. sight gauge (opt.)	Closed	Closed

FIGURE A.3.3.27.1 *Bladder Tank Proportioner.*

contained inside a pressure vessel. Operation is the same as a standard pressure proportioner, except that, because of the separation of the foam concentrate and water, this system can be used with all foam concentrates, regardless of specific gravity.

A.3.3.27.1 Bladder Tank Proportioner. See Figure A.3.3.27.1.

The bladder tank proportioner relies on water pressure externally applied to a bladder installed inside the tank as a means of expelling the foam concentrate. This arrangement allows relatively accurate proportioning over a wide range of flow. It is economical, since external power is not required, but is limited by size. See Figure A.3.3.27.1 for details of a bladder tank proportioner. Exhibit 3.6 shows a bladder tank proportioning system.

EXHIBIT 3.6 *Bladder Tank Proportioning System.*

3.3.27.2* In-Line Balanced Pressure Proportioner. This system is similar to a standard balanced pressure system, except the pumped concentrate pressure is maintained at a fixed preset value. Balancing of water and liquid takes place at individual proportioners located in the system riser or in segments of multiple systems.

A.3.3.27.2 In-Line Balanced Pressure Proportioner. See Figure A.3.3.27.2.

Balanced pressure proportioning relies on the use of a separate pump for pressurizing foam concentrate. This system can be used across a wide range of flow rates.

3.3.27.3* Line Proportioner. This system uses a venturi pickup-type device where water passing through the unit creates a vacuum, thereby allowing foam concentrate to be picked up from an atmospheric storage container.

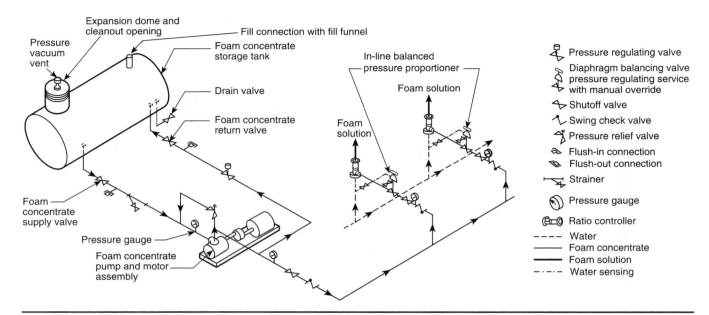

FIGURE A.3.3.27.2 In-Line Balanced Pressure Proportioner.

A.3.3.27.3 Line Proportioner. See Figure A.3.3.27.3.

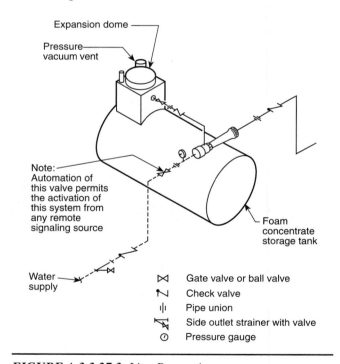

FIGURE A.3.3.27.3 Line Proportioner.

The line proportioner relies on venturi action. In other words, foam concentrate is drawn into the water stream as water flows through the pipe, to draft foam concentrate from a tank. Foam proportioning is sensitive to fluctuations in flow and pressure; variations in flow and pressure will result in incorrect concentrations of foam concentrate. See Figure A.3.3.27.3 for line proportioner details.

3.3.27.4* Standard Balanced Pressure Proportioner. This system utilizes a foam concentrate pump. Foam concentrate is drawn from an atmospheric storage tank, is pressurized by the pump, and passes back through a diaphragm balancing valve to the storage tank. Water- and foam concentrate-sensing lines are directed to the balancing valve and maintain the foam liquid at a pressure equal to that of the water pressure. The two equal pressures are fed to the proportioner proper and are mixed at a predetermined rate.

A.3.3.27.4 Standard Balanced Pressure Proportioner. See Figure A.3.3.27.4.

3.3.27.5* Standard Pressure Proportioner. This system uses a pressure vessel containing foam concentrate. Water is supplied to the proportioner, which directs an amount of the supply downward onto the contained concentrate, thereby pressurizing the tank. Pressurized concentrate then is forced through an orifice back into the flowing water stream. This type of system is applicable for use with foam concentrates having a specific gravity substantially higher than water. It is not applicable for use with foam concentrates with a specific gravity at or near that of water.

A.3.3.27.5 Standard Pressure Proportioner. See Figure A.3.3.27.5.

Similar to the bladder tank proportioner, the standard pressure proportioner relies on a viscosity of the foam concentrate that is higher than water. Water pressure is used to move the foam concentrate through a proportioner and mix it with the water stream. For details of this arrangement, see Figure A.3.3.27.5.

3.3.28 Qualified. A competent and capable person or company that has met the requirements and training for a given field acceptable to the AHJ. [**96,** 2008]

The use of the term *qualified* in this standard is intended to be general in nature. Nowhere in NFPA 25 does the standard require licensing of the individual or company performing the work, because many states do not require licensing and because licensing laws can and do vary considerably from one state to the next. As a result, NFPA 25 simply requires an inspector or service provider to possess the appropriate training in order to perform the work specified herein. It is important to note that many licensing laws specifically state that an individual who is *engaged in the business of providing this service for a fee* must be licensed. This stipulation means that for properties that have a qualified maintenance staff, licensing is not necessarily needed for work that is done "in-house." Before engaging in the business of performing any inspection, testing, or maintenance work for a fee, the local or state fire marshal's office should be contacted for any licensing requirements.

◀ FAQ
Does NFPA 25 require an inspector or service company to be licensed?

3.3.29 Reduced-Pressure Principle Backflow Prevention Assembly (RPBA). Two independently acting check valves together with a hydraulically operating, mechanically independent pressure differential relief valve located between the check valves and below the first check valve. These units are located between two tightly closed resilient-seated shutoff valves, as an assembly, and are equipped with properly located resilient-seated test cocks.

Exhibit 3.7 shows a reduced-pressure principle backflow prevention assembly (RPBA).

3.3.30 Sprinklers.

3.3.30.1 Corrosion-Resistant Sprinkler. A sprinkler fabricated with corrosion-resistant material, or with special coatings or platings, to be used in an atmosphere that would normally corrode sprinklers. [**13,** 2007]

Exhibit 3.8 shows a corrosion-resistant sprinkler.

3.3.30.2 Dry Sprinkler. A sprinkler secured in an extension nipple that has a seal at the inlet end to prevent water from entering the nipple until the sprinkler operates. [**13,** 2007]

Exhibit 3.9 shows a dry-type sprinkler. Dry sprinklers are available as upright, pendent, or sidewall types.

26 Chapter 3 • Definitions

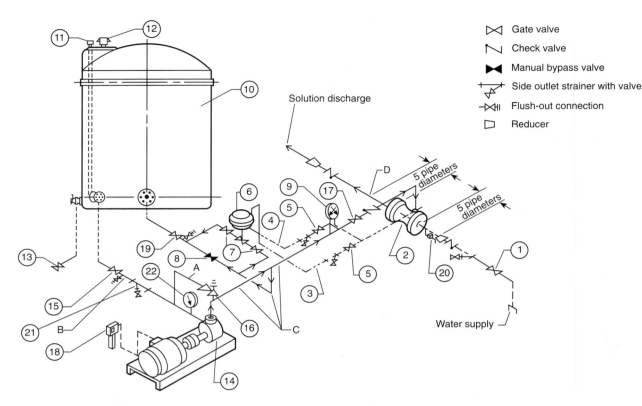

Legend:
1. Water supply valve (normally closed)
2. Ratio controller
3. Water balance line — minimum ³⁄₁₆ in. (5 mm) I.D. pipe or tubing recommended
4. Concentrate balance line — minimum ³⁄₁₆ in. (5 mm) I.D. pipe or tubing recommended
5. Sensing line valves (normally open)
6. Diaphragm control valve — automatic pressure balance — must be in vertical position
7. Block valves (normally open)
8. Manual bypass valve (normally open)
9. Water and concentrate pressure gauge (duplex)
10. Foam concentrate storage tank
11. Concentrate storage tank fill connection
12. Pressure vacuum vent
13. Concentrate storage tank drain valve (normally closed)
14. Foam concentrate pump and motor
15. Concentrate pump supply valve (normally open)
16. Pressure relief valve (setting as required by system)
17. Concentrate pump discharge valve (normally open)
18. Electric motor starter and switch
19. Concentrate return line valve (normally open)
20. Ball drip valve — ¾ in. (20 mm) (install in horizontal position)
21. Strainer with valved side outlet
22. Compound gauge

Operation:
Start concentrate pump (18). Open water supply valve (1). Open concentrate pump discharge valve (17). Equal gauge readings then maintained at (9) by the automatic valve (6). For manual operation, valves (7) can be closed and equal gauge readings maintained by regulating valve (8) manually.

System Automation:
By automating certain valves, the balanced pressure proportioning system can be activated from any remote signaling source.

- Water supply valve (1), normally closed, to be automatically operated;
- Concentrate pump discharge valve (17), normally closed, to be automatically operated;
- Electric motor starter switch (18) to be automatically operated.

FIGURE A.3.3.27.4 *Standard Balanced Pressure Proportioner.*

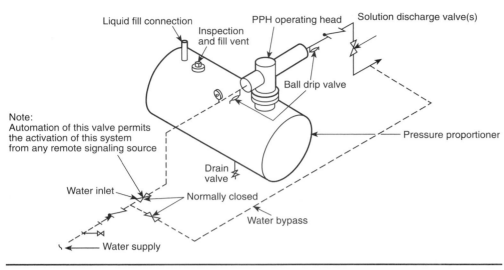

FIGURE A.3.3.27.5 *Standard Pressure Proportioner.*

EXHIBIT 3.7 *Reduced-Pressure Principle Backflow Prevention Assembly.*

EXHIBIT 3.8 *Corrosion-Resistant Sprinkler. (Source: Automatic Sprinkler Systems Handbook, 2007, Exhibit 3.30)*

EXHIBIT 3.9 Dry-Type Sprinkler.

3.3.30.3 Early Suppression Fast-Response (ESFR) Sprinkler. A type of fast-response sprinkler that meets the criteria of 3.6.1(a)(1) of NFPA 13, *Standard for the Installation of Sprinkler Systems*, and is listed for its capability to provide fire suppression of specific high-challenge fire hazards. [**13,** 2007]

Exhibit 3.10 shows an early suppression fast-response (ESFR) sprinkler.

3.3.30.4 Extended Coverage Sprinkler. A type of spray sprinkler with maximum coverage areas as specified in Sections 8.8 and 8.9 of NFPA 13. [**13,** 2007]

3.3.30.5 Large Drop Sprinkler. A type of specific application control mode sprinkler that is capable of producing characteristic large water droplets and that is listed for its capability to provide fire control of specific high-challenge fire hazards. [**13,** 2007]

3.3.30.6 Nozzles. A device for use in applications requiring special water discharge patterns, directional spray, or other unusual discharge characteristics. [**13,** 2007]

3.3.30.7 Old-Style/Conventional Sprinkler. A sprinkler that directs from 40 percent to 60 percent of the total water initially in a downward direction and that is designed to be installed with the deflector either upright or pendent. [**13,** 2007]

3.3.30.8 Open Sprinkler. A sprinkler that does not have actuators or heat-responsive elements. [**13,** 2007]

3.3.30.9 Ornamental/Decorative Sprinkler. A sprinkler that has been painted or plated by the manufacturer. [**13,** 2007]

3.3.30.10 Pendent Sprinkler. A sprinkler designed to be installed in such a way that the water stream is directed downward against the deflector. [**13,** 2007]

Exhibit 3.11 shows a pendent sprinkler.

 3.3.30.11 Quick-Response Early Suppression (QRES) Sprinkler. A type of quick-response sprinkler that meets the criteria of 3.6.1(a)(1) of NFPA 13 and is listed for its capability to provide fire suppression of specific fire hazards. [**13,** 2007]

3.3.30.12 Quick-Response Extended Coverage Sprinkler. A type of quick-response sprinkler that meets the criteria of 3.6.1(a)(1) of NFPA 13 and complies with the extended protection areas defined in Chapter 8 of NFPA 13. [**13,** 2007]

3.3.30.13 Quick-Response (QR) Sprinkler. A type of spray sprinkler that meets the fast response criteria of 3.6.1(a)(1) of NFPA 13 and is listed as a quick-response sprinkler for its intended use. [**13,** 2007]

Exhibit 3.12 shows a quick-response (QR) sprinkler. Note the smaller diameter of the glass bulb as compared to the standard-response upright sprinkler shown in Exhibit 3.14.

3.3.30.14 Recessed Sprinkler. A sprinkler in which all or part of the body, other than the shank thread, is mounted within a recessed housing. [**13,** 2007]

Exhibit 3.13 shows a recessed sprinkler.

3.3.30.15 Residential Sprinkler. A type of fast-response sprinkler that meets the criteria of 3.6.1(a)(1) of NFPA 13 that has been specifically investigated for its ability to enhance survivability in the room of fire origin and is listed for use in the protection of dwelling units. [**13,** 2007]

3.3.30.16 Special Sprinkler. A sprinkler that has been tested and listed as prescribed in 8.4.8 of NFPA 13. [**13,** 2007]

3.3.30.17 Spray Sprinkler. A type of sprinkler listed for its capability to provide fire control for a wide range of fire hazards. [**13,** 2007]

3.3.30.18 Standard Spray Sprinkler. A spray sprinkler with maximum coverage areas as specified in Sections 8.6 and 8.7 of NFPA 13. [**13,** 2007]

EXHIBIT 3.10 ESFR Sprinkler.

EXHIBIT 3.11 Pendent Sprinkler.

EXHIBIT 3.12 Quick-Response Sprinkler.

EXHIBIT 3.13 Recessed Sprinkler.

3.3.30.19 Upright Sprinkler. A sprinkler designed to be installed in such a way that the water spray is directed upwards against the deflector. [**13,** 2007]

Exhibit 3.14 shows a standard-response upright sprinkler. Note the larger diameter glass bulb as compared to the quick-response sprinkler shown in Exhibit 3.12.

3.3.31* Standpipe System. An arrangement of piping, valves, hose connections, and allied equipment installed in a building or structure, with the hose connections located in such a manner that water can be discharged in streams or spray patterns through attached hose and nozzles, for the purpose of extinguishing a fire, thereby protecting a building or structure and its contents in addition to protecting the occupants. [**14,** 2007]

These standpipe systems differ substantially from the hose connections defined previously in that they require a much higher flow and pressure to function properly. Therefore, standpipe systems require different inspection, testing, and maintenance procedures, as referenced in Chapter 6 of NFPA 25.

Exhibit 3.15 illustrates a combined sprinkler/standpipe system. The standpipe system is a single-zone type.

A.3.3.31 Standpipe System. This is accomplished by means of connections to water supply systems or by means of pumps, tanks, and other equipment necessary to provide an adequate supply of water to the hose connections.

3.3.31.1 Dry Standpipe. A standpipe system designed to have piping contain water only when the system is being utilized.

The application of a dry standpipe system is restricted to areas subject to freezing. Manual dry standpipe systems have the same limitations as manual wet systems. These limitations are as follows:

1. Manual standpipe systems must not be used in high-rise buildings.
2. Each hose connection for manual standpipes must be provided with a conspicuous sign that reads MANUAL STANDPIPE FOR FIRE DEPARTMENT USE ONLY.
3. Manual standpipes must not be used for Class II or Class III systems.

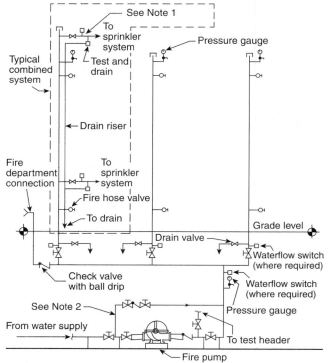

EXHIBIT 3.15 Combination Sprinkler/Standpipe System. [Source: NFPA 14, 2007, Figure A.7.1(a)]

EXHIBIT 3.14 Upright Sprinkler.

2008 Water-Based Fire Protection Systems Handbook

3.3.31.1.1 Class I System. A system that provides 2½ in. (65 mm) hose connections to supply water for use by fire departments and those trained in handling heavy fire streams. [**14,** 2007]

3.3.31.1.2 Class II System. A system that provides 1½ in. (40 mm) hose stations to supply water for use primarily by trained personnel or by the fire department during initial response. [**14,** 2007]

3.3.31.1.3 Class III System. A system that provides 1½ in. (40 mm) hose stations to supply water for use by trained personnel and 2½ in. (65 mm) hose connections to supply a larger volume of water for use by fire departments and those trained in handling heavy fire streams. [**14,** 2007]

Hose may or may not be installed in a Class III standpipe system. NFPA 14, *Standard for the Installation of Standpipe and Hose Systems*, permits the omission of hose on Class III systems if the building is completely sprinklered and a 2½ in. × 1½ in. reducer, cap, and chain are provided on the 2½ in. fire department valve. Where hose is provided, the hose must be inspected, tested, and maintained in accordance with the requirements of Chapter 6.

◄ **FAQ**
Is hose required to be installed in a Class III standpipe system?

3.3.31.2 Manual Standpipe. Standpipe system that relies exclusively on the fire department connection to supply the system demand.

3.3.31.3 Wet Standpipe System. A standpipe system having piping containing water at all times. [**14,** 2007]

Wet systems can be either automatic or manual depending on the application. Automatic systems must provide a water flow of 250 gpm (946 L/min) for each of the top two hose valves at the hydraulically most demanding riser at a pressure of 100 psi (6.9 bar).

A manual system simply has city water pressure applied to keep it full of water. The fire department is expected to supply the desired pressure while using a manual system. Manual systems are restricted to use in buildings that are not classified as high-rise and systems where occupant use is not intended (such as Class I).

3.3.32* Strainer. A device capable of removing from the water all solids of sufficient size that are obstructing water spray nozzles.

A.3.3.32 Strainer. There are two types of strainers. Pipeline strainers are used in water supply connections. These are capable of removing from the water all solids of sufficient size to obstruct the spray nozzles [⅛ in. (3.2 mm) perforations usually are suitable]. Pipeline strainer designs should incorporate a flushout connection or should be capable of flushing through the main drain.

Individual strainers for spray nozzles, where needed, are capable of removing from the water all solids of sufficient size to obstruct the spray nozzle that they serve.

3.3.33 Supervision. In water-based fire protection systems, a means of monitoring system status and indicating abnormal conditions.

3.3.34 Testing. A procedure used to determine the status of a system as intended by conducting periodic physical checks on waterbased fire protection systems such as waterflow tests, fire pump tests, alarm tests, and trip tests of dry pipe, deluge, or preaction valves. These tests follow up on the original acceptance test at intervals specified in the appropriate chapter of this standard.

3.3.35* Water Spray. Water in a form having a predetermined pattern, particle size, velocity, and density discharge from specially designed nozzles or devices. [**15,** 2007]

A.3.3.35 Water Spray. Water spray fixed systems are usually applied to special fire protection problems, since the protection can be specifically designed to provide for fire control, extinguishment, or exposure protection. Water spray fixed systems are permitted to be independent of, or supplementary to, other forms of protection.

3.3.36 Water Supply. A source of water that provides the flows [gal/min (L/min)] and pressures [psi (bar)] required by the water-based fire protection system.

FAQ ▶
What are some examples of acceptable water supplies?

Water supplies that are acceptable for use in fire protection systems include the following:

- Connections to water works systems
- Pumps
- Tanks
- Natural bodies of water, such as rivers and lakes

3.4 Deluge Foam-Water Sprinkler and Foam-Water Spray Systems Definitions

3.4.1 Foam-Water Spray System. A special system that is pipe-connected to a source of foam concentrate and to a water supply. The system is equipped with foam-water spray nozzles for extinguishing agent discharge (foam followed by water or in reverse order) and for distribution over the area to be protected. System operation arrangements parallel those for foam-water sprinkler systems as described in the definition of Foam-Water Sprinkler System. [**16,** 2003]

Foam-water spray systems are almost identical to water spray fixed systems; however, foam systems have foam-generating equipment attached to the feed main piping. The discharge device or nozzle must be listed for use with foam concentrate.

3.4.2 Foam-Water Sprinkler System. A special system that is pipe-connected to a source of foam concentrate and to a water supply. The system is equipped with appropriate discharge devices for extinguishing agent discharge and for distribution over the area to be protected. The piping system is connected to the water supply through a control valve that usually is actuated by operation of automatic detection equipment that is installed in the same areas as the sprinklers. When this valve opens, water flows into the piping system, foam concentrate is injected into the water, and the resulting foam solution discharging through the discharge devices generates and distributes foam. Upon exhaustion of the foam concentrate supply, water discharge follows and continues until shut off manually. Systems can be used for discharge of water first, followed by discharge of foam for a specified period, and then followed by water until manually shut off. Existing deluge sprinkler systems that have been converted to the use of aqueous film-forming foam or film-forming fluoroprotein foam are classified as foam-water sprinkler systems. [**16,** 2003]

Foam-water sprinkler systems and deluge systems are almost identical; they differ in that foam-water sprinkler systems have foam-generating equipment attached to the feed main piping. Like foam-water spray systems, foam-water sprinkler systems must employ discharge devices or sprinklers that are listed for use with foam concentrate. See the commentary in Chapter 11 for more information on foam-water sprinkler systems.

3.5 Valve Definitions

3.5.1* Control Valve. A valve controlling flow to water-based fire protection systems. Control valves do not include hose valves, inspector's test valves, drain valves, trim valves for dry pipe, preaction and deluge valves, check valves, or relief valves.

A.3.5.1 Control Valve. Experience has shown that closed valves are the primary cause of failure of water-based fire protection systems in protected occupancies.

As indicated in Commentary Table 1.1, shut control valves account for 35.4 percent of all sprinkler system failures. Therefore, the control valve requires attention through vigorous inspection and monitoring programs. NFPA 25 has devoted an entire chapter to valves primarily for this reason. Control valves require close scrutiny, particularly when there are no mechanical (chains with locks) or electronic (tamper switches) monitoring systems. When valves are provided with only a seal, the inspection frequency is weekly. With the aforementioned mechanical or electronic monitoring system in place, the inspection frequency is relaxed to monthly.

Further, whenever a valve is operated, a main drain flow test must be performed to verify that the valve has been opened fully. It is also good practice to confirm that any control valve is open by second-party verification. See the commentary in Chapter 13 for more information on valves and valve components.

3.5.2 Deluge Valve. A water supply control valve intended to be operated by actuation of an automatic detection system that is installed in the same area as the discharge devices. Each deluge valve is intended to be capable of automatic and manual operation.

3.5.3 Hose Valve. The valve to an individual hose connection. [**14,** 2007]

3.5.4 Pressure Control Valve. A pilot-operated pressure-reducing valve designed for the purpose of reducing the downstream water pressure to a specific value under both flowing (residual) and nonflowing (static) conditions. [**14,** 2007]

3.5.5 Pressure-Reducing Valve. A valve designed for the purpose of reducing the downstream water pressure under both flowing (residual) and nonflowing (static) conditions. [**14,** 2007]

Typically, pressure control and pressure-reducing valves are found on systems that are subjected to high pressures, usually in excess of 175 psi (12.1 bar). These valves are intended to protect system components from failure due to high pressure. The valves must be periodically tested to verify that they are functioning correctly, since failure of the valve can damage system components and may present a danger to personnel.

A typical application for these valves is a high-rise building where high pressures are needed to move water to the upper floors of the building. They can be used to control pressures for entire sprinkler or standpipe systems or individual fire department hose connections. In the case of valves controlling pressure to individual fire department valves, pressures may be either pre-set at the factory or field adjustable.

> **3.5.5.1* Master Pressure Reducing Valve.** A pressure reducing valve installed to regulate pressures in an entire fire protection system and/or standpipe system zone.

A master pressure reducing valve is used to control pressure to an entire sprinkler or standpipe system (see Exhibit 3.16). Because it controls water pressure to the entire system, the valve must be tested more frequently to verify that it is functioning properly. For more information regarding the inspection, testing, and maintenance of this valve, see Chapter 13.

> **A.3.5.5.1 Master Pressure Reducing Valve.** Master PRVs are typically found downstream of a fire pump's discharge.

3.6 Water-Based Fire Protection System Definitions

3.6.1 Combined Standpipe and Sprinkler System. A system where the water piping services both 2½ in. (65 mm) outlets for fire department use and outlets for automatic sprinklers.

A combined standpipe and sprinkler system (shown in Exhibit 3.15) is one that has common piping through the feed main and riser portion of the system. A sprinkler and standpipe system that shares a common header at the water service entrance to the building does not constitute a "combined system." In all cases, the sprinkler system is designed and installed in

EXHIBIT 3.16 Master Pressure Reducing Valve.

such a way that the sprinkler system can be isolated from the standpipe system, usually by means of a floor control assembly. With this configuration, manual fire fighting can continue even after the sprinkler system is shut off.

 3.6.2 Fire Pump. A pump that is a provider of liquid flow and pressure dedicated to fire protection. [20, 2007]

Whether taking suction from a natural body of water, a storage tank, or a municipal supply, the presence of a fire pump, as shown in Exhibit 3.17, indicates that the primary water source lacks sufficient pressure to meet the fire protection system demand. The fire pump must be properly maintained to be able to function in the event of a fire. Failure of the fire pump will result in inadequate pressure to meet the demands of the fire protection system. Therefore, frequent operation of the pump along with an annual evaluation of performance is necessary. Fire pumps and related equipment add a significant cost to the installation of a fire protection system and should be maintained to protect this investment. See the commentary in Chapter 8 for more information on fire pumps.

EXHIBIT 3.17 Horizontal Fire Pump Installation.

3.6.3* Private Fire Service Main. Private fire service main, as used in this standard, is that pipe and its appurtenances on private property (1) between a source of water and the base of the system riser for water-based fire protection systems, (2) between a source of water and inlets to foam-making systems, (3) between a source of water and the base elbow of private hydrants or monitor nozzles, and (4) used as fire pump suction and discharge piping, (5) beginning at the inlet side of the check valve on a gravity or pressure tank. [**24**, 2007]

See the commentary in Chapter 7 for more information on private fire service mains.

A.3.6.3 Private Fire Service Main. See Figure A.3.6.3.

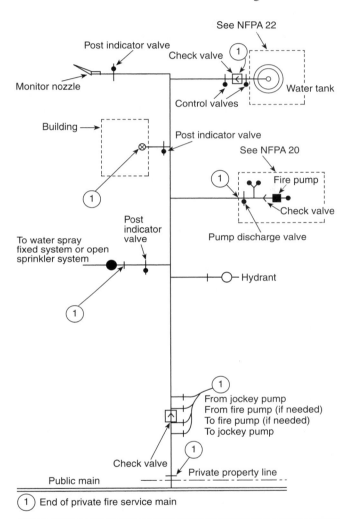

FIGURE A.3.6.3 *Typical Private Fire Service Main.* [*24:Figure A.3.3.11*]

3.6.4* Sprinkler System. For fire protection purposes, an integrated system of underground and overhead piping designed in accordance with fire protection engineering standards. The installation includes at least one automatic water supply which supplies one or more systems. The portion of the sprinkler system above ground is a network of specially sized or hydraulically designed piping installed in a building, structure, or area, generally overhead, and to which sprinklers are attached in a systematic pattern. Each system has a control valve located in the system riser or its supply piping. Each sprinkler system includes a device for actuating an alarm when the system is in operation. The system is usually activated by heat from a fire and discharges water over the fire area. [**13,** 2007]

A basic sprinkler system can be identified by the presence of either an alarm check valve or a dry pipe valve, a water supply control valve, a fire department connection, and a network of pipes and sprinklers. The NFPA 25 definition of *sprinkler system* also mentions underground pipe since the water supply is critical to the operation of the sprinkler system. See the commentary in Chapter 5 for more information on sprinkler systems.

A.3.6.4 Sprinkler System. A sprinkler system is considered to have a single system riser control valve. The design and installation of water supply facilities such as gravity tanks, fire pumps, reservoirs, or pressure tanks are covered by NFPA 20, *Standard for the Installation of Stationary Pumps for Fire Protection,* and NFPA 22, *Standard for Water Tanks for Private Fire Protection.*

> *3.6.4.1 Antifreeze Sprinkler System.* A wet pipe sprinkler system employing automatic sprinklers that are attached to a piping system that contains an antifreeze solution and that are connected to a water supply. The antifreeze solution is discharged, followed by water, immediately upon operation of sprinklers opened by heat from a fire. [**13,** 2007]

Antifreeze is one method of protecting a wet pipe system from freezing. Antifreeze systems are usually small, although there are no size limitations in NFPA 13 regarding the size of an antifreeze system, and require special attention due to the chemicals present and valves necessary for the fill connections.

Antifreeze systems must be tested to ensure that the antifreeze mixture will resist freezing at the temperatures to which the system is exposed. See Chapter 5 of NFPA 25 for details of the mixtures and test requirements for antifreeze systems. Exhibit 3.18 shows the arrangement of supply piping and valves for an antifreeze system.

EXHIBIT 3.18 Antifreeze System. (Source: NFPA 13, 2007, Figure 7.6.3.1)

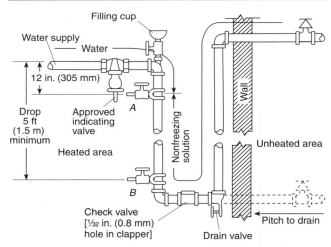

Notes:
1. Check valve shall be permitted to be omitted where sprinklers are below the level of valve *A*.
2. The 1/32 in. (0.8 mm) hole in the check valve clapper is needed to allow for expansion of the solution during a temperature rise, thus preventing damage to sprinklers.

> *3.6.4.2 Deluge Sprinkler System.* A sprinkler system employing open sprinklers that are attached to a piping system that is connected to a water supply through a valve that is opened by the operation of a detection system installed in the same areas as the sprin-

klers. When this valve opens, water flows into the piping system and discharges from all sprinklers attached thereto. [**13,** 2007]

Deluge sprinkler systems are typically used in high-hazard areas where water is required for the entire area, not only to extinguish or control a fire but also to provide exposure protection to areas adjacent to the fire. Since the deluge piping contains no water and is exposed to the atmosphere, the pipe is subject to the effects of corrosion resulting from condensation or improper draining of the system.

Deluge systems, such as the one shown in Exhibit 3.19, may require more attention than wet pipe systems due to the actuation systems used. These can be in the form of electrical detection systems or wet pilot or dry pilot sprinklers.

3.6.4.3 Dry Pipe Sprinkler System. A sprinkler system employing automatic sprinklers that are attached to a piping system containing air or nitrogen under pressure, the release of which (as from the opening of a sprinkler) permits the water pressure to open a valve known as a dry pipe valve, and the water then flows into the piping system and out the opened sprinklers. [**13,** 2007]

A dry pipe system (see Exhibit 3.20) is very similar to a wet pipe system in that the same type of pipe, fittings, hangers, and sprinklers are used. The dry pipe system, however, requires much more attention than a wet system since it is usually installed in a more hostile environment.

Exposure to fluctuating temperatures and corrosive environments can accelerate the corrosion of pipe, fittings, hangers, and sprinklers. The corrosion of the inner surfaces of pipe in a dry system is of particular concern because it can lead to the presence of obstructions, in the form of rust or scale, in the piping system. Careful observation of water discharge during testing of dry pipe systems is essential to the detection of this obstructing material.

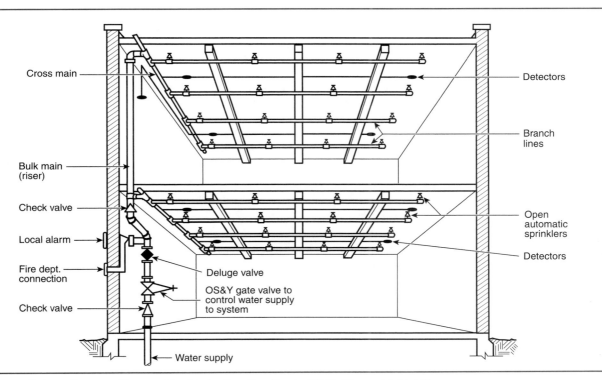

EXHIBIT 3.19 *Deluge System. (Source: Fire Protection Handbook, 2003, Figure 10.11.13)*

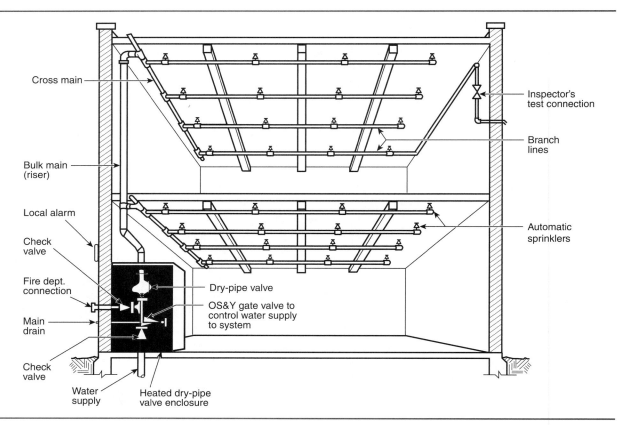

EXHIBIT 3.20 *Dry-Pipe System for Unheated Properties. (Source: Fire Protection Handbook, 2003, Figure 10.11.10)*

The dry pipe valve is a differential type valve, meaning that very little air pressure can hold back a great deal of water pressure, usually around 1 psi (0.07 bar) of air pressure to 5 psi (0.34 bar) or 6 psi (0.41 bar) of water pressure. Since a dry pipe system is prone to the formation of internal rust and scale, particular attention to the dry pipe valve is necessary to prevent the accumulation of this material in the valve seat. Accumulated rust and scale can prevent the clapper from sealing properly and result in malfunction of the dry pipe valve, thus necessitating an internal inspection.

The principle of a dry pipe valve is illustrated in Exhibit 3.21. Compressed air in the sprinkler system holds the dry valve closed, preventing water from entering the sprinkler piping until the air pressure has dropped below a predetermined point.

3.6.4.4 Preaction Sprinkler System. A sprinkler system employing automatic sprinklers that are attached to a piping system that contains air that might or might not be under pressure, with a supplemental detection system installed in the same areas as the sprinklers. [**13,** 2007]

Several types of preaction systems can be found in use, including the following:

- *Single interlock systems*, which allow water to enter the piping upon operation of detection devices
- *Non-interlock systems*, which allow water to enter the piping system upon activation of either detection devices or automatic sprinklers
- *Double interlock systems*, which only allow water to enter the piping system upon operation of both detection devices and automatic sprinklers

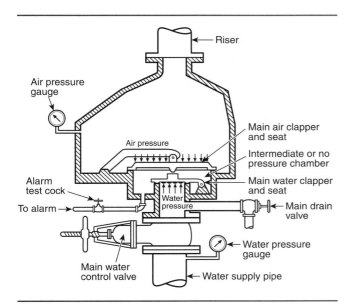

EXHIBIT 3.21 Dry Pipe Valve. (Source: Fire Protection Handbook, 2003, Figure 10.11.9)

A typical preaction sprinkler system is shown in the top portion of Exhibit 3.22. The bottom of Exhibit 3.22 compares the operating features of a single interlock preaction system with those of a double interlock preaction system.

***3.6.4.5* Wet Pipe Sprinkler System.** A sprinkler system employing automatic sprinklers attached to a piping system containing water and connected to a water supply so that water discharges immediately from sprinklers opened by heat from a fire. [**13**, 2007]

***A.3.6.4.5* Wet Pipe Sprinkler System.** Hose connections [1 ½ in. (40 mm) hose, valves, and nozzles] supplied by sprinkler system piping are considered components of the sprinkler system.

The wet pipe system (see Exhibit 3.23) is probably the most common of all sprinkler systems. This system is under water pressure at all times so that water is discharged immediately upon activation of a sprinkler. The alarm valve shown in Exhibit 3.23 causes a warning signal to sound when water flows through the system. (It must be noted here that sprinklers operate independently of each other; in many cases, fire is controlled by a single sprinkler.) One of the few limitations of a wet pipe system is that it must be protected from freezing temperatures.

3.6.5 Water Spray Fixed System. A special fixed pipe system connected to a reliable fire protection water supply and equipped with water spray nozzles for specific water discharge and distribution over the surface or area to be protected. The piping system is connected to the water supply through an automatically or manually actuated valve that initiates the flow of water. An automatic valve is actuated by operation of automatic detection equipment installed in the same areas as the water spray nozzles. (In special cases, the automatic detection system also is located in another area.)

Water spray fixed systems are usually controlled by a deluge valve supplying water to open or non-automatic discharge devices. For automatic systems, the actuating mechanism may be in the form of an electrical detection system or a system of hydraulic or pneumatic pilot sprinklers. These systems are usually more complex than either wet or dry pipe sprinkler systems. See the commentary in Chapter 10 for more information on water spray fixed systems.

3.6.6 Water Tank. A tank supplying water for water-based fire protection systems.

40 Chapter 3 • Definitions

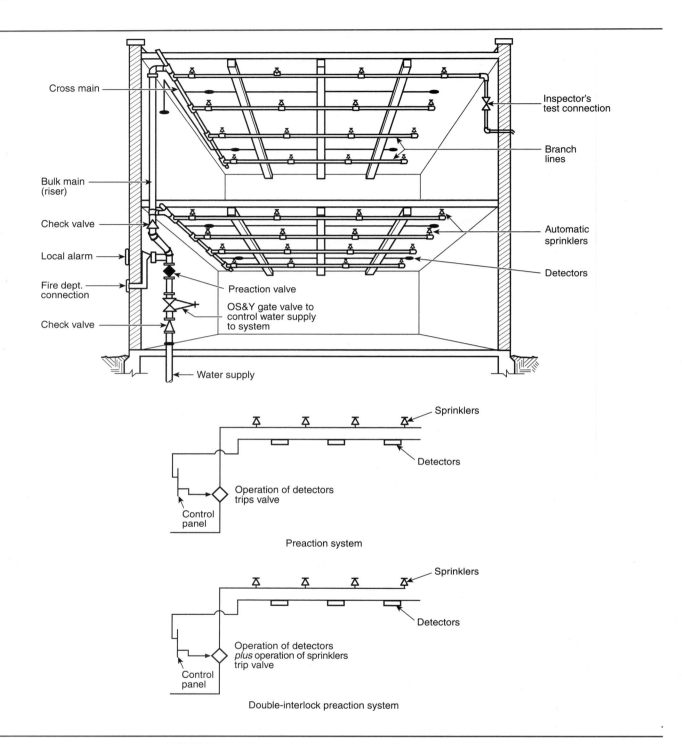

EXHIBIT 3.22 Typical Preaction Sprinkler System (top); Comparison of Operating Features for the Types of Preaction Systems (bottom). (Source: Fire Protection Handbook, 2003, Figures 10.11.11 and 10.11.12)

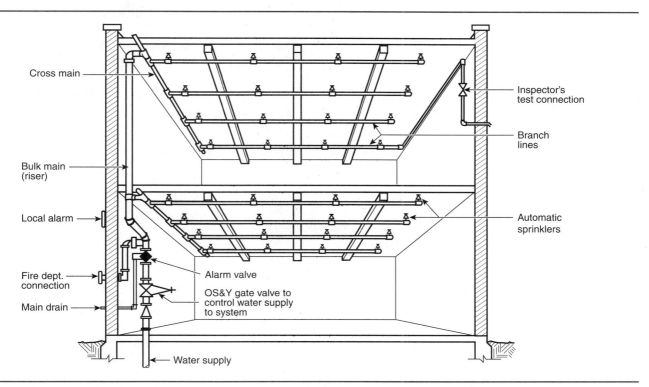

EXHIBIT 3.23 *Wet Pipe Sprinkler System. (Source: Fire Protection Handbook, 2003, Figure 10.11.8)*

A water tank is used primarily due to an insufficient water supply. Ground-mounted suction tanks and embankment-supported coated fabric suction tanks are used for this purpose. Pressure tanks and elevated tanks offer a pressure boost in addition to storage capacity. In most cases, tanks are the only source of water for a fire protection system and must be cared for to maintain system reliability. Water storage tanks are usually designed and installed in accordance with NFPA 22, *Standard for Water Tanks for Private Fire Protection*. See the commentary in Chapter 9 for more information on water tanks.

SUMMARY

Chapter 3 identifies the specialized terms and systems descriptions used in NFPA 25, as well as NFPA official definitions used throughout NFPA codes and standards. For definitions of terms not listed in Chapter 3, see *Merriam-Webster's Collegiate Dictionary*, 11th edition.

REFERENCES CITED IN COMMENTARY

National Fire Protection Association, 1 Batterymarch Park, Quincy, MA 02169-7471.

Cote, A. E., ed., *Fire Protection Handbook*®, 19th edition, 2003.
Dubay, C., ed., *Automatic Sprinkler Systems Handbook*, 2007 edition.
NFPA 13, *Standard for the Installation of Sprinkler Systems*, 2007 edition.
NFPA 14, *Standard for the Installation of Standpipe, Private Hydrant, and Hose Systems*, 2007 edition.

NFPA 16, *Standard for the Installation of Foam-Water Sprinkler and Foam-Water Spray Systems*, 2007 edition.

NFPA 22, *Standard for Water Tanks for Private Fire Protection*, 2003 edition.

NFPA 70E, *Standard for Electrical Safety in the Workplace*, 2004 edition.

NFPA 72®, National Fire Alarm Code®, 2007 edition.

Merriam-Webster, Inc., 7 Federal Street, P.O. Box 281, Springfield, MA 01102.

Merriam-Webster's Collegiate Dictionary, 11th edition, 2003.

General Requirements

CHAPTER 4

Chapter 4 of NFPA 25 deals with requirements that relate to all types of systems covered under the scope of the standard. The chapter outlines the administrative guidelines for compliance with the standard, which include such topics as impairments, corrective action, record keeping, and essential safety requirements for those performing the work specified in Chapters 5 through 14.

4.1 Responsibility of the Property Owner or Occupant

4.1.1* Responsibility for Inspection, Testing, and Maintenance.

The responsibility for properly maintaining a water-based fire protection system shall be that of the owner of the property.

A.4.1.1 The components are not required to be open or exposed. Doors, removable panels, or valve pits can be permitted to satisfy the need for accessibility. Such equipment should not be obstructed by features such as walls, ducts, columns, direct burial, or stock storage.

The requirement in 4.1.1 is intended to prevent the all-too-common practice of placing objects such as file cabinets, stock, and so on, in front of sprinkler risers and other control equipment. The property owner must maintain access to equipment for inspection, testing, and maintenance purposes. The standard refers to the property owner throughout the standard to distinguish between the property owner and the owner of the contracting firm. Ultimately, it is the property owner who is responsible for compliance with NFPA 25.

4.1.2* Accessibility.

The property owner or occupant shall provide ready accessibility to components of water-based fire protection systems that require inspection, testing, or maintenance.

CASE STUDY

Sprinkler System Not in Service at the Time of Fire in Electric Power Plant

This coal-fired electric power plant in Alabama was of unprotected noncombustible construction. The height of the plant and its ground floor area were not reported. The plant was operating at the time of the fire. This plant was reportedly equipped with a sprinkler system of unknown type and coverage. The system was not working at the time of the fire due to a malfunction.

Investigators for this federally owned plant reported that the fire started on a 400 ft conveyor belt supplying coal to the plant. The cause of the fire was thought to be a buildup of heat resulting from friction between the belt and a jammed roller. The fire spread up the belt and into the main plant, destroying the conveyor belt system, which had to be replaced. The fire disrupted plant operations for several weeks. Damages were estimated at $10,600,000.

A.4.1.2 Inspection, testing, and maintenance can be permitted to be contracted with an inspection, testing, and maintenance service.

Paragraph 4.1.2 clearly sets the tone for enforcement of the standard by indicating that the property owner is responsible for inspection, testing, and maintenance. Only in the case of an absentee owner can the responsibility for inspection, testing, and maintenance be transferred from the absentee owner to another person or management firm (see 4.1.2.3).

The standard does not specify who is responsible for heating a sprinklered building. However, the standard does establish responsibility for the proper maintenance of the fire protection systems in 4.1.2, which places the responsibility of proper maintenance on the property owner.

> **FAQ ▶** Who is responsible for the inspection, testing, and maintenance of the water-based fire protection system?

> **FAQ ▶** Who is responsible for keeping the building heated and the heating system operational?

4.1.2.1 Inspection, testing, and maintenance shall be implemented in accordance with procedures meeting those established in this document and in accordance with the manufacturer's instructions.

The statements in NFPA 25 are clearly minimum requirements. As such, any program meeting or exceeding these requirements is generally considered to be in compliance with this standard.

4.1.2.2 These tasks shall be performed by personnel who have developed competence through training and experience.

> **FAQ ▶** Is the property owner required to hire a contractor to perform the activities required by this standard or can the in-house maintenance staff perform these functions?

It is not the intent of NFPA 25 to establish the requirements for competency for service personnel, since qualification requirements can vary considerably from one jurisdiction to the next. However, it is important to point out that those who lack such qualification should not attempt some activities prescribed by this standard. Although training and experience are necessary for many of the prescribed tasks, property owners can perform some of the required inspection activities — for example, basic inspections of valve position or pressure gauge readings, which require a minimum of training and instruction. See 3.3.28 for a definition of the term *qualified*.

4.1.2.3 Where the property owner is not the occupant, the property owner shall be permitted to pass on the authority for inspecting, testing, and maintaining the fire protection systems to the occupant, management firm, or managing individual through specific provisions in the lease, written use agreement, or management contract.

4.1.2.4 Where an occupant, management firm, or managing individual has received the authority for inspection, testing, and maintenance, the occupant, management firm, or managing individual shall comply with the requirements identified for the owner or occupant throughout this standard.

4.1.3 Notification of System Shutdown.

The property owner or occupant shall notify the authority having jurisdiction, the fire department, if required, and the alarm-receiving facility before testing or shutting down a system or its supply.

Paragraph 4.1.3 establishes the requirement for a formal impairment procedure by requiring notification when systems are removed from service. Testing a system without proper notification might — and often does — result in a false alarm. False alarms must be avoided since they remove emergency services personnel from service at a time when their services may be needed for an actual emergency. In many jurisdictions, repeated false alarms can result in a fine or other penalty for property owners.

4.1.3.1 The notification of system shutdown shall include the purpose for the shutdown, the system or component involved, and the estimated time of shutdown.

4.1.3.2 The authority having jurisdiction, the fire department, and the alarm-receiving facility shall be notified when the system, supply, or component is returned to service.

The requirements in 4.1.3.2 assist the authority having jurisdiction (AHJ) or alarm company in establishing a follow-up program to ensure that systems are properly returned to service.

4.1.4* Corrections and Repairs.

The property owner or occupant shall promptly correct or repair deficiencies, damaged parts, or impairments found while performing the inspection, test, and maintenance requirements of this standard.

While NFPA 25 focuses on the specific requirements for inspection, testing, and maintenance of water-based fire protection systems, it cannot be overlooked that the ultimate goal of such a program is to maintain the operational status of a system. Paragraph 4.1.4 provides the enforcement tool to initiate corrective action to ensure that systems are functioning at all times.

A.4.1.4 Recalled products should be replaced or remedied. Remedies include entrance into a program for scheduled replacement. Such replacement or remedial product should be installed in accordance with the manufacturer's instructions and the appropriate NFPA installation standards. A recalled product is a product subject to a statute or administrative regulation specifically requiring the manufacturer, importer, distributor, wholesaler, or retailer of a product, or any combination of such entities, to recall the product, or a product voluntarily recalled by a combination of such entities.

4.1.4.1* Corrections and repairs shall be performed by qualified maintenance personnel or a qualified contractor.

A.4.1.4.1 System deficiencies not explained by normal wear and tear, such as hydraulic shock, can often be indicators of system problems and should be investigated and evaluated by a qualified person or engineer. Failure to address these issues could lead to catastrophic failure. Examples of deficiencies that can be caused by issues beyond normal wear and tear are as follows:

Pressure Gauge

(1) Gauge not returning to zero
(2) Gauge off scale
(3) Gauge with bent needle

Support Devices

(1) Bent hangers and/or rods
(2) Hangers pulled out/off structure
(3) Indication of pipe or hanger movement such as the following:
  (a) Hanger scrape marks on pipe, exposed pipe surface where pipe and hangers are painted
  (b) Fire stop material damaged at pipe penetration of fire rated assembly

Unexplained System Damage

(1) Unexplained system damage beyond normal wear and tear
(2) Bent or broken shafts on valves
(3) Bent or broken valve clappers

(4) Unexplained leakage at branch lines, cross main, or feed main piping
(5) Unexplained leakage at close nipples
(6) Loose bolts on flanges and couplings

Fire Pump

(1) Fire pump driver out of alignment
(2) Vibration of fire pump and/or driver
(3) Unusual sprinkler system piping noises (sharp report, loud bang)

 4.1.5* Changes in Occupancy, Use, Process, or Materials.

The property owner or occupant shall not make changes in the occupancy, the use or process, or the materials used or stored in the building without evaluation of the fire protection systems for their capability to protect the new occupancy, use, or materials.

CASE STUDY

Fire in Hotel/Conference Center Caused by Overloaded Electrical Switch

A fire in a hotel and conference center in New Jersey started when an overloaded electrical dimmer switch ignited paper-backed wall insulation. Voids in construction and sprinklers that were blocked by renovations allowed the fire to spread through concealed spaces before it was detected by an employee.

The two-story building, which had a basement, was of wood-frame construction and measured 100 ft by 50 ft. It contained a conference center with several meeting rooms, a restaurant, and sleeping rooms. The fire occurred in a wing that consisted of meeting rooms, a bar/lounge, a kitchen, and a basement-level game and exercise room.

The corridors, stairwells, and sleeping rooms contained automatic smoke detectors. Heat detectors were located in the bathrooms and mechanical rooms. A wet pipe, ordinary-hazard automatic sprinkler system that used 165°F sprinklers protected all occupied areas and the ceiling/floor concealed spaces. However, there were no sprinklers in the attic. All the systems and manual pull stations were supervised by in-house security.

At approximately 6:54 p.m. a bartender heard a crackling noise in a light dimmer switch on the wall in the first-floor kitchen preparation room, but he continued working. Several minutes later, he saw smoke near the ceiling and called security. Using an extinguisher, he then tried to put out the fire. Security called the fire department at 7:00 p.m. Five minutes later, another employee smelled smoke and activated a manual pull station.

Fire fighters arrived at 7:09 p.m. and found that the fire had spread to an open void above a mini-bar ceiling and had burned through an exterior wall, igniting the outside cedar siding. Because there were no draft stops or separations in the walls, the fire spread up the wall and across the roof. Nine sprinklers in the ceiling/floor void activated, but they could not completely control the fire. HVAC ducts and structural framing that had been added during renovations blocked the sprinklers, limiting their effectiveness.

Fire fighters used several handlines to extinguish flames on the exterior wall, on the ceiling over the bar, and in the bar area. A ladder crew opened up the roof and used a cellar nozzle to prevent further horizontal spread in the ceiling/roof void. Fire fighters then made a large ventilation hole in the roof and opened up the ceilings on the second floor to permit final extinguishment. Four engine companies, a truck company, and an 85 ft snorkel were needed to extinguish the fire. Damage to the building, which was valued at $4.2 million, was estimated at $900,000.

A.4.1.5 Fire protection systems should not be removed from service when the building is not in use; however, where a system that has been out of service for a prolonged period (such as in the case of idle or vacant properties) is returned to service, it is recommended that a responsible and experienced contractor be retained to perform all inspections and tests.

4.1.5.1 The evaluation shall consider factors that include, but are not limited to, the following:

(1) Occupancy changes such as converting office or production space into warehousing
(2) Process or material changes such as metal stamping to molded plastics
(3) Building revisions such as relocated walls, added mezzanines, and ceilings added below sprinklers
(4) Removal of heating systems in spaces with piping subject to freezing

Occupancy changes usually involve a building permit and subsequent review process by the AHJ. The more subtle changes listed in 4.1.5.1 items 3 and 4 may go unreported. Moving walls and adding mezzanines and ceilings can severely affect the spray pattern of sprinklers. Sprinkler protection in unheated spaces frequently causes damage to systems and water damage to buildings and their contents. By monitoring building alterations and performing inspections, costly damage to buildings and contents can be avoided.

4.1.6 Addressing Changes in Hazard.

Where changes in the occupancy, hazard, water supply, storage commodity, storage arrangement, building modification, or other condition that affects the installation criteria of the system are identified, the property owner or occupant shall promptly take steps, such as contacting a qualified contractor, consultant, or engineer, and the authority having jurisdiction, to evaluate the adequacy of the installed system in order to protect the building or hazard in question.

4.1.6.1 Where the evaluation reveals a deficiency causing a threat to life or property, the property owner shall make appropriate corrections. All requirements of the authority having jurisdiction shall be followed.

4.1.7 Valve Location.

The property owner shall ensure that responsible occupants are made aware of the location of the shutoff valves and the procedures for shutting down the system.

Where an impairment coordinator — that is, a person who is familiar with the building and its systems and who has the authority to manage a system shutdown — is employed (see Section 15.2), he or she should be responsible for control of all system water supply valves. Where an impairment coordinator is not on site, the property manager should be trained in the location and function of a water supply control valve. In many cases, time is wasted after a system begins to discharge water as the building occupants search for the location of the shutoff valve. This search results in additional water damage that could have been avoided.

4.1.8 Information Sign.

A permanently marked metal or rigid plastic information sign shall be placed at the system control riser supplying an antifreeze loop, dry system, preaction system, or auxiliary system control valve. Each sign shall be secured with a corrosion-resistant wire, chain, or other acceptable means and shall indicate the following information:

(1) Location of the area served by the system
(2) Location of auxiliary drains and low-point drains
(3) The presence and location of antifreeze or other auxiliary systems

The sign required in 4.1.8 is needed to inform anyone maintaining a system that auxiliary and low point drains are installed on the system and to indicate their number and location. Without this information, the system can be damaged due to freezing of the trapped sections of pipe. The sign is particularly useful where the service contractor is not the installing contractor. Exhibit 4.1 illustrates a sample information sign.

EXHIBIT 4.1 Sample Information Sign.

Fire Sprinkler Antifreeze System

This System Serves: _____

Auxiliary and Low Point Drains Are Located at: _____

Other Antifreeze/Auxiliary Systems in This Building Are Located at:

4.2 Impairments

4.2.1 Where an impairment to a water-based fire protection system occurs, the procedures outlined in Chapter 15 of this standard shall be followed, including the attachment of a tag to the impaired system.

4.2.2 Where a water-based fire protection system is returned to service following an impairment, the system shall be verified to be working properly by means of an appropriate inspection or test.

Recommissioning a fire protection system should be accomplished in accordance with the requirements of the appropriate installation standard. For example, the tests needed to recommission a sprinkler system would be performed in accordance with NFPA 13, *Standard for the Installation of Sprinkler Systems*. For the 2008 edition of the standard, component replacement and testing tables have been added for each system type to assist the user in determining the type and extent of recommissioning tests that are needed following a repair. See the appropriate systems chapter for the pertinent component replacement and testing table.

4.3 Corrective Action

Manufacturers shall be permitted to make modifications to their own listed product in the field with listed devices that restore the original performance as intended by the listing, where acceptable to the authority having jurisdiction.

4.4 Records

4.4.1* Records shall be made for all inspections, tests, and maintenance of the system and its components and shall be made available to the authority having jurisdiction upon request.

A.4.4.1 Typical records include, but are not limited to, valve inspections; flow, drain, and pump tests; and trip tests of dry pipe, deluge, and preaction valves.

Computer programs that file inspection and test results should provide a means of comparing current and past results and should indicate the need for corrective maintenance or further testing.

Acceptance test records should be retained for the life of the system or its special components. Subsequent test records should be retained for a period of 1 year after the next test. The comparison determines deterioration of system performance or condition and the need for further testing or maintenance.

The inspection, testing, and maintenance records referred to in 4.4.1 provide written verification of compliance with NFPA 25. These records also offer a service history of the installed systems by indicating their performance and/or maintenance history.

4.4.2 Records shall indicate the procedure performed (e.g., inspection, test, or maintenance), the organization that performed the work, the results, and the date.

Supplement 4 provides samples of some of the inspection, testing, and maintenance forms available to record the information specified in 4.4.2. Any form, whether produced commercially or developed independently, can be used to document inspection, testing, and maintenance activities provided that it details the activity sufficiently to produce a permanent record.

4.4.3* Records shall be maintained by the property owner.

A.4.4.3 See Section B.2 for information regarding sample forms.

4.4.4 As-built system installation drawings, hydraulic calculations, original acceptance test records, and device manufacturer's data sheets shall be retained for the life of the system.

4.4.5 Subsequent records shall be retained for a period of 1 year after the next inspection, test, or maintenance of that type required by the standard.

Compliance with 4.4.5 results in maintaining records for at least one year following successive inspection, testing, and maintenance activity. The requirement for records retention is intended to detect evidence of trending. For example, a main drain test performed on an annual basis should result in a total of three test reports: the original test, results for year 1, and results for year 2. With this information, fluctuations in the test results can be revealed.

4.5* Inspection

System components shall be inspected at intervals specified in the appropriate chapters.

A.4.5 Inspection and periodic testing determine what, if any, maintenance actions are required to maintain the operability of a water-based fire protection system. The standard establishes minimum inspection/testing frequencies, responsibilities, test routines, and reporting procedures but does not define precise limits of anomalies where maintenance actions are required.

Substandard conditions, such as a closed valve, subnormal water pressure, loss of building heat or power, or obstruction of sprinklers, nozzles, detectors, or hose stations, can delay or prevent system actuation and impede manual fire-fighting operations.

Section 4.5 refers to inspection intervals for specific system components that may be found in subsequent chapters of NFPA 25. Those chapters are as follows:

> Chapter 5 — Sprinkler Systems
>
> Chapter 6 — Standpipe and Hose Systems
>
> Chapter 7 — Private Fire Service Mains
>
> Chapter 8 — Fire Pumps
>
> Chapter 9 — Water Storage Tanks
>
> Chapter 10 — Water Spray Fixed Systems
>
> Chapter 11 — Foam-Water Sprinkler Systems
>
> Chapter 12 — Water Mist Systems
>
> Chapter 13 — Valves, Valve Components, and Trim

4.6 Testing

4.6.1* All components and systems shall be tested to verify that they function as intended.

A.4.6.1 As referred to in 4.4.4, original records should include, at a minimum, the contractor's material and test certificate, "as-built" drawings and calculations, and any other required or pertinent test reports. These documents establish the conditions under which the systems were first installed and offer some insight to the design intent, installation standards used, and water supply present at the time of installation. Original records are instrumental in determining any subsequent changes or modifications to the building or system.

4.6.1.1 The frequency of tests shall be in accordance with this standard.

FAQ ▶
Must a monthly test be conducted on the exact date each month, or can frequency be varied as long as the activity is carried out on or about the same day each month?

The inspection, testing, and maintenance frequencies established in this standard vary from weekly, monthly, quarterly, semiannually, annually, to every five years. The standard does not address the issue of any variance from the required frequency. The frequency requirements of NFPA 25 need some interpretation on the part of the AHJ. In the case of monthly inspections, it does not make sense to inspect a component on the 30th of the month and then again on the 2nd of the following month. Generally, the inspection, testing, or maintenance activity should be conducted within a few days of the target date for weekly frequencies and within approximately a week for monthly and semiannual frequencies and with a month of the annual and five year frequencies. It is impractical to expect any activity required by NFPA 25 to be performed on the exact day or date for each iteration.

4.6.1.1.1* As an alternative means of compliance, subject to the authority having jurisdiction, components and systems shall be permitted to be inspected, tested and maintained under a performance-based program.

A.4.6.1.1.1 Paragraph 4.6.1.1.1 provides the option to adopt a performance-based test and inspection method as an alternative means of compliance with 4.6.1.1. The prescriptive test and requirements contained in this standard are essentially qualitative. Equivalent or superior levels of performance can be demonstrated through quantitative performance-based analyses. This section provides a basis for implementing and monitoring a performance-based program acceptable under this option (providing approval is obtained by the AHJ).

The concept of a performance-based testing and inspection program is to establish the requirements and frequencies at which inspection must be performed to demonstrate an acceptable level of operational reliability. The goal is to balance the inspection/test frequency with proven reliability of the system or component. The goal of a performance-based inspection program is also to adjust test/inspection frequencies commensurate with historical documented equipment performance and desired reliability. Frequencies of test/inspection under a performance-based program can be extended or reduced from the prescriptive test requirements contained in this standard when continued testing has been documented indicating a higher or lower degree of reliability compared to the AHJ's expectations of performance. Additional program attributes that should be considered when adjusting test/inspection frequencies include the following:

(1) System/component preventive maintenance programs
(2) Consequences of system maloperation
(3) System/component repair history
(4) Building/service conditions

Fundamental to implementing a performance-based program is that adjusted test and inspection frequencies must be technically defensible to the AHJ and supported by evidence of higher or lower reliability. Data collection and retention must be established so that the data utilized to alter frequencies are representative, statistically valid, and evaluated against firm criteria. Frequencies should not arbitrarily be extended or reduced without a suitable basis and rationale. It must be noted that transitioning to a performance-based program might require additional expenditures of resources in order to collect and analyze failure data, coordinate review efforts, change program documents, and seek approval from the AHJ. The following factors should be considered in determining whether a transition to a performance-based test program as permitted in 4.6.1.1.1 is appropriate:

(1) Past system/component reliability — have problems routinely been identified during the performance of the prescriptive test requirements of 4.6.1.1, or have systems consistently performed with minimal discrepancies noted?
(2) Do the recurring resource expenditures necessary to implement the prescriptive test requirements in 4.6.1.1 justify conducting the detailed analysis needed to support a performance-based testing program?
(3) Is the increased administrative burden of implementing, documenting, and monitoring a performance-based program worthwhile?

Failure Rate Calculation

A performance-based program requires that a maximum allowable failure rate be established and approved by the AHJ in advance of implementation. The use of historical system/component fire system inspection records can be utilized to determine failure rates. One method of calculating the failure rate of a fire system is based on the following equation:

$$FSFR(t) = \frac{NF}{(NC)(t)}$$

where:

$FSFR(t)$ = fire system failure rate (failures per year)
NF = number of failures
NC = total number of fire systems inspected or tested
t = time interval of review in years

Example

Data are collected for 50 fire pump weekly tests over a 5-year period. The testing is conducted weekly, as described in 8.3.1. A review of the data has identified five failures:

Total components: 280

Data collection period: 5 years

Total failures: 5

$$FSFR = \frac{5}{280 \times 5} = 0.003/\text{year}$$

A fundamental requirement of a performance-based program is the continual monitoring of fire system/component failure rates and determining whether they exceed the maximum allowable failure rates as agreed upon with the AHJ. The process used to complete this review should be documented and repeatable.

Coupled with this ongoing review is a requirement for a formalized method of increasing or decreasing the frequency of testing/inspection when systems exhibit either a higher than expected failure rate or an increase in reliability as a result of a decrease in failures, or both.

A formal process for reviewing the failure rates and increasing or decreasing the frequency of testing must be well documented. Concurrence of the AHJ on the process used to determine test frequencies should be obtained in advance of any alterations to the test program. The frequency required for future tests might be reduced to the next inspection frequency and maintained there for a period equaling the initial data review or until the ongoing review indicates that the failure rate is no longer being exceeded — for example, going from annual to semiannual testing when the failure rate exceeds the AHJ's expectations or from annual to every 18 months when the failure trend indicates an increase in reliability.

References

Edward K. Budnick, P.E., "Automatic Sprinkler System Reliability," *Fire Protection Engineering*, Society of Fire Protection Engineers, Winter 2001.

Fire Protection Equipment Surveillance Optimization and Maintenance Guide, Electric Power Research Institute, July 2003.

William E. Koffel, P.E., *Reliability of Automatic Sprinkler Systems, Alliance for Fire Safety*.

NFPA's Future in Performance Based Codes and Standards, July 1995.

NFPA Performance Based Codes and Standards Primer, December 1999.

Paragraphs 4.6.1.1.1 and A.4.6.1.1.1 are new to the 2008 edition and provide a detailed procedure for determining the appropriate inspection/testing frequency for fire protection systems and equipment. Since its inception, NFPA 25 has included a provision allowing an alternate method of performing inspection, testing, and maintenance (see Section 1.3). But this provision does not detail exactly how such an alternate method should be implemented.

Paragraph 4.6.1.1.1 and its related annex material now explain in detail how to implement such a program. It is important to note that the performance-based option requires due diligence in conducting inspections and testing and demands a consistent approach to reporting and record keeping. Generally, if a component or system operates with few or no failures, the inspection and testing frequency could easily be decreased. Conversely, if a system or component continually breaks down or requires constant maintenance or replacement, the inspection and testing frequency must be increased. For example, a fire pump that experiences no failures could produce sufficient documentation to demonstrate or suggest a monthly running test (versus the prescribed weekly test in Chapter 8). However, a fire pump that is frequently unavailable — that is, one that does not start or experiences other types of failures — would produce documentation that suggests a weekly running test is necessary to verify optimal performance.

Performance-based options cannot be implemented without approval by the AHJ. The documentation submitted to the AHJ for review must be complete and sufficiently detailed to clearly justify the deviation in inspection and test frequencies from that specified in NFPA 25.

4.6.2 Fire protection system components shall be restored to full operational condition following testing, including reinstallation of plugs and caps for auxiliary drains and test valves.

4.6.3 During testing and maintenance, water supplies, including fire pumps, shall remain in service unless under constant attendance by qualified personnel or impairment procedures in Chapter 15 are followed.

The requirement in 4.6.3 is intended to avoid unnecessary system impairment by requiring impairment procedures for any water supply that is turned off for an extended period of time. Provided that water supplies, including fire pumps, are in constant attendance, impairment procedures are not needed for valves that are closed for short periods for testing. An example of such a situation would be closing the discharge isolation valve of a fire pump for the annual flow test. As illustrated in Commentary Table 1.1, a shut valve is the most common cause of system failure.

4.6.4* Test results shall be compared with those of the original acceptance test (if available) and with the most recent test results.

A.4.6.4 The types of tests required for each protection system and its components, and the specialized equipment required for testing, are detailed in the appropriate chapter.

4.6.5* When a major component or subsystem is rebuilt or replaced, the subsystem shall be tested in accordance with the original acceptance test required for that subsystem.

A.4.6.5 Examples of components or subsystems are fire pumps, drivers or controllers, pressure regulating devices, detection systems and controls, alarm check, and dry pipe, deluge, and preaction valves.

4.7* Maintenance

Maintenance shall be performed to keep the system equipment operable or to make repairs.

A.4.7 Preventive maintenance includes, but is not limited to, lubricating control valve stems; adjusting packing glands on valves and pumps; bleeding moisture and condensation from air compressors, air lines, and dry pipe system auxiliary drains; and cleaning strainers. Frequency of maintenance is indicated in the appropriate chapter.

Corrective maintenance includes, but is not limited to, replacing loaded, corroded, or painted sprinklers; replacing missing or loose pipe hangers; cleaning clogged fire pump impellers; replacing valve seats and gaskets; restoring heat in areas subject to freezing temperatures where water-filled piping is installed; and replacing worn or missing fire hose or nozzles.

Emergency maintenance includes, but is not limited to, repairs due to piping failures caused by freezing or impact damage; repairs to broken underground fire mains; and replacement of frozen or fused sprinklers, defective electric power, or alarm and detection system wiring.

4.8 Safety

Inspection, testing, and maintenance activities shall be conducted in a safe manner.

4.8.1 Confined Spaces.

Legally required precautions shall be taken prior to entering confined spaces such as tanks, valve pits, or trenches.

Paragraph 4.8.1 addresses precautions to be taken when working in confined spaces. A confined space is any space that is large enough for a person to enter and perform work and that has limited or restricted access (such as a water storage tank). At times, an inspector will find it necessary to enter a confined space, such as a valve pit, water storage tank, or trench.

The U.S. Department of Labor, Occupational Safety and Health Administration (OSHA) regulates confined space entry in 29 CFR 1910.146, "Permit-Required Confined Spaces." That standard establishes the requirement for a permit-required confined space entry program for controlling and protecting employees from confined space hazards and for regulating entry into these spaces. In areas not subject to OSHA regulations (e.g., outside the United States), other government regulations may apply. Exhibit 4.2 illustrates a warning sign identifying a confined space as part of a permit-required confined space entry program.

EXHIBIT 4.2 Sample Confined Space Entry Sign. (Source: NFPA 22, 2003, Figure A.4.8.2)

Prior to entering a confined space, the internal atmosphere must be tested for oxygen content, flammable gases and vapors, and potential toxic air contaminants. The space must also be ventilated to eliminate or control any hazardous atmosphere.

Access to any confined space must be coordinated with the property owner, contractor, or host employer. (A host employer is the customer of the contractor/inspector.) Controlled access to a confined space may be managed through the use of a permit. The permit should indicate that atmospheric testing has been completed and that proper safety equipment (such as ventilation, communications, personal protective equipment, lighting, ladders, and rescue and emergency equipment) is in place. The permit should also address any other requirements as established in the host employer's confined space entry program. Exhibit 4.3 illustrates a sample confined space entry permit.

Ongoing testing or monitoring of the confined space atmosphere is necessary to ensure that safe conditions are maintained for the duration of entry operations.

During confined space entry, at least one attendant, trained in summoning rescue and emergency services, rescuing entrants from the space, and providing emergency services to rescued employees, must remain outside of the permit space.

EXHIBIT 4.3 Sample Confined Space Entry Permit. (Source: OSHA, 29 CFR 1910.146, Appendix D)

Confined Space Entry Permit

Date and Time Issued: _____ Date and Time Expires: _____
Job Site/Space I.D.: _____ Job Supervisor: _____
Equipment to be worked on: _____ Work to be performed: _____

Stand-by personnel: _____

1. Atmospheric checks: Time _____
   Oxygen _____ %
   Explosive _____ % L•F•L•
   Toxic _____ PPM

2. Tester's signature: _____

3. Source isolation (No Entry):   N/A   Yes   No
   Pumps or lines blinded,   ( )   ( )   ( )
   disconnected, or blocked   ( )   ( )   ( )

4. Ventilation modification: N/A   Yes   No
   Mechanical   ( )   ( )   ( )
   Natural ventilation only   ( )   ( )   ( )

5. Atmospheric check after
   isolation and ventilation:
   Oxygen _____ %   >   19.5   %
   Explosive _____ % L•F•L•   <   10   %
   Toxic _____ PPM   <   10   PPM H(2)S
   Time _____
   Testers signature: _____

6. Communication procedures: _____

7. Rescue procedures: _____
   _____
   _____

8. Entry, standby, and backup persons: Yes   No
   Successfully completed required training?
   Is it current?   ( )   ( )

                                                          N/A   Yes   No
9. Equipment:
   Direct reading gas monitor - tested   ( )   ( )   ( )
   Safety harnesses and lifelines
   for entry and standby persons   ( )   ( )   ( )
   Hoisting equipment   ( )   ( )   ( )
   Powered communications   ( )   ( )   ( )
   SCBA's for entry and standby persons   ( )   ( )   ( )
   Protective Clothing   ( )   ( )   ( )
   All electric equipment listed
   Class I, Division I, Group D and
   Non-sparking tools   ( )   ( )   ( )

10. Periodic atmospheric tests:
    Oxygen _____ %   Time _____   Oxygen _____ %   Time _____
    Oxygen _____ %   Time _____   Oxygen _____ %   Time _____
    Explosive _____ % Time _____   Explosive _____ % Time _____
    Explosive _____ % Time _____   Explosive _____ % Time _____
    Toxic _____ %   Time _____   Toxic _____ %   Time _____
    Toxic _____ %   Time _____   Toxic _____ %   Time _____

We have reviewed the work authorized by this permit and the information contained here-in. Written instructions and safety procedures have been received and are understood. Entry cannot be approved if any squares are marked in the "No" column. This permit is not valid unless all appropriate items are completed.

Permit Prepared By: (Supervisor) _____
Approved By: (Unit Supervisor) _____
Reviewed By (Cs Operations Personnel): _____

_____   _____
       (printed name)   (signature)

This permit to be kept at job site. Return job site copy to Safety Office following job completion.

It is important for anyone involved in the inspection, testing, and maintenance of systems to recognize a confined space and be prepared to follow the host employer's confined space entry program. Paragraph 4.8.1 is not intended to outline a comprehensive confined space entry program.

4.8.2 Fall Protection.

Legally required equipment shall be worn or used to prevent injury from falls to personnel.

Fall protection, as referred to in 4.8.2, can include wearing safety harnesses as well as the proper use of ladders, staging, and powered lifts. The inspector should be familiar with any regulations in effect governing this type of equipment.

4.8.3 Hazards.

Precautions shall be taken to address any hazards, such as protection against drowning where working on the top of a filled embankment or a supported, rubberized fabric tank, or over open water or other liquids.

4.8.4* Hazardous Materials.

A.4.8.4 Most places using or storing hazardous materials have stations set up for employees where material safety data sheets (MSDSs) are stored. The inspector should be familiar with the types of materials present and the appropriate actions to take in an emergency.

4.8.4.1 Legally required equipment shall be used where working in an environment with hazardous materials present.

4.8.4.2 The property owner shall advise anyone performing inspection, testing, and maintenance on any system under the scope of this document, with regard to hazardous materials stored on the premises.

The hazardous materials referenced in 4.8.4.2 can be any material capable of causing harm or bodily injury. These materials can include cleaning agents, solvents, paint, adhesives, or other chemicals used for a variety of purposes. It is important that inspectors and service personnel be able both to recognize the presence of a hazardous material and to identify and classify the material to prevent accident, injury, or fire.

In 1983, OSHA established the Hazard Communication Standard, 29 CFR 1910, Subpart Z, "Toxic and Hazardous Substances." This standard requires manufacturers to provide information about hazardous materials to their customers in the form of Material Safety Data Sheets (MSDS), such as the one shown in Exhibit 4.4.

A completed MSDS provides chemical and toxicological data regarding hazardous materials and the need for any personal protective equipment. Prior to entering an area containing hazardous materials, an inspector or service personnel should request a copy of the MSDS for all substances in the area.

4.9* Electrical Safety

Legally required precautions shall be taken when testing or maintaining electric controllers for motor-driven fire pumps.

A.4.9 WARNING: NFPA 20, *Standard for the Installation of Stationary Pumps for Fire Protection*, includes electrical requirements that discourage the installation of a disconnect means in the power supply to electric motor-driven fire pumps. This is intended to ensure the availability of power to the fire pumps. Where equipment connected to those circuits is serviced or maintained, the service person could be subject to unusual exposure to electrical and other

EXHIBIT 4.4 OSHA Form 174 — Material Safety Data Sheet. (Source: OSHA, 29 CFR 1910.1200, Subpart Z)

Material Safety Data Sheet May be used to comply with OSHA's Hazard Communication Standard, 29 CFR 1910. 1200. Standard must be consulted for specific requirements.	U.S. Department of Labor Occupational Safety and Health Administration (Non-Mandatory Form) Form Approved OMB No. 1218-0072
IDENTITY (*As Used on Label and List*)	Note: Blank spaces are not permitted. If any item is not applicable, or no information is available, the space must be marked to indicate that.

Section I

Manufacturer's Name	Emergency Telephone Number
Address (*Number, Street, City, State, and ZIP Code*)	Telephone Number for Information
	Date Prepared
	Signature of Preparer (*optional*)

Section II—Hazardous Ingredients/Identity Information

Hazardous Components (*Specific Chemical Identity; Common Name(s)*) OSHA PEL   ACGIH TLV   Other Limits Recommended   % (*optional*)

Section III—Physical/Chemical Characteristics

Boiling Point		Specific Gravity ($H_2O = 1$)	
Vapor Pressure (mm Hg)		Melting Point	
Vapor Density (Air = 1)		Evaporation Rate (Butyl Acetate = 1)	
Solubility in Water			
Appearance and Odor			

Section IV—Fire and Explosion Hazard Data

Flash Point (*Method Used*)		Flammable Limits	LEL	UEL
Extinguishing Media				
Special Fire-Fighting Procedures				
Unusual Fire and Explosion Hazards				

hazards. It could be necessary to establish special safe work practices and to use safeguards or personal protective clothing, or both.

Safety-related work practices and procedures for persons who work on energized electrical equipment should be in place prior to engaging in such activity. As a minimum, procedures should be in accordance with 29 CFR 1926, "Safety and Health Regulations for Construction," and NFPA 70E, *Standard for Electrical Safety in the Workplace*. A fire pump controller is classified by NFPA 70E as a 600V Class Motor Control Center. Commentary Table 4.1, excerpted from NFPA 70E, defines the hazard/risk category and voltage-rated gloves and tools necessary for work specifically involving fire pump controllers.

EXHIBIT 4.4 *Continued*

Section V—Reactivity Data

Stability	Unstable		Conditions to Avoid	
	Stable			

Incompatibility (*Materials to Avoid*)

Hazardous Decomposition or By-products

Hazardous Polymerization	May Occur		Conditions to Avoid
	Will Not Occur		

Section VI—Health Hazard Data

Route(s) of Entry: Inhalation?   Skin?   Ingestion?

Health Hazards (*Acute and Chronic*)

Carcinogenicity: NTP?   IARC Monographs?   OSHA Regulated?

Signs and Symptoms of Exposure

Medical Conditions Generally Aggravated by Exposure

Emergency and First Aid Procedures

Section VII—Precautions for Safe Handling and Use

Steps to Be Taken in Case Material Is Released or Spilled

Waste Disposal Method

Precautions to Be Taken in Handling and Storing

Other Precautions

Section VIII—Control Measures

Respiratory Protection (*Specify Type*)

Ventilation	Local Exhaust		Special	
	Mechanical (*General*)		Other	
Protective Gloves			Eye Protection	
Other Protective Clothing or Equipment				
Work/Hygienic Practices				

*U.S.G.P.O.: 1986-491-529/45775

Once the hazard/risk categories for testing and maintenance of fire pump controllers have been determined from Commentary Table 4.1, the need for and type of protective clothing and personal protective equipment can be determined from Commentary Table 4.2, which is also excerpted from NFPA 70E.

Exhibit 4.5 illustrates the necessary personal protective equipment needed for service and maintenance of fire pump controllers.

In addition to electrical safety requirements and procedures, other precautions such as lock-out/tag-out procedures should be exercised to prevent injury from the accidental starting

COMMENTARY TABLE 4.1 Hazard/Risk Category Classifications

Task (Assumes Equipment Is Energized, and Work Is Done Within the Flash Protection Boundary)	Hazard/ Risk Category	V-rated Gloves	V-rated Tools
Panelboards Rated 240 V and Below — Notes 1 and 3			
Circuit breaker (CB) or fused switch operation with covers on	0	N	N
CB or fused switch operation with covers off	0	N	N
Work on energized parts, including voltage testing	1	Y	Y
Remove/install CBs or fused switches	1	Y	Y
Removal of bolted covers (to expose bare, energized parts)	1	N	N
Opening hinged covers (to expose bare, energized parts)	0	N	N
Panelboards or Switchboards Rated >240 V and up to 600 V (with molded case or insulated case circuit breakers) — Notes 1 and 3			
CB or fused switch operation with covers on	0	N	N
CB or fused switch operation with covers off	1	N	N
Work on energized parts, including voltage testing	2*	Y	Y
600 V Class Motor Control Centers (MCCs) — Notes 2 (except as indicated) and 3			
CB or fused switch or starter operation with enclosure doors closed	0	N	N
Reading a panel meter while operating a meter switch	0	N	N
CB or fused switch or starter operation with enclosure doors open	1	N	N
Work on energized parts, including voltage testing	2*	Y	Y
Work on control circuits with energized parts 120 V or below, exposed	0	Y	Y
Work on control circuits with energized parts >120 V, exposed	2*	Y	Y
Insertion or removal of individual starter "buckets" from MCC — Note 4	3	Y	N
Application of safety grounds, after voltage test	2*	Y	N
Removal of bolted covers (to expose bare, energized parts)	2*	N	N
Opening hinged covers (to expose bare, energized parts)	1	N	N
600 V Class Switchgear (with power circuit breakers or fused switches) — Notes 5 and 6			
CB or fused switch operation with enclosure doors closed	0	N	N
Reading a panel meter while operating a meter switch	0	N	N
CB or fused switch operation with enclosure doors open	1	N	N
Work on energized parts, including voltage testing	2*	Y	Y
Work on control circuits with energized parts 120 V or below, exposed	0	Y	Y
Work on control circuits with energized parts >120 V, exposed	2*	Y	Y
Insertion or removal (racking) of CBs from cubicles, doors open	3	N	N
Insertion or removal (racking) of CBs from cubicles, doors closed	2	N	N
Application of safety grounds, after voltage test	2*	Y	N
Removal of bolted covers (to expose bare, energized parts)	3	N	N
Opening hinged covers (to expose bare, energized parts)	2	N	N
Other 600 V Class (277 V through 600 V, nominal) Equipment — Note 3			
Lighting or small power transformers (600 V, maximum)	—	—	—
Removal of bolted covers (to expose bare, energized parts)	2*	N	N
Opening hinged covers (to expose bare, energized parts)	1	N	N
Work on energized parts, including voltage testing	2*	Y	Y
Application of safety grounds, after voltage test	2*	Y	N
Revenue meters (kW-hour, at primary voltage and current)	—	—	—
Insertion or removal	2*	Y	N
Cable trough or tray cover removal or installation	1	N	N
Miscellaneous equipment cover removal or installation	1	N	N
Work on energized parts, including voltage testing	2*	Y	Y
Application of safety grounds, after voltage test	2*	Y	N
NEMA E2 (fused contactor) Motor Starters, 2.3 kV Through 7.2 kV			
Contactor operation with enclosure doors closed	0	N	N
Reading a panel meter while operating a meter switch	0	N	N
Contactor operation with enclosure doors open	2*	N	N

(continues)

COMMENTARY TABLE 4.1 *Hazard/Risk Category Classifications (continued)*

Task (Assumes Equipment Is Energized, and Work Is Done Within the Flash Protection Boundary)	Hazard/ Risk Category	V-rated Gloves	V-rated Tools
Work on energized parts, including voltage testing	3	Y	Y
Work on control circuits with energized parts 120 V or below, exposed	0	Y	Y
Work on control circuits with energized parts >120 V, exposed	3	Y	Y
Insertion or removal (racking) of starters from cubicles, doors open	3	N	N
Insertion or removal (racking) of starters from cubicles, doors closed	2	N	N
Application of safety grounds, after voltage test	3	Y	N
Removal of bolted covers (to expose bare, energized parts)	4	N	N
Opening hinged covers (to expose bare, energized parts)	3	N	N
Metal Clad Switchgear, 1 kV and Above			
CB or fused switch operation with enclosure doors closed	2	N	N
Reading a panel meter while operating a meter switch	0	N	N
CB or fused switch operation with enclosure doors open	4	N	N
Work on energized parts, including voltage testing	4	Y	Y
Work on control circuits with energized parts 120 V or below, exposed	2	Y	Y
Work on control circuits with energized parts >120 V, exposed	4	Y	Y
Insertion or removal (racking) of CBs from cubicles, doors open	4	N	N
Insertion or removal (racking) of CBs from cubicles, doors closed	2	N	N
Application of safety grounds, after voltage test	4	Y	N
Removal of bolted covers (to expose bare, energized parts)	4	N	N
Opening hinged covers (to expose bare, energized parts)	3	N	N
Opening voltage transformer or control power transformer compartments	4	N	N
Other Equipment 1 kV and Above			
Metal clad load interrupter switches, fused or unfused	—	—	—
Switch operation, doors closed	2	N	N
Work on energized parts, including voltage testing	4	Y	Y
Removal of bolted covers (to expose bare, energized parts)	4	N	N
Opening hinged covers (to expose bare, energized parts)	3	N	N
Outdoor disconnect switch operation (hookstick operated)	3	Y	Y
Outdoor disconnect switch operation (gang-operated, from grade)	2	N	N
Insulated cable examination, in manhole or other confined space	4	Y	N
Insulated cable examination, in open area	2	Y	N

Legend:

V-rated Gloves are gloves rated and tested for the maximum line-to-line voltage upon which work will be done.

V-rated Tools are tools rated and tested for the maximum line-to-line voltage upon which work will be done.

2* means that a double-layer switching hood and hearing protection are required for this task in addition to the other Hazard/Risk Category 2 requirements of Table 130.7(C)(10) of NFPA 70E. [Shown as Commentary Table 4.2].

Y = yes (required)

N = no (not required)

Notes:

1. 25 kA short circuit current available, 0.03 second (2 cycle) fault clearing time.
2. 65 kA short circuit current available, 0.03 second (2 cycle) fault clearing time.
3. For < 10 kA short circuit current available, the hazard/risk category required may be reduced by one number.
4. 65 kA short circuit current available, 0.33 second (20 cycle) fault clearing time.
5. 65 kA short circuit current available, up to 1.0 second (60 cycle) fault clearing time.
6. For < 25 kA short circuit current available, the hazard/risk category required may be reduced by one number.

Source: NFPA 70E, 2004, Table 130.7(C)(9)(a).

COMMENTARY TABLE 4.2 *Protective Clothing and Personal Protective Equipment (PPE) Matrix*

Protective Clothing and Equipment	Protective Systems for Hazard/Risk Category					
Hazard/Risk Category Number	−1 (Note 3)	0	1	2	3	4
Non-melting (according to ASTM F 1506-00) or Untreated Natural Fiber						
a. T-shirt (short-sleeve)	X			X	X	X
b. Shirt (long-sleeve)		X				
c. Pants (long)	X	X	X (Note 4)	X (Note 6)	X	X
FR Clothing (Note 1)						
a. Long-sleeve shirt			X	X	X (Note 9)	X
b. Pants			X (Note 4)	X (Note 6)	X (Note 9)	X
c. Coverall			(Note 5)	(Note 7)	X (Note 9)	(Note 5)
d. Jacket, parka, or rainwear			AN	AN	AN	AN
FR Protective Equipment						
a. Flash suit jacket (multilayer)						X
b. Flash suit pants (multilayer)						X
c. Head protection						
1. Hard hat			X	X	X	X
2. FR hard hat liner					AR	AR
d. Eye protection						
1. Safety glasses	X	X	X	AL	AL	AL
2. Safety goggles				AL	AL	AL
e. Face and head area protection			—	—	—	—
1. Arc-rated face shield, or flash suit hood				X (Note 8)		
2. Flash suit hood					X	X
3. Hearing protection (ear canal inserts)				X (Note 8)	X	X
f. Hand protection			—		—	—
Leather gloves (Note 2)			AN	X	X	X
g. Foot protection						
Leather work shoes			AN	X	X	X

Legend:
AN = As needed
AL = Select one in group
AR = As required
X = Minimum required

Notes:
1. See Table 130.7(C)(11) of NFPA 70E. Arc rating for a garment is expressed in cal/cm^2.
2. If voltage-rated gloves are required, the leather protectors worn external to the rubber gloves satisfy this requirement.
3. Hazard/Risk Category Number "−1" is only defined if determined by Notes 3 or 6 of Table 130.7(C)(9)(a) of NFPA 70E.
4. Regular weight (minimum 12 oz/yd^2 fabric weight), untreated, denim cotton blue jeans are acceptable in lieu of FR pants. The FR pants used for Hazard/Risk Category 1 shall have a minimum arc rating of 4.
5. Alternate is to use FR coveralls (minimum arc rating of 4) instead of FR shirt and FR pants.
6. If the FR pants have a minimum arc rating of 8, long pants of non-melting or untreated natural fiber are not required beneath the FR pants.
7. Alternate is to use FR coveralls (minimum arc rating of 4) over non-melting or untreated natural fiber pants and T-shirt.
8. A faceshield with a minimum arc rating of 8, with wrap-around guarding to protect not only the face, but also the forehead, ears, and neck (or, alternatively, a flash suit hood), is required.
9. Alternate is to use two sets of FR coveralls (the inner with a minimum arc rating of 4 and outer coverall with a minimum arc rating of 5) over non-melting or untreated natural fiber clothing, instead of FR coveralls over FR shirt and FR pants over non-melting or untreated natural fiber clothing.

Source: NFPA 70E, 2004, Table 130.7(C)(10).

EXHIBIT 4.5 Personal Protective Equipment.

of mechanical equipment (such as motors and pumping equipment). For more information on system impairment and lock-out/tag-out procedures, see Chapter 15 of NFPA 25.

SUMMARY

Chapter 4 establishes the general requirements for the performance of inspections, testing, and maintenance prescribed throughout NFPA 25. Because the inspector or service personnel can be exposed to a multitude of hazards while performing the activities specified in this standard, the importance of satisfying the reporting and safety requirements cannot be overemphasized. It is the reporting that verifies compliance with the standard.

REVIEW QUESTIONS

1. Who is responsible for compliance with the requirements of NFPA 25?
2. How long must inspection, testing, and maintenance records be kept on file?
3. What four factors must be considered when evaluating the fire protection systems in a building that has changed occupancy, use, or processes?
4. What safety precautions should an inspector take prior to working in a building?

REFERENCES CITED IN COMMENTARY

National Fire Protection Association, 1 Batterymarch Park, Quincy, MA 02169-7471.

NFPA 13, *Standard for the Installation of Sprinkler Systems*, 2007 edition.
NFPA 22, *Standard for Water Tanks for Private Fire Protection*, 2003 edition.
NFPA 70E, *Standard for Electrical Safety in the Workplace*, 2004 edition.

U.S. Government Printing Office, Washington, DC 20402.

Title 29, Code of Federal Regulations, Part 1910.146, "Permit-Required Confined Spaces."

Title 29, Code of Federal Regulations, Part 1910, Subpart Z, "Toxic and Hazardous Substances."

Title 29, Code of Federal Regulations, Part 1926, "Safety and Health Regulations for Construction."

Title 29, Code of Federal Regulations, Part 1910.1200.

Sprinkler Systems

CHAPTER 5

Chapter 5 covers the inspection, testing, and maintenance of the most common type of fire protection: the fire sprinkler system. Sprinkler systems have proven to be exceptionally reliable, but they do require some attention to remain operational. Unlike other types of building systems, such as plumbing or HVAC, the sprinkler system is relatively static. In other words, it is only called on to activate when a fire emergency occurs. An ongoing inspection, testing, and maintenance program is necessary to ensure that the system and its components will function as intended in the event of a fire emergency.

5.1 General

This chapter shall provide the minimum requirements for the routine inspection, testing, and maintenance of sprinkler systems. Table 5.1 shall be used to determine the minimum required frequencies for inspection, testing, and maintenance.

In addition to the requirements of this chapter, many of the requirements of Chapter 13 (Valves, Valve Components, and Trim), Chapter 14 (Obstruction Investigation), and Chapter 15 (Impairments) apply to a sprinkler system. Table 5.1 summarizes the specific requirements of sprinkler system inspection, testing, and maintenance. The "Reference" column includes NFPA 25 paragraph numbers so that the reader can refer to details on the inspection, testing, or maintenance to be performed.

Many of the inspection activities listed in Table 5.1 can be performed by property owners or property maintenance personnel with a minimal amount of training. For example, the weekly/monthly inspection of gauges and control valves can be performed with some basic instruction, and the quarterly inspection for the hydraulic nameplate is simply intended to verify that it is attached to the system riser and is legible. The other inspection, testing, and maintenance activities required by Table 5.1 should be performed by a qualified contractor, since these activities require specialized training, or they can simply be performed by the owner or maintenance personnel while the contractor is present and supervising the activities.

5.1.1 Valves and Connections.

Valves and fire department connections shall be inspected, tested, and maintained in accordance with Chapter 13.

5.1.2 Impairments.

The procedures outlined in Chapter 15 shall be followed where an impairment to protection occurs.

5.1.3 Notification to Supervisory Service.

To avoid false alarms where a supervisory service is provided, the alarm receiving facility shall be notified by the property owner or designated representative as follows:

64 Chapter 5 • Sprinkler Systems

TABLE 5.1 Summary of Sprinkler System Inspection, Testing, and Maintenance

Item	Frequency	Reference
Inspection		
Gauges (dry, preaction, and deluge systems)	Weekly/monthly	5.2.4.2, 5.2.4.3
Control valves	Weekly/monthly	Table 13.1
Waterflow devices	Quarterly	5.2.6
Valve supervisory devices	Quarterly	5.2.6
Supervisory signal devices (except valve supervisory switches)	Quarterly	5.2.6
Gauges (wet pipe systems)	Monthly	5.2.4.1
Hydraulic nameplate	Quarterly	5.2.7
Buildings	Annually (prior to freezing weather)	5.2.5
Hanger/seismic bracing	Annually	5.2.3
Pipe and fittings	Annually	5.2.2
Sprinklers	Annually	5.2.1
Spare sprinklers	Annually	5.2.1.3
Fire department connections	Quarterly	Table 13.1
Valves (all types)		Table 13.1
Obstruction	5 years	14.2.2
Test		
Waterflow devices	Quarterly/semiannually	5.3.3
Valves supervisory devices	Semiannually	Table 13.1
Supervisory signal devices (except valve supervisory switches)	Semiannually	Table 13.1
Main drain	Annually	Table 13.1
Antifreeze solution	Annually	5.3.4
Gauges	5 years	5.3.2
Sprinklers — extra-high temperature	5 years	5.3.1.1.1.3
Sprinklers — fast-response	At 20 years and every 10 years thereafter	5.3.1.1.1.2
Sprinklers	At 50 years and every 10 years thereafter	5.3.1.1.1
Maintenance		
Valves (all types)	Annually or as needed	Table 13.1
Obstruction investigation	5 years or as needed	13.2.1, 13.2.2
Low-point drains (dry pipe system)	Annually prior to freezing and as needed	13.4.4.3.2
Investigation		
Obstruction	As needed	14.2.1

(1) Before conducting any test or procedure that could result in the activation of an alarm
(2) After such tests or procedures are concluded

 5.1.4 Hose connections shall be inspected, tested, and maintained in accordance with Chapters 6 and 13.

5.2* Inspection

A.5.2 The provisions of the standard are intended to apply to routine inspections. In the event of a fire, a post-fire inspection should be made of all sprinklers within the fire area. In situations where the fire was quickly controlled or extinguished by one or two sprinklers, it might be necessary only to replace the activated sprinklers. Care should be taken that the replacement sprinklers are of the same make and model or that they have compatible performance characteristics (*see 5.4.1.1*). Soot-covered sprinklers should be replaced because deposits can result in corrosion of operating parts. In the event of a substantial fire, special consideration should be given to replacing the first ring of sprinklers surrounding the operated sprinklers because of the potential for excessive thermal exposure, which could weaken the response mechanisms.

5.2.1 Sprinklers.

5.2.1.1* Sprinklers shall be inspected from the floor level annually.

Paragraph 5.2.1.1 requires a visual inspection of sprinklers to reveal obvious signs of damage. The inspection is conducted from floor level, because it is usually impractical (and costly) to get closer to the sprinklers for a more detailed inspection. Binoculars can assist in the inspection of sprinklers (or piping) in buildings with high ceilings. If problems are suspected, or when other work is being done at the ceiling level and access is not an issue, it may be desirable to do a more thorough inspection by getting closer to the sprinklers.

◄ **FAQ**
Is it necessary to inspect sprinklers and system piping above suspended ceilings?

A.5.2.1.1 The conditions described in this section can have a detrimental effect on the performance of sprinklers by affecting water distribution patterns, insulating thermal elements, delaying operation, or otherwise rendering the sprinkler inoperable or ineffectual.

Severely loaded or corroded sprinklers should be rejected as part of the visual inspection. Such sprinklers could be affected in their distribution or other performance characteristics not addressed by routine sample testing. Lightly loaded or corroded sprinklers could be permitted for continued use if samples are selected for testing based on worst-case conditions and the samples successfully pass the tests.

5.2.1.1.1* Sprinklers shall not show signs of leakage; shall be free of corrosion, foreign materials, paint, and physical damage; and shall be installed in the proper orientation (e.g., upright, pendent, or sidewall).

A.5.2.1.1.1 Sprinkler orientation includes the position of the deflector in relation to the ceiling slope. The deflector is generally required to be parallel to the slope of the ceiling. The inspection should identify any corrections made where deficiencies are noted, for example, pipe with welded outlets and flexible grooved couplings that have "rolled" out of position.

5.2.1.1.2 Any sprinkler shall be replaced that has signs of leakage; is painted, other than by the sprinkler manufacturer, corroded, damaged, or loaded; or in the improper orientation

NFPA 25 in 5.2.1.1.2 requires replacement of any sprinkler that has been painted or is corroded or exhibits any sign of damage. The only painting that is permitted is the cover plate of concealed sprinklers and the frame of fusible link type sprinklers as indicated in the listing for the sprinkler. The painting must be completed *only* by the sprinkler manufacturer.

◄ **FAQ**
Which NFPA standard contains requirements concerning the painting of sprinklers?

As required by 5.2.1.1.2, sprinklers that are leaking, painted, corroded, loaded, or installed in the wrong orientation (e.g., a pendent sprinkler mistakenly installed in the upright position) must be replaced to ensure that they will operate properly.

Sprinklers that are loaded with some type of contaminant such as dirt, dust, grease, or paint must be replaced. However, sprinklers that have a light coating of dust do not need to be replaced as long as they are cleaned by a blast of compressed air or by a vacuum (provided that the vacuum attachments do not touch the sprinkler, to avoid potential damage or weakening of the glass bulb or link). On the other hand, sprinklers that require wiping or scrubbing or the use of detergents or solvents to remove contaminants must be replaced. Painted sprinklers should never be cleaned and reinstalled, because the paint can get under the sprinkler cap and act like an adhesive, delaying or preventing sprinkler operation. A "light" overspray or loading can be tolerated when a representative sample is tested to verify the sprinklers will operate as intended (see A.5.2.1.1). Any residue that collects on the sprinkler's thermal elements can have an insulating effect that could also delay or prevent sprinkler operation. If corrosion is a significant problem, special corrosion-resistant sprinklers may be used.

◄ **FAQ**
Are there situations in which sprinklers can be cleaned of foreign material instead of being replaced?

Sprinklers that are leaking or that have been damaged must be replaced without testing. Dissolved minerals and other residues in the water can solidify as the sprinkler leaks, hampering the operation of the sprinkler by changing internal clearances or acting like an adhesive, preventing parts from moving as intended.

Exhibit 5.1 shows several examples of sprinklers that require replacement.

Exhibit 5.2 shows a sprinkler that was "repaired" and reinstalled in a sprinkler system. The machine bolt is not a fusible element, and this sprinkler will not operate in a fire emergency. Sprinklers should never be repaired but must be replaced instead.

5.2.1.1.3 Glass bulb sprinklers shall be replaced if the bulbs have emptied.

The requirement in 5.2.1.1.3 addresses the fact that glass bulb sprinklers operate when the liquid inside the bulb expands until the internal pressure is sufficient to break the glass. Missing liquid (or a crack in the bulb) will prevent the buildup of pressure inside the bulb and, therefore, prevent the sprinkler from operating.

In some cases, rough handling of glass bulb sprinklers during shipment or installation can cause minute fractures in the bulb, allowing the fluid to leak out. Exhibit 5.3 shows a glass bulb sprinkler that contains no liquid inside the glass bulb. This sprinkler will not operate until

EXHIBIT 5.1 Examples of Sprinklers Needing Replacement. (Top row, left to right): (1) Recessed pendent sprinkler that has been painted. (2) Brass upright sprinkler that has been painted. (3) Pendent sprinkler installed in the upright position. (Bottom row, left to right): (1) Upright sprinkler installed in the pendent position. (2) Loaded sprinkler from a candy factory coated with sugar. (3) Sprayed-on fireproofing covering an upright sprinkler.

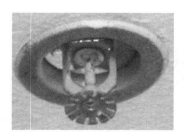

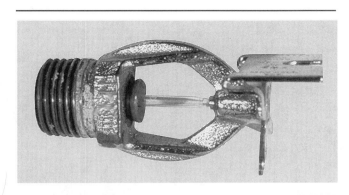

EXHIBIT 5.3 Glass Bulb Sprinkler Without Fluid. (Courtesy of National Fire Sprinkler Association)

the glass melts, which is well beyond the anticipated operating time and temperature intended for this type of sprinkler.

5.2.1.1.4* Sprinklers installed in concealed spaces such as above suspended ceilings shall not require inspection.

A.5.2.1.1.4 Examples include some floor/ceiling or roof/ceiling assemblies, areas under theater stages, pipe chases, and other inaccessible areas.

Paragraph 5.2.1.1.4 provides an exemption from the inspection requirements for sprinklers in concealed spaces on the basis of two factors:

1. It is extremely costly and impractical to inspect sprinklers in these spaces.
2. The sprinklers are in an inaccessible location, which means that they are less likely to be damaged, painted, or loaded than those in an open space.

However, when working in concealed areas, it is a good idea to conduct an inspection of the sprinklers in these spaces.

5.2.1.1.5 Sprinklers installed in areas that are inaccessible for safety considerations due to process operations shall be inspected during each scheduled shutdown.

5.2.1.1.6 Sprinklers that are subject to recall shall be replaced per the manufacturer's requirements.

5.2.1.2* The minimum clearance required by the installation standard shall be maintained below all sprinklers. Stock, furnishings, or equipment closer to the sprinkler than the clearance rules allow shall be corrected.

Obstructions to sprinkler distribution patterns, such as those referred to in 5.2.1.2, can hamper the effectiveness of sprinklers. Obstructions less than 18 in. (457 mm) below the sprinkler have greater impact than obstructions located further away. NFPA 13, *Standard for the Installation of Sprinkler Systems,* provides guidance on these types of obstructions.

Examples of obstructions include the installation of new (installed since the last inspection) equipment at the ceiling, the storage of stock on the top shelves in a retail store, and the installation of ceiling-mounted signs or decorations. Exhibit 5.4 shows a typical obstruction, and Exhibit 5.5 shows a very unusual obstruction.

A.5.2.1.2 NFPA 13, *Standard for the Installation of Sprinkler Systems,* allows stock furnishings and equipment to be as close as 18 in. (457 mm) to standard spray sprinklers or as close as 36 in. (914 mm) to other types of sprinklers such as ESFR and large drop sprinklers. Objects against walls are permitted to ignore the minimum spacing rules as long as the sprinkler is not directly above the object. Other obstruction rules are impractical to enforce under this standard. However, if obstructions that might cause a concern are present, the owner is advised to have an engineering evaluation performed.

EXHIBIT 5.2 Sprinkler Repaired and Returned to Service.

EXHIBIT 5.4 *Typical Obstruction.*

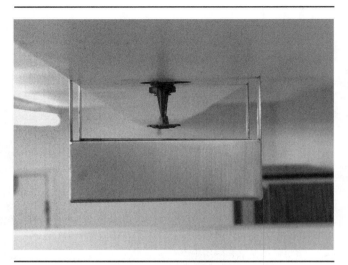

EXHIBIT 5.5 *Collection Device for a Leaking Sprinkler.*

CASE STUDY

Piled Storage Obscures Sprinklers

A fire in a brewery bottle-washing facility in Colorado caused $5 million in damage. This three-story building of fire-resistive construction had a ground-floor area of 400,000 ft^2. The facility was in operation. A complete wet pipe sprinkler system was present in the building. There was no automatic detection equipment.

Plant employees discovered a fire inside a palletized stack of boxed bottles on the third floor and notified the fire department. Fire fighters responded approximately 4 minutes later to find a deep-seated fire obscured by large quantities of dense smoke. They were unable to ventilate the structure due to the heavy concrete construction, and the fire continued to smolder and spread through the high-piled storage. Fire department officials convinced management to bring in a wrecking ball to breach a wall of the plant for ventilation 22 hours after the fire was first discovered. The fire was extinguished approximately 1 hour later. Damage to the structure was confined to the ventilation hole. The contents of the building were heavily damaged by smoke, fire, and water. Arson was determined to be the cause of the fire. Three fire fighters were injured. It was found that storage piled high directly under the sprinkler system limited the effect of sprinklers. Water could not reach the seat of the fire. In addition, lack of an effective means of ventilating the building prevented fire fighters from locating the main body of the fire early in the incident.

5.2.1.3 The supply of spare sprinklers shall be inspected annually for the following:

(1) The proper number and type of sprinklers
(2) A sprinkler wrench for each type of sprinkler

The inspection required in 5.2.1.3 is necessary to ensure that a supply of spare sprinklers, including all of the types and temperature ratings installed in the system, is available should a sprinkler operate or incur damage. With many types of sprinklers on the market, it is important to have the correct types of sprinklers immediately available to restore the system to service quickly. Local contractors may not be readily available or may not have the proper replacement sprinkler on hand. Sprinklers also require a special wrench to prevent damage to the sprinkler during installation, so spare sprinkler wrenches are also required. A pipe wrench or crescent wrench should not be used as these can stress the sprinkler frame, causing the sprinkler to leak. Exhibit 5.6 shows a spare sprinkler cabinet complete with two of each type of installed sprinkler and a sprinkler wrench.

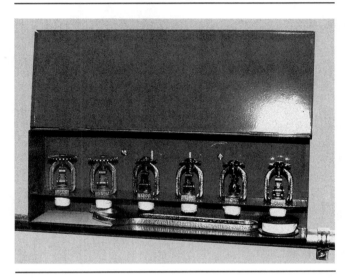

EXHIBIT 5.6 Spare Sprinkler Cabinet.

5.2.2* Pipe and Fittings.

Sprinkler pipe and fittings shall be inspected annually from the floor level.

The inspection required by 5.2.2 should include a check for conditions, described in 5.2.2.1 and 5.2.2.2, that can affect the performance of sprinkler pipe, fittings, and their supports. These conditions can be identified by a visual inspection conducted from floor level. It is impractical (and costly) in most cases to get closer to the piping for a more detailed inspection. Binoculars can assist in the inspection of piping in buildings with high ceilings. Where problems are suspected, or when other work is being done at the ceiling level and access is not an issue, it may be desirable to conduct a more thorough inspection by getting closer to the piping system.

Note that installation issues such as piping pitch are not covered by the annual inspection required by 5.2.2.

◀ **FAQ**
Is the inspector required to reverify piping pitch during the annual inspection?

A.5.2.2 The conditions described in 5.2.2 can have a detrimental effect on the performance and life of pipe by affecting corrosion rates or pipe integrity or otherwise rendering the pipe ineffectual.

5.2.2.1 Pipe and fittings shall be in good condition and free of mechanical damage, leakage, and corrosion.

5.2.2.2 Sprinkler piping shall not be subjected to external loads by materials either resting on the pipe or hung from the pipe.

The design of sprinkler piping, fittings, and their supports takes into account the weight of the pipe and water and the normal forces anticipated during sprinkler operation. Although there are safety factors included, the design does not consider the additional weight from other loads, such as those prohibited by 5.2.2.2. Equipment, signs, decorations, or other materials should therefore not be hung from or otherwise supported by the sprinkler piping. In addition to the potential for failure from the added load, attachments to sprinkler piping can also obstruct sprinklers or hinder their operation.

5.2.2.3* Pipe and fittings installed in concealed spaces such as above suspended ceilings shall not require inspection.

A.5.2.2.3 Examples include some floor/ceiling or roof/ceiling assemblies, areas under theater stages, pipe chases, and other inaccessible areas.

Paragraph 5.2.2.3 provides an exemption from the inspection requirements for pipe and fittings in concealed spaces on the basis of two factors:

1. It is extremely costly and impractical to inspect piping in these spaces.
2. The inaccessibility of such piping means it is less likely to be damaged or otherwise affected than piping in an open space.

When working in concealed spaces, however, it is a good idea to conduct an inspection of the pipe and fittings in the space.

5.2.2.4 Pipe installed in areas that are inaccessible for safety considerations due to process operations shall be inspected during each scheduled shutdown.

5.2.3* Hangers and Seismic Braces.

Sprinkler pipe hangers and seismic braces shall be inspected annually from the floor level.

The visual inspection of hangers and seismic braces is conducted from floor level, as specified in 5.2.3, because it is impractical (and costly) in most cases to get closer to the piping for a more detailed inspection. Binoculars can assist in the inspection of hangers and braces in buildings with high ceilings. Where problems are suspected, or when other work is being done at the ceiling level and access is not an issue, it may be desirable to do a more thorough inspection by getting closer to the piping system.

A.5.2.3 The conditions described in this section can have a detrimental effect on the performance of hangers and braces by allowing failures if the components become loose.

5.2.3.1 Hangers and seismic braces shall not be damaged or loose.

When hangers or seismic braces are loose, other hangers must support the load. If enough supports need repair, overloading and the failure of additional hangers or braces may occur. Damaged hangers can also fail. When such conditions are found, the damaged or loose supports must be replaced or refastened as appropriate.

5.2.3.2 Hangers and seismic braces that are damaged or loose shall be replaced or refastened.

See the commentary following 5.2.3.1.

5.2.3.3* Hangers and seismic braces installed in concealed spaces such as above suspended ceilings shall not require inspection.

A.5.2.3.3 Examples of hangers and seismic braces installed in concealed areas include some floor/ceiling or roof/ceiling assemblies, areas under theater stages, pipe chases, and other inaccessible areas.

5.2.3.4 Hangers installed in areas that are inaccessible for safety considerations due to process operations shall be inspected during each scheduled shutdown.

5.2.4 Gauges.

5.2.4.1* Gauges on wet pipe sprinkler systems shall be inspected monthly to ensure that they are in good condition and that normal water supply pressure is being maintained.

The effectiveness of sprinklers depends on the availability and reliability of the water supply. An inspection of the gauge on the sprinkler system riser, as required by 5.2.4.1, can identify potential problems with the water supply (see Exhibit 5.7). For example, if the gauge reads significantly less than normal, there may be a problem with the water supply such as an obstruction or closed valve. If the gauge reads close to zero, there may be a shut valve or other problem that is preventing water from reaching the system. Excessive pressures on the top side of an alarm check valve may indicate a problem with the relief valve on a gridded system. Abnormal conditions should be investigated and damaged gauges repaired or replaced. Note that a normal pressure does not guarantee that no problems exist. A valve can be almost totally shut and the gauge will still indicate a normal static pressure. A partially shut valve is usually discovered when the main drain is flowed and the residual pressure falls dramatically below normal.

EXHIBIT 5.7 Pressure Gauge.

A.5.2.4.1 Due to the high probability of a buildup of excess pressure, gridded wet pipe systems should be provided with a relief valve not less than $\frac{1}{4}$ in. (6.3 mm) in accordance with NFPA 13, *Standard for the Installation of Sprinkler Systems*.

It is normal, though, that the pressure above the alarm or system check valve is typically higher than that of the water supply as a result of trapped pressure surges.

5.2.4.2 Gauges on dry, preaction, and deluge systems shall be inspected weekly to ensure that normal air and water pressures are being maintained.

5.2.4.3 Where air pressure supervision is connected to a constantly attended location, gauges shall be inspected monthly.

5.2.4.4* For dry pipe or preaction systems protecting freezers, in accordance with Figure A.5.2.4.4 the air pressure gauge near the compressor shall be compared weekly to the pressure gauge above the dry pipe or preaction valve. When the gauge near the compressor is reading higher than the gauge near the dry pipe valve, the air line in service shall be taken out of service, and the alternate air line opened to equalize the pressure. The air line taken out of service shall be internally inspected, shall have all ice blockage removed, and shall be reassembled for use as a future alternate air line.

A.5.2.4.4 See Figure A.5.2.4.4.

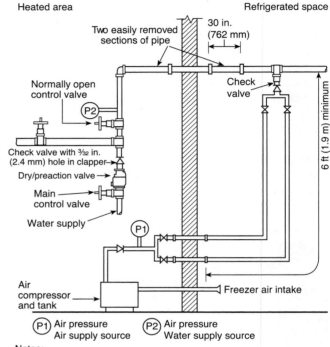

FIGURE A.5.2.4.4 *Refrigerator Area Sprinkler System Used to Minimize the Chances of Developing Ice Plugs.*

The arrangement of the air supply for a dry pipe system, such as the one depicted in Figure A.5.2.4.4, is designed to prevent moisture in the make-up air for the dry system from condensing and freezing in the main sprinkler piping by allowing the moisture (ice) to accumulate in the air supply line. When the gauge near the compressor reads higher than the air pressure in the system, there is usually an ice blockage in the air line. The alternate line maintains the correct air pressure in the system, allowing the blocked line to be disconnected, the ice removed, and the line reinstalled without removing the dry pipe valve from service.

5.2.5 Buildings.

Annually, prior to the onset of freezing weather, buildings with wet pipe systems shall be inspected to verify that windows, skylights, doors, ventilators, other openings and closures, blind spaces, unused attics, stair towers, roof houses, and low spaces under buildings do not expose water-filled sprinkler piping to freezing and to verify that adequate heat [minimum 40°F (4.4°C)] is available.

The rationale for the requirement in 5.2.5 is that wet pipe sprinkler systems require an ambient temperature of 40°F (4.4°C) to prevent freezing and the potential for broken pipes or fittings and an inoperable system. During warm weather, freezing is seldom a problem, so problems with the building enclosure often go unnoticed until the weather grows colder.

Prior to the onset of cold weather, it is important to check for areas in the building where cold air could enter. Particular attention should be paid to unoccupied areas such as openings and closures, blind spaces, unused attics, and so on, because problems in unattended areas are most likely to go unnoticed. Heating equipment should also be inspected to ensure it operates as needed when cold weather arrives.

Sprinklers located near unit heaters should be of intermediate- or high-temperature rating as shown in Exhibit 5.8 to avoid an unwanted activation. Before heating systems are needed, the presence of intermediate- or high-temperature sprinklers should be verified. For other requirements for intermediate- and high-temperature sprinklers, see NFPA 13, 2007 edition, Tables 8.3.2.5(a), 8.3.2.5(b), and 8.3.2.5(c).

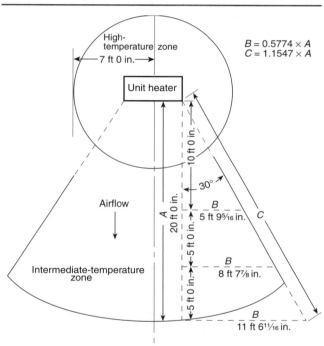

EXHIBIT 5.8 High-Temperature and Intermediate-Temperature Zones at Unit Heaters. (Source: NFPA 13, 2007, Figure 8.3.2.5)

CASE STUDY

Sprinkler System Turned Off Due to Concerns for Freezing

Fire in a baled paper warehouse that was part of a large mill complex containing multiple occupancies was started intentionally, causing $5 million in damage. The buildings were of heavy timber construction with a combined ground floor area of approximately 100,000 ft^2. The three-story warehouse was closed at the time of the fire.

A nearby resident discovered the fire in the complex and reported it by calling 911. Fire fighters arrived within 4 minutes to find a large fire on the first floor in the baled paper storage area. The fire spread vertically and laterally to more than 60,000 ft^2 of the complex before it was contained.

> A dry pipe sprinkler system should have been used in this type of environment. The wet pipe system had been turned off because the baled paper storage area did not require heat and the wet pipe system would have frozen if left unheated.
>
> The private hydrant system for the complex could not deliver the required flow for a fire of this size. Supplemental water had to be brought in on a tanker, and water was drafted from a nearby river.

5.2.6 Alarm Devices.

Alarm devices shall be inspected quarterly to verify that they are free of physical damage.

5.2.7* Hydraulic Nameplate.

The hydraulic nameplate for hydraulically designed systems shall be inspected quarterly to verify that it is attached securely to the sprinkler riser and is legible.

A.5.2.7 The hydraulic nameplate should be secured to the riser with durable wire, chain, or equivalent. *(See Figure A.5.2.7.)*

```
This system as shown on _____ company
print no _____ dated _____
for _____
at _____ contract no _____
is designed to discharge at a rate of _____
Gpm per ft² (L/min per m²) of floor area over a maximum
area of _____ ft² (m²) when supplied
with water at a rate of _____ gpm (L/min)
at _____ psi (bar) at the base of the riser.
Hose stream allowance of _____
gpm (L/min) is included in the above.
```

FIGURE A.5.2.7 *Sample Hydraulic Nameplate.*

Paragraph 5.2.7 requires that the hydraulic nameplate be inspected on a quarterly basis. The presence of a hydraulically designed system cannot be determined by visual inspection because the piping is sized based on the water supply available and the desired density and area of operation. NFPA 13 requires a hydraulic nameplate (or placard) on hydraulically designed systems so that the design can be readily determined. The details are also documented on the approved plans and hydraulic calculations, but these plans can be misplaced and may not be available when the property changes owners. A nameplate, securely fastened to the riser, can provide the details when these other data are missing (see Exhibit 5.9). If the placard becomes loose, or is difficult to read, it must be repaired or replaced.

A hydraulic nameplate will *not* be present on systems where the piping diameter was determined by the pipe schedule method.

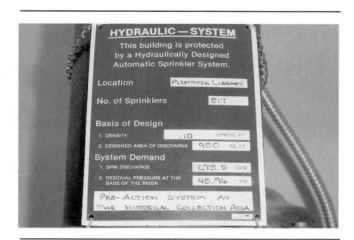

EXHIBIT 5.9 Hydraulic Data Nameplate.

5.3 Testing

5.3.1* Sprinklers.

A.5.3.1 The sprinkler field service testing described in this section is considered routine testing. Non-routine testing should be conducted to address unusual conditions not associated with the routine test cycles mandated within this standard. Due to the nature of non-routine testing, specific tests cannot be identified in this standard. The type of tests to be conducted and the number and location of samples to be submitted should be appropriate to the problem discovered or being investigated and based on consultation with the manufacturer, listing agency, and the authority having jurisdiction.

Where documentation of the installation date is not available, the start date for the in-service interval should be based upon the sprinkler's manufacture date.

5.3.1.1* Where required by this section, sample sprinklers shall be submitted to a recognized testing laboratory acceptable to the authority having jurisdiction for field service testing.

Sprinklers are extremely reliable devices that can last as long as the building in which they are installed. However, unlike plumbing, electrical, and HVAC systems, sprinkler systems sit idle for years (possibly forever) if no fire emergency occurs. As a result of this idleness, sprinkler system operation and the operation of the sprinkler itself cannot be verified without proper testing.

Paragraph 5.3.1.1 outlines the routine testing that is needed to identify potential problems that would otherwise go unnoticed. Sprinklers that have visible signs of damage, that are leaking, or that are loaded, corroded, or painted should be replaced without testing. The test frequency will vary based on the sprinkler type.

As sprinklers age, the frequency with which they should be tested increases. For example, standard response sprinklers must be tested after 50 years of service. They must then be re-tested every 10 years.

As sprinklers age, they may fail more frequently. As stated in 5.3.1.3, where one sprinkler fails the test, all sprinklers represented by that sample must be replaced. With the improvements being made in sprinkler technology, it may be desirable to replace the aging sprinklers rather than continue testing. A cost-benefit analysis is necessary to determine if replacement is the best strategy.

The sprinklers being tested undergo a procedure known as a plunge test. The sprinkler is inserted (or plunged) into a device known as a plunge test apparatus where it is exposed to an

◀ **FAQ**
What is the basis for determining whether sprinklers should be tested or simply replaced?

airflow that has a controlled velocity and temperature (see Exhibit 5.10). The temperature in the device is considerably higher than the operating temperature of the sprinkler. The sprinkler is pressurized with 5 psi of air pressure. The amount of time taken for the fusible element or glass bulb to activate is measured. If the sprinkler fails to operate in the specified amount of time, or if the sprinkler does not operate at all, the sprinkler fails the test and all sprinklers represented by that test sprinkler have to be replaced.

EXHIBIT 5.10 Plunge Test Apparatus (top) and Test Sprinkler (bottom).

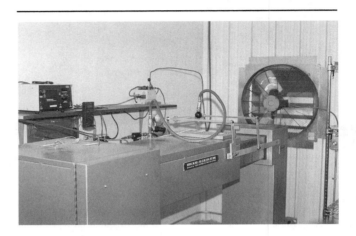

A.5.3.1.1 Sprinklers should be first given a visual inspection for signs of mechanical damage, cleaning, painting, leakage in service, or severe loading or corrosion, all of which are considered causes for immediate replacement. Devices that have passed the visual inspection should then be laboratory tested for sensitivity and functionality. The waterway should clear when sensitivity/functionality tested at 5 psi (0.4 bar) or the minimum listed operating pressure for dry sprinklers.

Thermal sensitivity should be not less than that permitted in post-corrosion testing of new sprinklers of the same type.

Sprinklers that have been in service for a number of years should not be expected to have all of the performance qualities of a new sprinkler. However, if there is any question about their continued satisfactory performance, the sprinklers should be replaced.

See Figure A.5.3.1.1.

Where are the requirements located for testing 1% of piping and sprinklers when they reach 50 years old?

FAQ ▶

5.3.1.1.1 Where sprinklers have been in service for 50 years, they shall be replaced or representative samples from one or more sample areas shall be tested. Test procedures shall be repeated at 10-year intervals.

Paragraph 5.3.1.1.1 deals with the removal and testing of sprinklers after a specific service period (50 years for standard spray sprinklers). Piping is not involved in this particular test.

FIGURE A.5.3.1.1 *Sprinkler Operating Element Identification.*

The sprinklers should be removed in the correct quantity and sent to a testing lab for evaluation of the response time index (RTI). Subsection 5.3.1 should be reviewed carefully as the test frequency varies based on the sprinkler type. Paragraph 5.3.1.2 addresses the number of sprinklers that must be removed for testing.

5.3.1.1.1.1 Sprinklers manufactured prior to 1920 shall be replaced.

5.3.1.1.1.2 Sprinklers manufactured using fast-response elements that have been in service for 20 years shall be replaced, or representative samples shall be tested. They shall be retested at 10-year intervals.

Quick-response sprinklers, as defined in 3.3.30.13, are a relatively recent advancement in sprinkler technology. The 20-year (rather than typical 50-year) testing frequency requirement in 5.3.1.1.1.2 is included in the standard because the long-term performance of these sprinklers is unknown. A reduced frequency could identify potential problems early in the life cycle of quick-response sprinklers, rather than wait for the 50-year test required of sprinklers whose performance is well known. As quick-response sprinklers begin to be tested over the next few years, the Committee on Inspection, Testing, and Maintenance of Water-Based Systems intends to monitor their performance and adjust the test frequency requirement if no significant problems are identified.

To determine the age of sprinklers in a system, the inspector would normally examine the sprinklers in the spare sprinkler cabinet. However, if a building has undergone renovations, or if the spare sprinklers have been used and replaced over the years, they may not have the same date as those in the system. Therefore, the inspector should not rely on examining the spare sprinklers to determine the age of the sprinklers in the system.

Although glass bulb sprinklers have a different operating mechanism, there is no exception for testing glass bulb type sprinklers at any other testing frequency because other types

Are glass bulb sprinklers an exception to the requirement in 5.3.1.1.1.2?
◄ **FAQ**

of failure for these sprinklers are possible, such as corrosion around the seat, contamination, and loading.

5.3.1.1.1.3* Representative samples of solder-type sprinklers with a temperature classification of extra high 325°F (163°C) or greater that are exposed to semicontinuous to continuous maximum allowable ambient temperature conditions shall be tested at 5-year intervals.

A.5.3.1.1.1.3 Due to solder migration caused by the high temperatures to which these devices are exposed, it is important to test them every 5 years. Because of this phenomenon, the operating temperature can vary over a wide range.

5.3.1.1.1.4 Where sprinklers have been in service for 75 years, they shall be replaced or representative samples from one or more sample areas shall be submitted to a recognized testing laboratory acceptable to the authority having jurisdiction for field service testing. Test procedures shall be repeated at 5-year intervals.

5.3.1.1.1.5* Dry sprinklers that have been in service for 10 years shall be replaced, or representative samples shall be tested. They shall be retested at 10-year intervals.

FAQ ▶
Do dry sprinklers have to be tested or can they be replaced?

Dry sprinklers have experienced a much higher failure rate than standard sprinklers and are more susceptible to corrosion both internally (when moisture condenses inside the device) and externally. Corrosion at the water seal and at the weep hole at the bottom of the sprinkler has been reported in sprinklers older than 10 years. These sprinklers are usually installed in harsh environments, further compounding this problem.

The failure rate of dry-type sprinklers installed for 10 years is approximately 50 percent. Because of this high failure rate, it is very important that these sprinklers be identified and tested as required. Dry-type sprinklers are custom made in exact lengths, and therefore NFPA 13 does not require spare dry-type sprinklers in the spare sprinkler cabinet. The inspector has to look for typical applications for dry-type sprinklers and perform a close inspection of a sampling of the sprinklers themselves to determine the date on the sprinklers.

The 10-year threshold requirement noted in 5.3.1.1.1.5 was added to NFPA 25 in the 2002 edition because all of the conditions that cause failure are not well understood, and the frequency of failure is higher for sprinklers that have been in service for more than 10 years. In cases where only a few dry sprinklers are installed or where corrosion is noted, it may be more cost-effective to replace the sprinklers rather than test them.

FAQ ▶
Does 5.3.1.1.1.5 apply to listed dry sprinklers, or does it apply to all sprinklers installed in a dry pipe system?

Paragraph 5.3.1.1.1.5 is referring to the listed dry-type sprinkler. This paragraph does not apply to standard spray sprinklers installed on a dry pipe system.

A.5.3.1.1.1.5 See 3.3.30.3.

5.3.1.1.2* Where sprinklers are subjected to harsh environments, including corrosive atmospheres and corrosive water supplies, on a 5-year basis, sprinklers shall either be replaced or representative sprinkler samples shall be tested.

A.5.3.1.1.2 Examples of these environments are paper mills, packing houses, tanneries, alkali plants, organic fertilizer plants, foundries, forge shops, fumigation areas, pickle and vinegar works, stables, storage battery rooms, electroplating rooms, galvanizing rooms, steam rooms of all descriptions including moist vapor dry kilns, salt storage rooms, locomotive sheds or houses, driveways, areas exposed to outside weather, around bleaching equipment in flour mills, all portions of cold storage areas, and portions of any area where corrosive vapors prevail. Harsh water environments include water supplies that are chemically reactive.

FAQ ▶
Are walk-in refrigerators and walk-in freezers considered a "harsh environment" and therefore subject to this standard?

Paragraph A.5.3.1.1.2 provides examples of areas that are considered to be harsh environments and includes "cold storage area." While the annex does not further define the description of a cold storage area, a walk-in refrigerator or walk-in freezer could fall into this category.

2008 Water-Based Fire Protection Systems Handbook

5.3.1.1.3 Where historical data indicate, longer intervals between testing shall be permitted.

5.3.1.2* A representative sample of sprinklers for testing per 5.3.1.1.1 shall consist of a minimum of not less than four sprinklers or 1 percent of the number of sprinklers per individual sprinkler sample, whichever is greater.

The requirement in 5.3.1.2 for a minimum of four samples or 1 percent of the total of sprinklers installed is intended to balance the cost of testing with the likelihood of identifying a possible problem. The sample should be somewhat random and should be representative of the sprinklers installed in the system. For example, sprinklers should be selected from different floors or areas of the building and not selected simply because they are more accessible than other sprinklers. In addition, the selection should take into consideration the age and types of sprinklers as well as environmental conditions to which they are subjected.

The inspector and/or the owner can determine the groups of sprinklers that the sprinkler sample represents. Keep in mind that if a single sprinkler from the sample fails the plunge test, all the sprinklers that the sample represents must be replaced. The sample can represent an entire system or one floor of a multi-story building.

Note that only sprinklers that have been exposed to service conditions must be tested. The sprinklers in the spare sprinkler cabinet, for example, have not been exposed to service conditions and may not reveal any deficiencies.

◄ **FAQ**
Can the sprinklers in the spare sprinkler cabinet be used to comply with 5.3.1.2?

A.5.3.1.2 Within an environment, similar sidewall, upright, and pendent sprinklers produced by the same manufacturer could be considered part of the same sample, but additional sprinklers would be included within the sample if produced by a different manufacturer.

5.3.1.3 Where one sprinkler within a representative sample fails to meet the test requirement, all sprinklers within the area represented by that sample shall be replaced.

The sample size specified in 5.3.1.2 is not very large and will not guarantee that all problems will be discovered. As a result, when a single sprinkler fails, there is a very high probability that many other sprinklers will also fail. Therefore, 5.3.1.3 requires that all of the sprinklers represented by the sample be replaced.

5.3.1.3.1 Manufacturers shall be permitted to make modifications to their own sprinklers in the field with listed devices that restore the original performance as intended by the listing, where acceptable to the authority having jurisdiction.

5.3.2* Gauges.

Gauges shall be replaced every 5 years or tested every 5 years by comparison with a calibrated gauge. Gauges not accurate to within 3 percent of the full scale shall be recalibrated or replaced.

Subsection 5.3.2 of NFPA 25 requires gauges to be tested every five years by comparison with a calibrated gauge. Gauges not accurate to within 3 percent of the full scale are required to be recalibrated or replaced.

Many of the inspections and tests prescribed in NFPA 25 rely on the accuracy of pressure gauges. It is not the intent of 5.3.2 to require that each and every gauge be individually calibrated. It is acceptable to compare the reading of one gauge with that of a calibrated gauge. Other gauges on systems that are installed in similar positions and elevations (for example, on an adjacent riser) that show similar readings are acceptable.

◄ **FAQ**
Does NFPA 25 require that sprinkler system pressure gauges be tested for accuracy or calibration? If so, how often does it have to be done?

A.5.3.2 The normal life expectancy of a gauge is between 10 and 15 years. A gauge can be permitted to have a reading with an error of ±3 percent of the maximum (full scale) gauge reading. For example, a gauge having 200 psi (13.8 bar) maximum radius installed on a system with 60 psi (4.1 bar) normal pressure can be permitted if the gauge reads from 54 psi to 66 psi (3.7 bar to 4.5 bar).

The gauge pictured in Exhibit 5.11 has a maximum radius of 300 psi (20.7 bar). If the normal pressure on the system into which it is installed is 125 psi (8.6 bar) as shown, the actual pressure in the system can vary from 116 psi to 134 psi (8 bar to 9.2 bar). This range is acceptable.

5.3.3 Alarm Devices.

5.3.3.1 Mechanical waterflow devices including, but not limited to, water motor gongs, shall be tested quarterly.

5.3.3.2* Vane-type and pressure switch–type waterflow devices shall be tested semiannually.

A.5.3.3.2 Data concerning reliability of electrical waterflow switches indicate no appreciable change in failure rates for those tested quarterly and those tested semiannually. Mechanical motor gongs, however, have additional mechanical and environmental failure modes and need to be tested more often.

The testing frequency for waterflow alarms was changed from quarterly to semi-annually for the 2002 edition of NFPA 25 but included vane-type devices only. For the 2008 edition of NFPA 25, the technical committee changed the test frequency for pressure switch–type waterflow devices to semi-annual to match that of vane-type devices. The initial change to a semi-annual frequency was made to correlate with the testing frequency for waterflow switches with that of *NFPA 72®*, *National Fire Alarm Code®*. In 1996, *NFPA 72* changed the test frequency for waterflow devices from quarterly to semi-annually based on a survey of test results for these devices. Commentary Table 5.1 contains the results of this survey and clearly indicates that semi-annual testing does not illustrate a significant difference in failure rates from that of quarterly testing. Note, however, that testing on an annual versus semi-annual basis results in an 85 percent increase in failures of waterflow devices. Based on these results, semi-annual testing is appropriate.

EXHIBIT 5.11 System Pressure Gauge.

COMMENTARY TABLE 5.1 Testing Frequency Survey

Frequency of Testing	Devices Tested	Devices Failed	Percent of Failure
Quarterly	13,455	161	1.20
Semiannual	3,355	51	1.52
Annual	1,772	50	2.82
Total	**18,582**	**262**	**1.41**

The waterflow alarm is a key part of the sprinkler system and must be tested. Paragraphs 5.3.3.1 and 5.3.3.2 take into account that each type of alarm has its own failure modes and that some are more susceptible than others to environmental conditions. For example, water motor gongs (see Exhibit 5.12) may be affected more by environmental conditions (bird or wasp nests or freezing) than an electric alarm initiating device such as a pressure switch or vane-type flow switch.

FAQ ▶
Is a licensed fire alarm technician permitted to test waterflow alarms on sprinkler systems?

NFPA 25 does not specify who may perform the required tests other than to require a qualified individual. The definition of *qualified* in 3.3.28 is very general, and the licensing laws of the local jurisdiction need to be consulted in order to determine who is permitted to conduct the tests.

For more information on alarm devices, see the commentary on 13.2.6.

5.3.3.3 Testing the waterflow alarms on wet pipe systems shall be accomplished by opening the inspector's test connection.

EXHIBIT 5.12 Water Motor Alarm Gong (top) and Water Motor Gong with Wasp Nests (bottom).

5.3.3.3.1 Where freezing weather conditions or other circumstances prohibit use of the inspector's test connection, the bypass connection shall be permitted to be used.

5.3.3.4 Fire pumps shall not be turned off during testing unless all impairment procedures contained in Chapter 15 are followed.

As required in 5.3.3.4, it is vital that water supplies remain in service during testing, for several reasons:

1. Turning off fire pumps creates an impairment.
2. Sprinkler systems designed to be supplied by a fire pump may not perform properly without the flow and pressure the pump provides.
3. Keeping the pump in service tests the alarm under the conditions in which it is expected to operate.

Testing with the fire pumps turned on exercises the pump and ensures that adequate water will be available should a fire occur during the test. It is important that a fire pump operator be present in the pump room during operation. When it is necessary to shut down water supplies during testing, extra care and coordination is required to ensure that the pump or other water supply can be promptly returned to operation in an emergency. All of the requirements in Chapter 15 of NFPA 25 should be followed. See NFPA 20, *Standard for the Installation of Stationary Pumps for Fire Protection*, for more information on fire pump operation.

5.3.3.5* Testing the waterflow alarm on dry pipe, preaction, or deluge systems shall be accomplished by using the bypass connection.

A.5.3.3.5 Opening the inspector's test connection can cause the system to trip accidentally.

5.3.4* Antifreeze Systems.

The freezing point of solutions in antifreeze shall be tested annually by measuring the specific gravity with a hydrometer or refractometer and adjusting the solutions if necessary.

A.5.3.4 Listed CPVC sprinkler pipe and fittings should be protected from freezing with glycerin only. The use of diethylene, ethylene, or propylene glycols is specifically prohibited. Where inspecting antifreeze systems employing listed CPVC piping, the solution should be verified to be glycerin based.

Many refractometers are calibrated for a single type of antifreeze solution and will not provide accurate readings for the other types of solutions.

New antifreeze solutions are being introduced for use in early suppression fast-response (ESFR) systems, and NFPA 13 restricts the use of antifreeze in ESFR systems to those that are referenced in NFPA 13 and those that are listed specifically for that use. Further, the ESFR sprinkler must be listed for use with the antifreeze solutions listed in NFPA 13.

Controlling antifreeze concentrations is relatively easy for small systems; however, additional inspection, testing, or maintenance activities may be needed to ensure uniform mixtures when antifreeze systems are used to protect large areas. An example of such additional activities would be circulating the mixture to prevent the antifreeze from settling out of solution.

Subsection 5.3.4 provides needed guidance on the steps required in testing the antifreeze solution in an antifreeze system. Since concentrations can be higher at different points in the system, it is important to evaluate more than one point in the system. The danger of too little antifreeze is obvious, while too much antifreeze may present the same potential to freeze. Too much antifreeze may also present problems of flammability as the type of chemical used in most antifreeze solutions is defined as a Class IIIB Combustible Liquid [a liquid having a flashpoint at or above 200°F (93°C)]. Further, large antifreeze systems should be tested at multiple points to verify a consistent concentration of antifreeze solution throughout. Exhibit 5.13 illustrates antifreeze testing.

FAQ ▶
Is an antifreeze system required to be drained for the annual test?

NFPA 25 does not require the antifreeze system to be drained for the annual test. However, to measure the antifreeze solution properly in smaller systems, the system should be drained and the solution remixed prior to taking a measurement. The requirement in NFPA 25 is to test at the most remote portion of the system. If a test port is not available at that point, then the system must be drained.

5.3.4.1* Solutions shall be in accordance with Table 5.3.4.1(a) and Table 5.3.4.1(b).

A.5.3.4.1 See Figure A.5.3.4.1.

5.3.4.2 The use of antifreeze solutions shall be in accordance with any state or local health regulations.

5.3.4.3 The antifreeze solution shall be tested at its most remote portion and where it interfaces with the wet-pipe system. When antifreeze systems have a capacity larger than 150 gal (568 L), tests at one additional point for every 100 gal (379 L) shall be made. If the test results indicate an incorrect freeze point at any point in the system, the system shall be drained,

EXHIBIT 5.13 *Antifreeze Testing. (1) Obtaining an antifreeze sample by draining a small quantity of antifreeze from the antifreeze system (top). (2) Applying the antifreeze test sample to a refractometer (middle). (3) Reading the refractometric index of the antifreeze sample by holding the refractometer up to a natural light source (bottom).*

TABLE 5.3.4.1(a) *Antifreeze Solutions To Be Used If Nonpotable Water Is Connected to Sprinklers*

Material	Solution (by volume)	Specific Gravity at 60°F (15.6°C)	Freezing Point °F	Freezing Point °C
Glycerine*				
Diethylene glycol	50% water	1.078	−13	−25.0
	45% water	1.081	−27	−32.8
	40% water	1.086	−42	−41.1
	Hydrometer scale 1.000 to 1.120 (subdivisions 0.002)			
Ethylene glycol	61% water	1.056	−10	−23.3
	56% water	1.063	−20	−28.9
	51% water	1.069	−30	−34.4
	47% water	1.073	−40	−40.0
	Hydrometer scale 1.000 to 1.120 (subdivisions 0.002)			
Propylene glycol*				
Calcium chloride 80% "flake"	lb CaCl$_2$/gal of water			
Fire protection grade** Add corrosion inhibitor of sodium bichromate $^3/_4$ oz/gal water	2.83	1.183	0	−17.8
	3.38	1.212	−10	−23.3
	3.89	1.237	−20	−28.9
	4.37	1.258	−30	−34.4
	4.73	1.274	−40	−40.0
	4.93	1.283	−50	−45.6

*If used, see Table 5.3.4.1(b).
**Free from magnesium chloride and other impurities.

TABLE 5.3.4.1(b) *Antifreeze Solutions To Be Used If Potable Water Is Connected to Sprinklers*

Material	Solution (by volume)	Specific Gravity at 60°F (15.6°C)	Freezing Point °F	Freezing Point °C
Glycerine C.P. or U.S.P. grade*	50% water	1.145	−20.9	−29.4
	40% water	1.171	−47.3	−44.1
	30% water	1.197	−22.2	−30.1
	Hydrometer scale 1.000 to 1.200			
Propylene glycol	70% water	1.027	+9	−12.8
	60% water	1.034	−6	−21.2
	50% water	1.041	−26	−32.2
	40% water	1.045	−60	−51.1
	Hydrometer scale 1.000 to 1.200 (subdivisions 0.0002)			

*C.P. = Chemically pure; U.S.P. = United States Pharmacopoeia 96.9%.

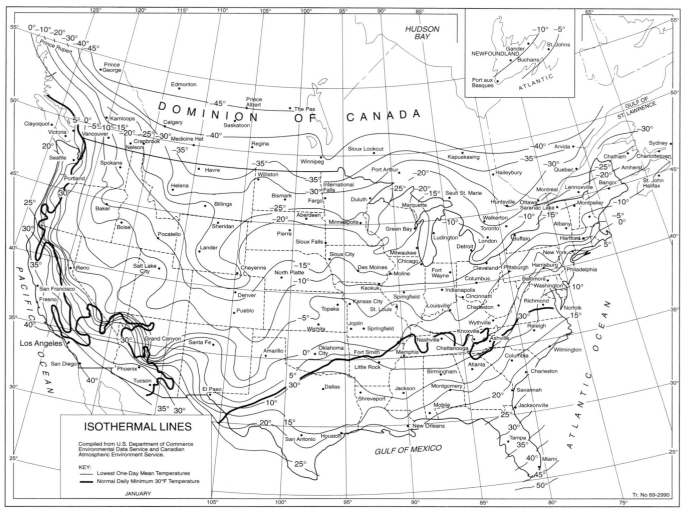

FIGURE A.5.3.4.1 *Isothermal Lines — Lowest One-Day Mean Temperature (°F).*
[24:Figure A.10.5.1]

the solution adjusted, and the systems refilled. For premixed solutions, the manufacturer's instructions shall be permitted to be used with regard to the number of test points and refill procedure.

5.4 Maintenance

5.4.1 Sprinklers.

5.4.1.1* Replacement sprinklers shall have the proper characteristics for the application intended. These shall include the following:

(1) Style
(2) Orifice size and *K*-factor
(3) Temperature rating

(4) Coating, if any
(5) Deflector type (e.g., upright, pendent, sidewall)
(6) Design requirements

There are many types of sprinklers, a number of which have a similar appearance yet have different performance characteristics. Paragraph 5.4.1.1 identifies sprinkler characteristics that must be matched to the application.

Care must be taken to ensure that replacement sprinklers have the same characteristics as those they are replacing so that the design performance of the system is maintained. This is especially important for special sprinklers, as noted in 5.4.1.3. When the same make and model of a special sprinkler is no longer manufactured, persons with appropriate expertise and knowledge should be consulted when selecting a replacement sprinkler.

A.5.4.1.1 To help in the replacement of like sprinklers, unique sprinkler identification numbers (SINs) are provided on all sprinklers manufactured after January 1, 2001. The SIN accounts for differences in orifice size, deflector characteristics, pressure rating, and thermal sensitivity.

5.4.1.1.1* Spray sprinklers shall be permitted to replace old-style sprinklers.

A.5.4.1.1.1 Old-style sprinklers are permitted to replace existing old-style sprinklers. Old-style sprinklers should not be used to replace standard sprinklers without a complete engineering review of the system. The old-style sprinkler is the type manufactured before 1953. It discharges approximately 40 percent of the water upward to the ceiling, and it can be installed in either the upright or pendent position.

5.4.1.1.2 Replacement sprinklers for piers and wharves shall comply with NFPA 307, *Standard for the Construction and Fire Protection of Marine Terminals, Piers, and Wharves*.

5.4.1.2 Only new, listed sprinklers shall be used to replace existing sprinklers.

FAQ ▶
Can a building owner reinstall an existing sprinkler after removing the sprinkler and/or the drop nipple from the branch piping for maintenance or repairs?

Paragraph 5.4.1.2 prohibits the installation of used sprinklers in a new installation or to replace sprinklers that have either activated or become damaged or those that are leaking. This paragraph does not prohibit the relocation of a sprinkler in a system due to a new ceiling configuration, for example. This paragraph also does not prevent the relocation of existing sprinklers where system piping must be relocated due to interferences. However, sprinklers that are removed from one system should never be used in another system, building, or construction project.

5.4.1.3* Special and quick-response sprinklers as defined by NFPA 13, *Standard for the Installation of Sprinkler Systems*, shall be replaced with sprinklers of the same orifice, size, temperature range and thermal response characteristics, and *K*-factor.

Previously, the standard required that sprinklers be replaced by the exact make and model in addition to the same performance characteristics. This requirement has been deleted in the 2008 edition, and the standard simply requires that sprinklers be replaced with types of equal performance characteristics such as orifice size, *K*-factor, temperature rating, and thermal response.

A.5.4.1.3 It is imperative that any replacement sprinkler have the same characteristics as the sprinkler being replaced. If the same temperature range, response characteristics, spacing requirements, flow rates, and *K*-factors cannot be obtained, a sprinkler with similar characteristics should be used, and the system should be evaluated to verify the sprinkler is appropriate for the intended use. With regard to response characteristics, matching identical Response Time Index (RTI) and conductivity factors is not necessary unless special design considerations are given for those specific values.

5.4.1.4* A supply of spare sprinklers (never fewer than six) shall be maintained on the premises so that any sprinklers that have operated or been damaged in any way can be promptly replaced.

Following an incident, it is important to minimize system impairments. Thus, 5.4.1.4 specifies that a supply of spare sprinklers must be available so sprinklers can be replaced quickly following a small fire or accidental discharge. Note that the supply must be "on the premises," but need not be in each building of a complex. For example, a supply of spare sprinklers for a complex can be stored in a central location, provided that the supply is accessible and does not substantially delay the replacement of sprinklers following an incident. (A sprinkler cabinet with spare sprinklers is shown in Exhibit 5.6.)

A.5.4.1.4 A minimum of two sprinklers of each type and temperature rating installed should be provided.

5.4.1.4.1 The sprinklers shall correspond to the types and temperature ratings of the sprinklers in the property.

5.4.1.4.2 The sprinklers shall be kept in a cabinet located where the temperature in which they are subjected will at no time exceed 100°F (38°C).

Spare sprinklers must be maintained at normal temperatures to prevent their operation and also to prevent the phenomenon of "cold flow," which can result in premature operation. The 100°F (38°C) temperature specified in 5.4.1.4.2 is a safe storage temperature for sprinklers of all temperature ratings.

5.4.1.4.2.1 Where dry sprinklers of different lengths are installed, spare dry sprinklers shall not be required, provided that a means of returning the system to service is furnished.

5.4.1.5 The stock of spare sprinklers shall include all types and ratings installed and shall be as follows:

(1) For protected facilities having under 300 sprinklers — no fewer than 6 sprinklers
(2) For protected facilities having 300 to 1000 sprinklers — no fewer than 12 sprinklers
(3) For protected facilities having over 1000 sprinklers — no fewer than 24 sprinklers

5.4.1.6* A special sprinkler wrench shall be provided and kept in the cabinet to be used in the removal and installation of sprinklers. One sprinkler wrench shall be provided for each type of sprinkler installed.

Because modern sprinklers are delicate devices that can be damaged by mishandling, 5.4.1.6 mandates that a special wrench be used to prevent damage during installation. Special wrenches must be kept in the sprinkler cabinet so that sprinklers can be properly replaced following an incident or in the event inspection or testing indicates damage.

A.5.4.1.6 Other types of wrenches could damage the sprinklers.

5.4.1.7 Sprinklers protecting spray coating areas shall be protected against overspray residue.

As noted in the commentary for 5.2.1.1.2, paint or overspray can prevent the proper operation of sprinklers. Paragraph 5.4.1.7 requires sprinklers to be protected from such overspray.

5.4.1.7.1 Sprinklers subject to overspray accumulations shall be protected using plastic bags having a maximum thickness of 0.003 in. (0.076 mm) or shall be protected with small paper bags.

5.4.1.7.2 Coverings shall be replaced when deposits or residue accumulate.

Testing has shown that lightweight plastic or paper bags will not adversely affect the operation of the sprinkler. Sprinklers protected by lightweight plastic or paper bags may require more frequent inspection than the annual inspection outlined in 5.2.1.1.2 to prevent excessive buildup on the bags. Exhibit 5.14 illustrates a sprinkler protected from overspray by a plastic bag.

5.4.1.8* Sprinklers shall not be altered in any respect or have any type of ornamentation, paint, or coatings applied after shipment from the place of manufacture.

A.5.4.1.8 Corrosion-resistant or specially coated sprinklers should be installed in locations where chemicals, moisture, or other corrosive vapors exist.

5.4.1.9 Sprinklers and automatic spray nozzles used for protecting commercial-type cooking equipment and ventilating systems shall be replaced annually.

5.4.1.9.1 Where automatic bulb-type sprinklers or spray nozzles are used and annual examination shows no buildup of grease or other material on the sprinklers or spray nozzles, such sprinklers and spray nozzles shall not be required to be replaced.

5.4.2* Dry Pipe Systems.

Dry pipe systems shall be kept dry at all times.

 A.5.4.2 Conversion of dry pipe systems to wet pipe systems on a seasonal basis causes corrosion and accumulation of foreign matter in the pipe system and loss of alarm service.

Moisture is one of the main causes of corrosion in sprinkler piping. Alternating from a dry system to a wet system allows moisture to enter the system and can accelerate corrosion rates. In addition, the waterflow alarm on most dry systems will need to be removed from service if the system is left wet. To minimize corrosion and prevent freezing of sprinkler piping, it is recommended that these systems always remain dry and that they be reset and drained following operation to minimize moisture inside the system. Exhibit 5.15 shows rust and scale removed from a dry pipe system.

5.4.2.1 During nonfreezing weather, a dry pipe system shall be permitted to be left wet if the only other option is to remove the system from service while waiting for parts or during repair activities.

EXHIBIT 5.14 Sprinkler Protected from Overspray with Plastic Bag.

EXHIBIT 5.15 Rust and Scale Removed from Dry Pipe System.

5.4.2.2 Air driers shall be maintained in accordance with the manufacturer's instructions.

5.4.2.3 Compressors used in conjunction with dry pipe sprinkler systems shall be maintained in accordance with the manufacturer's instructions.

5.4.3* Installation and Acceptance Testing.

Where maintenance or repair requires the replacement of sprinkler system components affecting more than 20 sprinklers, those components shall be installed and tested in accordance with NFPA 13, *Standard for the Installation of Sprinkler Systems*.

A.5.4.3 Where pressure testing listed CPVC piping, the sprinkler systems should be filled with water and air should be bled from the highest and farthest sprinkler before test pressure is applied. Air or compressed gas should never be used for pressure testing.

For repairs affecting the installation of less than 20 sprinklers, a test for leakage should be made at normal system working pressure.

5.4.4* Marine Systems.

Sprinkler systems that are normally maintained using fresh water as a source shall be drained and refilled, then drained and refilled again with fresh water following the introduction of raw water into the system.

A.5.4.4 Certain sprinkler systems, such as those installed aboard ships, are maintained under pressure by a small fresh-water supply but are supplied by a raw water source following system activation. In these systems, the effects of raw water are minimized by draining and refilling with freshwater. For systems on ships, flushing within 45 days or the vessel's next port of call, whichever is longer, is considered acceptable.

5.5 Component Action Requirements

Component replacement tables were added in the 2008 edition to offer guidance to the user of the standard when system components are adjusted, repaired, rebuilt, or replaced. It is not necessary in each case to require a complete acceptance test for each component when maintenance is performed.

5.5.1 Whenever a component in a sprinkler system is adjusted, repaired, reconditioned, or replaced, the actions required in Table 5.5.1 shall be performed.

5.5.1.1 Where the original installation standard is different from the cited standard, the use of the appropriate installing standard shall be permitted.

5.5.1.2 A main drain test shall be required if the system control or other upstream valve was operated in accordance with 13.3.3.4.

5.5.1.3 These actions shall not require a design review, which is outside the scope of this standard.

SUMMARY

As was illustrated in Commentary Table 1.1, inadequate maintenance accounts for 8.4 percent of sprinkler system failures. Chapter 5 of NFPA 25 addresses the components that require periodic testing and maintenance to avoid such failures. The inspections required by this chapter can be considered preventive maintenance in that they can reveal potential problems before those problems compromise system performance. As mentioned in the commentary at

TABLE 5.5.1 Summary of Component Replacement Action Requirements

Component	Adjust	Repair/ Recondition	Replace	Required Action
Water Delivery Components				
Pipe and fittings affecting less than 20 sprinklers	X	X	X	Check for leaks at system working pressure
Pipe and fittings affecting more than 20 sprinklers	X	X	X	Hydrostatic test in conformance with NFPA 13, *Standard for the Installation of Sprinkler Systems*
Sprinklers, less than 20	X		X	Check for leaks at system working pressure
Sprinklers, more than 20	X		X	Hydrostatic test in conformance with NFPA 13
Fire department connections	X	X	X	See Chapter 13
Antifreeze solution	X		X	Check freezing point of solution
				Check for leaks at system working pressure
Valves				See Chapter 13
Fire pump				See Chapter 8
Alarm and Supervisory Components				
Vane-type waterflow	X	X	X	Operational test using inspector's test connection
Pressure switch–type waterflow	X	X	X	Operational test using inspector's test connection
Water motor gong	X	X	X	Operational test using inspector's test connection
High and low air pressure switch	X	X	X	Operational test of high and low settings
Valve supervisory device	X	X	X	Test for conformance with NFPA 13 and/or NFPA 72, *National Fire Alarm Code*
Detection system (for deluge or preaction system)	X	X	X	Operational test for conformance with NFPA 13 and/or *NFPA 72*
Status-Indicating Components				
Gauges			X	Verify at 0 bar (0 psi) and system working pressure
Testing and Maintenance Components				
Air compressor	X	X	X	Operational test for conformance with NFPA 13
Automatic air maintenance device	X	X	X	Operational test for conformance with NFPA 13
Main drain	X	X	X	Main drain test
Auxiliary drains	X	X	X	Check for leaks at system working pressure
				Main drain test
Inspector's test connection	X	X	X	Check for leaks at system working pressure
				Main drain test
Structural Components				
Hanger/seismic bracing	X	X	X	Check for conformance with NFPA 13
Pipe stands	X	X	X	Check for conformance with NFPA 13
Informational Components				
Identificaion Signs	X	X	X	Check for conformance with NFPA 13
Hydraulic Placards	X	X	X	Check for conformance with NFPA 13

the beginning of this chapter, many of the basic weekly and monthly inspection activities can be performed by the property owner.

Care of the sprinklers themselves is important because it is their temperature-sensing and water-distribution capabilities that make a sprinkler system so effective. This chapter pays particular attention to the inspection of sprinklers to verify that the physical clearance of storage required under a sprinkler is maintained, to allow development of the water spray pattern for proper water distribution. This chapter of the standard also calls attention to the condition of the sprinkler and operating element for proper temperature sensing, by requiring laboratory testing to ensure that the sprinkler will operate at its designed temperature rating and sensitivity.

REVIEW QUESTIONS

1. What quantity and type of spare sprinklers must be provided for a sprinkler system having 150 brass upright, 165°F (74°C), and 149 brass upright, 286°F (141°C), sprinklers?
2. How often should extra high–temperature sprinklers be tested?
3. How often are sprinkler alarms required to be tested?
4. What action should be taken if obstructing material is observed in the discharge from the inspector's test connection?
5. How can sprinklers be protected from paint overspray?
6. Can a loaded sprinkler be cleaned?
7. Is a license required to inspect, test, or maintain sprinkler systems?
8. Does the property owner need a license to inspect, test, or maintain the property's sprinkler system?

REFERENCES CITED IN COMMENTARY

National Fire Protection Association, 1 Batterymarch Park, Quincy, MA 02169-7471.

NFPA 13, *Standard for the Installation of Sprinkler Systems*, 2007 edition.
NFPA 20, *Standard for the Installation of Stationary Pumps for Fire Protection*, 2007 edition.
NFPA 72®, *National Fire Alarm Code®*, 2007 edition.

Standpipe and Hose Systems

CHAPTER 6

Chapter 6 of NFPA 25 covers the inspection, testing, and maintenance of standpipe systems. Standpipe systems may or may not include hose and nozzles. When hose and nozzles are part of the system, these critical components must be tested and maintained properly. Hose must be dried and folded correctly before being placed in a rack. It must also be hydrostatically tested periodically to verify that it is still capable of holding water pressure without failure. Failure of standpipe equipment can result in unsafe conditions for the user of the equipment.

Exhibit 6.1 illustrates a hose rack assembly in poor condition. The hose is not racked properly and shows no signs of proper care. Exhibit 6.2 shows a properly racked hose and nozzle. This installation demonstrates a properly maintained standpipe system hose rack assembly.

6.1 General

This chapter shall provide the minimum requirements for the routine inspection, testing, and maintenance of standpipe and hose systems. Table 6.1 shall be used to determine the minimum required frequencies for inspection, testing, and maintenance.

EXHIBIT 6.1 Hose Rack Assembly in Poor Condition.

EXHIBIT 6.2 Hose Rack Assembly in Proper Condition.

TABLE 6.1 Summary of Standpipe and Hose Systems Inspection, Testing, and Maintenance

Item	Frequency	Reference
Inspection		
Control valves	Weekly/monthly	Table 13.1
Pressure regulating devices	Quarterly	Table 13.1
Piping	Annually	6.2.1
Hose connections	Annually	Table 13.1
Cabinet	Annually	NFPA 1962, *Standard for the Inspection, Care, and Use of Fire Hose, Couplings, and Nozzles and the Service Testing of Fire Hose*
Hose	Annually	NFPA 1962
Hose storage device	Annually	NFPA 1962
Hose nozzle	Annually and after each use	NFPA 1962
Test		
Waterflow devices	Quarterly/semiannually	Table 13.1
Valve supervisory devices	Semiannually	Table 13.1
Supervisory signal devices (except valve supervisory switches)	Semiannually	Table 13.1
Hose storage device	Annually	NFPA 1962
Hose	5 years/3 years	NFPA 1962
Pressure control valve	5 years	Table 13.1
Pressure reducing valve	5 years	Table 13.1
Hydrostatic test	5 years	6.3.2
Flow test	5 years	6.3.1
Main drain test	Annually	Table 13.1
Maintenance		
Hose connections	Annually	Table 6.2.2
Valves (all types)	Annually/as needed	Table 13.1

The inspections required by Table 6.1 can be performed with minimal training and in many cases can be conducted by the property owner or his or her representative. These inspections are as follows:

- Control valves — verify operating position.
- Piping — look for signs of obvious damage or leaks, and so on.
- Hose, hose connections, and cabinets — inspect to ensure that the equipment has not been vandalized, is free of debris, and exhibits no evidence of mildew, rot, or damage from chemicals, burns, cuts, abrasion, or vermin.

Except for those items listed above, a qualified inspector or contractor should perform all testing and maintenance activities listed in Table 6.1.

6.1.1 Valves and Connections.

Valves and fire department connections shall be inspected, tested, and maintained in accordance with Chapter 13.

6.1.2 Impairments.

Where the inspection, testing, and maintenance of standpipe and hose systems results or involves a system that is out of service, the procedures outlined in Chapter 15 shall be followed.

6.2 Inspection

6.2.1 Components of standpipe and hose systems shall be visually inspected annually or as specified in Table 6.1.

Inspection of piping and hose connections has been changed from quarterly to annually in the 2008 edition because the committee feels that this equipment does not require such rigorous inspections. The inspection frequencies for pressure regulating devices and alarms were not changed because these devices are addressed in Chapter 13 and require different inspection frequencies.

The piping in Exhibit 6.3 is severely corroded and could fail if pressurized. Such a failure under pressure could present a safety hazard to the end user or to nearby building occupants. It is obvious from this photo that this system has not been inspected for some time.

6.2.2 Table 6.2.2 shall be used for the inspection, testing, and maintenance of all classes of standpipe and hose systems.

6.2.3 Checkpoints and corrective actions outlined in Table 6.2.2 shall be followed to determine that components are free of corrosion, foreign material, physical damage, tampering, or other conditions that adversely affect system operation.

◄ FAQ
Why was the inspection of components changed from quarterly to annually?

6.3 Testing

Where water damage is a possibility, an air test shall be conducted on the system at 25 psi (1.7 bar) prior to introducing water to the system.

The rationale for Section 6.3 is that extreme care is needed when testing with pressurized air due to the amount of energy stored in the compressed air that can be released if a portion of the system fails catastrophically. This section relates to standpipe systems that are normally

EXHIBIT 6.3 Severely Corroded Manual Dry Standpipe.

TABLE 6.2.2 Standpipe and Hose Systems

Component/Checkpoint	Corrective Action
Hose Connections	
Cap missing	Replace
Fire hose connection damaged	Repair
Valve handles missing	Replace
Cap gaskets missing or deteriorated	Replace
Valve leaking	Close or repair
Visible obstructions	Remove
Restricting device missing	Replace
Manual, semiautomatic, or dry standpipe — valve does not operate smoothly	Lubricate or repair
Piping	
Damaged piping	Repair
Control valves damaged	Repair or replace
Missing or damaged pipe support device	Repair or replace
Damaged supervisory devices	Repair or replace
Hose	
Inspect	Remove and inspect the hose, including gaskets, and rerack or rereel at intervals in accordance with NFPA 1962, *Standard for the Care, Use, and Service Testing of Fire Hose Including Couplings and Nozzles*
Mildew, cuts, abrasions, and deterioration evident	Replace with listed lined, jacketed hose
Coupling damaged	Replace or repair
Gaskets missing or deteriorated	Replace
Incompatible threads on coupling	Replace or provide thread adapter
Hose not connected to hose rack nipple or valve	Connect
Hose test outdated	Retest or replace in accordance with NFPA 1962
Hose Nozzle	
Hose nozzle missing	Replace with listed nozzle
Gasket missing or deteriorated	Replace
Obstructions	Remove
Nozzle does not operate smoothly	Repair or replace
Hose Storage Device	
Difficult to operate	Repair or replace
Damaged	Repair or replace
Obstruction	Remove
Hose improperly racked or rolled	Remove
Nozzle clip in place and nozzle correctly contained?	Replace if necessary
If enclosed in cabinet, will hose rack swing out at least 90 degrees?	Repair or remove any obstructions
Cabinet	
Check overall condition for corroded or damaged parts	Repair or replace parts; replace entire cabinet if necessary
Difficult to open	Repair
Cabinet door will not open fully	Repair or move obstructions
Door glazing cracked or broken	Replace
If cabinet is break-glass type, is lock functioning properly?	Repair or replace
Glass break device missing or not attached	Replace or attach
Not properly identified as containing fire equipment	Provide identification
Visible obstructions	Remove
All valves, hose, nozzles, fire extinguisher, etc., easily accessible	Remove any material not related

dry, such as a manual dry or semiautomatic dry standpipe systems. An automatic dry system is already pressurized with air. Wet systems of all types do not require an air test as leaks will be readily apparent in these systems. Pressures in excess of the 25 psi (1.7 bar) recommended should therefore be avoided.

6.3.1 Flow Tests.

6.3.1.1* A flow test shall be conducted every 5 years at the hydraulically most remote hose connections of each zone of an automatic standpipe system to verify the water supply still provides the design pressure at the required flow.

A.6.3.1.1 The hydraulically most remote hose connections in a building are generally at a roof manifold, if provided, or at the top of a stair leading to the roof. In a multizone system, the testing means is generally at a test header at grade or at a suction tank on higher floors.

A flow test is conducted to verify that the water supply for the system is still available and that all of the devices that could restrict flow in the standpipe are operating properly.

The test is intended to verify the design flow and pressure. Design pressure will vary depending on the age of the system. For example, systems installed prior to 1993 should be tested to verify 65 psi (4.5 bar) at the topmost outlet. Systems installed after 1993 should be tested to verify 100 psi (6.9 bar) (at the topmost outlet. The test flow must be based on the calculated flow of 500 gpm (1892 L/min) for the most demanding riser and 250 gpm (946 L/min) for each additional riser up to a maximum of 1250 gpm (4731 L/min) for buildings that are partially sprinklered and 1000 gpm (3785 L/min) for buildings that are fully sprinklered. Note that NFPA 14, *Standard for the Installation of Standpipe and Hose Systems*, requires an additional outlet at the top of the most demanding standpipe for testing purposes. This requirement is intended to facilitate testing of the standpipe system at the design flow of 500 gpm (1892 L/min) [250 gpm × 2 = 500 gpm (946 L/min × 2 = 1892 L/min)] from NFPA 14.

Exhibit 6.4 illustrates the five-year flow test for a manual dry standpipe system with three risers or standpipes. Since the system is manual dry, there is no permanently attached water supply, and a fire department pumper (in this case) must be used to perform the test.

Usually, the most remote flow device, as referred to in 6.3.1.1, is located on the roof or upper floors. Testing of devices in such locations requires coordination to conduct the test safely, prevent water damage, and ensure proper water disposal. In some cases, special drain risers may be needed to remove the water generated by the test. If the test cannot be conducted at the most remote outlet, other arrangements can be made when acceptable to the AHJ.

6.3.1.2 Where a flow test of the hydraulically most remote outlet(s) is not practical, the authority having jurisdiction shall be consulted for the appropriate location for the test.

6.3.1.3 All systems shall be flow tested and pressure tested at the requirements for the design criteria in effect at the time of the installation.

Paragraph 6.3.1.3 points out the significant differences in design pressure based on the date of installation of the standpipe. In the 1950s, standpipe systems were designed to flow water at a pressure of 50 psi (3.5 bar). The 65 psi (4.5 bar) pressure requirement was introduced into NFPA 14 in the early 1970s and was subsequently revised to the current pressure of 100 psi (6.9 bar). It is important to remember these pressure differences and to determine the date of installation of the standpipe system in order to test to the proper pressure. Obviously, a standpipe system that was designed to operate at 65 psi (4.5 bar) will most likely not flow water at 100 psi (6.9 bar).

◄ **FAQ**
What are some of the possible design pressures based on date of installation?

6.3.1.3.1 The actual test method(s) and performance criteria shall be discussed in advance with the authority having jurisdiction.

EXHIBIT 6.4 Five-Year Flow Test. (Top left) Flow test from most remote standpipe. (Top right) Flow test from second most remote standpipe. (Bottom left) Flow test from third most remote standpipe. (Bottom right) Fire department pump used for standpipe system flow test.

6.3.1.4 Standpipes, sprinkler connections to standpipes, or hose stations equipped with pressure reducing valves or pressure regulating valves shall have these valves inspected, tested, and maintained in accordance with the requirements of Chapter 13.

> **CASE STUDY**
>
> **Internal Fire Suppression Systems Hampered by Inadequate Fire Attack Hose Stream Pressure**
>
> Fire in a 38-story high-rise building in Philadelphia resulted in the death of three fire fighters, fire extension to nine floors, and severe structural damage to the building. In addition, 24 fire fighters and 1 civilian were injured in this mostly unoccupied office building.
>
> The fire department received the initial alarm from a person located outside the building just before 8:30 p.m. Upon arrival at One Meridian Plaza, the fire fighters observed heavy smoke at the mid-height of the building. During the $18\frac{1}{2}$ hour effort

to control the blaze, interior fire suppression activities were hampered by the loss of electrical power (including emergency power) and inadequate fire attack hose stream pressure. As a result, the fire spread from the floor of origin, the 22nd floor, to the 29th floor by various spread mechanisms. Vertical fire spread was eventually stopped by the 30th floor automatic sprinkler system supplied by fire department pumpers through the siamese connection. One of the major factors contributing to the loss of life and the severity of the fire was found to be inadequate pressures for fire attack hose lines due to the improper setting of the standpipe pressure regulating valves.

6.3.1.5 A main drain test shall be performed on all standpipe systems with automatic water supplies in accordance with the requirements of Chapter 13.

6.3.1.5.1 The test shall be performed at the low point drain for each standpipe or the main drain test connection where the supply main enters the building (when provided).

6.3.1.5.2 Pressure gauges shall be provided for the test and shall be maintained in accordance with 5.3.2.

6.3.2 Hydrostatic Tests.

6.3.2.1 Hydrostatic tests of not less than 200 psi (13.8 bar) pressure for 2 hours, or at 50 psi (3.4 bar) in excess of the maximum pressure, where maximum pressure is in excess of 150 psi (10.3 bar), shall be conducted every 5 years on manual standpipe systems and automatic-dry standpipe systems, including piping in the fire department connection.

Paragraph 6.3.2.1 requires a hydrostatic test every five years for manual standpipe systems and automatic dry standpipe systems to detect failures before they become catastrophic in nature.

◀ **FAQ**
Why is a hydrostatic test required for a manual dry standpipe system?

Problems with piping integrity in wet systems can be detected by leaks. However, it is not possible to detect leakage within dry standpipes, which are more susceptible to corrosion due to the combination of air and moisture in the system. Undetected leaks can lead to failures when the dry standpipe is pressurized.

6.3.2.2* Hydrostatic tests shall be conducted in accordance with 6.3.2.1 on any system that has been modified or repaired.

A.6.3.2.2 The intent of 6.3.2.2 is to ascertain whether the system retains its integrity under fire conditions. Minimum leakage existing only under test pressure is not cause for repair.

6.3.2.2.1 Manual wet standpipes that are part of a combined sprinkler/standpipe system shall not be required to be tested in accordance with 6.3.2.1.

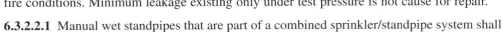

It is not the intent to require a hydrostatic test on manual wet standpipes that are part of a combined system since leaks in this type of system are usually detected immediately.

6.3.2.3 The hydrostatic test pressure shall be measured at the low elevation point of the individual system or zone being tested. The inside standpipe piping shall show no leakage.

6.3.3 Alarm Devices.

Where provided, waterflow alarm and supervisory devices shall be tested in accordance with 13.2.6 and 13.3.3.5.

6.3.3.1 Where freezing conditions necessitate a delay in testing, tests shall be performed as soon as weather allows.

TABLE 6.5.1 Summary of Component Replacement Action Requirements

Component	Adjust	Repair	Replace	Required Action
Water Delivery Components				
Control valves	X	X	X	See Chapter 13
Hose valve pressure regulating devices	X	X	X	See Chapter 13
System pressure regulating devices	X	X	X	See Chapter 13
Piping	X	X	X	Hydrostatic test in conformance with NFPA 14, *Standard for the Installation of Standpipe and Hose Systems*
Fire hose			X	
Hose valve	X	X	X	See Chapter 13
Fire department connections	X	X	X	See Chapter 13
Backflow prevention device	X	X	X	See Chapter 13
Valves				See Chapter 13
Fire pump				See Chapter 8
Alarm and Supervisory Components				
Vane-type waterflow	X	X		Operational test using inspector's test connection
Vane-type waterflow			X	Operational test using inspector's test connection
Pressure switch–type waterflow	X	X	X	Operational test using inspector's test connection
Water motor gong	X	X	X	Operational test using inspector's test connection
Valve supervisory device	X	X	X	Operational test for receipt of alarms and verification of conformance with NFPA 14 and/or *NFPA 72, National Fire Alarm Code*
Status-Indicating Components				
Gauges			X	Verify at 0 psi and system working pressure
System Housing and Protection Components				
Cabinet	X	X	X	Verify compliance with NFPA 14
Hose storage rack	X	X	X	Verify compliance with NFPA 14
Testing and Maintenance Components				
Drain riser	X	X	X	Check for leaks while flowing from connection above the repair
Auxiliary drains	X	X	X	Check for leaks at system working pressure
Main drain	X	X	X	Check for leaks and residual pressure during main drain test
Structural Components				
Hanger/seismic bracing	X	X	X	Verify conformance with NFPA 14
Pipe stands	X	X	X	Verify conformance with NFPA 14
Informational Components				
Identification signs	X	X	X	Verify conformance with NFPA 14
Hydraulic placards	X	X	X	Verify conformance with NFPA 14

6.4 Maintenance

Maintenance and repairs shall be in accordance with 6.2.3 and Table 6.2.2.

6.4.1 Equipment that does not pass the inspection or testing requirements shall be repaired and tested again or replaced.

This requirement relates to hose only. Following operation, hose must be cleaned, dried, and inspected, as specified in 6.4.1, prior to being returned to service. If not dried, wet hose can deteriorate (especially unlined hose) and fail the next time it is used. Hose that is cut or otherwise damaged during use can also fail during subsequent use. Hose that cannot be repaired should be replaced.

6.5 Component Action Requirements

Component replacement tables were added in the 2008 edition to offer guidance to the user of the standard when system components are adjusted, repaired, rebuilt, or replaced. It is not necessary in each case to require a complete acceptance test for each component when maintenance is performed.

6.5.1 Whenever a component in a standpipe and hose system is adjusted, repaired, reconditioned or replaced, the action required in Table 6.5.1, Summary of Component Replacement Action Requirements, shall be performed.

6.5.1.1 Where the original installation standard is different from the cited standard, the use of the appropriate installing standard shall be permitted.

6.5.1.2 A main drain test shall be required if the control valve or other upstream valve was operated in accordance with 13.3.3.4.

6.5.1.3 These actions shall not require a design review, which is outside the scope of this standard.

SUMMARY

Chapter 6 of NFPA 25 covers the inspection, testing, and maintenance of all types of standpipe systems covered in NFPA 14, including Class I standpipes (for fire department use), Class II standpipes (for building occupant use), and Class III standpipes (intended for use by both fire departments and building occupants). Failure of this equipment can create an unsafe condition for the user; therefore, it is extremely important to verify the operation of a standpipe system.

REVIEW QUESTIONS

1. When are hydrostatic tests required on a standpipe system?
2. What design pressure and required flow must be measured during the five-year flow test?
3. When conducting an air test of a standpipe system, how much air pressure should be used?

REFERENCE CITED IN COMMENTARY

National Fire Protection Association, 1 Batterymarch Park, Quincy, MA 02169-7471.

NFPA 14, *Standard for the Installation of Standpipe and Hose Systems,* 2007 edition.

Private Fire Service Mains

CHAPTER 7

Chapter 7 of NFPA 25 deals with the inspection, testing, and maintenance of private underground mains and their appurtenances. The term *private* indicates that the piping and equipment is located on private property. The flow testing discussed in this chapter is intended to apply to hydrants located on private property, although the procedures are the same for testing hydrants on public property.

Exhibit 7.1 illustrates the boundary between public and private equipment and systems. Generally, when piping enters private property, the property owner is responsible for inspection, testing, and maintenance of piping and related equipment. Piping and equipment on the public side of the property line generally is the responsibility of the water purveyor.

7.1 General

This chapter shall provide the minimum requirements for the routine inspection, testing, and maintenance of private fire service mains and their appurtenances. Table 7.1 shall be used to determine the minimum required frequencies for inspection, testing, and maintenance.

The purpose of Section 7.1 mirrors the basic purpose of the standard; that is, "to provide requirements that ensure a reasonable degree of protection for life and property from fire through minimum inspection, testing, and maintenance methods. . . ." It should be recognized that when inspecting equipment such as private fire service mains, a visual examination of the equipment is usually not possible. Much of the inspection, therefore, centers on the inspection, testing, and maintenance of the attached equipment, such as hydrants, hose houses, and monitor nozzles. Testing of this equipment is necessary to assess the condition of the underground piping.

Section 7.1 specifically states that the chapter applies to private fire service mains and appurtenances. In this case, this is the equipment that has been installed in accordance with NFPA 24, *Standard for the Installation of Private Fire Service Mains and Their Appurtenances*. Neither standard is intended to apply to public systems.

◄ **FAQ**
Is the intent of NFPA 25 to require the inspection, testing, and maintenance requirements of fire hydrants (specifically wet barrel) to apply to both private and public hydrants?

7.1.1 Valves and Connections.

Valves and fire department connections shall be inspected, tested, and maintained in accordance with Chapter 13.

7.1.2 Fire Hose.

Fire hose shall be maintained in accordance with NFPA 1962, *Standard for the Inspection, Care, and Use of Fire Hose, Couplings, and Nozzles and the Service Testing of Fire Hose*.

7.1.3 Impairments.

The procedures outlined in Chapter 15 shall be followed wherever such an impairment to protection occurs.

EXHIBIT 7.1 Public/Private Equipment Boundary. (Source: NFPA 24, 2007, Figure A.3.3.11)

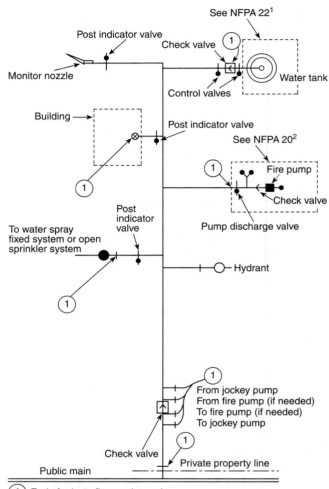

① End of private fire service main

Note: The piping (aboveground or buried) shown is specific as to the end of the private fire service main, and this schematic is only for illustrative purposes beyond the end of the fire service main. Details of valves and their location requirements are covered in the specific standard involved.

1. See NFPA 22, *Standard for Water Tanks for Private Fire Protection*, 2003.
2. See NFPA 20, *Standard for the Installation of Stationary Pumps for Fire Protection*, 2007.

Impairments such as those referred to in 7.1.3 involve removing a system from service. Impairments can be of an emergency nature, such as a broken fire service main, or they can be preplanned, as in cases of a physical modification to the system or routine maintenance of the equipment. In either case, an impairment program must be in place to minimize the length of time the system is impaired and to verify that systems and valves are properly returned to service upon completion of the work.

An impairment program is critical for any work associated with a private fire service main because impairment of these systems involves shutting water supply control valves to one or more water-based fire protection systems. As mentioned in the commentary on Chapter 1, shut valves are the most common cause of sprinkler system failure (see Commentary Table 1.1).

TABLE 7.1 Summary of Private Fire Service Main Inspection, Testing, and Maintenance

Item	Frequency	Reference
Inspection		
Hose houses	Quarterly	7.2.2.7
Hydrants (dry barrel and wall)	Annually and after each operation	7.2.2.4
Monitor nozzles	Semiannually	7.2.2.6
Hydrants (wet barrel)	Annually and after each operation	7.2.2.5
Mainline strainers	Annually and after each significant flow	7.2.2.3
Piping (exposed)	Annually	7.2.2.1
Piping (underground)	See 7.2.2.2	7.2.2.2
Test		
Monitor nozzles	Flow, annually (range and operation)	7.3.3
Hydrants	Flow, annually	7.3.2
Piping (exposed and underground) (flow test)	5 years	7.3.1
Maintenance		
Mainline strainers	Annually and after each operation	7.2.2.3
Hose houses	Annually	7.2.2.7
Hydrants	Annually	7.4.2
Monitor nozzles	Annually	7.4.3

7.1.4 Notification to Supervisory Service.

To avoid false alarms where a supervisory service is provided, the alarm receiving facilities always shall be notified by the property owner or designated representative as follows:

(1) Before conducting any test or procedure that could result in the activation of an alarm
(2) After such tests or procedures are concluded

When flowing an inspector's test connection, it may be obvious that an alarm is intended to sound upon waterflow and that the supervisory service must be notified, as prescribed by 7.1.4. Perhaps not so obvious is the fact that other tests, such as a fire pump test, hydrant test, or main drain test, can also lead to the tripping of a waterflow switch, sounding an alarm and potentially sending a signal to the fire department. At the conclusion of such tests, the alarm-receiving facility should be notified that any alarms received from that point on are not test-generated and should be responded to appropriately.

7.2 Inspection

7.2.1 General.

Private fire service mains and their appurtenances shall be inspected at the intervals specified in Table 7.1.

Generally, as required by 7.2.1, inspectors should be checking each inspection point as listed in Table 7.1. The intent is to perform a quick visual examination to verify that the system and components appear to be in good working order. As referenced in Chapter 1, it is important to verify that water supply control valves are in the full open position.

7.2.2* Procedures.

All procedures shall be carried out in accordance with the manufacturer's instructions, where applicable.

A.7.2.2 The requirements in 7.2.2 outline inspection intervals, conditions to be inspected, and corrective actions necessary for private fire service mains and associated equipment.

7.2.2.1 Exposed Piping.

7.2.2.1.1 Exposed piping shall be inspected annually.

7.2.2.1.2 Piping shall be inspected, and the necessary corrective action shall be taken as shown in Table 7.2.2.1.2.

TABLE 7.2.2.1.2 Exposed Piping

Condition	Corrective Action
Leaks	Repair
Physical damage	Repair or replace
Corrosion	Clean or replace and coat with corrosion protection
Restraint methods	Repair or replace

7.2.2.1.3 Piping installed in areas that are inaccessible for safety considerations due to process operations shall be inspected during each scheduled shutdown.

7.2.2.2 Underground Piping. Generally, underground piping cannot be inspected on a routine basis. However, flow testing can reveal the condition of underground piping and shall be conducted in accordance with Section 7.3.

CASE STUDY

Broken Water Main Disables Wet Pipe Sprinkler System and Hydrants

Fire at a sawmill complex in Louisiana destroyed a storage shed. Damage to the property was estimated at $1.3 million, a total loss. The building materials storage shed was located on a 52-acre complex that included a sawmill that manufactured plywood. The single-floor building measured 300 ft by 360 ft and was two-and-a-half stories high. It was constructed of heavy timber on a metal frame and was covered by metal siding. A wet pipe sprinkler system was in place to protect the storage shed, but the connecting water main was broken, disabling the system and two nearby hydrants.

The building was closed for the night when a security guard making his rounds detected the fire at 8:29 p.m. He notified the police department, but the call was incorrectly dispatched as a house fire to the wrong address, causing a delay of approximately 10 minutes in fire fighter response.

By the time fire fighters arrived, the blaze had spread through the roof, and fire fighters were forced to take a defensive attack to protect exposures. They tried to use a pump hose, but its proximity to the fire and failure to operate further limited the water supply. The cause of the blaze was undetermined. An employee who had been on site about an hour before the fire had detected nothing unusual. One fire fighter received minor injuries.

7.2.2.3* Mainline Strainers. Mainline strainers shall be inspected and cleaned after each system flow exceeding that of a nominal 2 in. (50 mm) orifice and shall be removed and inspected annually for failing, damaged, and corroded parts with the necessary corrective action taken as shown in Table 7.2.2.3.

TABLE 7.2.2.3 Mainline Strainers

Condition	Corrective Action
Plugging or fouling	Clean
Corrosion	Replace or repair

A.7.2.2.3 Any flow in excess of the flow through the main drain connection should be considered significant.

A strainer, such as those addressed in 7.2.2.3, is installed on a system where foreign material is likely to be present in the water supply and might obstruct an orifice. For example, the presence of rocks, pebbles, leaves, and sediment in a raw water source, such as a pond, would necessitate a strainer.

Strainers must be installed when sprinkler or nozzle orifices are smaller than $\frac{3}{8}$ in. (10 mm) in diameter. The concern is increased at the higher volumes of water used for fire protection purposes, as opposed to those typically encountered in a plumbing system. In fire pump installations, the velocity of the water may be sufficient to dislodge and carry material to an orifice. In this case, the strainer serves to "filter" the water. During inspection the strainer flushing valve should be opened and allowed to flow until the water is clear. In cases of severe sediment, the strainer basket should be removed and cleaned. Exhibit 7.2 illustrates a typical fire protection system strainer.

7.2.2.4 Dry Barrel and Wall Hydrants. Dry barrel and wall hydrants shall be inspected annually and after each operation with the necessary corrective action taken as shown in Table 7.2.2.4.

The dry barrel hydrants in 7.2.2.4 (see Figure A.3.3.9.1) are also known as frostproof hydrants. These hydrants are used where there is a chance that temperatures will drop below freezing. The valve controlling the water is located below the frost line between the foot piece and barrel of the hydrant. The barrel of the hydrant is dry and water is admitted upon opening the operating nut. A drain valve at the base of the barrel is open when the main valve is closed, allowing residual water in the barrel to drain out.

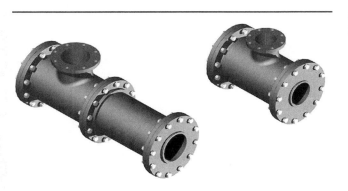

EXHIBIT 7.2 *Fire Protection Mainline Strainer.*

TABLE 7.2.2.4 *Dry Barrel and Wall Hydrants*

Condition	Corrective Action
Inaccessible	Make accessible
Barrel contains water or ice (presence of water or ice could indicate a faulty drain, a leaky hydrant valve, or high groundwater table)	Repair and drain; for high groundwater it could be necessary to plug the drain and pump out the barrel after each use
Improper drainage from barrel	Repair drain
Leaks in outlets or at top of hydrant	Repair or replace gaskets, packing, or parts as necessary
Cracks in hydrant barrel	Repair or replace
Tightness of outlet caps	Lubricate if necessary; tighten if necessary
Worn outlet threads	Repair or replace
Worn hydrant operating nut	Repair or replace
Availability of operating wrench	Make sure wrench is available

7.2.2.5 Wet Barrel Hydrants. Wet barrel hydrants shall be inspected annually and after each operation with the necessary corrective action taken as shown in Table 7.2.2.5.

TABLE 7.2.2.5 *Wet Barrel Hydrants*

Condition	Corrective Action
Inaccessible	Make accessible
Leaks in outlets or at top of hydrant	Repair or replace gaskets, packing, or parts as necessary
Cracks in hydrant barrel	Repair or replace
Tightness of outlet caps	Lubricate if necessary; tighten if necessary
Worn outlet threads	Repair or replace
Worn hydrant operating nut	Repair or replace
Availability of operating wrench	Make sure wrench is available

EXHIBIT 7.3 Wet Barrel Hydrant.

Paragraph 7.2.2.5 addresses the inspection of wet barrel hydrants, also known as "California" hydrants, such as the one shown in Exhibit 7.3. These hydrants are sometimes used where the temperature remains above freezing. A wet barrel hydrant usually has a compression valve at each outlet, but it may have another valve in the bonnet that controls the water flow to all outlets.

7.2.2.6 Monitor Nozzles. Monitor nozzles shall be inspected semiannually with the necessary corrective action taken as shown in Table 7.2.2.6.

Monitor nozzles are provided where large amounts of combustible materials, such as log piles, lumber piles, flammable or combustible liquids, or railway cars are stored in yards and

TABLE 7.2.2.6 *Monitor Nozzles*

Condition	Corrective Action
Leakage	Repair
Physical damage	Repair or replace
Corrosion	Clean or replace, and lubricate or protect as necessary

it is necessary to provide a means of quickly delivering large volumes of water to control fires. See Exhibit 7.4 for an example of a monitor nozzle.

7.2.2.7 Hose Houses. Hose houses shall be inspected quarterly with the necessary corrective action taken as shown in Table 7.2.2.7.

TABLE 7.2.2.7 Hose Houses

Condition	Corrective Action
Inaccessible	Make accessible
Physical damage	Repair or replace
Missing equipment	Replace equipment

A hose house is typically located in an industrial environment to provide manual fire-fighting equipment outside to protect buildings or process equipment that may be exposed to a fire hazard. Such equipment generally includes hose, nozzles, wrenches, gated wyes, and hose couplings.

EXHIBIT 7.4 Monitor Nozzle.

7.3 Testing

7.3.1* Underground and Exposed Piping Flow Tests.

Underground and exposed piping shall be flow tested to determine the internal condition of the piping at minimum 5-year intervals.

A.7.3.1 Full flow tests of underground piping can be accomplished by methods including, but not limited to, flow through yard hydrants, fire department connections once the check valve has been removed, main drain connections, and hose connections.

A hydrant flow test, which determines the internal condition of piping, should be conducted on the underground pipe as required by 7.3.1 and compared to both the original test and the most recent test conducted. If deterioration is noted, additional testing, such as a hydraulic gradient analysis, should be conducted. All data gathered during either test should be recorded on report forms. Exhibit 7.5 illustrates different steps of a hydrant flow test.

HYDRANT FLOW TEST EQUIPMENT

◄ **FAQ**
What equipment is needed for a hydrant flow test?

Conducting a flow test using fire hydrants requires the following equipment, calibrated in customary inch/pound units or appropriate SI units:

1. A 6 in. (52 mm.) steel ruler with $\frac{1}{16}$ in. (1 mm) divisions.
2. A Pitot tube (see Exhibit 7.6), together with a test-pressure gauge, suitable for the pressures to be expected. [Usually, a 60 psi (4 bar) gauge is satisfactory.]
3. A $2\frac{1}{2}$ in. (65 mm) hydrant cap with fittings (see Exhibit 7.7) [assuming $2\frac{1}{2}$ in. (65 mm) outlets], together with a test-pressure gauge suitable for the pressures to be expected, with 1 psi (0.07 bar) graduations. [Usually, a 200 psi (13.8 bar) gauge is satisfactory.]
4. A form for recording test data, including a sketch of the test location showing hydrants and any other salient features.

A Pitot-tube-and-gauge assembly (see Exhibit 7.6) is indispensable in conducting flow tests from hydrants and nozzles. The small opening at the end of the tube, not over $\frac{1}{16}$ in.

EXHIBIT 7.5 Hydrant Flow Test. (Top left) A calibrated test gauge attached to the pressure hydrant. (Top right) Measuring outlet diameter on flow hydrant. (Bottom left) Flow hydrant at full flow. (Bottom right) Taking Pitot measurement at flow hydrant.

(1.6 mm) in diameter, is inserted in the center of the stream, in a direct line with the flow, at a distance in front of the opening equal to one-half the opening diameter. Velocity pressure is registered on the gauge attached to the tube.

Note in Exhibit 7.6 the petcock blow-off. Its operation permits air from the hydrant barrel to be vented when the hydrant is opened and allows air to re-enter when the hydrant is closed. Failure to open the petcock during hydrant closing may subject the pressure gauge to a partial vacuum, which could introduce errors in future gauge readings. Also note the $\frac{3}{4}$ in. (20 mm) garden-hose thread on the $2\frac{1}{2}$ in. (65 mm) hydrant cap.

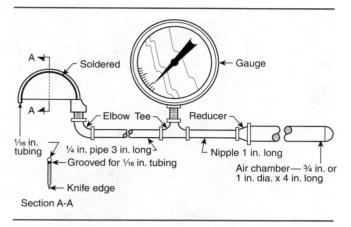

EXHIBIT 7.6 *Typical Pitot Tube Assembly. (Source: Fire Protection Handbook, 2003, Figure 10.6.1)*

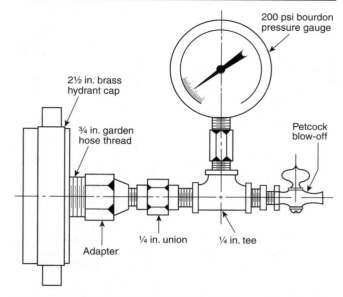

EXHIBIT 7.7 *Typical Hydrant Cap and Gauge. (Source: Fire Protection Handbook, 2003, Figure 10.6.2)*

Test-quality gauges should be used in accordance with ASME B40.1, *Gauges—Pressure Indicating Dial Type—Elastic Element, Grade AA*. The use of good quality test gauges has produced results that are considered reasonably accurate within the scope of the test procedure. However, care should be taken to protect the gauges from rough handling.

Gauges should be tested periodically by means of a dead-weight tester throughout the range of operation. Calibration sheets should be kept for each gauge and correction factors affixed to the back of each gauge before a test series is begun. All test data should be recorded in a neat, systematic fashion, along with any system operating conditions that might affect the test results.

HYDRANT FLOW TEST PROCEDURE

The procedure for conducting a hydrant flow test is rather straightforward. The tester should identify the test (non-flowing) hydrant and the flowing hydrant. In a single-direction feed water supply, the test hydrant should be the one nearest the source. In a gridded-type system, where the water flow comes from multiple directions, the location of the test and flow hydrants is not critical. In any case, the test should be conducted in the vicinity of the required point of connection.

Based on the underground piping configuration, Exhibit 7.8 provides the recommended arrangement of flow (*F*) and residual (*R*) hydrants. The hydrants used to measure flow and pressure vary based on the piping configuration.

After identifying the hydrants to be used, the tester should remove the hydrant cap and flush the test hydrant. Flushing the hydrant reduces the possibility of damage to the test equipment or the tester. Once a clean steady stream of water is observed, the hydrant valve should be slowly closed. The test hydrant is then prepared by installing the hydrant cap and gauge (see Exhibit 7.7) with the petcock open. The hydrant is opened and the petcock is closed. The pressure with no hydrants flowing — the static pressure — is recorded.

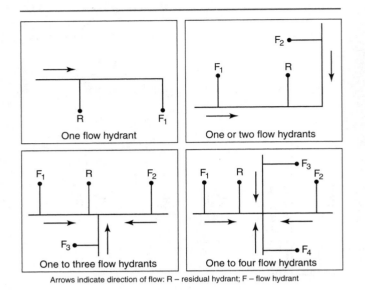

EXHIBIT 7.9 Verifying Hydrant Outlet Coefficient.

EXHIBIT 7.8 Suggested Flow Test Arrangements. (Source: NFPA 291, 2007, Figure 4.3.4)

The tester proceeds to the flow hydrant and measures the opening (assuming no flow aid is attached, such as stream straighteners or playpipes). At this point, the tester also should determine the hydrant coefficient of discharge (see Exhibit 7.9). That determination can be made by inserting one hand into the hydrant and comparing the shape of the outlet to those shown in Exhibit 7.10.

After obtaining the size of the hydrant opening and the hydrant coefficient of discharge, the tester should open up the flow hydrant. When a clean steady stream of water is present, the tester should measure the flowing pressure by inserting the Pitot tube and gauge assembly (see Exhibit 7.6) into the middle of the stream, one-half the opening diameter away from the edge of the opening.

The tester should record the Pitot gauge reading while someone else simultaneously records the gauge pressure on the test hydrant. The pressure recorded by the Pitot tube assembly is used to calculate flow. Flow can be found listed in the flow tables in NFPA 291, *Recommended Practice for Fire Flow Testing and Marking of Hydrants*, or can be calculated by inserting the outlet coefficient (c), outlet diameter (d), and velocity pressure (p) into the following formula:

$$Q = 29.83\ cd^2\sqrt{p}$$

EXHIBIT 7.10 Three General Types of Hydrant Outlets and Their Coefficients of Discharge. (Source: NFPA 291, 2007, Figure 4.7.1)

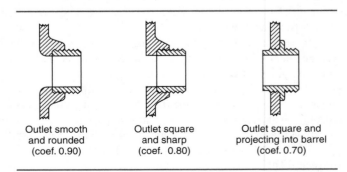

where:

Q = flow in gpm

c = coefficient of discharge

d = diameter of outlet (inches)

$\sqrt{p}$ = flowing pressure (psi)

In metric units, the formula is as follows:

$$Q = 0.0666\, cd^2\sqrt{p}$$

Where:

Q = flow in L/min

c = coefficient of discharge

d = diameter of outlet (mm)

$\sqrt{p}$ = flowing pressure (bar)

To obtain satisfactory test results, sufficient discharge should be achieved to cause a drop in pressure at the test hydrant of at least 25 percent of static pressure or to flow the total system demand. Should additional flow rates be desired, other outlets or hydrants may be opened and pressures measured accordingly. After the flowing hydrants have been shut off slowly, the static pressure should be observed and compared with the initial reading.

Pitot readings of less than 10 psi (0.7 bar) or over 30 psi (2.1 bar) at any open hydrant should be avoided. To keep within these pressure limits, the rate of flow can be controlled by throttling the hydrant or opening a second outlet, or both. However, the flow hydrants should be opened sufficiently so that hydrant drains are closed. Water continuously discharged through the drains tends to erode the soil from the base of the hydrant.

The use of the larger pumper connection on a hydrant for testing should be avoided unless flow and pressure are strong enough to produce a full stream. When pumper outlets are used, a proper coefficient of discharge must be determined, based on the extent to which the orifice is completely filled with water. On occasion, it may be desirable to obtain an average velocity pressure by moving the Pitot tube through the entire vertical dimension of the orifice.

7.3.1.1 Flow tests shall be made at flows representative of those expected during a fire, for the purpose of comparing the friction loss characteristics of the pipe with those expected for the particular type of pipe involved, with due consideration given to the age of the pipe and to the results of previous flow tests.

If the flow test required by 7.3.1.1 indicates a potential problem, additional information can be obtained by conducting an engineering analysis or hydraulic gradient analysis. For further information regarding how to conduct a hydraulic gradient analysis, see the *Fire Protection Handbook*, 2003, Section 10.

7.3.1.2 Any flow test results that indicate deterioration of available waterflow and pressure shall be investigated to the complete satisfaction of the authority having jurisdiction to ensure that the required flow and pressure are available for fire protection.

7.3.1.3 Where underground piping supplies individual fire sprinkler, standpipe, water spray, or foam-water sprinkler systems and there are no means to conduct full flow tests, tests generating the maximum available flows shall be permitted.

7.3.2 Hydrants.

Hydrants shall be tested annually to ensure proper functioning.

7.3.2.1 Each hydrant shall be opened fully and water flowed until all foreign material has cleared.

7.3.2.2 Flow shall be maintained for not less than 1 minute.

7.3.2.3 After operation, dry barrel and wall hydrants shall be observed for proper drainage from the barrel.

Proper drainage, as required in 7.3.2.3, is important. The dry barrel is often provided due to concerns regarding freezing temperatures. If the water does not drain out of the barrel promptly, the potential for freezing increases. (See the commentary following 7.2.2.4 for more information.)

CASE STUDY

Frozen Hydrants Fail to Operate

Fire in a steam power plant in Virginia caused $45 million in damage. This three- and four-story steam power plant was of protected, noncombustible construction. It had a ground floor area of 240,000 ft^2. The plant was operating with a skeleton crew at the time of the fire.

The type and coverage of the detector system was unknown, but it was effective. There were two automatic suppression systems — a wet pipe system and a deluge system — but they became inoperative when metal fragments ruptured the water supply pipe. Plant employees and fire fighters used dry chemical extinguishers to control and extinguish the fire.

A turbine broke loose without warning and started to come apart, immediately producing smoke and fire, which an employee tried to fight with an extinguisher. The turbine came to an abrupt stop, and metal shrapnel ruptured the fuel line and the water supply lines for the sprinkler system. The hot metal also ignited the fuel. Fire spread to the large air intake unit and to the air filter and coolant unit. The area in which the fire occurred was not easily accessible.

No injuries were reported. The fact that on-site hydrants were frozen or inoperative was found to be a contributing factor to the fire.

7.3.2.4 Full drainage shall take no longer than 60 minutes.

7.3.2.5 Where soil conditions or other factors are such that the hydrant barrel does not drain within 60 minutes, or where the groundwater level is above that of the hydrant drain, the hydrant drain shall be plugged and the water in the barrel shall be pumped out.

7.3.2.6 Dry barrel hydrants that are located in areas subject to freezing weather and that have plugged drains shall be identified clearly as needing pumping after operation.

7.3.3 Monitor Nozzles.

7.3.3.1 Monitor nozzles that are mounted on hydrants shall be tested as specified in 7.3.2.

7.3.3.2 All monitor nozzles shall be oscillated and moved throughout their full range annually to ensure proper operability.

A monitor nozzle may be a fixed-position device or a directional device capable of being manually or automatically rotated either from side to side or up and down. Paragraph 7.3.3.2 addresses the fact that it is necessary to ensure that the device maintains full movement in all directions. (See the commentary following Table 7.2.2.6 for more information.)

7.4 Maintenance

7.4.1 General.

All equipment shall be maintained in proper working condition, consistent with the manufacturer's recommendations.

7.4.2 Hydrants.

7.4.2.1 Hydrants shall be lubricated annually to ensure that all stems, caps, plugs, and threads are in proper operating condition.

7.4.2.2* Hydrants shall be kept free of snow, ice, or other materials and protected against mechanical damage so that free access is ensured.

A.7.4.2.2 The intent of 7.4.2.2 is to maintain adequate space for use of hydrants during a fire emergency. The amount of space needed depends on the configuration as well as the type and size of accessory equipment, such as hose, wrenches, and other devices that could be used.

The requirement in 7.4.2.2 that hydrants be kept free from snow is sometimes difficult to meet. For example, a hydrant near the street runs the risk of being buried by a snowplow. Where the potential exists for snow burial, extra measures should be taken to ensure that hydrants can be easily located. Such measures may include the provision of flags, indicating poles, banners, and so on. (See Exhibit 7.11.) Note that in addition to snow, vegetation may also obscure a hydrant location.

◀ **FAQ**
What measures can be taken to indicate the locations of hydrants in case of snow burial?

7.4.3 Monitor Nozzles.

Monitor nozzles shall be lubricated annually to ensure proper operating condition.

EXHIBIT 7.11 Hydrant with Indicating Pole.

7.5 Component Action Requirements

Component replacement tables were added in the 2008 edition to offer guidance to the user of the standard when system components are adjusted, repaired, rebuilt, or replaced. It is not necessary in each case to require a complete acceptance test for each component when maintenance is performed.

7.5.1 Whenever a component in a private fire service system is adjusted, repaired, reconditioned, or replaced, the action required in Table 7.5.1 shall be performed.

7.5.1.1 Where the original installation standard is different from the cited standard, the use of the appropriate installing standard shall be permitted.

7.5.1.2 A main drain test shall be required if the system control or other upstream valve was operated in accordance with 13.3.3.4.

7.5.1.3 These actions shall not require a design review, which is outside the scope of this standard.

TABLE 7.5.1 Summary of Component Replacement Action Requirements

Component	Adjust	Repair/ Recondition	Replace	Test Criteria
Water Delivery Components				
Pipe and fittings (exposed and underground)	X	X	X	Hydrostatic test in conformance with NFPA 24, *Standard for the Installation of Private Fire Service Mains and Their Appurtenances*
Hydrants	X	X	X	Hydrostatic test in conformance with NFPA 24
				Water flow in conformance with NFPA 24
				Check for proper drainage
Monitor nozzles	X	X	X	Hydrostatic test in conformance with NFPA 24
				Flush in conformance with NFPA 24
Mainline strainers	X	X	X	Flow test downstream of strainer
Fire department connection	X	X	X	See Chapter 13
Valves				See Chapter 13
Fire pump				See Chapter 8
Alarm and Supervisory Components				
Valve supervisory device	X	X	X	Operational test for conformance with NFPA 24 and/or NFPA 72, *National Fire Alarm Code*
System-Indicating Components				
Gauges			X	Verify at 0 psi and system working pressure
System Housing and Protection Components				
Hose houses	X	X	X	Verify integrity of hose and hose house components
Structural Components				
Thrust blocks	X	X	X	Test at system working pressure
Tie rods	X	X	X	Test at system working pressure
Retainer glands	X	X	X	Test at system working pressure
Informational Components				
Identification signs	X	X	X	Verify conformance with NFPA 24

SUMMARY

Chapter 7 of NFPA 25 covers the requirements for private water supplies. This type of system is often difficult to maintain because much of the pipe and other components is usually several feet underground. It is therefore necessary to conduct the annual hydrant flushing test and the periodic flow test, which are the only methods available to determine the condition of the system. Other components, such as hydrants, strainers, and monitor nozzles, can be visually inspected without difficulty.

REVIEW QUESTIONS

1. A flow test of a $2\frac{1}{2}$ in. (65 mm) hydrant butt (coefficient of discharge = 0.8) produced a Pitot reading of 27 psi (186 kPa). How much water was flowing?
2. A $1\frac{1}{8}$ in. (29 mm) nozzle with a coefficient of discharge of 0.97 is attached to the hydrant and the Pitot now reads 48 psi (331 kPa). How much water is flowing?
3. Who should be notified when conducting an inspector's test connection trip test?
4. A dry barrel hydrant is not draining properly due to the water table in the area. Since it is July, what should be done?

REFERENCES CITED IN COMMENTARY

National Fire Protection Association, 1 Batterymarch Park, Quincy, MA 02169-7471.

Cote, A. E. ed., *Fire Protection Handbook®*, 19th edition, 2003.

NFPA 24, *Standard for the Installation of Private Fire Service Mains and Their Appurtenances*, 2007 edition.

NFPA 291, *Recommended Practice for Fire Flow Testing and Marking of Hydrants*, 2007 edition.

American Society of Mechanical Engineers, Three Park Avenue, New York, NY 10016-5990.

ASME B40.1, *Gauges—Pressure Indicating Dial Type—Elastic Element, Grade AA,* 2000 edition.

Fire Pumps

CHAPTER 8

Chapter 8 covers the inspection, testing, and maintenance of fire pumps and related equipment. In many cases, fire pumps are the sole source of water and pressure for a fire protection system and therefore must be properly maintained to be sure they are available at all times. Fire pumps and their associated equipment also constitute a major investment, ranging in cost from $5,000 to $25,000. Failure to properly maintain fire pumps and related equipment can result in costly repairs.

8.1* General

This chapter shall provide the minimum requirements for the routine inspection, testing, and maintenance of fire pump assemblies. Table 8.1 shall be used to determine the minimum required frequencies for inspection, testing, and maintenance.

A.8.1 A fire pump assembly provides waterflow and pressure for private fire protection. The assembly includes the water supply suction and discharge piping and valving; pump; electric, diesel, or steam turbine driver and control; and the auxiliary equipment appurtenant thereto.

A stationary fire pump must be subject to a running test on a weekly basis and a flow test annually. These tests are independent of system type or building served.

Exhibit 8.1 shows a fire pump assembly. Exhibit 8.2 shows the major internal components of a fire pump, including the shaft, bearings, packing glands, and impeller.

◄ **FAQ**
What is the fire pump testing frequency for a pump that services a standpipe system in a municipal building?

TABLE 8.1 Summary of Fire Pump Inspection, Testing, and Maintenance

Item	Frequency	Reference
Inspection		
Pump house, heating ventilating louvers	Weekly	8.2.2(1)
Fire pump system	Weekly	8.2.2(2)
Test		
Pump operation		
No-flow condition	Weekly	8.3.1
Flow condition	Annually	8.3.3.1
Maintenance		
Hydraulic	Annually	8.5
Mechanical transmission	Annually	8.5
Electrical system	Varies	8.5
Controller, various components	Varies	8.5
Motor	Annually	8.5
Diesel engine system, various components	Varies	8.5

EXHIBIT 8.1 Fire Pump Assembly.

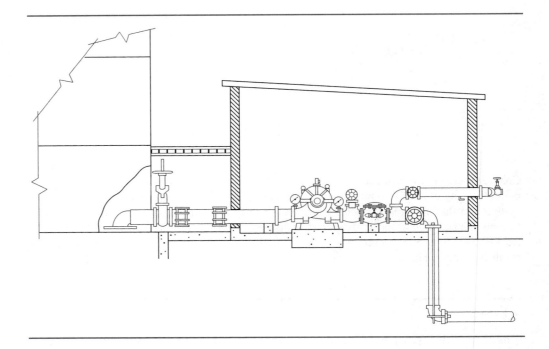

EXHIBIT 8.2 Fire Pump Internal Components. (Courtesy of John Jensen)

8.1.1 Valves and Connections.

Valves and fire department connections shall be inspected, tested, and maintained in accordance with Chapter 13.

8.1.2* Auxiliary Equipment.

The pump assembly auxiliary equipment shall include the following:

(1) Pump accessories as follows:
    (a) Pump shaft coupling
    (b) Automatic air release valve

The automatic air relief valve removes the air that may become trapped inside the pump casing. On some old installations, the air release valve may be manually operated. Exhibit 8.3 shows an air release valve mounted on top of the fire pump casing.

EXHIBIT 8.3 Air Release Valve. (Courtesy of John Jensen)

(c) Pressure gauges

Pressure gauges, such as the one in Exhibit 8.4, are located on the suction, discharge, and cooling water lines for diesel engine drivers.

EXHIBIT 8.4 Pressure Gauge.

The suction gauge is a compound gauge having a range of –30 psi (–2.1 bar) to +100 psi (6.9 bar) that is used on horizontal split-case, vertical in-line, and end suction pumps. The water level in a well or below-grade reservoir is found using the air line method explained in A.7.3.5.3 of the 2007 edition of NFPA 20, *Standard for the Installation of Stationary Pumps for Fire Protection.* The water level must be known in order to calculate the net pump pressure for a vertical turbine pump.

The cooling water pressure gauge has a range of twice the rating of the cooling water heat exchanger. This heat exchanger rating is normally 30 psi to 60 psi (2.1 bar to 4.1 bar). Because the cooling water is taken off of the pump discharge, it is easy to overpressurize the cooling water heat exchanger, resulting in failure of the heat exchanger.

(d) Circulation relief valve (not used in conjunction with diesel engine drive with heat exchanger)

The circulation relief valve is designed to open and discharge a small amount of water when the pump is running at churn. The valve should be $\frac{3}{4}$ in. (19 mm) for pump capacities up to 2500 gpm (9462 L/min) and 1 in. (25 mm) for larger pumps. This valve prevents the pump casing from overheating and damaging the impeller, packings, and bearings during extended periods of running at churn. When the fire pump is operating while discharging water through the fire protection system, the circulation relief valve is not needed for cooling and should close. The circulation relief valve is usually spring-operated and can fail with the slightest bit of obstructing material in the valve. Discharge from this valve should be piped to a drain or to the outdoors, and the discharge should be observed during the weekly test of the fire pump. Exhibit 8.5 shows a typical circulation relief valve.

EXHIBIT 8.5 Circulation Relief Valve.

(2) Pump test device(s)

As indicated in 8.1.2(2), each pump must have the means to be flow tested. The most common method to meet that requirement is to install a fire pump test header (see Exhibit 8.6), with one hose valve for each 250 gpm (946 L/min), which is to be used for the annual test and can also be used as a hydrant in an emergency situation.

The second most common method is to install a flowmeter (see Exhibit 8.7) on a bypass or with the discharge back to a water storage tank. This method is preferable because the testing can be completed without making elaborate provisions, and wasting the water to ground. This method also reduces the time and resources necessary to conduct the testing.

If a test header or flowmeter cannot be installed, a plant fire hydrant or standpipe system can be used to connect hoses and nozzles to conduct the testing.

(3) Pump relief valve and piping (where maximum pump discharge pressure exceeds the rating of the system components or the driver is of variable speed)

A relief valve is designed to open at a predetermined pressure [usually 175 psi (1.2 bar)] to protect the piping and other components in the system from overpressurization. See Exhibit 13.24 for an illustration of a pressure relief valve.

EXHIBIT 8.6 *Fire Pump Test Header.*

EXHIBIT 8.7 *Fire Pump Flowmeter.*

The need for a main pump relief valve is determined by the following considerations:

1. The working pressure of the piping components downstream of the pump is 175 psi (1.2 bar). If the shutoff pressure of the fire pump is less than 175 psi (1.2 bar), a main pressure relief valve is not needed.
2. If the pump driver is a variable-speed drive (such as a diesel engine), the pressure that can be developed during engine overspeed until the governor shuts down the engine can exceed the rating of the system components. In such a case, a pressure relief valve is needed.
3. If the pump is sized properly, sound engineering practice does not require the installation of a main pressure relief valve to control the pump discharge pressure.
4. NFPA 20 is very clear that a pressure relief valve is only needed when a diesel engine fire pump is installed and where a total of 121 percent of the net rated shutoff (churn) pressure plus the maximum static suction pressure, adjusted for elevation, exceeds the pressure for which the system components are rated. Pressure relief valves are also required for variable-speed drive fire pumps. Pressure relief valves should never be used on electric drive (constant speed) fire pumps.

(4) Alarm sensors and indicators

The alarm sensors in 8.1.2(4) are really supervisory signals. They must be transmitted to a constantly attended location and received by a knowledgeable person who can respond to the signal and determine what is wrong with the fire pump installation. That location, which preferably should be on premises, need not be the fire alarm monitoring service.

(5) Right-angle gear sets (for engine-driven vertical shaft turbine pumps)

Right-angle gear drives are built with a cooling oil reservoir. That reservoir must be checked to make sure that the oil level is where it should be and that the drive is not overheating. Oil temperatures should be kept below 200°F (93.3°C), and preferably below 135°F (57°C). If these temperatures are exceeded, cooling coils may be used to address the overheating.

(6) Pressure maintenance (jockey) pump and accessories

The pressure maintenance, or jockey, pump is a low-flow, high-pressure pump, such as the one shown in Exhibit 8.8. It is designed to maintain a constant pressure on the system, accounting for minor pressure fluctuations. The discharge of the pressure maintenance pump is not intended to keep pace with the discharge of a single sprinkler, thus allowing the system pressure to decrease when one or more sprinklers activate. Under these circumstances, as the system pressure drops, the pressure switch (see Exhibit 8.9) in the pressure maintenance pump controller senses the pressure drop and activates the pump. If the pressure maintenance pump cannot keep pace with the flow of one or more sprinklers and the system pressure continues to drop, then the pressure switch in the fire pump controller senses the pressure drop and starts the fire pump. There is usually at least a 5 psi (0.3 bar) pressure differential between the pressure maintenance pump and the fire pump. Pressure switch settings closer than 5 psi (0.3 bar) may cause unintentional starting of the fire pump. This situation should be avoided as it presents false alarms.

EXHIBIT 8.8 Jockey Pump.

EXHIBIT 8.9 Pressure Switch.

The pressure switch shown in Exhibit 8.9 is located in the fire pump controller and the jockey pump controller and has high- and low-pressure settings to start and stop each pump. The pressure switch is internally mounted in each controller and has specific settings based on the water supply. An example of pressure switch settings can be found in NFPA 20 as follows:

Examples of fire pump settings follow (for SI units, 1 psi = 0.0689 bar):

i. Pump: 1000 gpm, 100 psi pump with churn pressure of 115 psi
ii. Suction supply: 50 psi from city — minimum static; 60 psi from city — maximum static
iii. Jockey pump stop = 115 psi + 50 psi = 165 psi

iv. Jockey pump start = 165 psi − 10 psi = 155 psi
v. Fire pump stop = 115 psi + 50 psi = 165 psi
vi. Fire pump start = 155 psi − 5 psi = 150 psi
vii. Fire pump maximum churn = 115 psi + 60 psi = 175 psi

The jockey pump can be of any commercially available type. It should be able to maintain the system pressure above the starting pressure of the main fire pump while accounting for small leaks in the system. The jockey pump is used on automatic controller fire pump systems (see Exhibit 8.10). The pump should be equipped with a running period timer of at least 10 minutes to obtain the maximum life from the pump by not overheating the motor coils from excessive starting and stopping. There is no need to conduct flow tests of the jockey pump and measure the water discharge. The pump should be started and run through its normal cycle during the weekly and annual fire pump test.

Each pressure switch in each controller is connected by means of a pressure-sensing line. This pipe consists of a direct connection from the fire protection system piping and the pressure switch inside the controller. Two check valves or ground-face unions are installed in the sensing line to protect the pressure switch from momentary pressure surges. It is important to note that each pump, including the jockey pump, must have its own sensing line. Frequently, sensing lines are grouped together, which is in direct violation of the installation requirement. If this situation is found during an inspection, a notation should be made in the inspection report. Exhibits 8.11 and 8.12 illustrate the correct installation of a pressure-sensing line.

A.8.1.2 Types of centrifugal fire pumps include single and multistage units of horizontal or vertical shaft design. Listed fire pumps have rated capacities of 25 gpm to 5000 gpm (95 L/min to 18,925 L/min), with a net pressure range from approximately 40 psi to 400 psi (2.75 bar to 27.6 bar).

(1) *Horizontal Split Case.* This pump has a double suction impeller with an inboard and outboard bearing and is used with a positive suction supply. A variation of this design can be mounted with the shaft in a vertical plane. *[See Figure A.8.1.2(a).]*

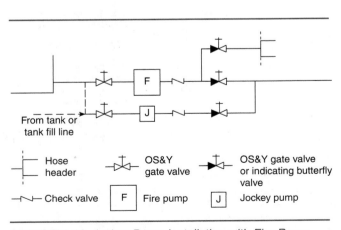

EXHIBIT 8.10 *Jockey Pump Installation with Fire Pump.* (Source: NFPA 20, 2007, Figure A.5.24.4)

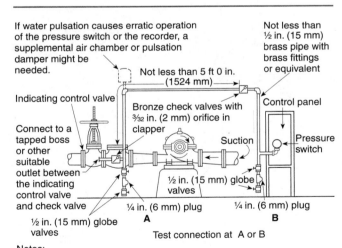

EXHIBIT 8.11 *Piping Connection for Each Automatic Pressure Switch (for Electric Fire Pump and Jockey Pumps).* [Source: NFPA 20, 2007, Figure A.10.5.2.1(a)]

EXHIBIT 8.12 Piping Connection for Pressure-Sensing Line (Electric Fire Pump). [Source: NFPA 20, 2007, Figure A.10.5.2.1(b)]

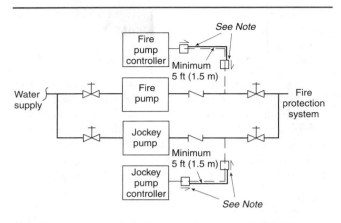

Note: Check valves or ground-face unions complying with 10.5.2.1.

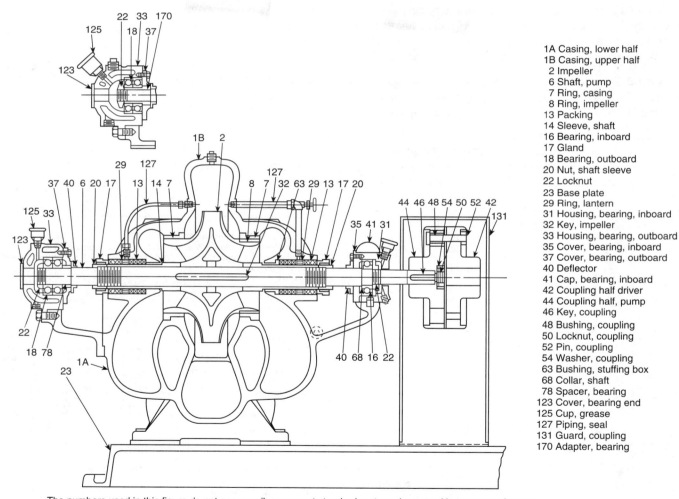

1A Casing, lower half
1B Casing, upper half
2 Impeller
6 Shaft, pump
7 Ring, casing
8 Ring, impeller
13 Packing
14 Sleeve, shaft
16 Bearing, inboard
17 Gland
18 Bearing, outboard
20 Nut, shaft sleeve
22 Locknut
23 Base plate
29 Ring, lantern
31 Housing, bearing, inboard
32 Key, impeller
33 Housing, bearing, outboard
35 Cover, bearing, inboard
37 Cover, bearing, outboard
40 Deflector
41 Cap, bearing, inboard
42 Coupling half driver
44 Coupling half, pump
46 Key, coupling
48 Bushing, coupling
50 Locknut, coupling
52 Pin, coupling
54 Washer, coupling
63 Bushing, stuffing box
68 Collar, shaft
78 Spacer, bearing
123 Cover, bearing end
125 Cup, grease
127 Piping, seal
131 Guard, coupling
170 Adapter, bearing

The numbers used in this figure do not necessarily represent standard part numbers used by any manufacturer.

FIGURE A.8.1.2(a) Impeller Between Bearings, Separately Coupled, Single-Stage Axial (Horizontal) Split Case. (Courtesy of Hydraulic Institute Standard for Centrifugal, Rotary and Reciprocating Pumps.)

(2) *End Suction and Vertical In-Line.* This pump can have either a horizontal or vertical shaft with a single suction impeller and a single bearing at the drive end. *[See Figure A.8.1.2(b).]*
(3) *Vertical Shaft, Turbine Type.* This pump has multiple impellers and is suspended from the pump head by a column pipe that also serves as a support for the shaft and bearings. This pump is necessary where a suction lift is needed, such as from an underground reservoir, well, river, or lake. *[See Figure A.8.1.2(c).]*

The 2007 edition of NFPA 20 allows positive displacement pumps for water mist and foam system applications. For details on the inspection, testing, and maintenance of these pumps, contact the pump manufacturer.

8.1.3 Water Supply to Pump Suction.

The suction supply for the fire pump shall provide the required flow at a gauge pressure of zero (0) psi [zero (0) bar] or higher at the pump suction flange to meet the system demand.

As indicated in 8.1.3, the pressure observed at the suction gauge while the pump is operating should be positive, even when the fire pump is operating at its overload condition. The suction gauge is permitted to drop to -3 psi (-0.2 bar) when the supply is a suction tank with its base at or above the same elevation as the pump.

8.1.3.1 Those installations for which NFPA 20, *Standard for the Installation of Stationary Pumps for Fire Protection*, permitted negative suction gauge pressures at the time of pump installation, where the system demand still can be met by the pump and water supply, shall be considered to be in compliance with 8.1.3.

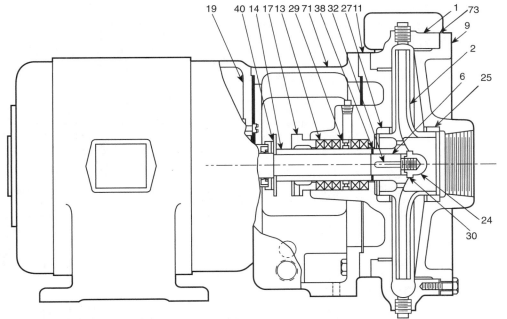

The numbers used in this figure do not necessarily represent standard part numbers used by any manufacturer.

FIGURE A.8.1.2(b) *Overhung Impeller, Close-Coupled, Single-Stage, End Suction. (Courtesy of Hydraulic Institute Standard for Centrifugal, Rotary and Reciprocating Pumps.)*

128 Chapter 8 • Fire Pumps

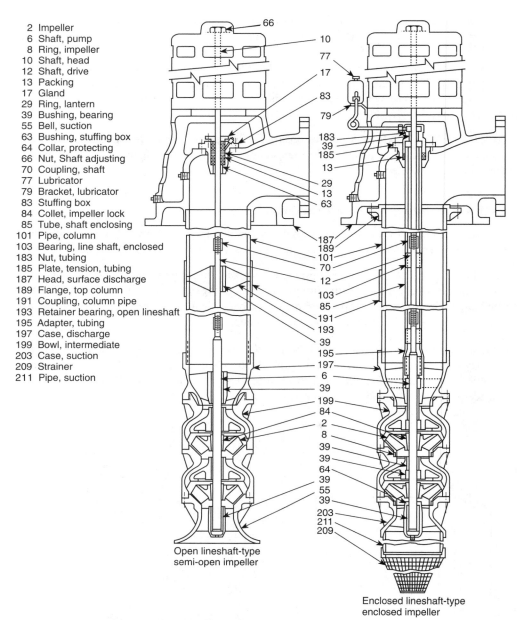

FIGURE A.8.1.2(c) *Turbine-Type, Vertical, Multistage, Deep Well. (Courtesy of Hydraulic Institute Standard for Centrifugal, Rotary and Reciprocating Pumps.)*

8.1.4 Energy Source.

The energy sources for the pump driver shall supply the necessary brake horsepower of the driver so that the pump meets system demand.

The energy sources in 8.1.4 are electric power, steam, and diesel fuel. Other fuels are currently not allowed for newly installed fire pump installations.

8.1.5 Driver.

The pump driver shall not overload beyond its rating (including any service factor allowance) when delivering the necessary brake horsepower.

8.1.6* Controller.

Automatic and manual controllers for applying the energy source to the driver shall be capable of providing this operation for the type of pump used.

There are two types of controllers listed or approved for electric fire pump service. For installations requiring less than 30 horsepower, there is a limited-service unit. The other type is a full-service controller.

A.8.1.6 Controllers include air-, hydraulic-, or electric-operated units. These units can take power from the energy source for their operation, or the power can be obtained elsewhere. Controllers used with electric power sources can apply the source to the driver in one (across-the-line) or two (reduced voltage or current) steps. Controllers can be used with automatic and manual transfer switches to select the available electric power source where more than one is provided.

The most common reason for installing multiple electric power supplies, as referred to in A.8.1.6, is that one or more of the supplies do not offer the required reliability. When testing fire pumps, each power source must be tested to determine its adequacy for supplying the fire pump.

8.1.7 Impairments.

The procedures outlined in Chapter 15 shall be followed where an impairment to protection occurs.

8.1.8 Notification to Supervisory Service.

To avoid false alarms where a supervisory service is provided, the alarm receiving facility always shall be notified by the property owner or designated representative as follows:

(1) Before conducting any test or procedure that could result in the activation of an alarm
(2) After such tests or procedures are concluded

8.2 Inspection

8.2.1 The purpose of inspection shall be to verify that the pump assembly appears to be in operating condition and is free from physical damage.

The inspection specified in Section 8.2 is usually conducted during the required weekly test. It is not necessary for separate visits to be made for the inspection and test. The inspection should also include the system components, such as valves, tanks, underground mains, and so on, covered in other chapters.

8.2.2* The pertinent visual observations specified in the following checklists shall be performed weekly:

(1) Pump house conditions:
 (a) Heat is adequate, not less than 40°F (4.4°C) [70°F (21°C)] for pump room with diesel pumps without engine heaters.

The pump house should be checked not only for heating, as detailed in 8.2.2(1)(a), but also for cooling. In the case of diesel engines, the engine is derated 1 percent for each 10°F (5.6°C) above 77°F (25°C) ambient temperature.

(b) Ventilating louvers are free to operate.

Paragraph 8.2.2(1)(b) requires that weekly pump house inspections confirm that ventilating louvers are free to operate because the louvers, shown in Exhibit 8.13, are needed to maintain fresh air for diesel systems and to control air temperature. For purposes of fresh air intake, the inlet ventilation must be automatically opened and have the capacity to provide adequate air for combustion purposes as well as ventilation. In cold climates, the inlet air may need to be heated to prevent small piping and gauges from freezing.

EXHIBIT 8.13 Ventilation Louvers.

(2) Pump system conditions:
   (a) Pump suction and discharge and bypass valves are fully open.

It is suggested that all valves be supervised in their normal position by electronic means or by chaining and locking. In cold climates, the hose header control valve should be in the closed position and the hose header drained of any water. Frequently, the hose valves are found broken from freezing if they are not left cracked open when the hose header valve is closed in the fall.

   (b) Piping is free of leaks.
   (c) Suction line pressure gauge reading is within acceptable range.

As required by 8.2.2(2)(c), an inspection of pump system conditions must include a check to ensure that the suction line pressure gauge reading is within an acceptable range. Where the water supply is taken from a tank, the gauge should be positive at a pressure equal to the water height in the tank or should equal the public water supply pressure, adjusted for elevation.

   (d) System line pressure gauge reading is within acceptable range.

The pump only boosts pressure when it is running. Therefore, when it is not running, the pressure on the system line pressure gauge should be the same as on the suction pressure gauge.

(e) Suction reservoir is full.

An altitude float located on the exterior of the tank is used to check the pressure in the suction reservoir. In some installations, a pressure gauge that reads in feet of water or psi is installed in the pump room. See Chapter 9 for inspection, testing, and maintenance requirements for the suction tanks.

In the case of a fire well or underground reservoir a different method is used to check the water level. It involves using an air line that extends below the water surface and a pressure gauge that, when air is blown into the tubing, reads how far below grade the water is. See A.7.3.5.3 and Figure A.7.3.5.3 of NFPA 20, 2007, for a detailed description of this method.

(f) Wet pit suction screens are unobstructed and in place.

The size and shape of the wet pit are governed by the Hydraulic Institute standards. The following must be checked during the inspection required by 8.2.2(2)(f):

- The trash rack
- Two sets of screens
- The pump suction strainer at the bottom of the submerged pump assembly

A small amount of debris can seriously affect pump operation, either by damaging the pump impeller or by causing obstruction of the pump and piping around the pump.

(3) Electrical system conditions:
   (a) Controller pilot light (power on) is illuminated.
   (b) Transfer switch normal pilot light is illuminated.

The transfer switch is only installed with electric motors that have an alternative electrical power supply installed. Exhibit 8.14 illustrates a fire pump controller with an attached power transfer switch.

EXHIBIT 8.14 Fire Pump Controller with Transfer Switch.

   (c) Isolating switch is closed — standby (emergency) source.

There is an isolating switch on each source of power to the electric motor drive. Both switches must be closed when the pump is in the operating or standby condition.

   (d) Reverse phase alarm pilot light is off, or normal phase rotation pilot light is on.

Once the electric motor/controller installation is completed and the phases have been checked, there is little chance that reverse phase will be a problem unless work is conducted on the electrical supply system.

 (e) Oil level in vertical motor sight glass is within acceptable range.

In addition to the items listed in 8.2.2(3)(a) through 8.2.2(3)(e), the general condition of the electrical components should be observed and recorded. Potential problems that can be discovered include rodent nesting, plugged motor ventilation, broken parts, unlocked controller doors, and improperly labeled electrical panelboards. If the pump room is equipped with electric heat, the thermostat should be checked to determine that it is operating in cold weather and that the temperature is set at normal room temperature. Likewise, any electric controls for ventilation should be in operation.

(4) Diesel engine system conditions:
 (a) Fuel tank is two-thirds full.

The fuel tank is designed to hold an eight-hour supply of fuel for the diesel engine, based on a formula of one gallon per horsepower. If an engine does not consume this quantity of fuel — and many do not — the fuel storage can be adjusted for the actual demand. Fuel in the tank should be consumed within one year. Fuel kept longer than one year has an increased risk of becoming contaminated with biological growth and clogging the engine fuel filters, preventing the engine from running.

The bottom of the fuel tank is reserved for collecting water or other contaminants. The bottom 5 percent of fuel should therefore be removed on an annual basis. Also, the bottom of the fuel tank should be observed to make sure that it is above the level of the fuel injectors so that if the fuel pump fails, the engine can still operate. Exhibit 8.15 illustrates a diesel fuel tank.

EXHIBIT 8.15 Fuel Tank for a Diesel Fire Pump.

(b) Controller selector switch is in auto position.
 (c) Batteries' (2) voltage readings are within acceptable range.
 (d) Batteries' (2) charging current readings are within acceptable range.
(e) Batteries' (2) pilot lights are on or battery failure (2) pilot lights are off.
(f) All alarm pilot lights are off.
(g) Engine running time meter is reading.
 (h) Oil level in right angle gear drive is within acceptable range.
(i) Crankcase oil level is within acceptable range.
(j) Cooling water level is within acceptable range.
 (k) Electrolyte level in batteries is within acceptable range.

The electrolyte level in a battery is normal when the battery is full of water up to the ring under the cell cap. A lower water level exposes the battery to an accumulation of hydrogen in the cell, which can lead to an explosion when the pump is started and there is a spark in the battery. Proper personal protective equipment should be used whenever filling the battery with water.

Diesel fire pumps are required to have two battery banks per 11.2.5.2.2.1 of NFPA 20. During the cranking cycle, the fire pump controller will alternate cranking from one bank to the other with an appropriate rest period until the engine starts. Both batteries must be maintained fully charged at all times. Exhibit 8.16 illustrates the two battery banks for a diesel fire pump.

EXHIBIT 8.16 Diesel Fire Pump Batteries.

   (l) Battery terminals are free from corrosion.
   (m) Water-jacket heater is operating.
(5)* Steam system conditions: Steam pressure gauge reading is within acceptable range.

A.8.2.2(5) Visual indicators other than pilot lights can be used for the same purpose.

A.8.2.2 See Table A.8.2.2 and Figure A.8.2.2.

TABLE A.8.2.2 Weekly Observations — Before Pumping

Item	Before Pump Is Operated
Horizontal pumps	1. Check drip pockets under packing glands for proper drainage. Standing water in drip pockets is the most common cause of bearing failure.
	2. Check packing adjustment — approximately one drop per second is necessary to keep packing lubricated.
	3. Observe suction and discharge gauges. Readings higher than suction pressure indicate leakage back from system pressure through either the fire pump or jockey pump.

Frequently, inexperienced personnel may view the water in the drip pocket as a leak. If the packing glands are tightened to the point where no water is allowed to drip, then the packing gland will dry out and fail. Exhibit 8.17 illustrates a packing gland that has been properly adjusted, because water can be seen in the drip pocket.

134 Chapter 8 • Fire Pumps

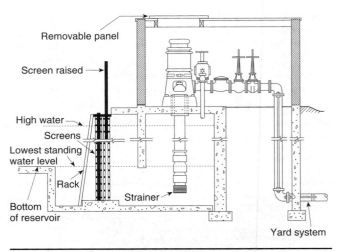

FIGURE A.8.2.2 *Wet Pit Suction Screen Installation.*

EXHIBIT 8.17 *Fire Pump Packing Gland and Drip Pocket.*

8.3* Testing

CASE STUDY

Fire Pump Fails During Fire

Fire caused over $32 million in damage to a New York electric power generating station of unprotected noncombustible construction. The station, which was operating at the time of the fire, is 10 stories high with a group-floor area of 115,000 ft^2.

The facility was partially protected with combination deluge/preaction sprinkler systems located over transformers.

An electrical fault occurred in the vault of the Unit 2 main transformer. Fire spread inside the generator station, extensively damaging the mezzanine, basement, and turbine gallery, as well as the boiler towers in each unit. The local fire department was notified immediately and successfully rescued 12 plant personnel trapped inside the control room by the fire.

Contributing factors included the following:

- Fire spread from the transformer vault and into the turbine building through bus ductwork.
- Hydrogen and natural gas lines contributed fuel to the fire until the control valves were located and shut.
- Concerns over energized electrical equipment and the rescue of the trapped plant workers delayed fire suppression operations.
- Fire spreading from unprotected areas overwhelmed the partial sprinkler system. According to officials, the fire pump was not large enough to supply both the hydrant and sprinkler systems. The fire pump failed during the fire.

Although several factors contributed to this loss, the failure of the fire pump is of particular concern because without a water supply, sprinkler systems cannot be expected to control or extinguish a fire. Had proper maintenance procedures, such as weekly and annual testing, been followed, failure of the pump could have been avoided.

A.8.3 The purpose of testing the pump assembly is to ensure automatic or manual operation upon demand and continuous delivery of the required system output. An additional purpose is to detect deficiencies of the pump assembly not evident by inspection.

8.3.1 A weekly test of fire pump assemblies shall be conducted without flowing water.

The reason that 8.3.1 references the weekly test "without flowing water" is to indicate that it is not the intent of the standard to require a flow test on a weekly basis. Subsection 8.3.1 is intended to verify that the pump will start and will not overheat. Paragraph 8.3.1.4 allows the circulation relief valve to open to flow water as a cooling measure. Allowing additional water flow to prevent overheating is not a requirement of the standard. Flow from the circulation relief valve should be sufficient to prevent overheating of the pump. The valve should be at least $3/4$ in. in size as required by NFPA 20.

Subsection 8.3.1 is intended to distinguish between weekly and annual performance tests. The weekly testing verifies that the pump will start and not overheat when operating without flowing water.

The weekly testing requirement has engendered much discussion in the industry. Weekly testing of fire pumps can be costly due to the time and frequency involved. Further, it is difficult to apply a one-size-fits-all approach to testing such a critical piece of equipment, given the various configurations of the equipment such as electric drive, diesel engine drive, and fixed speed versus variable speed drive systems. Few data are currently available on failure rates and loss history to suggest another test frequency for fire pumps at this time. The new performance-based provision in Chapter 4, however, can be very useful when applied to testing fire pumps. In fact, A.4.6.1.1.1 uses a fire pump as an example of how a performance-based program can be implemented.

Another consideration for testing frequencies has been that emergency generators are required by NFPA 110, *Standard for Emergency and Standby Power Systems*, to be tested at least monthly. While this requirement appears to set a precedent for testing this type of equipment, a fire pump has different operating characteristics and related equipment and cannot and should not be compared to an emergency power supply system (EPSS).

The critical factor in determining the appropriate testing frequency for a fire pump is the accumulation of operating data for as long a period of time as possible. These data will establish an operating history for the specific installation and should indicate operating trends that will enable the system operator to determine an appropriate testing frequency based on

◀ **FAQ**
Is cracking open a test header valve to provide enough waterflow to keep the pump casing cool an acceptable procedure, even though 8.3.1 specifically states "without flowing water"?

◀ **FAQ**
Must pumps be tested every week or is there another approach to properly test a fire pump at frequencies other than weekly?

the number of system failures over a given time period. This provision of the standard should not be viewed as avoiding weekly testing for economic reasons, since the accumulation and analysis of test data can be time consuming. Further, the test data and proposal to deviate from the prescriptive requirements of this standard must be approved by the AHJ.

8.3.1.1 The weekly test shall be conducted by starting the pump automatically.

FAQ ▶
What is meant by an automatic start?

Paragraph 8.3.1.1 requires that pumps be started automatically, rather than by the use of the start button on the front panel of the fire pump controller. The pump must be started by drawing off water from the sensing line to simulate a pressure drop in the system. As the pressure drops, the pressure switch (see Exhibit 8.9) will sense the drop in pressure and start the pump automatically. Using the "start" button on the fire pump controller does not constitute an automatic start.

Exhibit 8.18 shows a pressure-sensing line connection that can be used to draw off water to simulate a pressure drop in the pressure-sensing line. Doing so will cause the jockey pump, and ultimately the fire pump, to start automatically.

EXHIBIT 8.18 Pressure-Sensing Line.

8.3.1.2 The electric pump shall run a minimum of 10 minutes.

When a pump is started, a great deal of heat is generated from the energy needed to bring the pump up to speed. Paragraph 8.3.1.2 requires that the electric motor be run for 10 minutes so the motor windings can cool down after starting across the line. Another reason for the 10-minute requirement is that it allows time to check the pump packing and bearings to determine if they are overheating or leaking excessively.

8.3.1.3 The diesel pump shall run a minimum of 30 minutes.

FAQ ▶
Why does NFPA 25 require the pump to run at churn for $\frac{1}{2}$ hour first?

Paragraph 8.3.1.3 requires that a diesel fire pump be operated for 30 minutes. This requirement is intended to allow the pump and driver to reach operating temperature and will reveal any overheating problems. The 30 minute operating time is also intended to consume fuel to prevent the fuel from stagnating and to prevent wet stacking in the exhaust system.

8.3.1.4 A valve installed to open as a safety feature shall be permitted to discharge water.

The valve referred to in 8.3.1.4 is the main relief valve (see Exhibit 13.24), which may be installed to limit the pressure on the system downstream of the pump. It is important that this valve be set so that it only discharges water at or above the rating of the piping downstream of the pump. This valve, if present, should cause water to discharge while the pump is operating at churn, meaning that water will be discharging for 30 minutes each week.

Water is intended to discharge from the cooling water line when a heat exchanger is supplied with water from the pump discharge. This allows cooler water to enter the cooling line while the pump is running.

8.3.1.5 The automatic weekly test timer shall be permitted to be substituted for the starting procedure.

8.3.2 Weekly Tests.

8.3.2.1 Qualified operating personnel shall be in attendance during the weekly pump operation.

8.3.2.2* The pertinent visual observations or adjustments specified in the following checklists shall be conducted while the pump is running:

(1) Pump system procedure:

    (a) Record the system suction and discharge pressure gauge readings.

The system suction and discharge pressure gauge readings are taken both when the pump is running and when the pump is shut off. For diesel engine–driven pumps, it is recommended that the cooling water pressure be recorded to ensure that the heat exchanger is not overpressurized.

    (b) Check the pump packing glands for slight discharge.
    (c) Adjust gland nuts if necessary.

The pump should run a minimum of 10 minutes to allow time for the motor windings to cool down. Running the motor for less than 10 minutes will shorten the motor's life.

Exhibit 8.19 shows a stone in the impeller of a fire pump. The stone was revealed by weekly testing.

    (d) Check for unusual noise or vibration.
    (e) Check packing boxes, bearings, or pump casing for overheating.
    (f) Record the pump starting pressure.

(2) Electrical system procedure:

    (a) Observe the time for motor to accelerate to full speed.
    (b) Record the time controller is on first step (for reduced voltage or reduced current starting).
    (c) Record the time pump runs after starting (for automatic stop controllers).

EXHIBIT 8.19 Stone in Impeller (left) and Size of Stone (right). (Courtesy of John Jensen)

(3) Diesel engine system procedure:
  (a) Observe the time for engine to crank.

The engine will alternate battery supplies with each attempt-to-start sequence. There should be a pilot light on the controller that indicates which bank of batteries is being used for the present start cycle. If the engine is difficult to start, maintenance or repair may be necessary to correct the problem because excessive cranking time causes the batteries to discharge.

  (b) Observe the time for engine to reach running speed.

The engine should come up to speed within 10 seconds. This delay has been another reason that property owners resist weekly testing, as it does not permit time for the engine to warm up. Excessive starting can shorten the operating life of the equipment. A tachometer, such as the one shown in Exhibit 8.20, is used to measure engine running speed.

EXHIBIT 8.20 Handheld Tachometer Used to Measure Motor Speed in rpm.

  (c) Observe the engine oil pressure gauge, speed indicator, water, and oil temperature indicators periodically while engine is running.
  (d) Record any abnormalities.

Paragraph 8.3.2.2(3)(d) requires the recording of any abnormalities such as low rpm, low oil pressure, high temperature, high cooling water pressure for diesel engines using discharge water for cooling, leaking hoses, and so on.

  (e) Check the heat exchanger for cooling waterflow.

The heat exchanger should have a sight glass or other location where the water can be observed directly. The temperature, determined by touching the pipe or heat exchanger, should be slightly above the normal temperature of the water source. There should not be any antifreeze or discharge water discoloration in the heat exchanger.

(4) Steam system procedure:
  (a) Record the steam pressure gauge reading.
  (b) Observe the time for turbine to reach running speed.

A.8.3.2.2 See Table A.8.3.2.2.

While the pump is running, observations including the items listed in Table A.8.3.2.2 should be made and a record of the test prepared. Exhibit 8.21 shows operating personnel recording test data during the weekly test.

TABLE A.8.3.2.2 Weekly Observations — While Pumping

Item	While Pump Is Operating
Horizontal pumps	1. Read suction and discharge gauges — difference between these readings indicates churn pressure, which should match churn pressure as shown on fire pump nameplate. 2. Observe packing glands for proper leakage for cooling of packing. 3. Observe discharge from casing relief valve — adequate flow keeps pump case from overheating.
Vertical pumps	1. Read discharge gauge — add distance to water level in feet (or meters) and divide by 2.31 to compute psi (30.47 to compute bar). This total must match churn pressure as shown on fire pump nameplate. 2. Observe packing glands for proper leakage for cooling of packing. 3. Observe discharge from casing relief valve — adequate flow keeps pump case from overheating.
Diesel engines	1. Observe discharge of cooling water from heat exchanger — if not adequate, check strainer in cooling system for obstructions. If still not adequate, adjust pressure reducing valve for correct flow. 2. Check engine instrument panel for correct speed, oil pressure, water temperature, and ammeter charging rate. 3. Check battery terminal connections for corrosion and clean if necessary. 4. After pump has stopped running, check intake screens, if provided; change diesel system pressure recorder chart and rewind if necessary.

EXHIBIT 8.21 Weekly Test.

8.3.3 Annual Tests.

8.3.3.1* An annual test of each pump assembly shall be conducted under minimum, rated, and peak flows of the fire pump by controlling the quantity of water discharged through approved test devices.

As with the acceptance test, three points must be measured and plotted on graph paper: churn, rated capacity, and overload.

The minimum flow condition, sometimes called shutoff or churn, is the point where very little, if any, water is flowing. The normal water flowing at churn is restricted to diesel engine cooling water, casing relief, or the main relief valve. The main relief valve is keeping the discharge pressure below the pipe rated pressure of 175 psi (1.2 bar).

At rated capacity, the pump should be flowing the amount of water at the pressure indicated on the pump nameplate. For overload, the pump should be flowing 150 percent of rated flow at 65 percent of rated pressure.

Exhibit 8.22 illustrates the annual test at churn. Note that calibrated test gauges have been installed. Exhibit 8.23 shows measuring flow at rated capacity. Exhibit 8.24 illustrates the annual test at overload. Note the quantity of water versus that of the annual test at rated capacity.

EXHIBIT 8.22 Annual Test at Churn.

EXHIBIT 8.23 Discharge from Annual Fire Pump Test at Rated Capacity.

EXHIBIT 8.24 Discharge from Annual Test at Overload.

A.8.3.3.1 Peak flow for a fire pump is 150 percent of the rated flow. Minimum flow for a pump is the churn pressure.

8.3.3.1.1 If available suction supplies do not allow flowing of 150 percent of the rated pump capacity, the fire pump shall be permitted to operate at maximum allowable discharge.

8.3.3.1.2* The annual test shall be conducted as described in 8.3.3.1.2.1, 8.3.3.1.2.2, or 8.3.3.1.2.3.

A.8.3.3.1.2 The method described in 8.3.3.1.2.3 is not considered as complete as those in 8.3.3.1.2.1 and 8.3.3.1.2.2, because it does not test the adequacy of the water supply for compliance with the requirements of 8.1.3 at the suction flange.

The following is a description of the test procedures necessary to conduct the annual test required in 8.3.3.1.2.

ANNUAL TEST PROCEDURES

Test Equipment. Test equipment should be provided to determine net pump pressures, rate of flow through the pump, volts and amperes for electric motor–driven pumps, and speed.

It is normal to replace the suction and discharge gauges with test gauges that have been recently calibrated in accordance with nationally recognized standards. Gauges used with Pitot tubes also should be recently calibrated. The flow characteristics of the test nozzles should be known. All flowmeters require a known factory calibration.

All test volt or amp meters should be calibrated.

Flow Tests. The pump flow for positive displacement pumps should be tested and determined to meet the specified rated performance criteria. One performance point is required to establish positive displacement pump performance.

For other types of pumps, such as centrifugal, the flow is measured, and suction and discharge pressure is recorded at the churn, rated, and peak flow points. For diesel engines, the engine speed in rpm is recorded.

Measurement Procedure. The quantity of water discharging from the fire pump assembly should be determined and stabilized. Immediately thereafter, the operating conditions of the fire pump and driver should be measured. Foam concentrate pumps should be permitted to be tested with water. However, water flow rates can be lower than expected foam flow rates because of viscosity.

The pump flow test for positive displacement pumps should be accomplished using a flowmeter or orifice plate installed in a test loop back to the foam concentrate tank or the inlet side of a water pump. The flowmeter reading or discharge pressure should be recorded and should be in accordance with the pump manufacturer's flow performance data. If orifice plates are used, the orifice size and corresponding discharge pressure to be maintained on the upstream side of the orifice plate should be made available to the authority having jurisdiction (AHJ). Flow rates should be as specified while operating at the system design pressure. Tests should be performed in accordance with HI 3.6, *Rotary Pump Tests*, 1994 edition.

Fire Pump Operation. The fire pump operation is as follows.

Motor-Driven Pump. To start a motor-driven pump, complete the following steps:

1. See that pump is completely primed.
2. Close isolating switch and then close circuit breaker.
3. Verify that the automatic controller will start pump if system demand is not satisfied (e.g., pressure low, deluge tripped, etc.).
4. For manual operation, activate switch or pushbutton, or manual start handle.

The circuit breaker-tripping mechanism should be set so that it will not operate when current in the circuit is excessively large. The circuit breaker and any protective devices are factory-set at stalled rotor condition or six times full load amps. These settings should not be changed.

Steam-Driven Pump. A steam turbine driving a fire pump should always be kept warmed up to permit instant operation at full rated speed. The automatic starting of the turbine should not be dependent on any manual valve operation or period of low-speed operation. If the pop safety valve on the casing blows, steam should be shut off and the exhaust piping examined for a possible closed valve or an obstructed portion of piping. Steam turbines are provided with governors to maintain a predetermined speed, with some adjustment for higher or lower speeds. Desired speeds below this range can be obtained by throttling the main throttle valve.

Diesel Engine-Driven Pump. Only an operator who has studied the instruction books issued by the engine and control manufacturer and is familiar with the operation of diesel engine–driven pumps should start this type of equipment.

The storage batteries should always be maintained in good order to ensure prompt satisfactory operation of this equipment. Maintenance should include checking electrolyte level and specific gravity, inspecting cable conditions, checking for corrosion, and so on.

Fire Pump Settings. The fire pump system, when started by pressure drop, should be arranged as follows:

1. The jockey pump stop point should equal the pump churn pressure plus the minimum static supply pressure.
2. The jockey pump start point should be at least 10 psi (0.68 bar) less than the jockey pump stop point.
3. The fire pump start point should be 5 psi (0.34 bar) less than the jockey pump start point. Use 10 psi (0.68 bar) increments for each additional pump.
4. Where minimum run timers are provided, the pump will continue to operate after attaining these pressures. The final pressures should not exceed the pressure rating of the system.
5. Where the operating differential of pressure switches does not permit these settings, the settings should be as close as equipment will permit. The settings should be established by pressures observed on test gauges.

An example of fire pump settings given in inch/pound units follows.

Pump: 1000 gpm, 100 psi pump with churn pressure of 115 psi.

Suction supply: 50 psi from city — minimum static; 60 psi from city — maximum static

Jockey pump stop = 115 + 50 = 165 psi

Jockey pump start = 165 – 10 = 155 psi

Fire pump stop = 115 + 50 = 165 psi

Fire pump start = 155 – 5 = 150 psi

Fire pump maximum churn = 115 + 60 = 175 psi

An example of fire pump settings given in SI units follows.

Pump: 3785 L/min, 7 bar pump with churn pressure of 7.9 bar.

Suction supply: 3.45 bar from city — minimum static; 4.13 bar from city — maximum static

Jockey pump stop = 7.9 + 3.45 = 11.35 bar

Jockey pump start = 11.35 – 0.65 = 10.7 bar

Fire pump stop = 7.9 + 3.45 = 11.35 bar

Fire pump start = 10.7 – 0.35 = 10.35 bar

Fire pump maximum churn = 7.9 + 4.1 = 12 bar

Automatic Recorder. The performance of all fire pumps should be indicated on an automatic pressure recorder to provide a record of pump operation. Such a record would be of assistance in fire loss investigation.

Whether the test equipment is furnished by the AHJ, the installing contractor, or the pump manufacturer is determined by prevailing arrangements made among those parties. The equipment should include, but need not be limited to, the following:

1. Use with test valve header: 50 ft (15 m) lengths, 2½ in. (65 mm) lined hose, and Underwriters Laboratories play pipe nozzles as needed to flow required volume of water. Where test meter is provided, hose may not be needed.

2. Instrumentation: the following test instruments should be of high quality, accurate, and in good repair:
   a. Clamp on volt/ammeter
   b. Test gauges
   c. Tachometer
   d. Pitot tube with gauge (for use with hose and nozzle)
3. Instrumentation calibration: all test instrumentation should have been calibrated by an approved testing and calibration facility within the previous 12 months. Calibration documentation should be available for review by the AHJ.

A majority of the test equipment used for acceptance and annual testing has never been calibrated. This equipment can have errors of 15 to 30 percent in readings, leading to inaccurate test results.

8.3.3.1.2.1 Use of the Pump Discharge via the Hose Streams. Pump suction and discharge pressures and the flow measurements of each hose stream shall determine the total pump output. Care shall be taken to prevent water damage by verifying there is adequate drainage for the high-pressure water discharge from hoses.

8.3.3.1.2.2 Use of the Pump Discharge via the Bypass Flowmeter to Drain or Suction Reservoir. Pump suction and discharge pressures and the flowmeter measurements shall determine the total pump output.

8.3.3.1.2.3 Use of the Pump Discharge via the Bypass Flowmeter to Pump Suction (Closed-Loop Metering). Pump suction and discharge pressures and the flowmeter measurements shall determine the total pump output.

8.3.3.1.3 Where the annual test is conducted periodically in accordance with 8.3.3.1.2.3, a test shall be conducted every 3 years in accordance with 8.3.3.1.2.1 or 8.3.3.1.2.2 in lieu of the method described in 8.3.3.1.2.3.

The purpose of the flow test requirement every three years is to verify the accuracy of the flowmeter. Any outlet having the capacity to discharge water at the peak flow should be acceptable for the pump test.

◀ **FAQ**
Why is a flow test required every 3 years? Is it permitted to use the roof manifold for the test?

8.3.3.1.4 Where 8.3.3.1.2.2 or 8.3.3.1.2.3 is used, the flow meter shall be adjusted immediately prior to conducting the test in accordance with the manufacturer's instructions. If the test results are not consistent with the previous annual test, 8.3.3.1.2.1 shall be used. If testing in accordance with 8.3.3.1.2.1 is not possible, a flowmeter calibration shall be performed and the test shall be repeated.

8.3.3.2 The pertinent visual observations, measurements, and adjustments specified in the following checklists shall be conducted annually while the pump is running and flowing water under the specified output condition:

(1) At no-flow condition (churn):

  (a) Check the circulation relief valve for operation to discharge water.

The circulation relief valve opens when the pump is running to allow a small amount of water to discharge (usually outside of the pump room). The intent is to allow additional, cooler water to enter the pump casing for cooling purposes. Circulation relief valves are usually pressure operated (spring-loaded) and can fail easily when a small amount of obstructing material or corrosion enters the valve.

The circulation relief valve should be set to open at the pump rated pressure.

  (b) Check the pressure relief valve (if installed) for proper operation.

The pressure relief valve mentioned in 8.3.3.2(1)(b) refers to the main relief valve that is used to maintain the system pressure below the rated pressure. In the case of a diesel engine, this valve is used to control the pressure if the engine reaches a runaway or overspeed condition. The overspeed governor is set at 120 percent of rated engine speed. Using the pump affinity laws, the formula to determine the overspeed governor setting is as follows:

$$H_1 = H_2 \left(\frac{N_1}{N_2}\right)^2$$

where:

H_1 = Head at test speed (m or ft)
H_2 = Head at rated speed (m or ft)
N_1 = Test speed (rpm)
N_2 = Rated speed (rpm)

For example, a rated speed of 1750 rpm, a rated pressure of 125 psi (8.6 bar), a test speed of 2100 rpm gives us the following:

$$H_1 = 125 \left(\frac{2100}{1750}\right)^2 = 180 \text{ psi}$$

A pressure of 180 psi (12.4 bar) is above the rating of the pipe.

The overspeed governor should be tested annually to make sure it will operate in an emergency situation. For pipe and fittings rated for 175 psi (12.1 bar), the setting of the main relief valve should be no higher than that.

(c) Continue the test for $1/2$ hour.

The entire test should take 30 minutes, including the churn and the flow tests.

(2) At each flow condition:

(a) Record the electric motor voltage and current (all lines).

The electric motor voltage and current data should only be recorded by someone trained and qualified in electrical hazards and equipped with needed safety equipment as outlined in NFPA 70E, *Standard for Electrical Safety in the Workplace*. Voltage and amperage should be recorded for each phase of the electrical circuits at each flow condition, including shutoff.

(b) Record the pump speed in rpm.

The pump speed can be recorded using a calibrated strobe tachometer or handheld rpm counter placed on the end of the pump shaft. The newer engine electronic tachometers are as accurate as any portable unit and are considered adequate for fire pump testing purposes. Electric motors run at their synchronous speed and, unlike a diesel engine, will not vary with load.

(c) Record the simultaneous (approximately) readings of pump suction and discharge pressures and pump discharge flow.

Calibrated test equipment improves the quality of the test results. A pressure dampener will minimize the gauge needle movement during testing, improving the accuracy of the pressure reading.

8.3.3.3* For installations having a pressure relief valve, the operation of the relief valve shall be closely observed during each flow condition to determine whether the pump discharge pressure exceeds the normal operating pressure of the system components.

A.8.3.3.3 A pressure relief valve that opens during a flow condition is discharging water that is not measured by the recording device(s). It can be necessary to temporarily close the pressure relief valve to achieve favorable pump test results. At the conclusion of the pump test, the pressure relief valve must be readjusted to relieve pressures in excess of the normal operating pressure of the system components.

If the pressure relief valve is open during the flowing conditions due to the fact that the pressure is too high for the components in the fire protection system, the discharge control valve should be closed prior to closing the pressure relief valve to make sure that the fire protection system is not overpressurized. After the test, the valve must be opened again.

The pressure relief valve should be set to open at a pressure that is just below the rated pressure of the piping system.

8.3.3.3.1* The pressure relief valve shall also be observed during each flow condition to determine whether the pressure relief valve closes at the proper pressure.

A.8.3.3.3.1 A pressure relief valve that is open during a flow condition will affect test results.

8.3.3.3.2 The pressure relief valve shall be closed during flow conditions if necessary to achieve minimum rated characteristics for the pump, and reset to normal position at the conclusion of the pump test.

The pressure setting of the pressure relief valve must be set at the working pressure of the system [usually 175 psi (12.1 bar) or less].

8.3.3.4 For installations having an automatic transfer switch, the following test shall be performed to ensure that the overcurrent protective devices (i.e., fuses or circuit breakers) do not open:

(1) Simulate a power failure condition while the pump is operating at peak load.
(2) Verify that the transfer switch transfers power to the alternate power source.
(3) Verify that the pump continues to perform at peak load.
(4) Remove the power failure condition, and verify that, after a time delay, the pump is reconnected to the normal power source.

8.3.3.5 Alarm conditions shall be simulated by activating alarm circuits at alarm sensor locations, and all such local or remote alarm indicating devices (visual and audible) shall be observed for operation.

The conditions for alarming purposes as required by 8.3.3.5 are supervisory signals that require local attention and should not summon the fire department.

8.3.3.6 Safety. Section 4.8 shall be followed for safety requirements while working near electric motor-driven fire pumps.

Work near electric motor–driven pumps requires a licensed and qualified electrician using the appropriate safety equipment such as gloves, protective clothing, and a face shield.

Due to the voltages present in a typical fire pump controller, NFPA 70E considers it a motor control center (MCC). Therefore, protective equipment in the form of flash protection for the face, rated gloves, and noncombustible shirt and trousers must be worn when opening the controller. (See Exhibit 4.5 for the safety equipment needed to work on a fire pump controller.)

8.3.3.7* Suction Screens. After the waterflow portions of the annual test or fire protection system activations, the suction screens shall be inspected and cleared of any debris or obstructions.

A.8.3.3.7 During periods of unusual water supply conditions such as floods, inspection should be on a daily basis.

The "unusual water supply conditions" mentioned in A.8.3.3.7 include floods and times of low water. In times of low water, the suction crib should be inspected to make sure there is adequate water to cover the pump bowls so that cavitation does not occur.

If the pump takes suction from a well, the water level needs to be checked on a regular basis during times of drought or unusual water demand to make sure there is adequate water available for the pump. If the water level is too low, it may be necessary to extend the well deeper or add additional sections of pipe in the well. In extreme cases, additional bowel sections and a larger pump driver may have to be added.

8.3.3.8* Where engines utilize electronic fuel management control systems, the backup electronic control module (ECM), and the primary and redundant sensors for the ECM, shall be tested annually.

A.8.3.3.8 *ECM and Sensor Testing.* To verify the operation of the alternate ECM with the stop, the ECM selector switch should be moved to the alternate ECM position. Repositioning of this should cause an alarm on the fire pump controller. Then the engine is started; it should operate normally with all functions. Next, the engine is shut down, switched back to the primary ECM, and restarted briefly to verify that correct switchback has been accomplished.

To verify the operation of the redundant sensor, with the engine running, the wires are disconnected from the primary sensor. There should be no change in the engine operation. The wires are then reconnected to the sensor, then disconnected from the redundant sensor. There should be no change in the engine operation. The wires should next be reconnected to the sensor. This process is repeated for all primary and redundant sensors on the engines. It should be noted whether disconnecting and reconnecting of wires to the sensors can be done while the engine is not running, then starting the engine after each disconnecting and reconnecting of the wires to verify engine operation.

8.3.4 Other Tests.

8.3.4.1 Engine generator sets supplying emergency or standby power to fire pump assemblies shall be tested routinely in accordance with NFPA 110, *Standard for Emergency and Standby Power Systems*.

8.3.4.2 Automatic transfer switches shall be tested routinely and exercised in accordance with NFPA 110, *Standard for Emergency and Standby Power Systems*.

8.3.4.3 Tests of appropriate environmental pump room space conditions (e.g., heating, ventilation, illumination) shall be made to ensure proper manual or automatic operation of the associated equipment.

8.3.4.4* Parallel and angular alignment of the pump and driver shall be checked during the annual test. Any misalignment shall be corrected.

When inspecting and testing fire pumps, the inspector should closely examine beneath the coupling guard for filings, which can be indicative of a misaligned coupling. When the pump is running, excessive vibration can also be an indication of coupling misalignment. Exhibit 8.25 shows a coupling alignment check during an annual test.

A.8.3.4.4 If pumps and drivers were shipped from the factory with both machines mounted on a common base plate, they were accurately aligned before shipment. All base plates are flexible to some extent and, therefore, must not be relied on to maintain the factory alignment. Realignment is necessary after the complete unit has been leveled on the foundation and again after the grout has set and foundation bolts have been tightened. The alignment should be checked after the unit is piped and rechecked periodically. To facilitate accurate field alignment, most manufacturers either do not dowel the pumps or drivers on the base plates before shipment or, at most, dowel the pump only.

After the pump and driver unit has been placed on the foundation, the coupling halves should be disconnected. The coupling should not be reconnected until the alignment operations have been completed.

EXHIBIT 8.25 Coupling Alignment During Annual Test.

The purpose of the flexible coupling is to compensate for temperature changes and to permit end movement of the shafts without interference with each other while transmitting power from the driver to the pump.

There are two forms of misalignment between the pump shaft and the driver shaft:

(1) *Angular misalignment.* Shafts with axes concentric but not parallel
(2) *Parallel misalignment.* Shafts with axes parallel but not concentric

The faces of the coupling halves should be spaced within the manufacturer's recommendations and far enough apart so that they cannot strike each other when the driver rotor is moved hard over toward the pump. Due allowance should be made for wear of the thrust bearings. The necessary tools for an approximate check of the alignment of a flexible coupling are a straight edge and a taper gauge or a set of feeler gauges.

A check for angular alignment is made by inserting the taper gauge or feelers at four points between the coupling faces and comparing the distance between the faces at four points spaced at 90 degree intervals around the coupling [*see Figure A.8.3.4.4(a)*]. The unit will be in angular alignment when the measurements show that the coupling faces are the same distance apart at all points.

A check for parallel alignment is made by placing a straight edge across both coupling rims at the top, bottom, and at both sides [*see Figure A.8.3.4.4(b)*]. The unit will be in parallel alignment when the straight edge rests evenly on the coupling rim at all positions. Allowance may be necessary for temperature changes and for coupling halves that are not of the same outside diameter. Care must be taken to have the straight edge parallel to the axes of the shafts.

Angular and parallel misalignment are corrected by means of shims under the motor mounting feet. After each change, it is necessary to recheck the alignment of the coupling halves. Adjustment in one direction may disturb adjustments already made in another direction. It should not be necessary to adjust the shims under the pump.

The permissible amount of misalignment will vary with the type of pump and driver; and coupling manufacturer, model, and size. [20: A.6.5]

8.3.5 Test Results and Evaluation.

8.3.5.1* Interpretation.

A.8.3.5.1 Where the information is available, the test plot should be compared with the original acceptance test plot. It should be recognized that the acceptance test plot could exceed

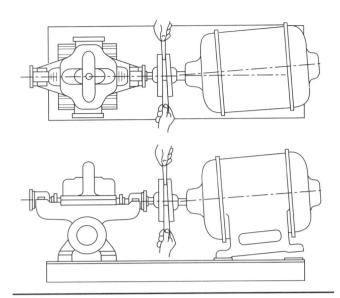

FIGURE A.8.3.4.4(a) *Checking Angular Alignment. (Courtesy of Hydraulic Institute Standard for Centrifugal, Rotary and Reciprocating Pumps.)*

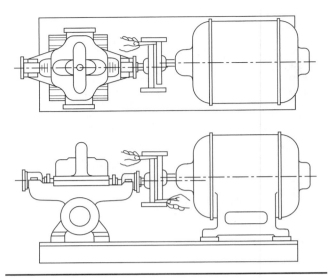

FIGURE A.8.3.4.4(b) *Checking Parallel Alignment. (Courtesy of Hydraulic Institute Standard for Centrifugal, Rotary and Reciprocating Pumps.)*

the minimum acceptable pump requirements as indicated by the rated characteristics for the pump. While a reduction in output is a matter of concern, this condition should be evaluated in light of meeting the rated characteristics for the pump. [*See Figure A.8.3.5.3(1).*]

The test equipment should be of high quality and accuracy. All equipment should have been calibrated within the last 12 months by an approved calibration facility. Where possible, the calibration facility should provide documentation indicating the instrument reading against the calibrated reading. Instruments that pass the calibration test should be labeled by the calibration facility with the name of the facility and the date of the test.

Pressure gauges should have an accuracy not greater than 1 percent of full scale. To prevent damage to a pressure gauge utilizing a Bourdon tube mechanism, it should not be used where the expected test pressure is greater than 75 percent of the test gauge scale. Some digital gauges can be subjected to twice the full scale pressure without damage. The manufacturer's recommendations should be consulted for the proper use of the gauge. To be able to easily read an analog gauge, the diameter of the face of the analog gauge should be greater than 3 in. (76 mm). Pressure snubbers should be used for all gauges to minimize needle fluctuation. All gauges used in the test should be such that a gauge with the lowest full scale pressure is used. For example, a 300 psi (20.7 bar) gauge should not be used to measure a 20 psi (1.4 bar) pitot pressure.

Equipment other than pressure gauges, such as volt/ammeters, tachometers, and flowmeters, should be calibrated to the manufacturer's specifications. The readings from equipment with this level of accuracy and calibration can be used without adjustment for accuracy.

8.3.5.1.1 The interpretation of the test results shall be the basis for determining performance of the pump assembly.

8.3.5.1.2 Qualified individuals shall interpret the test results.

8.3.5.2 Engine Speed.

8.3.5.2.1 Theoretical factors for correction to the rated speed shall be applied where determining the compliance of the pump per the test.

As with the acceptance test requirements from NFPA 20, NFPA 25 now permits the use of affinity laws for correction to the rated pump speed. Paragraph 8.3.5.2.1 was revised for the 2008 edition to allow for correction because the committee felt that prohibiting such correction would cause most pumps to fail the test when comparing test data from the annual test to the original, unadjusted acceptance test curve, because of the 5 percent degradation requirement of 8.3.5.4.

8.3.5.2.2 Increasing the engine speed beyond the rated speed of the pump at rated condition is not an acceptable method for meeting the rated pump performance.

8.3.5.3 The fire pump assembly shall be considered acceptable if either of the following conditions is shown during the test:

(1)*The test is no less than 95 percent of the pressure at rated flow and rated speed of the initial unadjusted field acceptance test curve, provided that the original acceptance test curve matches the original certified pump curve by using theoretical factors.

A.8.3.5.3(1) See Figure A.8.3.5.3(1).

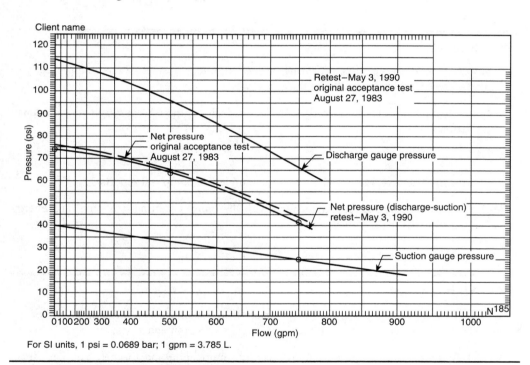

FIGURE A.8.3.5.3(1) Fire Pump Retest.

(2) The fire pump is no less than 95 percent of the performance characteristics as indicated on the pump nameplate.

The annual fire pump test is intended to evaluate the net fire pump performance. This net performance includes flow and pressure. Figure A.8.3.5.3(1) illustrates that the pump is evaluated based on discharge minus suction pressure. Variations of more than 5 percent must be investigated as to the cause.

8.3.5.4* Degradation in excess of 5 percent of the pressure of the initial unadjusted acceptance test curve or nameplate shall require an investigation to reveal the cause of degraded performance.

◀ **FAQ**
Where in NFPA 25 is the requirement located that states that the net head is the performance parameter for a fire pump test?

FAQ ▶
Does degraded pump performance require corrective action?

Although 8.3.5.4 requires that only degradation in excess of 5 percent be investigated, degradation of less than that amount should also be noted and compared to previous tests to determine if a steady deterioration in performance is occurring. For example, if pump performance were to decrease by a factor of 2 percent per year for several years, the requirement of 8.3.5.4 would not come into effect. However, comparing test results to the original test curve would reveal a steady decline in pump performance. Corrective action would be appropriate in this situation also.

A.8.3.5.4 See Annex C.

 8.3.5.5 Current and voltage readings whose product does not exceed the product of the rated voltage and rated full-load current multiplied by the permitted motor service factor shall be considered acceptable.

 8.3.5.6 Voltage readings at the motor within 5 percent below or 10 percent above the rated (i.e., nameplate) voltage shall be considered acceptable.

8.4 Reports

8.4.1 Any abnormality observed during inspection or testing shall be reported promptly to the person responsible for correcting the abnormality.

8.4.2* Test results shall be recorded and retained for comparison purposes in accordance with Section 4.4.

A.8.4.2 See 8.3.3.4.

See Forms S4.4 and S4.5 in Supplement 4 for sample fire pump test reports.

8.4.2.1 All time delay intervals associated with the pump's starting, stopping, and energy source transfer shall be recorded.

8.5 Maintenance

8.5.1* A preventive maintenance program shall be established on all components of the pump assembly in accordance with the manufacturer's recommendations.

A.8.5.1 It is important to provide proper bearing lubrication and to keep bearings clean. Some bearings are the sealed type and need no relubrication. Couplings with rubber drive parts do not need lubrication; other types generally do. The following practices are recommended:

(1) Lubricant fittings should be cleaned before relubricating with grease.
(2) The proper amount of lubricant should be used. Too much lubricant results in churning, causing excessive power loss and overheating.
(3) The correct lubricant should be used.

Engine Maintenance. Engines should be kept clean, dry, and well lubricated. The proper oil level in the crankcase should be maintained.

Battery Maintenance. Only distilled water should be used in battery cells. Plates should be kept submerged at all times. An automatic battery charger is not a substitute for proper maintenance of the battery and charger. Periodic inspection ensures that the charger is operating correctly, the water level in the battery is adequate, and the battery is holding its proper charge.

An adequate water level in the battery cells is up to the ring at the bottom of the cell fill connection. If there is any space in the top of the cell, hydrogen may accumulate in that space. If that occurs, an explosion could result when the pump starts.

Fuel Supply Maintenance. The fuel storage tank should be kept at least two-thirds full. Fuel should be maintained free of water and foreign material by draining water and foreign material from the tank sump annually. This necessitates draining approximately 5 gal (19 L).

Fuel that is kept for longer than one year may have biological growth in it that can clog the engine fuel filters. Therefore, fuel should be rotated or filtered annually and new biological inhibitors added.

Temperature Maintenance. The temperature of the pump room, pump house, or area where engines are installed should never be less than the minimum recommended by the engine manufacturer. The manufacturer's temperature recommendations for water and oil heaters should be followed.

8.5.2 Records shall be maintained on all work performed on the pump, driver, controller, and auxiliary equipment.

8.5.3 In the absence of manufacturers recommendations for preventive maintenance, Table 8.5.3 shall be used for alternative requirements.

Table 8.5.3 is intended to be used only in the absence of the pump manufacturer's recommendations for maintenance. It is best to obtain the manufacturer's operations and maintenance manuals for a fire pump, because some manufacturers may have special requirements particular to their equipment. Failure to follow such requirements could result in damage to the pump or its components and may void any warranties.

8.5.4 The preventive maintenance program shall be initiated immediately after the pump assembly has passed acceptance tests.

Paragraph 8.5.4 reflects the fact that, as with every installation standard, once commissioning activity has been completed, the scope of the installation standard is no longer applicable and the scope (and frequencies) of the maintenance standard begins.

8.6 Component Replacement Testing Requirements

Component replacement tables were added in the 2008 edition to offer guidance to the user of the standard when system components are adjusted, repaired, rebuilt, or replaced. It is not necessary in each case to require a complete acceptance test for each component when maintenance is performed.

8.6.1 Whenever a component in a fire pump is adjusted, repaired, rebuilt, or replaced, the tests required to restore the system to service shall be performed in accordance with Table 8.6.1.

8.6.2 NFPA 20, *Standard for the Installation of Stationary Pumps for Fire Protection*, shall be consulted for the minimum requirements for design and installation, including acceptance testing.

TABLE 8.5.3 Summary of Fire Pump Inspection, Testing, and Maintenance

Complete as Applicable	Visual Inspection	Check	Change	Clean	Test	Frequency
Pump System						
Lubricate pump bearings			X			Annually
Check pump shaft end play		X				Annually
Check accuracy of pressure gauges and sensors		X	X			Annually (change or recalibrate when 5% out of calibration)
Check pump coupling alignment		X				Annually
Wet pit suction screens		X		X		After each pump operation
Mechanical Transmission						
Lubricate coupling			X			Annually
Lubricate right-angle gear drive			X			Annually
Electrical System						
Exercise isolating switch and circuit breaker					X	Monthly
Trip circuit breaker (if mechanism provided)					X	Annually
Operate manual starting means (electrical)					X	Semiannually
Inspect and operate emergency manual starting means (without power)	X				X	Annually
Tighten electrical connections as necessary		X				Annually
Lubricate mechanical moving parts (excluding starters and relays)		X				Annually
Calibrate pressure switch settings		X				Annually
Grease motor bearings			X			Annually
Diesel Engine System						
Fuel						
Tank level	X	X				Weekly
Tank float switch	X				X	Weekly
Solenoid valve operation	X				X	Weekly
Strainer, filter, or dirt leg, or combination thereof				X		Quarterly
Water and foreign material in tank				X		Annually
Water in system		X			X	Weekly
Flexible hoses and connectors	X					Weekly
Tank vents and overflow piping unobstructed		X			X	Annually
Piping	X					Annually
Lubrication system						
Oil level	X	X				Weekly
Oil change			X			50 hours or annually
Oil filter(s)			X			50 hours or annually
Lube oil heater		X				Weekly
Crankcase breather	X		X	X		Quarterly
Cooling system						
Level	X	X				Weekly
Antifreeze protection level					X	Semiannually
Antifreeze		X				Annually
Adequate cooling water to heat exchanger		X				Weekly

TABLE 8.5.3 Continued

Complete as Applicable	Visual Inspection	Check	Change	Clean	Test	Frequency
Rod out heat exchanger				X		Annually
Water pump(s)	X					Weekly
Condition of flexible hoses and connections	X	X				Weekly
Jacket water heater		X				Weekly
Inspect duct work, clean louvers (combustion air)	X	X	X			Annually
Water strainer				X		Quarterly
Exhaust system						
Leakage	X	X				Weekly
Drain condensate trap		X				Weekly
Insulation and fire hazards	X					Quarterly
Excessive back pressure					X	Annually
Exhaust system hangers and supports	X					Annually
Flexible exhaust section	X					Semiannually
Battery system						
Electrolyte level		X				Weekly
Terminals clean and tight	X	X				Quarterly
Case exterior clean and dry	X	X				Monthly
Specific gravity or state of charge					X	Monthly
Charger and charge rate	X					Monthly
Equalize charge		X				Monthly
Clean terminals				X		Annually
Electrical system						
General inspection	X					Weekly
Tighten control and power wiring connections		X				Annually
Wire chafing where subject to movement	X	X				Quarterly
Operation of safeties and alarms		X			X	Semiannually
Boxes, panels, and cabinets				X		Semiannually
Circuit breakers or fuses	X	X				Monthly
Circuit breakers or fuses			X			Biennially

TABLE 8.6.1 Summary of Component Replacement Testing Requirements

Component	Adjust	Repair	Rebuild	Replace	Test Criteria
Fire Pump System					
Entire pump assembly				X	Perform acceptance test in accordance with NFPA 20, *Standard for the Installation of Stationary Pumps for Fire Protection*
Impeller/rotating assembly		X		X	Perform acceptance test in accordance with NFPA 20
Casing		X		X	Perform acceptance test in accordance with NFPA 20
Bearings				X	Perform annual test in accordance with NFPA 25
Sleeves				X	Perform annual test in accordance with NFPA 25
Wear rings				X	Perform annual test in accordance with NFPA 25
Main shaft		X		X	Perform annual test in accordance with NFPA 25
Packing	X			X	Perform weekly test in accordance with NFPA 25
Mechanical Transmission					
Gear right angle drives		X	X	X	Perform acceptance test in accordance with NFPA 20
Drive coupling	X	X	X	X	Perform weekly test in accordance with NFPA 25
Electrical System/Controller					
Entire controller		X	X	X	Perform acceptance test in accordance with NFPA 20
Isolating switch				X	Perform weekly test in accordance with NFPA 25 and exercise 6 times
Circuit breaker	X				Perform six momentary starts in accordance with NFPA 20
Circuit breaker				X	Perform a one-hour full-load current test
Electrical connections	X				Perform weekly test in accordance with NFPA 25
Main contactor			X		Perform weekly test in accordance with NFPA 25
Main contactor				X	Perform acceptance test in accordance with NFPA 20
Power monitor				X	Perform weekly test in accordance with NFPA 25
Start relay				X	Perform weekly test in accordance with NFPA 25
Pressure switch	X			X	Perform acceptance test in accordance with NFPA 20
Pressure transducer	X			X	Perform acceptance test in accordance with NFPA 20
Manual start or stop switch				X	Perform six operations under load
Transfer switch — load carrying parts		X	X	X	Perform a one-hour full-load current test, and transfer from normal power to emergency power and back one time
Transfer switch — non-load parts		X	X	X	Perform six no-load operations of transfer of power
Electric Motor Driver					
Electric motor		X	X	X	Perform acceptance test in accordance with NFPA 20
Motor bearings				X	Perform annual test in accordance with NFPA 25
Incoming power conductors				X	Perform a one-hour full-load current test
Diesel Engine Driver					
Entire engine			X	X	Perform annual test in accordance with NFPA 25
Fuel transfer pump	X		X	X	Perform weekly test in accordance with NFPA 25

TABLE 8.6.1 Continued

Component	Adjust	Repair	Rebuild	Replace	Test Criteria
Fuel injector pump	X			X	Perform weekly test in accordance with NFPA 25
Fuel system filter		X		X	Perform weekly test in accordance with NFPA 25
Combustion air intake system		X		X	Perform weekly test in accordance with NFPA 25
Fuel tank		X		X	Perform weekly test in accordance with NFPA 25
Cooling system		X	X	X	Perform weekly test in accordance with NFPA 25
Batteries		X		X	Perform a start/stop sequence in accordance with NFPA 25
Battery charger		X		X	Perform weekly test in accordance with NFPA 25
Electric system		X		X	Perform weekly test in accordance with NFPA 25
Lubrication filter/oil service		X		X	Perform weekly test in accordance with NFPA 25
Steam Turbines					
Steam turbine		X		X	Perform annual test in accordance with NFPA 20
Steam regulator or source upgrade		X		X	Perform annual test in accordance with NFPA 20
Positive Displacement Pumps					
Entire pump				X	Perform annual test in accordance with NFPA 20
Rotors				X	Perform annual test in accordance with NFPA 25
Plungers				X	Perform annual test in accordance with NFPA 25
Shaft				X	Perform annual test in accordance with NFPA 25
Driver		X	X	X	Perform annual test in accordance with NFPA 25
Bearings				X	Perform weekly test in accordance with NFPA 25
Seals				X	Perform weekly test in accordance with NFPA 25
Pump House and Miscellaneous Components					
Base plate		X		X	Perform weekly test in accordance with NFPA 25 with alignment check
Foundation		X	X	X	Perform weekly test in accordance with NFPA 25 with alignment check
Suction/discharge pipe		X		X	Perform visual inspection in accordance with NFPA 25
Suction/discharge fittings		X		X	Perform visual inspection in accordance with NFPA 25
Suction/discharge valves		X	X	X	Perform operational test in accordance with NFPA 25

SUMMARY

Chapter 8 of NFPA 25 provides inspection, testing, and maintenance requirements and recommendations for fire pumps and related equipment. Fire pumps are considered a critical component of fire protection systems, because they are frequently the only source of water supply. With proper care and maintenance, fire pumps can provide reliable service for many years. Lack of proper maintenance can lead to failure of the pump or components and result in expensive repairs.

REVIEW QUESTIONS

1. During the weekly running test for fire pumps, how long must the pump operate?
2. During the annual fire pump performance test, what points along the performance curve must be tested?

3. What is the proper adjustment for packing glands on a fire pump?
4. Describe the three methods of flow testing a fire pump during the annual test.
5. When testing a fire pump by means of discharge through the bypass flowmeter to the pump suction, how often must the pump be tested by means of actual water discharge?

REFERENCES CITED IN COMMENTARY

National Fire Protection Association, 1 Batterymarch Park, Quincy, MA 02169-7471.

NFPA 20, *Standard for the Installation of Stationary Pumps for Fire Protection,* 2007 edition.
NFPA 70E, *Standard for Electrical Safety in the Workplace,* 2004 edition.
NFPA 110, *Standard for Emergency and Standby Power Systems,* 2005 edition.

Hydraulics Institute, 1230 Keith Building, Cleveland, OH 44115.

HI 3.6, *Rotary Pump Tests,* 1994 edition.

Water Storage Tanks

CHAPTER 9

Water storage tanks are frequently used where an adequate quantity or volume of water is not readily available. Water storage tanks can be made of various materials such as wood, steel, concrete, fiberglass-reinforced plastic, and rubberized fabric.

Some tanks, such as pressure tanks, serve dual purposes by providing the necessary quantity of water and the needed pressure to propel the water through a fire protection system. In 2007, NFPA 20, *Standard for the Installation of Stationary Pumps for Fire Protection*, introduced the concept of a "break tank," which has basically three purposes:

1. As a backflow prevention device between the city water supply and the fire pump suction
2. To eliminate pressure fluctuations in the city water supply and provide a steady suction pressure to the fire pump
3. To augment the city water supply when the volume of water available from the city is inadequate for the fire protection system demand

Other tanks, such as those made of wood and elevated steel, rely on gravity to provide the 0.433 psi/ft (0.09 bar/m) of pressure necessary. Suction tanks rely on fire pumps to pressurize fire protection systems. Regardless of the arrangement, water storage tanks are a critical part of the water-based fire protection system and must be properly maintained if they are to function when needed.

Chapter 9 of NFPA 25 prescribes the inspection, testing, and maintenance activities necessary to ensure that water storage tanks will provide the proper quantity, and, in some cases, pressure, in the event of a fire. The requirements in Chapter 9, as is the case with those in NFPA 22, *Standard for Water Tanks for Private Fire Protection,* apply to private tanks installed on private property only. Storage tanks installed on public property serving public domestic water systems are not within the scope of this standard.

9.1* General

This chapter shall provide the minimum requirements for the routine inspection, testing, and maintenance of water storage tanks dedicated to fire protection use. Table 9.1 shall be used to determine the minimum required frequencies for inspection, testing, and maintenance.

Section 9.1 was revised for the 2008 edition to specifically point out that the private fire service tanks referenced in this chapter refer only to those tanks that are dedicated to fire protection use. The concern is that this chapter has been inappropriately applied to municipal water storage tanks and private tanks that supply both domestic and fire protection water. The inspection, testing, and maintenance of domestic, potable water is not within the scope of this standard.

A.9.1 One source of information on the inspection and maintenance of steel gravity and suction tanks is the AWWA *Manual of Water Supply Practices — M42 Steel Water-Storage Tanks*, Part III and Annex C.

157

TABLE 9.1 Summary of Water Storage Tank Inspection, Testing, and Maintenance

Item	Frequency	Reference
Inspection		
Condition of water in tank	Monthly/quarterly*	9.2.1
Water temperature	Daily/weekly*	9.2.4
Heating system	Daily/weekly*	9.2.6.6
Control valves	Weekly/monthly	Table 13.1
Water — level	Monthly/quarterly	9.2.1
Air pressure	Monthly/quarterly	9.2.2
Tank — exterior	Quarterly	9.2.5.1
Support structure	Quarterly	9.2.5.1
Catwalks and ladders	Quarterly	9.2.5.1
Surrounding area	Quarterly	9.2.5.2
Hoops and grillage	Annually	9.2.5.4
Painted/coated surfaces	Annually	9.2.5.5
Expansion joints	Annually	9.2.5.3
Interior	5 years/3 years	9.2.6
Check valves	5 years	Table 13.1
Test		
Temperature alarms	Quarterly*	9.2.4.2, 9.2.4.3
High temperature limit switches	Monthly*	9.3.4
Water level alarms	Semiannually	9.3.5
Level indicators	5 years	9.3.1
Pressure gauges	5 years	9.3.6
Maintenance		
Water level	—	9.4.1
Drain silt	Semiannually	9.4.5
Control valves	Annually	Table 13.1
Embankment-supported coated fabric (ESCF)	—	9.4.6
Check valves	—	13.4.2.2

*Cold weather/heating season only.

In many cases, water storage tanks are the primary — if not the only — source of water for fire-fighting purposes. Requirements for the design, construction, and operation of water storage tanks can be found in NFPA 22.

The appropriate size for a tank is determined by multiplying the water demand of the fire protection system it serves by the discharge duration specified in the installation standard. The total capacity is measured from the discharge pipe to the overflow pipe.

Exhibit 9.1 shows a steel ground-level tank that constitutes the sole source of water for a small manufacturing plant. The building in the foreground is a fire pump house.

Exhibit 9.2 shows an elevated storage tank with a riser more than 3 ft (0.9 m) in diameter. In some areas of the United States, the size of the riser will determine the need for heating.

The use of wooden gravity (elevated) tanks for fire protection has decreased significantly over the past 100 years, though they are still found in many older cities (see Exhibit 9.3). These wooden tanks were installed high in the building (usually on or above the roof), feeding a sprinkler system by gravity. Exhibit 9.4 shows a wooden gravity tank under construction. Exhibit 9.5 shows a fiberglass-reinforced plastic (FRP) tank.

Exhibit 9.6 illustrates a ground-level suction tank and the discharge pipe connected to the bottom of the tank in a valve pit.

The use of steel pressure tanks, as shown in Exhibit 9.7, has declined largely due to space considerations. These tanks require a large volume of space to provide sufficient water

Section 9.1 • General 159

EXHIBIT 9.1 Steel Ground-Level Tank.

EXHIBIT 9.2 Elevated Storage Tank.

EXHIBIT 9.3 Wooden Gravity Tanks. (Courtesy of Hall-Woolford Tank Co., Inc.)

EXHIBIT 9.4 Wooden Gravity Tank Under Construction. (Courtesy of Hall-Woolford Tank Co., Inc.)

EXHIBIT 9.5 Fiberglass-Reinforced Plastic Tank.

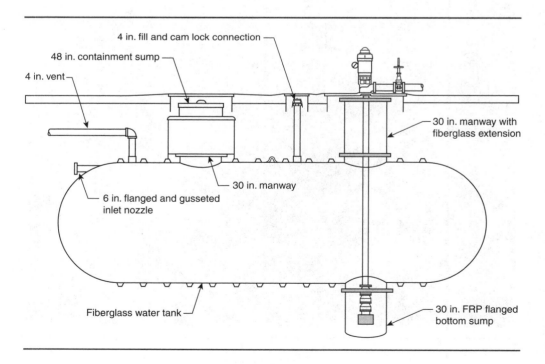

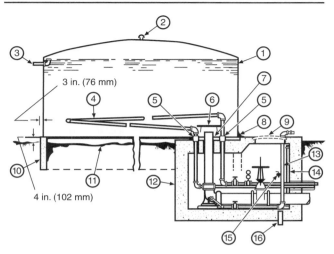

EXHIBIT 9.7 Steel Pressure Tank.

1. Pump suction tank
2. Screened vent
3. Stub overflow pipe
4. Steam coil for heating
5. Extra-heavy couplings welded to tank bottom
6. Vortex plate
7. Watertight lead slip joint
8. Flashing around tank
9. Manhole with cover
10. Concrete ring wall
11. Sand or concrete pad (depending on soil condition)
12. Valve pit
13. Drain pipe
14. Ladder
15. Drain cock
16. Valve pit drain

EXHIBIT 9.6 Ground-Level Suction Tank. (Source: Fire Protection Handbook, 2003, Figure 10.2.1)

capacity for a sprinkler system. Smaller-volume pressure tanks are being provided, in part due to the introduction of residential sprinkler systems and the corresponding lower duration values they typically require when installed in accordance with NFPA 13D, *Standard for the Installation of Sprinkler Systems in One- and Two-Family Dwellings and Manufactured Homes*.

An embankment-supported fabric suction tank, as shown in Exhibit 9.8, can be thought of as a waterbed. Water is enclosed in fabric that is given its shape by berms located around the tank. The fabric reduces the likelihood of marine life attaching to the walls of the tank.

9.1.1 Valves and Connections.

Valves and fire department connections shall be inspected, tested, and maintained in accordance with Chapter 13.

9.1.2 Impairments.

The procedures outlined in Chapter 15 shall be followed where an impairment to protection occurs.

9.1.3* Notification to Supervisory Service.

To avoid false alarms where a supervisory service is provided, the alarm receiving facility always shall be notified by the property owner or designated representative as follows:

(1) Before conducting any test or procedure that could result in the activation of an alarm
(2) After such tests or procedures are concluded

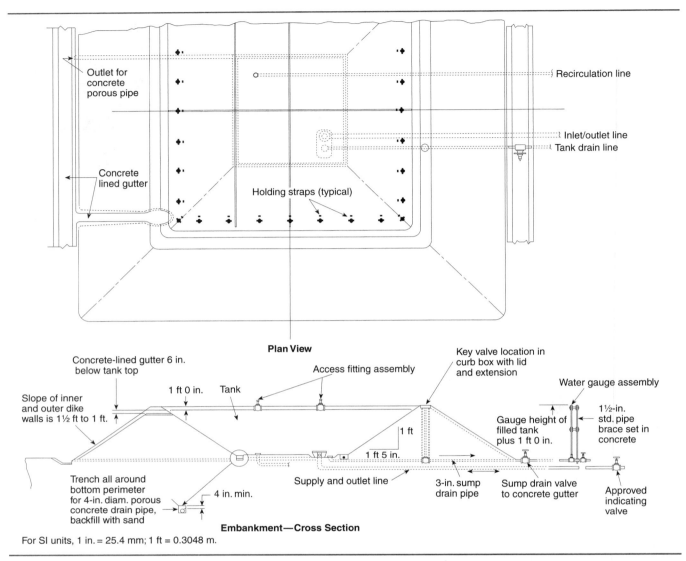

EXHIBIT 9.8 Embankment-Supported Fabric Suction Tank. [Source: NFPA 22, 2003, Figure B.1(e)]

When flowing an inspector's test connection, it may be obvious that an alarm is intended to sound upon waterflow and that the supervisory service must be notified, as prescribed by 9.1.3(2). Perhaps not so obvious is the fact that other tests, such as a fire pump test, hydrant test, main drain test, and so on, may also lead to the tripping of a waterflow switch. All of these may trip a switch, sounding an alarm and potentially sending a signal to the fire department.

A.9.1.3 The inspection, testing, and maintenance of water storage tanks can involve or result in a system that is out of service. In cases where a tank is the sole source of supply to a fire protection system, it is recommended that an alternate water supply be arranged while maintenance is performed on the tank.

FAQ ▶
What are some alternate sources of water?

An alternate source of water, as referred to in A.9.1.3, can take the form of temporary hose connections to a nearby hydrant, the temporary placement of a tanker truck, or hose connections and portable fire pump to a body of water, whether natural or manmade.

9.2 Inspection

9.2.1 Water Level.

9.2.1.1* Tanks equipped with supervised water level alarms that are connected to a constantly attended location shall be inspected quarterly.

A.9.2.1.1 More frequent inspections should be made where extreme conditions, such as freezing temperatures or arid climate, can increase the probability of adversely affecting the stored water.

Supervisory water level alarms installed on tanks provide notification that the tank water level is above or below an acceptable level. The water level of the tank is the main concern as opposed to the condition of the water. For convenience, inspection of the condition of the water can take place concurrently with the water level inspection.

The water level in a tank is measured by the capacity of the tank between the overflow and discharge outlet. This quantity is determined by the calculated flow of the fire protection system for a specified duration (e.g., 30, 60, or 90 minutes). The level should never be lower than 3 in. to 4 in. (76 mm to 102 mm) below the designated fire service level.

9.2.1.2 Tanks not equipped with supervised water level alarms connected to a constantly attended location shall be inspected monthly.

CASE STUDY

Water Tank Found Empty During Fire Investigation

Fire in a furniture manufacturing plant in Indiana caused over $15 million in damage. The one- and two-story, 375,000 ft^2 plant of unprotected, noncombustible construction was used for assembling chairs. At the time of the fire, one person was operating a forklift truck in an area separated from the point of fire origin by a fire wall.

The type and coverage of the automatic detection system in this building was not reported, but it failed to operate due to poor maintenance. The building had wet and dry pipe sprinkler systems, but not in the room of origin. These systems also failed to work due to a lack of maintenance. A fire wall between the two parts of the building did slow fire spread, even though it did not extend to the roof.

This incendiary fire was started with accelerant that was poured on foam seat cushions and seat backs in the production area on the east side of the building. The fire wall on the north end of the production area slowed the fire spread so the fire department could stop it at that point.

When fire fighters arrived, they made an exterior deluge attack, stopping the fire when it threatened to jump a street. Some siding suffered minor damage, and homes across the street had minor roof burns. One fire fighter and one civilian were injured.

The contributing factors were found to be the following:

- Sprinkler system maintenance was lacking — the 200,000 gal tank was empty, and 8 of 11 post indicator valves and the fire pump were shut off.
- A 6 in. riser broke when a wall in the room of fire origin collapsed, causing low water pressure.
- The fire wall did not extend to the roof, but it did contain the fire.

9.2.2 Air Pressure.

9.2.2.1 Pressure tanks that have their air pressure source supervised in accordance with *NFPA 72, National Fire Alarm Code,* shall be inspected quarterly.

According to NFPA 22 a pressure tank must be kept two-thirds full of water and a minimum air pressure of at least 75 psi (5.2 bar) must be maintained. As the last of the water leaves the pressure tank, the residual pressure shown on the gauge cannot be less than zero. The air pressure must be sufficient to provide not less than 15 psi (1.03 bar) at the highest sprinkler under the main roof of the building.

9.2.2.2 The air pressure in pressure tanks with a nonsupervised air pressure source shall be inspected monthly.

9.2.3 Heating System.

NFPA 22 recognizes the following heating systems:

- Steam water heaters
- Gas-fired water heaters
- Oil-fired water heaters
- Coal-burning water heaters
- Electric water heaters
- Vertical steam radiators
- Hot water
- Steam coils inside tanks
- Direct discharge of steam
- Solar heating

Regardless of which method is used to heat the tank, the intent is to ensure reliability. An ice plug in a riser pipe, for example, can obstruct flow. In the worst case, it may break the pipe. In addition, ice in, or even on, the tank structure can cause collapse of the tank.

Generally, only tanks that are subject to freezing will have heating systems. The need for a heating system is based on the lowest one-day mean temperature as determined by isothermal lines (see Exhibit 9.9) and the type of tank construction. For example, a suction tank in Atlanta, Georgia, does not need to be heated. An elevated tank with a riser of 3 ft (0.9 m) or less in diameter in the same city must be heated in the riser portion only. See NFPA 22 to determine if a heating system is needed for a particular tank.

9.2.3.1 Tank heating systems installed on tanks equipped with a supervised low water temperature alarm that are connected to a constantly attended location shall be inspected weekly.

9.2.3.2 Tank heating systems without a supervised low temperature alarm connected to a constantly attended location shall be inspected daily during the heating season.

9.2.4 Water Temperature.

9.2.4.1 The temperature of water tanks shall not be less than 40°F (4.4°C).

9.2.4.2 The temperature of water in tanks with low temperature alarms connected to a constantly attended location shall be inspected and recorded monthly during the heating season when the mean temperature is less than 40°F (4.4°C).

9.2.4.3 The temperature of water in tanks without low temperature alarms connected to a constantly attended location shall be inspected and recorded weekly during the heating season when the mean temperature is less than 40°F (4.4°C).

9.2.5 Exterior Inspection.

9.2.5.1* The exterior of the tank, supporting structure, vents, foundation, and catwalks or ladders, where provided, shall be inspected quarterly for signs of obvious damage or weakening.

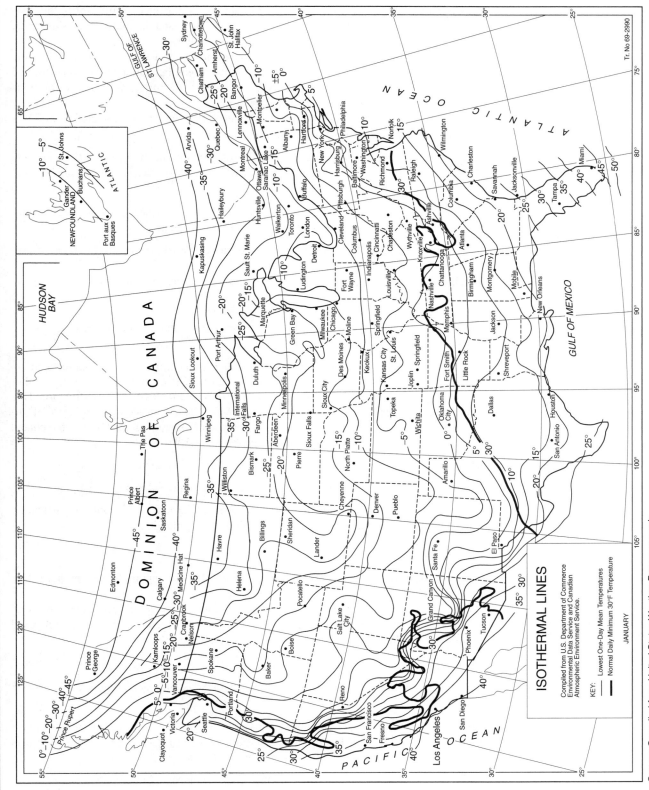

EXHIBIT 9.9 Isothermal Lines — Lowest One-Day Mean Temperature. (Source: NFPA 22, 2003, Figure 15.1.4)

The inspection required by 9.2.5.1 is a visual inspection, as indicated by the words "obvious damage." An exterior visual inspection should uncover items such as loose or missing bolts, excessive corrosion, cracking, and peeling paint, among others.

The roof vent provides ventilation above the maximum water level while at the same time keeping birds and other animals out of the tank. The vent may be equipped with a perforated plate having $3/8$ in. (9.5 mm) perforations or a corrosion-resistant screen. The perforations must be kept clean to maintain proper ventilation.

For FRP tanks, the exterior inspection should verify that the tank is protected from freezing, mechanical, and ultraviolet (UV) damage.

A.9.2.5.1 Lightning protection systems, where provided, should be inspected, tested, and maintained in accordance with NFPA 780, *Standard for the Installation of Lightning Protection Systems*.

9.2.5.2 The area surrounding the tank and supporting structure, where provided, shall be inspected quarterly to ensure that the following conditions are met:

(1) The area is free of combustible storage, trash, debris, brush, or material that could present a fire exposure hazard.
(2) The area is free of the accumulation of material on or near parts that could result in accelerated corrosion or rot.
(3) The tank and support are free of ice buildup.
(4) The exterior sides and top of embankments supporting coated fabric tanks are free of erosion.

An inspection of the area surrounding the tank and supporting structure is analogous to an exposure evaluation provided during an insurance inspection. It must be determined whether any surrounding material presents an exposure hazard to the tank or support structure. This exposure hazard can be from fires, as discussed in 9.2.5.2(1), corrosion [see 9.2.5.2(2)], increased loads due to ice [see 9.2.5.2(3)], and, in the case of an embankment-supported coated fabric tank, erosion of the berm or embankment [see 9.2.5.2(4)].

9.2.5.3 Expansion joints, where provided, shall be inspected annually for leaks and cracks.

9.2.5.4 The hoops and grillage of wooden tanks shall be inspected annually.

9.2.5.5 Exterior painted, coated, or insulated surfaces of the tank and supporting structure, where provided, shall be inspected annually for signs of degradation.

9.2.6 Interior Inspection.

When an interior inspection of a tank is performed, it is critical to comply with proper safety precautions, including confined space entry requirements.

FAQ ▶
Is it necessary to drain a tank to perform an interior inspection?

It is not necessary to drain the tank for the inspection. The standard recognizes the use of a certified commercial diver in 9.2.6.2. Alternatively, remote video equipment could be used if acceptable to the AHJ, also in accordance with 9.2.6.2.

9.2.6.1 Frequency.

9.2.6.1.1* The interior of steel tanks without corrosion protection shall be inspected every 3 years.

A.9.2.6.1.1 To aid in the inspection and evaluation of test results, it is a good idea for property owners to stencil the last known date of an interior paint job on the exterior of the tank in a conspicuous place. A typical place is near one of the manways at eye level.

What is the inspection requirement for tanks used for fire protection?

FAQ ▶

9.2.6.1.2 The interior of all other types of tanks shall be inspected every 5 years.

A three-year internal inspection is required for tanks that do not have corrosion protection (i.e., cathodic protection or equal); a five-year inspection is required for all other tanks.

9.2.6.2 Where interior inspection is made by means of underwater evaluation, silt shall first be removed from the tank floor.

Underwater interior inspections were introduced into NFPA 25 during the 2002 revision cycle. Underwater inspections are desirable because draining, inspecting, and refilling a tank are time consuming and costly and create a lengthy system impairment. Over time, silt will accumulate on the bottom of a tank. This accumulation not only poses a threat to a diver, but also may hide flaws in the tank.

It is important to note that NFPA 22 requires a fill connection that is designed to fill the tank within 8 hours. Therefore, draining, inspection, and refilling can take longer than a normal work shift. Section 15.5 of NFPA 25 requires supplemental protection and special precautions for any impairment in excess of 10 hours in a 24-hour period. As a result, interior inspections can involve considerable planning and expense.

Exhibit 9.10 illustrates the presence of a potential obstruction — a fish — inside a steel tank. The photo was taken during an internal inspection.

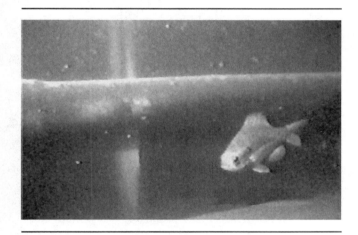

EXHIBIT 9.10 Potential Obstruction Inside Steel Tank. (Courtesy of Conrady Consultant Services)

9.2.6.3 The tank interior shall be inspected for signs of pitting, corrosion, spalling, rot, other forms of deterioration, waste materials and debris, aquatic growth, and local or general failure of interior coating.

9.2.6.4 Steel tanks exhibiting signs of interior pitting, corrosion, or failure of coating shall be tested in accordance with 9.2.7.

9.2.6.5* Tanks on ring-type foundations with sand in the middle shall be inspected for evidence of voids beneath the floor.

A.9.2.6.5 This inspection can be performed by looking for dents on the tank floor. Additionally, walking on the tank floor and looking for buckling of the floor will identify problem areas.

Ground-level suction tanks are typically set on crushed stone, sand, or concrete foundations. A concrete ring wall at least $2\frac{1}{2}$ ft (760 mm) high and 10 in. (250 mm) thick surrounds the tank foundation. This ring typically projects 6 in. (152 mm) above grade. A shifting of the sand beneath the tank, which could create a void, remains possible. Leaks from the tank bottom can also cause subsurface erosion. Without adequate support at the void, failure of the tank bottom can occur.

9.2.6.6 The heating system and components including piping shall be inspected.

9.2.6.7 The anti-vortex plate shall be inspected for deterioration or blockage.

An anti-vortex plate, such as the one illustrated in Exhibit 9.11, is provided on a tank to reduce the likelihood of introducing air pockets into the suction line. The introduction of an air pocket into a pump suction line could result in cavitation of the pump. Cavitation of the pump, in turn, can damage the pump casing and cause a reduction in pump performance.

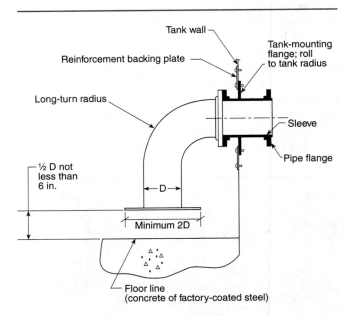

EXHIBIT 9.11 Anti-Vortex Plate. [Source: NFPA 22, 2003, Figure B.1(p)]

For SI units, 1 in. = 25.4 mm.

Note: Large, standard size anti-vortex plates (48 in. × 48 in.) are recommended, as they are adequate for all sizes of pump suction pipes normally used. Smaller plates may be used; however, they should comply with 13.2.13 of NFPA 22.

9.2.7 Interior Inspection.

Where a drained interior inspection of a steel tank is conducted in accordance with 9.2.6.4, the following tests shall be conducted:

(1) Evaluation of tank coatings shall be made in accordance with the adhesion test of ASTM D 3359, *Standard Test Methods for Measuring Adhesion by Tape Test*, generally referred to as the "cross-hatch test."
(2) Dry film thickness measurements shall be taken at random locations to determine the overall coating thickness.
(3) Nondestructive ultrasonic readings shall be taken to evaluate the wall thickness where there is evidence of pitting or corrosion.
(4) Interior surfaces shall be spot wet-sponge tested to detect pinholes, cracks, or other compromises in the coating. Special attention shall be given to sharp edges such as ladder rungs, nuts, and bolts.
(5) Tank bottoms shall be tested for metal loss and/or rust on the underside by use of ultrasonic testing where there is evidence of pitting or corrosion. Removal, visual inspection, and replacement of random floor coupons shall be an acceptable alternative to ultrasonic testing.
(6) Tanks with flat bottoms shall be vacuum-box tested at bottom seams in accordance with test procedures found in NFPA 22, *Standard for Water Tanks for Private Fire Protection*.

All the tests required by 9.2.7 are intended to identify failure points or weakened areas of the tank.

9.3 Testing

9.3.1* Level indicators shall be tested every 5 years for accuracy and freedom of movement.

A.9.3.1 The testing procedure for listed mercury gauges is as follows.

To determine that the mercury gauge is accurate, the gauge should be tested every 5 years as follows [steps (1) through (7) coincide with Figure A.9.3.1]:

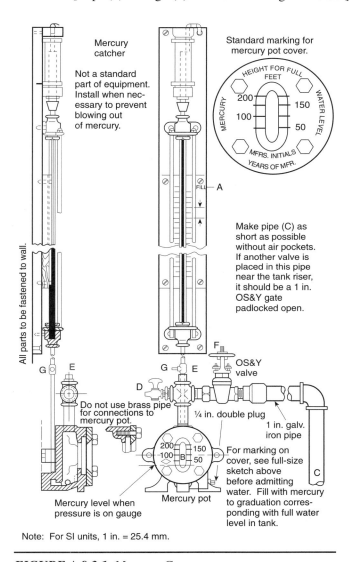

FIGURE A.9.3.1 *Mercury Gauge.*

(1) Overflow the tank.
(2) Close valve F. Open test cock D. The mercury will drop quickly into the mercury pot. If it does not drop, there is an obstruction that needs to be removed from the pipe or pot between the test cock and the gauge glass.
(3) If the mercury does lower at once, close cock D and open valve F. If the mercury responds immediately and comes to rest promptly opposite the "FULL" mark on the gauge board, the instrument is functioning properly.

(4) If the mercury column does not respond promptly and indicate the correct reading during the test, there probably are air pockets or obstructions in the water connecting pipe. Open cock D. Water should flow out forcibly. Allow water to flow through cock D until all air is expelled and rusty water from the tank riser appears. Close cock D. The gauge now likely will read correctly. If air separates from the water in the 1 in. (25 mm) pipe due to being enclosed in a buried tile conduit with steam pipes, the air can be removed automatically by installing a $\frac{3}{4}$ in. (20 mm) air trap at the high point of the piping. The air trap usually can be installed most easily in a tee connected by a short piece of pipe at E, with a plug in the top of the tee so that mercury can be added in the future, if necessary, without removing the trap. If there are inaccessible pockets in the piping, as where located below grade or under concrete floors, the air can be removed only through petcock D.

(5) If, in step (4), the water does not flow forcibly through cock D, there is an obstruction that needs to be removed from the outlet of the test cock or from the water pipe between the test cock and the tank riser.

(6) If there is water on top of the mercury column in the gauge glass, it will provide inaccurate readings and should be removed. First, lower the mercury into the pot as in step (2). Close cock D and remove plug G. Open valve F very slowly, causing the mercury to rise slowly and the water above it to drain through plug G. Close valve F quickly when mercury appears at plug G, but have a receptacle ready to catch any mercury that drains out. Replace plug G. Replace any escaped mercury in the pot.

(7) After testing, leave valve F open, except under the following conditions: If it is necessary to prevent forcing mercury and water into the mercury catcher, the controlling valve F can be permitted to be closed when filling the tank but should be left open after the tank is filled. In cases where the gauge is subjected to continual fluctuation of pressure, it could be necessary to keep the gauge shut off except when it needs to be read. Otherwise, it could be necessary to remove water frequently from the top of the mercury column as in step (5).

Mercury gauges are no longer permitted on new installations. They will therefore only be found on older tanks. A mercury gauge should not be adjusted until a Material Safety Data Sheet (MSDS) (see Exhibit 4.4) is obtained from the property owner and all precautions followed.

Electronic tank water level indicators should be tested in accordance with the manufacturer's instructions. Exhibit 9.12 illustrates an electronic water level indicator.

EXHIBIT 9.12 Electronic Water Level Indicator. *(Courtesy of Potter Electric Signal Co.)*

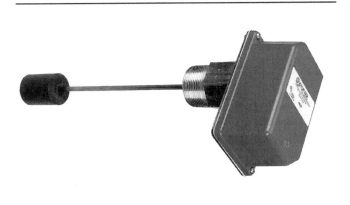

9.3.2 The tank heating system, where provided, shall be tested prior to the heating season to make certain it is in the proper working order.

9.3.3 Low water temperature alarms, where provided, shall be tested monthly (cold weather only).

As the water level decreases, the likelihood of freezing increases; hence the need for increased testing during cold weather. Exhibit 9.13 shows a water temperature sensor.

9.3.4* High water temperature limit switches on tank heating systems, where provided, shall be tested monthly whenever the heating system is in service.

A.9.3.4 The manufacturer's instructions should be consulted for guidance on testing. In some situations, it might not be possible to test the actual initiating device. In such cases, only the circuitry should be tested.

9.3.5* High and low water level alarms shall be tested semiannually.

Low water level alarms are tested more frequently during cold weather (see 9.3.3 and related commentary).

A.9.3.5 See A.9.3.4.

9.3.6 Pressure gauges shall be tested every 5 years with a calibrated gauge in accordance with the manufacturer's instructions. Gauges not accurate to within 3 percent of the scale of the gauge being tested shall be recalibrated or replaced.

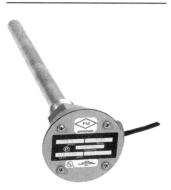

EXHIBIT 9.13 Electronic Water Temperature Sensor. (Courtesy of Potter Electric Signal Co.)

9.4 Maintenance

9.4.1 Voids discovered beneath the floors of tanks shall be filled by pumping in grout or accessing the sand and replenishing.

Voids can result from slow leaks. If left untreated, such leaks can cause further subsoil erosion, resulting in failure of the tank bottom.

9.4.2 The tank shall be maintained full or at the designed water level.

The water level should never be lower than 3 in. to 4 in. (76 mm to 102 mm) below the designated fire service level.

9.4.3 The hatch covers in the roofs and the door at the top of the frostproof casing shall always be kept securely fastened with substantial catches as a protection against freezing and windstorm damage.

9.4.4 No waste materials, such as boards, paint cans, trim, or loose material, shall be left in the tank or on the surface of the tank.

Any material left in the tank, or even on the tank, may block the waterway. Steel tanks are easy to inspect for the presence of such material. However, certain wooden tanks have space below the roof (see Exhibit 9.14), in which old paint cans or similar items were commonly placed.

9.4.5 Silt shall be removed during interior inspections or more frequently as needed to avoid accumulation to the level of the tank outlet.

The actual frequency required for silt removal, as required by 9.4.5, may be identified over time. For example, if a raw water source is used to fill the tank, the introduction of silt into the line presents a problem that must be dealt with more frequently than during the interior inspection.

EXHIBIT 9.14 Wooden Tank with Space Below the Roof. [Source: NFPA 22, 2003, Figure B.1(i)]

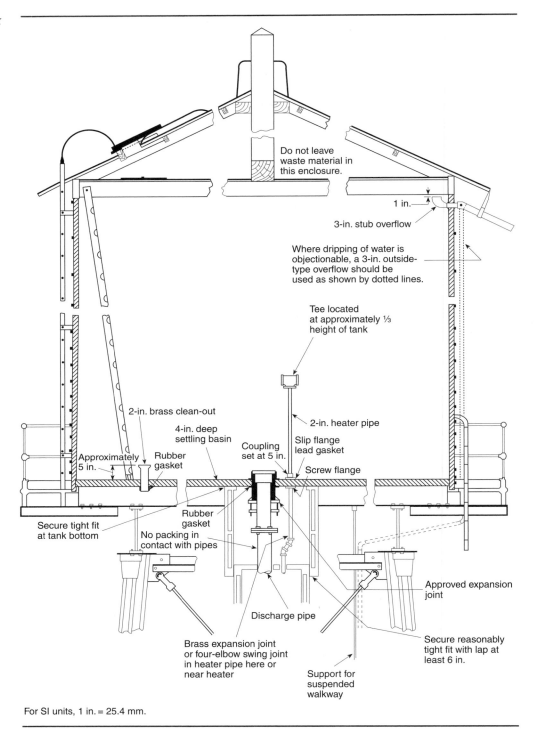

9.4.6 Maintenance of Embankment-Supported Coated Fabric (ESCF) Suction Tanks.

Embankment-supported coated fabric tanks present a unique installation. The tank manufacturer's maintenance requirements must be followed, in addition to those prescribed by this standard. Standard inspection requirements for these tanks, such as no waste materials, not allowing large accumulations of ice to collect on top of the tank, and inspection of the berms for erosion, are outlined in this standard. However, specific maintenance requirements, such as painting or refurbishing the top surface of the tank and interior inspection frequencies and methods, must come from the manufacturer.

9.4.6.1 The maintenance of ESCF tanks shall be completed in accordance with this section and the tank manufacturer's instructions.

9.4.6.2 The exposed surfaces of embankment-supported coated fabric (ESCF) tanks shall be cleaned and painted every 2 years or in accordance with the manufacturer's instructions.

9.5 Automatic Tank Fill Valves

Automatic tank fill valves are usually spring-loaded diaphragm-type valves that sense pressure in the tank and open and close to maintain a constant water level in the tank. The valve manufacturer's instructions should be followed for testing, maintenance, and adjustment. Exhibit 9.15 illustrates an automatic tank fill valve.

EXHIBIT 9.15 Automatic Tank Fill Valve.

9.5.1 Inspection.

9.5.1.1 Automatic tank fill valves shall be inspected weekly to ensure that the OS&Y isolation valves are in the normal open position in accordance with Table 9.5.1.1.

TABLE 9.5.1.1 Summary of Automatic Tank Fill Valve Inspection and Testing

Item	Frequency	Reference
Inspection		
Strainers, filters, orifices (inspect/clean)	Quarterly	13.4.1.2
Enclosure (during cold weather)	Daily/weekly	13.4.3.1.1
Exterior	Monthly	13.4.3.1.6
Interior	Annually/5 years	13.4.3.1.7
Test		
Automatic tank fill valve	Annually	

9.5.1.2 Valves secured with locks or electrically supervised in accordance with applicable NFPA standards shall be inspected monthly.

9.5.1.3 The enclosure shall be inspected to verify that it is heated and secured.

9.5.2 Maintenance.

9.5.2.1 Maintenance of all automatic tank fill valves shall be conducted by a qualified person following the manufacturer's instructions in accordance with the procedure and policies of the authority having jurisdiction.

9.5.2.2 Rubber parts shall be replaced in accordance with the frequency required by the authority having jurisdiction and the manufacturer's instructions.

9.5.2.3 Strainers shall be cleaned quarterly.

9.5.3 Testing.

All automatic tank fill valves shall be tested yearly in accordance with the following:

(1) The valve shall be actuated automatically by lowering the water level in the tank.
(2) The refill rate shall be measured and recorded.

9.6 Component Action Requirements

Component replacement tables were added in the 2008 edition to offer guidance to the user of the standard when system components are adjusted, repaired, rebuilt, or replaced. It is not necessary in each case to require a complete acceptance test for each component when maintenance is performed.

9.6.1 Whenever a component in a water storage tank is adjusted, repaired, reconditioned, or replaced, the action required in Table 9.6.1, Summary of Component Replacement Action Requirements, shall be performed.

9.6.1.1 Where the original installation standard is different from the cited standard, the use of the appropriate installing standard shall be permitted.

9.6.1.2 A main drain test shall be required if the system control or other upstream valve was operated in accordance with 13.3.3.4.

9.6.1.3 These actions shall not require a design review, which is outside the scope of this standard.

TABLE 9.6.1 *Summary of Component Replacement Action Requirements*

Component	Adjust	Repair/ Recondition	Replace	Test Criteria
Tank Components				
Tank interior		X	X	Remove debris Verify integrity in conformance with NFPA 22, *Standard for Water Tanks for Private Fire Protection*
Tank exterior		X	X	Verify integrity in conformance with NFPA 22
Support structure		X	X	Verify integrity in conformance with NFPA 22
Heating system	X	X	X	Verify heating system is in conformance with NFPA 22
Catwalks and ladders	X	X	X	Verify integrity in conformance with NFPA 22
Hoops and grillage	X	X	X	Verify integrity in conformance with NFPA 22
Expansion joints	X	X	X	Verify integrity in conformance with NFPA 22
Overflow piping	X	X	X	Verify integrity in conformance with NFPA 22
Insulation		X	X	Verify integrity in conformance with NFPA 22
Valves				See Chapter 13
Alarm and Supervisory Components				
High and low water level	X	X	X	Operational test for conformance with NFPA 22 and/or NFPA 72, *National Fire Alarm Code,* and the design water levels
Water temperature	X	X	X	Operational test for conformance with NFPA 22 and/or NFPA 72
Enclosure temperature	X	X	X	Operational test for conformance with NFPA 22 and/or NFPA 72
Valve supervision	X	X	X	Operational test for conformance with NFPA 22, and/or NFPA 72
Fill and Discharge Components				
Automatic fill valves				See Chapter 13
Valves	X	X	X	See Chapter 13
Status Indicators				
Level indicators	X	X	X	Verify conformance with NFPA 22
Pressure gauges			X	Verify at 0 psi (0 bar) and at system working pressure

SUMMARY

Although storage tanks are relatively simple components of a fire protection system, they are usually equipped with accessories that do require inspection, testing, and maintenance. For example, failure to maintain a storage tank heating system can result in the formation of ice, which can, in turn, obstruct the discharge outlet and cause failure of the tank shell. Also, corrosion or a failing paint system can result in costly repair or eventual tank failure. Following the requirements of this chapter can improve the reliability of water storage tanks, minimize maintenance costs, and reduce the need for expensive repairs.

REVIEW QUESTIONS

1. Identify four items to be addressed when conducting an inspection of the interior of a tank. What would be the specific concerns and, if left unchecked, how could they lead to failure?

2. What section of NFPA 25 requires an owner to remove excessive weeds growing near a water storage tank?
3. Wood tanks on the roofs of building used to be common. Why are they disappearing, and what is being provided in lieu of these tanks?
4. Water temperature in a storage tank must be maintained at what value?
5. Prior to entering a tank for internal inspection, what precautions (if any) should be taken?

REFERENCES CITED IN COMMENTARY

National Fire Protection Association, 1 Batterymarch Park, Quincy, MA 02169-7471.

Cote, A. E., ed., *Fire Protection Handbook*™, 19th edition, 2003.

NFPA 13D, *Standard for the Installation of Sprinkler Systems in One- and Two-Family Dwellings and Manufactured Homes,* 2007 edition.

NFPA 22, *Standard for Water Tanks for Private Fire Protection,* 2003 edition.

NFPA 20, *Standard for the Installation of Stationary Pumps for Fire Protection,* 2007 edition.

Water Spray Fixed Systems

CHAPTER 10

Water spray fixed systems are highly specialized systems that are typically installed under the scope of NFPA 15, *Standard for Water Spray Fixed Systems for Fire Protection*. These systems are used to protect equipment such as cable trays, belt conveyors, pumps, compressors, vessels, transformers, structures, and miscellaneous equipment. Water spray systems are also used on flammable and combustible liquid pool fires and are frequently used for exposure protection.

Water spray fixed systems use different types of detection systems, such as electronic (including smoke, heat, and infrared detection), pneumatic rate-of-rise, and wet and dry pilot detection systems for actuation. Depending on the detection system used, inspection, testing, and maintenance of water spray fixed systems may require specialized training beyond that required for conventional sprinkler systems.

The requirements of Chapter 10 must be followed to avoid the very intense fires that can result from the hazards typically protected by these systems.

The inspection and testing of a water spray fixed system, as prescribed by this chapter, requires a certain knowledge of NFPA 15. To determine if test results are acceptable, the original installation drawings and hydraulic calculations should be available for comparison. If these documents are not available, or if the inspector does not have sufficient knowledge of NFPA 15, the interpretation of test results may require the assistance of a professional engineer or an engineering technician.

10.1* General

This chapter shall provide the minimum requirements for the routine inspection, testing, and maintenance of water spray protection from fixed nozzle systems only. Table 10.1 shall be used to determine the minimum required frequencies for inspection, testing, and maintenance.

Many of the inspection, testing, and maintenance requirements in Table 10.1 reference other chapters within NFPA 25 or other NFPA documents. Those components in a water spray fixed system that are common to other water-based fire protection systems, such as waterflow alarms, fire pumps, water storage tanks, and valves, are addressed in other chapters. The detection systems are addressed in *NFPA 72®, National Fire Alarm Code®*. All other components specific to water spray fixed systems, such as the nozzles, pipe, fittings, hangers, supports, strainers, and manual releases, are addressed in the specific Chapter 10 paragraphs indicated in Table 10.1.

A.10.1 The effectiveness and reliability of water spray fixed systems depends on maintenance of the integrity of hydraulic characteristics, water control valves, deluge valves and their fire detection/actuation systems, pipe hangers, and prevention of obstructions to nozzle discharge patterns.

Water spray fixed systems are most commonly used to protect processing equipment and structures, flammable liquid and gas vessels, piping, and equipment such as transformers, oil switches, and motors. They also have been shown to be effective on many combustible solids.

TABLE 10.1 Summary of Water Spray Fixed System Inspection, Testing, and Maintenance

Item	Frequency	Reference
Inspection		
Backflow preventer		Chapter 13
Check valves		Chapter 13
Control valves	Weekly (sealed)	Chapter 13
Control valves	Monthly (locked, supervised)	Chapter 13
Deluge valve		10.2.2, Chapter 13
Detection systems		NFPA 72, *National Fire Alarm Code*
Detector check valves		Chapter 13
Drainage	Quarterly	10.2.8
Electric motor		10.2.9, Chapter 8
Engine drive		10.2.9, Chapter 8
Fire pump		10.2.9, Chapter 8
Fittings	Quarterly	10.2.4, 10.2.4.1
Fittings (rubber-gasketed)	Quarterly	10.2.4.1, A.10.2.4.1
Gravity tanks		10.2.10, Chapter 9
Hangers	Annually and after each system activation	10.2.4.2
Heat (deluge valve house)	Daily/weekly	10.2.1.5, Chapter 13
Nozzles	Annually and after each system activation	10.2.1.1, 10.2.1.2, 10.2.1.6, 10.2.5.1, 10.2.5.2
Pipe	Annually and after each system activation	10.2.1.1, 10.2.1.2, 10.2.4, 10.2.4.1
Pressure tank		10.2.10, Chapter 9
Steam driver		10.2.9, Chapter 8
Strainers	Mfg. instruction	10.2.7
Suction tanks		10.2.10, Chapter 9
Supports	Quarterly	10.2.1.1, 10.2.1.2, 10.2.4.2
Water supply piping		10.2.6.1, 10.2.6.2
UHSWSS — detectors	Monthly	10.4.2
UHSWSS — controllers	Each shift	10.4.3
UHSWSS — valves	Each shift	10.4.4
Operational Test		
Backflow preventer		Chapter 13
Check valves		Chapter 13
Control valves	Annually	13.3.3.1
Deluge valve		10.2.2, Chapter 13
Detection systems		*NFPA 72*
Detector check valve		Chapter 13
Electric motor		10.2.9, Chapter 8
Engine drive		10.2.9, Chapter 8
Fire pump		10.2.9, Chapter 8
Flushing	Annually	10.2.1.3, Section 10.3 (flushing of connection to riser, part of annual test)
Gravity tanks		10.2.10, Chapter 9
Main drain test	Annually	13.3.3.4
Manual release	Annually	10.2.1.3, 10.3.6
Nozzles	Annually	10.2.1.3, 10.2.1.6, Section 10.3
Pressure tank		Section 10.2, Chapter 9
Steam driver		10.2.9, Chapter 8
Strainers	Annually	10.2.1.3, 10.2.1.7, 10.2.7
Suction tanks		10.2.10, Chapter 9
Water-flow alarm	Quarterly	Chapter 5
Water spray system test	Annually	Section 10.3, Chapter 13

TABLE 10.1 Continued

Item	Frequency	Reference
Operational Test (*continued*)		
Water supply flow test		7.3.2
UHSWSS	Annually	Section 10.4
Maintenance		
Backflow preventer		Chapter 13
Check valves		Chapter 13
Control valves	Annually	10.2.1.4, Chapter 13
Deluge valve		10.2.2, Chapter 13
Detection systems		NFPA 72
Detector check valve		Chapter 13
Electric motor		10.2.9, Chapter 8
Engine drive		10.2.9, Chapter 8
Fire pump		10.2.9, Chapter 8
Gravity tanks		10.2.10, Chapter 9
Pressure tank		10.2.6, Chapter 9
Steam driver		10.2.9, Chapter 8
Strainers	Annually	10.2.1.4, 10.2.1.7, 10.2.7
Strainers (baskets/screen)	5 years	10.2.1.4, 10.2.1.8, A.10.2.7
Suction tanks		10.2.10, Chapter 9
Water spray system	Annually	10.2.1.4, Chapter 13

Many of the components and subsystems found in a water spray system require the same inspection, test, and maintenance procedures where they are used in automatic sprinkler systems and other fixed water-based fire protection systems. Other chapters of this standard should be consulted for particulars on required inspection and maintenance.

Water spray fixed systems are designed and installed in accordance with the requirements of NFPA 15. These systems normally employ an open piping network with open spray nozzles that discharge water in a predetermined pattern on a hazard area, structure, or vessel (see Exhibits 10.1 through 10.4). Exhibit 10.2 illustrates a typical water spray system protecting a horizontal chemical storage tank. Note that two levels of nozzles are used to completely engulf the tank in water spray. This type of system design is typically used for fire extinguishment as well as exposure protection.

Exhibit 10.3 illustrates water spray protection for a pipe rack assembly that is usually located in a chemical processing plant or refinery. The water spray nozzles are intended to spray directly on process piping and are also used in some cases to protect the structural steel of the pipe rack itself. Note that in this illustration, water spray nozzles are also employed to protect a process pump.

Exhibit 10.4 illustrates the use of water spray nozzles to protect a coal or other type of materials conveyor. In this application, nozzles are located along the conveyor belt to spray water either across the belt or along the length of the belt. Note the detector at the roof of this enclosed conveyor. The detector in this case is used to activate the deluge valve in this water spray system.

Because the piping network is open to the atmosphere through the open spray nozzles, these systems are subject to interior corrosion. Scale from the inside of the pipes can clog the water spray nozzles, affecting both the discharge pattern and density being applied to the surface of the hazard.

EXHIBIT 10.1 Discharging System.

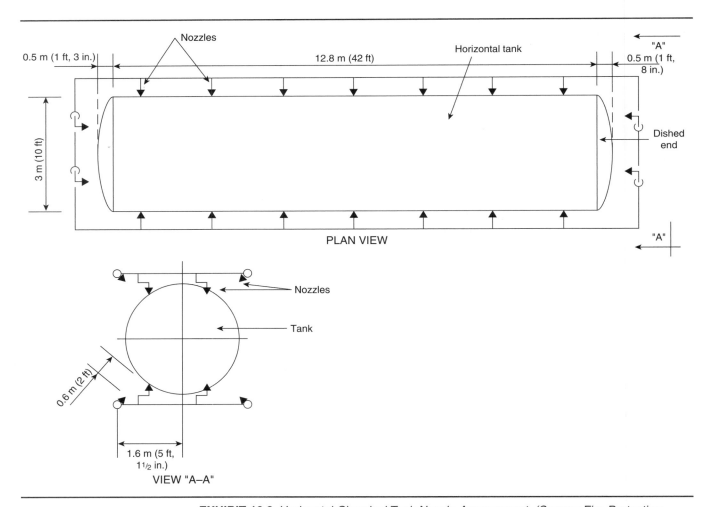

EXHIBIT 10.2 Horizontal Chemical Tank Nozzle Arrangement. (Source: Fire Protection Systems for Special Hazards, 2004, Figure 6.3)

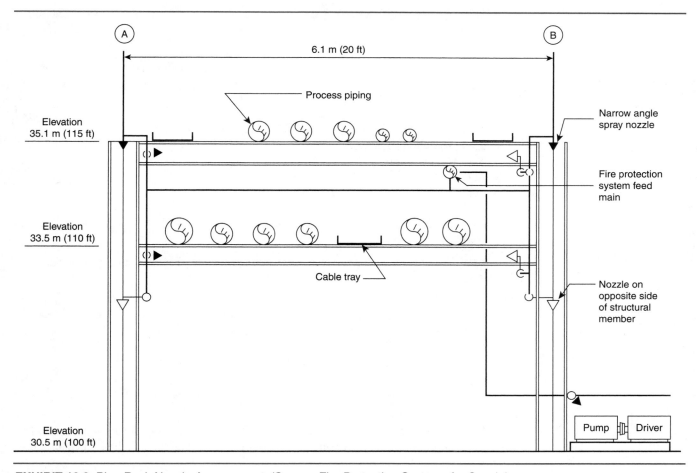

EXHIBIT 10.3 Pipe Rack Nozzle Arrangement. (Source: Fire Protection Systems for Special Hazards, 2004, Figure 6.16)

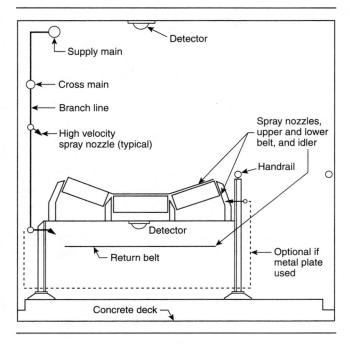

EXHIBIT 10.4 Typical Conveyor Belt Protection. [Source: NFPA 15, 2007, Figure A.7.2.3.3.1(a)]

Water spray fixed systems protect special hazards that require higher densities and direct impingement of water over the surface area of the equipment or structure. These hazards may include the following:

- Storage vessels containing hazardous chemicals or flammable and combustible liquids
- Piping and pumps involved in the processing or transfer of hazardous chemicals or flammable and combustible liquids
- Structural steel supporting vessels, piping, and other equipment used in the processing of hazardous chemicals or flammable and combustible liquids
- Cable trays and cable runs
- Transformers (see Exhibit 10.5)
- Belt conveyors
- Turbine bearings
- Boiler fronts
- Other similar equipment involving the use of hazardous chemicals or flammable and combustible liquids

10.1.1 This chapter shall not cover water spray protection from portable nozzles, sprinkler systems, monitor nozzles, or other means of application.

As indicated in 10.1.1, Chapter 10 does not specifically describe the inspection, testing, and maintenance requirements applicable to systems with portable nozzles, monitor nozzles, or other means of application. Nonetheless, many of the requirements in this chapter can be applied in the absence of another inspection, testing, and maintenance standard.

For example, monitor nozzles used in an aircraft hangar should be inspected, tested, and maintained in accordance with NFPA 409, *Standard on Aircraft Hangars*. However, monitor nozzles that are fed from a fire service underground main are covered in Chapter 7 of this document.

10.1.2* NFPA 15, *Standard for Water Spray Fixed Systems for Fire Protection*, shall be consulted to determine the requirements for design and installation, including acceptance testing.

FAQ ▶
Is the inspector required to have in-depth knowledge of NFPA 15?

The inspector of a water-based fire protection system is not required to have complete knowledge of the design requirements of the installation standard, NFPA 15. However, that standard should be consulted, as required by 10.1.2, and a certain level of knowledge is necessary to properly inspect and test a water spray fixed system.

For example, an inspector may not be able to determine if the nozzles are aligned correctly unless he knows whether direct impingement over the entire surface of the structure, vessel, or equipment is required or if the system is intended to protect the structure, vessel, or equipment from an exposure fire. (When protecting vessels to control burning, the nozzles are to be positioned to impinge directly on the vessel and on all areas around the vessel where a spill is likely to spread or accumulate. When protecting structural steel, water spray from the nozzles is only required to impinge on one side of the structural member and rundown is anticipated.)

A.10.1.2 Insulation acting in lieu of water spray protection is expected to protect a vessel or structure for the duration of the exposure. The insulation is to prevent the temperature from exceeding 850°F (454°C) for structural members and 650°F (393°C) for vessels. If the insulation is missing, the structure or vessel is not considered to be protected, regardless of water spray protection or insulation on other surfaces. To re-establish the proper protection, the insulation should be replaced or the water spray protection should be extended, using the appropriate density.

FAQ ▶
Why is it important to check insulation?

NFPA 15 allows encasing structural steel in AHJ-approved fire-resistant insulating material in lieu of protecting the steel with a water spray fixed system. Missing or damaged insulation has the same impact on the protection of the structure as a totally clogged nozzle. In cases of

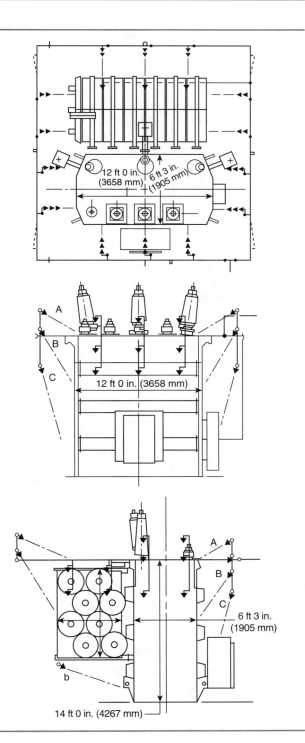

EXHIBIT 10.5 Typical Transformer Water Spray System. (Source: NFPA 15, 2007, Figure A.7.4.4.1)

damaged or missing insulation, immediate corrective action should be taken. Therefore, it is important that the inspection of the water spray fixed system include a check of the integrity of any such insulation.

10.1.3 Valves and Connections.

Valves and fire department connections shall be inspected, tested, and maintained in accordance with Chapter 13.

10.1.4* Impairments.

The procedures outlined in Chapter 15 shall be followed where an impairment to protection occurs.

A.10.1.4 The inspection, testing, and maintenance of water spray fixed systems can involve or result in a system that is out of service. Also see Chapter 15.

10.1.4.1 When a water spray fixed system or any portion thereof is out of service for any reason, notice shall be given to facility management, the local fire department, the on-site fire brigade, and other authorities having jurisdiction, as applicable.

CASE STUDY

Sprinkler Not in Service

A 1993 fire in an Alabama coal-fired electric power plant of unprotected noncombustible construction resulted in a dollar loss of over $10 million.

The federally owned plant, which was operating at the time of the fire, was reportedly equipped with a sprinkler system of unknown type and coverage. The system was not working at the time of the fire due to a malfunction.

Investigators report that the fire started on a 400 ft conveyor belt supplying coal to the plant. The cause of the fire is thought to be a buildup of heat resulting from friction between the belt and a jammed roller. The fire spread up the belt and into the main plant, destroying the conveyor belt system, which had to be replaced. The fire disrupted plant operations for several weeks.

Water spray fixed systems are typically employed in "ordinary-hazard" to "extra-hazard" areas. These are areas where the potential exists for fires of higher intensity than ordinarily encountered in light hazard occupancies such as office buildings or residential occupancies. These areas are also subject to more hazardous activities, such as welding, and the use of flammable or combustible liquids.

In addition to the system impairment requirements prescribed by Chapter 15, controls over hazardous activities and operations must be considered during any system impairment.

10.1.4.2 A sign shall be posted at each fire department connection or system control valve indicating which portion of the system is out of service.

10.2 Inspection and Maintenance Procedures

10.2.1 The components described in this section shall be inspected and maintained at the frequency specified in Table 10.1 and in accordance with this standard and the manufacturer's instructions.

As is the case with other chapters of NFPA 25 that address systems, a table (Table 10.1) summarizes this chapter's inspection, testing, and maintenance requirements. It is recommended that the user refer to the table to review the requirements, then refer to the referenced sections for more specific information.

10.2.1.1 Items in areas that are inaccessible for safety considerations due to factors such as continuous process operations and energized electrical equipment shall be inspected during each scheduled shutdown but not more than every 18 months.

The facility manager is obligated to notify the inspector when a shutdown occurs. In some cases, repairs or unscheduled maintenance will necessitate a shutdown, during which inspection, testing, and/or maintenance activities can be conducted. When processes are very rarely shut down, the facility manager must use reasonable judgment and take advantage of shutdowns as they occur. If inspection, testing, and maintenance have not been conducted for almost 18 months because no shutdown has occurred, a shutdown must be scheduled to allow these activities to take place.

10.2.1.2 Inspections shall not be required for items in areas with no provision for access and that are not subject to the conditions noted in 10.2.4.1, 10.2.4.2, and 10.2.5.1.

10.2.1.3 Items in areas that are inaccessible for safety considerations shall be tested at longer intervals in accordance with 13.4.3.2.2.2.

◄ **FAQ**
Why does NFPA 25 require longer intervals between discharge tests?

Because water spray fixed systems frequently protect equipment in industrial occupancies and are intended to discharge a considerable amount of water, performing a functional test requires coordination with the property owner or operator. Discharge tests must be planned carefully to avoid disruption of building or process operations with further consideration to potential damage to the building, equipment, or product. The standard allows a longer interval between discharge tests to facilitate this coordination.

10.2.1.4 Other maintenance intervals shall be permitted, depending on the results of the visual inspection and operating tests.

10.2.1.5 Deluge valve enclosures shall be inspected in accordance with the provisions of Chapter 13.

10.2.1.6 Nozzle strainers shall be removed, inspected, and cleaned during the flushing procedure for the mainline strainer.

Nozzle strainers are only installed with certain types of nozzles. When a nozzle strainer is present, it will usually need cleaning as required by 10.2.1.6, since it is intended to capture obstructing material and prevent the plugging of the nozzle. Exhibit 10.6 shows a typical fire protection strainer with a basket-type screen of corrosion-resistant metal.

10.2.1.7 Mainline strainers shall be removed and inspected every 5 years for damaged and corroded parts.

10.2.2 Deluge Valves.

Deluge valves shall be inspected, tested, and maintained in accordance with Chapter 13.

10.2.3 Automatic Detection Equipment.

Detection systems, which include smoke, heat, and infrared systems, are used as an actuating means for fixed water spray systems. Detection systems are not covered in detail in NFPA 25. They are usually installed in accordance with *NFPA 72* and should be maintained by a qualified electrician in accordance with that standard, as required by 10.2.3.1.

An electronic detection system functions as an actuation mechanism by sending a signal via a local fire alarm or deluge system panel to a solenoid valve mounted on the deluge system trim piping. Signaling the solenoid to open and release water pressure from the alarm line and deluge valve diaphragm allows the deluge valve to open. The solenoid can be configured

EXHIBIT 10.6 Strainer with Basket-Type Screen.

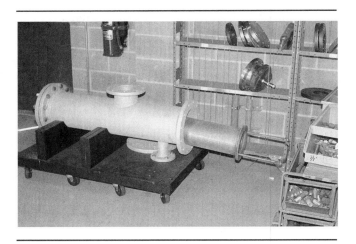

to either energize to open or de-energize to open (providing fail-safe operation). Exhibit 10.7 shows a detection system being tested.

EXHIBIT 10.7 Testing a Heat Detector on a Water Spray System.

A wet pilot system is composed of a system of $\frac{1}{2}$ in. (15 mm) pipe with sprinklers attached, similar to a wet pipe sprinkler system. The system is pressurized to normal static pressure. Sprinklers in this case are used as detectors. When a sprinkler activates, pressure in the system is released to the atmosphere. This reduced pressure allows the deluge valve diaphragm to open the deluge valve.

Wet pilot systems are usually provided with an inspector's test connection, in the form of a $\frac{1}{2}$ in. (15 mm) ball valve. Opening the inspector's test connection releases pressure and activates the system. Wet pilot systems are used where freezing is not a concern.

Dry pilot systems operate in a similar manner except that the system is filled with air, similar to a dry pipe sprinkler system. When a sprinkler activates in a dry pilot system, air pressure is released and a device such as a dry pilot actuator opens and bleeds water from the diaphragm on the deluge valve, allowing the deluge valve to open.

A dry pilot system is normally equipped with an inspector's test connection in the form of a $\frac{1}{2}$ in. (15 mm) ball valve. Dry pilot systems are typically provided where freezing is a concern.

Inspection, testing, and maintenance of pilot sprinkler systems should be conducted according to the requirements for sprinkler systems presented in Chapter 5 of NFPA 25. With either a wet or dry pilot system, spare sprinklers should be provided, as discussed in Chapter 5.

The air pressure for a dry pilot system can be supplied from either plant (or owner's) air or a nitrogen supply. Exhibit 10.8 illustrates the proper connections for air supplies.

EXHIBIT 10.8 Dry Pilot Actuation Trim Options.

10.2.3.1 Automatic detection equipment shall be inspected, tested, and maintained in accordance with *NFPA 72, National Fire Alarm Code*.

10.2.3.2 Automatic fire detection equipment not covered by *NFPA 72, National Fire Alarm Code*, shall be inspected, tested, and maintained to ensure that the detectors are in place, securely fastened, and protected from corrosion, weather, and mechanical damage and that the communication wiring, control panels, or tubing system is functional.

10.2.4* System Components.

System piping, fittings, hangers, and supports shall be inspected and maintained to ensure continuity of water delivery to the spray nozzles at full water-flow and design pressure.

Water spray fixed systems, due to the forces exerted on the pipe and pipe supports from the water spray coming from a system of open nozzles, are prone to water hammer and pipe vibration. It is therefore extremely important to inspect the integrity of the piping, the hanger or supports, and the nozzles.

A.10.2.4 The operation of the water spray system is dependent on the integrity of the piping, which should be kept in good condition and free of mechanical damage. The pipe should not be used for support of ladders, stock, or other material. Where piping is subject to a corrosive atmosphere, a protective corrosion-resistant coating should be provided and maintained. Where the age or service conditions warrant, an internal examination of the piping should be made. Where it is necessary to flush all or part of the piping system, this work should be done by sprinkler contractors or other qualified workers.

In many cases, pipe in water spray fixed systems is installed near equipment, rather than in a more protected spot near the roof or ceiling, as in the case of sprinkler systems. Therefore, an inspection of piping is essential to ensure that pipe, fittings, and pipe supports are not damaged.

Usually, where piping is installed in corrosive atmospheres, the pipe, fittings, and pipe supports are galvanized. The galvanization protects against corrosion, which could lead to the eventual failure of the component. Inspections should be made to ensure that the galvanizing has not been removed through abrasion or any other means.

As with sprinkler systems, water spray fixed systems must be supported in accordance with NFPA 13, *Standard for the Installation of Sprinkler Systems*. NFPA 13 requires that pipe supports other than those illustrated in the standard be designed to support five times the weight of the water-filled pipe plus a 250 lb (113.4 kg) safety factor. Additional weight from ladders, stock, or other material could place a load beyond that designed for the pipe support.

10.2.4.1* Piping and Fittings. System piping and fittings shall be inspected for the following:

(1) Mechanical damage (e.g., broken piping or cracked fittings)

Piping in a water spray fixed system is installed primarily in areas where production processes, machinery, and other equipment can come in contact with it. It is therefore more susceptible to mechanical damage and misalignment than piping in most other water-based fire protection systems.

Supporting detection system pipe or conduit is a common practice in the installation of water spray fixed systems. This method is acceptable because the detection system is a component of the overall fire protection system. A release line hanger is typically used for hanging detection piping from the water piping. Angle iron or strut is also used for this application.

No other items may be attached to or hanging from the water spray fixed system piping.

(2) External conditions (e.g., missing or damaged paint or coatings, rust, and corrosion)

Many water spray fixed systems are installed outside or in atmospheres where corrosion is a common problem. Although it is required that piping be protected, rust, damaged paint, or other signs of corrosion frequently will be found.

(3) Misalignment or trapped sections

In many installations, water spray fixed system piping that surrounds equipment or vessels is installed in sections. This configuration allows the piping to be disassembled if the equipment or vessels need to be removed for repair or replacement. When such disassembly takes place, piping sections can be reinstalled incorrectly. The inspection should assess whether this has happened.

(4) Low-point drains (automatic or manual)

Water spray fixed systems may be installed outdoors. If exposed to freezing temperatures, systems may freeze. Low point drains and/or trapped sections of pipe should be inspected, particularly prior to the onset of cold weather.

(5) Location of rubber-gasketed fittings

Mechanical damage to fittings may be more prevalent in a water spray system than in a wet or dry pipe system, because the pipe and fittings may be exposed to the hazard that is protected by the water spray system. For example, if pipe and fittings are exposed to heat prior to system actuation, the sudden contraction of cool water entering a hot pipe may cause fracture of the pipe or fitting. Inspection of a water spray system should include a check for this type of failure.

A.10.2.4.1 Rubber-gasketed fittings in the fire areas are inspected to determine whether they are protected by the water spray or other approved means. Unless properly protected, fire could cause loss of the rubber gasket following excessive leakage in a fire situation.

NFPA 15 allows the use of rubber-gasketed fittings in fire areas if those fittings are also protected by a water spray fixed system automatically controlled through the use of a detection system.

Spray from the open nozzles on the system is expected to keep the gaskets from melting or disfiguring during a fire scenario. These fittings may be protected by an adjacent system, provided that system is activated by a detection system in the same fire area as the fittings.

Rubber gaskets are normally found in two types of fittings that can be installed in a water spray fixed system: flanged and grooved.

When using flanged fittings and/or joints, a rubber gasket, usually red or black, is sandwiched between every two flanges to prevent leakage. These gaskets can be either the ring type that will be visible inside the ring of bolts holding the flanges together or the full-face type, which extends beyond the bolts to the outer edge of the flanges. There is a metallic alternative to rubber gaskets that can be used with flanges. These metallic gaskets, which are normally silver or gray, do not need protection when installed in a fire area.

Grooved fittings use rubber gaskets that are installed around the outside of the pipe in a coupling. These gaskets are not normally visible during an inspection. To date, there is no alternative to rubber gaskets in grooved fittings. Therefore, a water spray fixed system, or other method in accordance with NFPA 15, must protect all grooved fittings or couplings in a fire area.

10.2.4.2* Hangers and Supports. Hangers and supports shall be inspected for the following and repaired or replaced as necessary:

(1) Condition (e.g., missing or damaged paint or coating, rust, and corrosion)
(2) Secure attachment to structural supports and piping
(3) Damaged or missing hangers

A.10.2.4.2 Hangers and supports are designed to support and restrain the piping from severe movement when the water supply operates and to provide adequate pipe slope for drainage of water from the piping after the water spray system is shut down. Hangers should be kept in good repair. Broken or loose hangers can put undue strain on piping and fittings, cause pipe breaks, and interfere with proper drainage of the pipe. Broken or loose hangers should be replaced or refastened.

10.2.5* Water Spray Nozzles.

A.10.2.5 Systems need inspection to ensure water spray nozzles effectively discharge water unobstructed onto surfaces to be protected from radiant heat (exposure protection) or onto flaming surfaces to extinguish or control combustion. Factors affecting the proper placement of water spray nozzles include the following:

(1) Changes or additions to the protected area that obstruct existing nozzles or require additional coverage for compliance
(2) Removal of equipment from the protected area that results in nozzle placement at excessive distances from the hazard
(3) Mechanical damage or previous flow tests that have caused nozzles to be misdirected
(4) A change in the hazard being protected that requires more or different nozzles to provide adequate coverage for compliance

Spray nozzles can be permitted to be placed in any position necessary to obtain proper coverage of the protected area. Positioning of nozzles with respect to surfaces to be protected, or to fires to be controlled or extinguished, should be guided by the particular nozzle design and the character of water spray produced. In positioning nozzles, care should be taken that the water spray does not miss the targeted surface and reduce the efficiency or calculated discharge rate.

Two basic types of nozzles are used on fixed water spray systems and so are governed by the requirements of 10.2.5: automatic and non-automatic (open) water spray nozzles.

Automatic water spray nozzles are intended to open automatically by operation of a heat-responsive element. That element keeps the discharge orifice closed by means such as the exertion of force on a cap (button or disc). When discharging water under pressure, the nozzle will distribute the water in a specific, directional pattern, very similar to a sprinkler. In fact, automatic water spray nozzles can be mistaken for sprinklers. Exhibit 10.9 illustrates typical automatic spray nozzles.

EXHIBIT 10.9 Automatic Spray Nozzles.

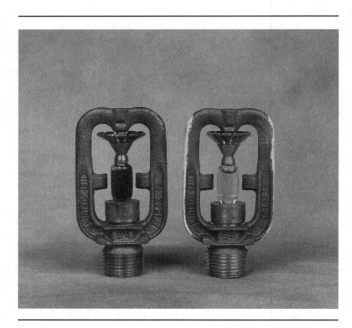

A non-automatic, or open, water spray nozzle is an open water discharge device that, when discharging water under pressure, will distribute the water in a specific, directional pattern. Some non-automatic nozzles resemble an operated sprinkler, while others appear very different. Exhibit 10.10 illustrates different types of non-automatic spray nozzles.

10.2.5.1 Water spray nozzles shall be inspected and maintained to ensure that they are in place, continue to be aimed or pointed in the direction intended in the system design, and are free from external loading and corrosion.

The requirements of 10.2.5.1 are important because water spray nozzles are subject to corrosion from the effects of weather or exposure to corrosive chemicals. When water spray nozzles are installed in harsh environments, corrosion protection in the form of wax or lead coatings, applied by the nozzle manufacturer, may be used.

A problem commonly seen with the use of grooved fittings is the misalignment of piping sections due to the thrust from the nozzles, which causes the pipe to rotate. To correct this condition, nozzles should be realigned and additional restraints added.

10.2.5.2 Where caps or plugs are required, the inspection shall confirm they are in place and free to operate as intended.

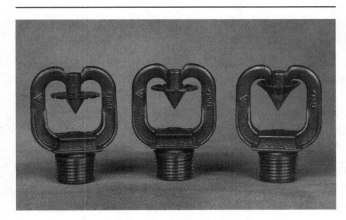

EXHIBIT 10.10 Low Velocity Non-Automatic Water Spray Nozzles (top) and High Velocity Non-Automatic Spray Nozzles (bottom).

Bees and hornets have been known to build nests inside open nozzles and sprinklers, creating obstructions. Dust and dirt can also find their way into nozzles and piping. If nests or other obstructing material are found during an inspection, consideration should be given to the use of listed plastic plugs or nozzle caps in 10.2.5.2. Only plastic plugs or caps listed for use with the nozzle may be used. Exhibit 10.11 shows an example of a nozzle with such a cap.

10.2.5.3 Misaligned water spray nozzles shall be adjusted (aimed) by visual means, and the discharge patterns shall be checked at the next scheduled flow test.

When adjusting misaligned water spray nozzles as required by 10.2.5.3, if it is not obvious in what direction the nozzle should be aimed, that information should be obtained from the as-built drawings for the system.

10.2.6 Water Supply.

10.2.6.1 The dependability of the water supply shall be ensured by regular inspection and maintenance, whether furnished by a municipal source, on-site storage tanks, a fire pump, or private underground piping systems.

10.2.6.2* Water supply piping shall be maintained free of internal obstructions.

A.10.2.6.2 Water supply piping should be free of internal obstructions that can be caused by debris (e.g., rocks, mud, tubercles) or by closed or partially closed control valves. See Chapter 5 for inspection and maintenance requirements.

The effective cross-sectional area of many water spray nozzles requires the installation of mainline strainers. Because nozzles can be obstructed easily, the requirements of Chapter 7 are necessary to ensure that water supplies remain free of obstructing material.

EXHIBIT 10.11 Nozzle with Blow-Off Cap. (Source: Fire Protection Handbook, 2003, Figure 10.15.11)

10.2.7* Strainers.

A.10.2.7 Mainline strainers should be removed and inspected for damaged and corroded parts every 5 years.

10.2.7.1 Mainline strainers (basket or screen) shall be flushed until clear after each operation or flow test.

Listed strainers are equipped with flushing connections that allow back-flushing of the strainer basket (see Exhibit 10.6). If flushing, as required by 10.2.7.1, reveals excessive obstructing material, the basket can be removed easily for visual inspection and cleaning.

10.2.7.2 Individual water spray nozzle strainers shall be removed, cleaned, and inspected after each operation or flow test.

10.2.7.3 All strainers shall be inspected and cleaned in accordance with the manufacturer's instructions.

10.2.7.4 Damaged or corroded parts shall be replaced or repaired.

10.2.8 Drainage.

The area beneath and surrounding a water spray fixed system shall be inspected visually on a quarterly basis to ensure that drainage facilities, such as trap sumps and drainage trenches, are not blocked and retention embankments or dikes are in good repair.

10.2.9 Fire Pumps.

Chapter 8 shall be followed for inspection and maintenance requirements.

10.2.10 Water Tanks (Gravity, Pressure, or Suction Tanks, or Reservoirs).

Chapter 9 shall be followed for inspection and maintenance requirements.

10.3 Operational Tests

10.3.1 Performance.

Water spray fixed systems shall require competent and effective care and maintenance to ensure they perform as designed.

10.3.1.1 Frequency of system tests shall be in accordance with Table 10.1.

10.3.1.2 Water spray fixed systems shall be serviced in accordance with this standard and with the manufacturer's instructions.

10.3.2 Notification.

10.3.2.1 To avoid false alarms where a supervisory service is provided, the alarm receiving facility always shall be notified by the property owner or designated representative as follows:

(1) Before conducting any test or procedure that could result in the actuation of an alarm
(2) After such tests or procedures are concluded

10.3.2.2 All personnel whose operations could be affected by the system operation shall be notified.

10.3.3* Test Preparation.

Precautions shall be taken to prevent damage to property during the test.

A.10.3.3 The property owner's representative should take care to prevent damage to equipment or the structure during the test. Damage could be caused by the system discharge or by runoff from the test site. It should be verified that there is adequate and unobstructed drainage. Equipment should be removed or covered as necessary to prevent damage. Means such as curbing or sandbagging should be used to prevent entry of the water.

A test of a water spray system can result in the discharge of water into the environment. Proper procedures for the disposal of test water must be followed based on local, state, and national environmental policies.

10.3.4 Operational Test Performance.

Operational tests shall be conducted to ensure that the water spray fixed systems respond as designed, both automatically and manually.

For procedures necessary to satisfy the operational test requirement in 10.3.4, see 13.4.3.2.

10.3.4.1* Response Time.

A.10.3.4.1 Test methods are as follows:

(1) Some detection circuits can be permitted to be deliberately desensitized in order to override unusual ambient conditions. In such cases, the response required in 10.3.4.1 can be permitted to be exceeded.
(2) Testing of integrating tubing systems can be permitted to be related to this test by means of a standard pressure impulse test specified by the listing laboratory.
(3) One method of testing heat detection uses a radiant heat surface at a temperature of 300°F (149°C) and a capacity of 350 watts at a distance of 1 in. (25 mm) but not more than 2 in. (50 mm) from the nearest part of the detector. This method of testing with an electric test set should not be used in hazardous locations. Other test methods can be permitted to be employed, but the results should be obtained under these conditions.

10.3.4.1.1 Under test conditions, the heat detection systems, where exposed to a heat test source, shall operate within 40 seconds.

10.3.4.1.2 Under test conditions, the flammable gas detection system, where exposed to a standard test gas concentration, shall operate within the time frame specified in the system design.

10.3.4.1.3 These response times shall be recorded.

10.3.4.2 Discharge Time. The time lapse between operation of detection systems and water delivery time to the protected area shall be recorded.

10.3.4.3* Discharge Patterns.

A.10.3.4.3 Spray nozzles can be of different sizes and types. Some are more subject to internal obstructions than others.

10.3.4.3.1* The water discharge patterns from all of the open spray nozzles shall be observed to ensure that patterns are not impeded by plugged nozzles, to ensure that nozzles are correctly positioned, and to ensure that obstructions do not prevent discharge patterns from wetting surfaces to be protected.

The observation of water discharge patterns during the discharge test required by 10.3.4.3.1 are intended to verify not only that there are no obstructions, but also that the water spray pattern attains complete impingement on the protected surface.

A.10.3.4.3.1 See 13.4.3.2.2.1.

10.3.4.3.1.1 Where the nature of the protected property is such that water cannot be discharged, the nozzles shall be inspected for proper orientation and the system tested with air to ensure that the nozzles are not obstructed.

10.3.4.3.2 Where obstructions occur, the piping and nozzles shall be cleaned and the system retested.

10.3.4.4 Pressure Readings.

10.3.4.4.1 Pressure readings shall be recorded at the hydraulically most remote nozzle to ensure the waterflow has not been impeded by partially closed valves or by plugged strainers or piping.

The proper functioning of a water spray nozzle is based on the nozzle pressure. To achieve the intended water spray pattern, pressures of either 15 psi (1 bar) or 30 psi (2 bar) are necessary for most nozzles. Evaluation of the most hydraulically demanding nozzle ensures that all other nozzles in the system will be subject to at least equivalent — if not higher — pressures.

10.3.4.4.2 A second pressure reading shall be recorded at the deluge valve to ensure the water supply is adequate.

10.3.4.4.3 Readings shall be compared to the hydraulic design pressures to ensure the original system design requirements are met and the water supply is adequate to meet the design requirements.

The residual pressure measured at the deluge valve should be compared to the calculated pressure as indicated in the hydraulic calculations. Any pressure in excess of that calculated is cause for concern and should be sufficient to initiate a complete evaluation of the piping system.

10.3.4.4.3.1 Where the hydraulically most remote nozzle is inaccessible, nozzles shall be permitted to be checked visually without taking a pressure reading on the most remote nozzle.

10.3.4.4.3.2 Where the reading taken at the riser indicates that the water supply has deteriorated, a gauge shall be placed on the hydraulically most remote nozzle and the results compared with the required design pressure.

10.3.5 Multiple Systems.

The maximum number of systems expected to operate in case of fire shall be tested simultaneously to check the adequacy of the water supply.

10.3.6 Manual Operation.

Manual actuation devices shall be operated annually.

10.3.7 Return to Service.

After the full flow test, the water spray system shall be maintained and returned to service in accordance with the manufacturer's instructions.

10.3.7.1 Main Drain Tests.

10.3.7.1.1 Main drain tests shall be conducted at the main riser to determine whether there has been any change in the condition of the water supply piping and controlling valves.

10.3.7.1.2 Static and residual water pressures shall be recorded respectively before, during, and after the operation of the fully opened drain valve.

10.3.7.1.3 Readings shall be compared with those made at the time of the original acceptance tests or with those made at the time of the last test to determine whether there has been any deterioration of the water supply.

10.3.7.2 Low Point Drains.

10.3.7.2.1 To prevent freezing and corrosion, all low point drains in aboveground piping shall be opened, the pipe drained, and the valves closed and plugs replaced.

10.3.7.2.2 Where weep holes are provided in lieu of low-point drains, they shall be inspected to ensure they are clear and unobstructed.

The weep holes mentioned in 10.3.7.2.2 are usually $\frac{3}{16}$ in. diameter holes drilled into fittings to create an automatic drain for trapped sections of pipe. These drains, the presence of which may not be obvious, should be carefully inspected.

10.4 Ultra-High-Speed Water Spray System (UHSWSS) Operational Tests

An ultra-high-speed water spray system, covered in Section 10.4, is a type of automatic water spray system in which water spray is rapidly applied to protect specific hazards in areas where deflagrations are anticipated. Ultra-high-speed water spray systems are intended to operate in less than 100 milliseconds. Specialized training and experience are necessary to work on these systems.

10.4.1 A full operational test, including measurements of response time, shall be conducted at intervals not exceeding 1 year.

10.4.1.1 Systems out of service shall be tested before being placed back in service.

10.4.2 All detectors shall be tested and inspected monthly for physical damage and accumulation of deposits on the lenses of optical detectors.

10.4.3 Controllers shall be inspected for faults at the start of each working shift.

10.4.4 Valves.

10.4.4.1 Valves on the water supply line shall be inspected at the start of each working shift to verify they are open.

10.4.4.2 Valves secured in the open position with a locking device or monitored by a signaling device that sounds a trouble signal at the deluge system control panel or other central location shall not require inspection.

10.4.5 Response Time.

10.4.5.1 The response time shall be verified during the operational test.

10.4.5.2 The response time shall be in accordance with the requirements of the system but not more than 100 milliseconds.

10.5 Component Action Requirements

Component replacement tables were added in the 2008 edition to offer guidance to the user of the standard when system components are adjusted, repaired, rebuilt, or replaced. It is not necessary in each case to require a complete acceptance test for each component when maintenance is performed.

10.5.1 Whenever a component in a water spray fixed system is adjusted, repaired, reconditioned, or replaced, the action required in Table 10.5.1 shall be performed.

10.5.1.1 Where the original installation standard is different from the cited standard, the use of the appropriate installing standard shall be permitted.

10.5.1.2 A main drain test shall be required if the system control or other upstream valve was operated in accordance with 13.3.3.4.

10.5.1.3 These actions shall not require a design review, which is outside the scope of this standard.

TABLE 10.5.1 Summary of Component Replacement Action Requirements

Component	Adjust	Repair/ Recondition	Replace	Required Action
Water Delivery Components				
Pipe and fittings	X	X	X	(1) Operational flow test
Nozzles	X	X	X	(1) Operational flow test
Manual release	X	X	X	(1) Operational test
				(2) Check for leaks at system working pressure
				(3) Test all alarms
Fire department connections				See Chapter 13
Valves	X	X	X	See Chapter 13
Fire pump	X	X	X	See Chapter 8
Alarm and Supervisory Components				
Pressure switch–type waterflow	X	X	X	(1) Operational test using inspector's test connection
Water motor gong	X	X	X	(1) Operational test using inspector's test connection
Valve supervisory device	X	X	X	Test for conformance with NFPA 15, *Standard for Water Spray Fixed Systems for Fire Protection*, and/or NFPA 72, *National Fire Alarm Code*
Detection system	X	X	X	Operational test for conformance with NFPA 15 and/or NFPA 72
Status-Indicating Components				
Gauges			X	Verify at 0 psi (0 bar) and system working pressure
Testing and Maintenance Components				
Main drain	X	X	X	Full flow main drain test
Auxiliary drains	X	X	X	(1) Check for leaks at system working pressure
				(2) Main drain test 2
Structural Components				
Hanger/seismic bracing	X	X	X	Check for conformance with NFPA 15 and/or NFPA 13, *Standard for the Installation of Sprinkler Systems*
Pipe stands	X	X	X	Check for conformance with NFPA 15 and /or NFPA 13
Informational Components				
Identification signs	X	X	X	Check for conformance with NFPA 15

SUMMARY

Chapter 10 of NFPA 25 covers the inspection, testing, and maintenance of highly specialized fire protection equipment. The applications, devices, and design of these systems are critical to protection of high-hazard areas or buildings. Water spray fixed systems must be maintained to ensure adequate protection and personnel safety.

REVIEW QUESTIONS

1. What is the purpose of performing the annual discharge test?
2. What types of equipment are water spray fixed systems used to protect?
3. What types of detection systems are typically used to actuate water spray fixed systems?
4. What is the difference between an automatic directional spray nozzle and a non-automatic one?

REFERENCES CITED IN COMMENTARY

National Fire Protection Association, 1 Batterymarch Park, Quincy, MA 02169-7471.

Cote, A. E., ed., *Fire Protection Handbook®*, 19th edition, 2003.
Hague, D. R., *Fire Protection Systems for Special Hazards,* 2004.
NFPA 13, *Standard for the Installation of Sprinkler Systems,* 2007 edition.
NFPA 15, *Standard for Water Spray Fixed Systems for Fire Protection,* 2007 edition.
NFPA 72®, National Fire Alarm Code®, 2007 edition.
NFPA 409, *Standard on Aircraft Hangars,* 2004 edition.

Foam-Water Sprinkler Systems
CHAPTER 11

Chapter 11 of NFPA 25 covers the inspection, testing, and maintenance of foam-water sprinkler and foam-water spray systems. Foam-water systems are used in extra hazard areas containing flammable or combustible liquids. As is the case with water spray fixed systems, the proper inspection, testing, and maintenance of foam-water systems require specialized training and experience. In addition to the requirements in this chapter, the manufacturer's instructions for inspection, testing, and maintenance should be followed.

Foam-water sprinkler systems are installed to protect structures, equipment, and facilities where a potential hazard for two-dimensional flammable liquid fire exists. These systems can provide prevention, extinguishment, control, and/or exposure protection, depending on the application. Typically, foam-water sprinkler systems are installed in aircraft hangars, petrochemical plants, tank farms, fuel loading facilities, and power plants.

The uses of foam (or mechanical foam, as it was initially called) for fire protection have increased greatly since it was first used in the 1930s. Original applications of this agent utilized a proteinaceous-type liquid foam-forming concentrate delivered in water solution to a turbulence-producing foam generator or nozzle that then directed the mechanically formed foam to a burning fuel tank or area of burning flammable fuel.

As technology advanced, new film-forming liquid concentrates were developed, and new systems and devices for applying the foam were proven useful for fire protection purposes. For example, one early (circa 1954) development was the application of foam from overhead sprinkler systems using specially designed foam-making nozzles. The nozzles were capable of either forming foam from protein-type foam concentrate solutions or, where supplied with water only, delivering a satisfactory water discharge pattern.

Protein, fluoroprotein, and aqueous film-forming concentrates or film-forming fluoroprotein foam concentrates are suitable for use with foam-water sprinklers. This latter type of foam concentrate has also proven suitable for use with standard sprinklers of the type referred to in NFPA 13, *Standard for the Installation of Sprinkler Systems*, where the system is provided with the necessary foam concentrate proportioning equipment. Care should be exercised to ensure that the chosen concentrate and discharge device are listed for use together.

11.1 General

This chapter shall provide the minimum requirements for the routine inspection, testing, and maintenance of foam-water systems. Table 11.1 shall be used to determine the minimum required frequencies for inspection, testing, and maintenance.

11.1.1 Fire pumps, water storage tanks, and valves common to other types of water-based fire protection systems shall be inspected, tested, and maintained in accordance with Chapters 8, 9, and 13, respectively, and as specified in Table 11.1.

11.1.2 Foam-Water Systems.

11.1.2.1 This section shall apply to foam-water systems as specified in NFPA 16, *Standard for the Installation of Foam-Water Sprinkler and Foam-Water Spray Systems*.

TABLE 11.1 Summary of Foam-Water Sprinkler System Inspection, Testing, and Maintenance

System/Component	Frequency	Reference
Inspection		
Discharge device location (sprinkler)	Annually	11.2.5
Discharge device location (spray nozzle)	Monthly	11.2.5
Discharge device position (sprinkler)	Annually	11.2.5
Discharge device position (spray nozzle)	Monthly	11.2.5
Foam concentrate strainer(s)	Quarterly	11.2.7.2
Drainage in system area	Quarterly	11.2.8
Proportioning system(s) — all	Monthly	11.2.9
Pipe corrosion	Annually	11.2.3
Pipe damage	Annually	11.2.3
Fittings corrosion	Annually	11.2.3
Fittings damage	Annually	11.2.3
Hangers/supports	Annually	11.2.4
Waterflow devices	Quarterly	11.2.1
Water supply tank(s)		Chapter 9
Fire pump(s)		Chapter 8
Water supply piping		11.2.6.1
Control valve(s)	Weekly/monthly	—
Deluge/preaction valve(s)		11.2.1, Chapter 13
Detection system	See *NFPA 72, National Fire Alarm Code*	11.2.2
Test		
Discharge device location	Annually	11.3.3.6
Discharge device position	Annually	11.3.3.6
Discharge device obstruction	Annually	11.3.3.6
Foam concentrate strainer(s)	Annually	11.2.7.2
Proportioning system(s) — all	Annually	11.2.9
Complete foam-water system(s)	Annually	11.3.3
Foam-water solution	Annually	11.3.6
Manual actuation device(s)	Annually	11.3.5
Backflow preventer(s)	Annually	Chapter 13
Fire pump(s)	See Chapter 8	—
Waterflow devices	Quarterly/Semi-annually	11.3.1
Water supply piping	Annually	Chapter 10
Control valve(s)	See Chapter 13	—
Strainer(s) — mainline	See Chapter 10	11.2.7.1
Deluge/preaction valve(s)	See Chapter 13	11.2.1
Detection system	See *NFPA 72*	11.2.2
Backflow preventer(s)	See Chapter 13	—
Water supply tank(s)	See Chapter 9	—
Water supply flow test	See Chapter 4	11.2.6
Maintenance		
Foam concentrate pump operation	Monthly	11.4.6.1, 11.4.7.1
Foam concentrate strainer(s)	Quarterly	Section 11.4
Foam concentrate samples	Annually	11.2.10
Proportioning system(s) standard pressure type		
Ball drip (automatic type) drain valves	5 years	11.4.3.1
Foam concentrate tank — drain and flush	10 years	11.4.3.2
Corrosion and hydrostatic test	10 years	11.4.3.3
Bladder tank type		
Sight glass	10 years	11.4.4.1
Foam concentrate tank — hydrostatic test	10 years	11.4.4.2

TABLE 11.1 Continued

System/Component	Frequency	Reference
Maintenance *(continued)*		
Line type		
Foam concentrate tank — corrosion and pickup pipes	10 years	11.4.5.1
Foam concentrate tank — drain and flush	10 years	11.4.5.2
Standard balanced pressure type		
Foam concentrate pump(s)	5 years *(see Note)*	11.4.6.2
Balancing valve diaphragm	5 years	11.4.6.3
Foam concentrate tank	10 years	11.4.6.4
In-line balanced pressure type		
Foam concentrate pump(s)	5 years *(see Note)*	11.4.7.2
Balancing valve diaphragm	5 years	11.4.7.3
Foam concentrate tank	10 years	11.4.7.4
Pressure vacuum vents	5 years	11.4.8
Water supply tank(s)	See Chapter 9	—
Fire pump(s)	See Chapter 8	—
Water supply	Annually	11.2.6.1
Backflow preventer(s)	See Chapter 13	—
Detector check valve(s)	See Chapter 13	—
Check valve(s)	See Chapter 13	—
Control valve(s)	See Chapter 13	—
Deluge/preaction valves	See Chapter 13	11.2.1
Strainer(s) — mainline	See Chapter 10	—
Detection system	See *NFPA 72*	11.2.2

Note: Also refer to manufacturer's instructions and frequency. Maintenance intervals other than preventive maintenance are not provided, as they depend on the results of the visual inspections and operational tests. For foam-water systems in aircraft hangars, refer to the inspection, test, and maintenance requirements of NFPA 409, *Standard on Aircraft Hangars*, Table 11.1.1.

According to 11.1.2.1, Chapter 11 of NFPA 25 addresses the foam-water system types and applications covered in NFPA 16, *Standard for the Installation of Foam-Water Sprinkler and Foam-Water Spray Systems*. These systems are very different from those addressed in NFPA 11, *Standard for Low-, Medium-, and High-Expansion Foam*.

It is important to note that the foam-water sprinkler systems addressed in this chapter are very similar to the standard wet, dry, preaction, and deluge systems covered in NFPA 13. The only difference is that the systems addressed in this chapter are also equipped with a foam concentrate connection, a proportioning device, and a foam concentrate storage tank. The discharge devices are similar to those used in sprinkler systems but are listed for use with foam concentrate.

◀ **FAQ**
What is the difference between the systems covered in NFPA 13 and the systems covered by Chapter 11 of NFPA 25?

The foam-water spray systems addressed in this chapter are identical to those addressed by NFPA 15, *Standard for Water Spray Fixed Systems for Fire Protection*. Foam-water spray systems are equipped with a foam concentrate connection, a proportioning device, and a foam concentrate storage tank. These systems can either discharge foam in spray form or discharge water in a satisfactory pattern for fire protection purposes.

The types of foam concentrates used in these systems are protein, fluoroprotein, or aqueous film-forming foam or film-forming fluoroprotein foam concentrates. They are low-expansion types only. For medium- and high-expansion foam, see NFPA 11. The types of concentrates used in the systems covered in this chapter are described below.

TYPES OF FOAM CONCENTRATES

Protein-Foam Concentrate. Protein-foam concentrate consists primarily of products from a protein hydrolysate, plus stabilizing additives and inhibitors to protect against freezing, prevent corrosion of equipment and containers, resist bacterial decomposition, control viscosity, and otherwise ensure readiness for use under emergency conditions. Protein-foam concentrates are diluted with water to form 3 percent to 6 percent solutions, depending on the type. These concentrates are compatible with certain dry chemicals.

Aqueous Film-Forming Foam (AFFF) Concentrate. AFFF concentrate is based on fluorinated surfactants plus foam stabilizers. They are usually diluted with water to a 1 percent, 3 percent, or 6 percent solution. The foam formed acts as a barrier both to exclude air or oxygen and to develop an aqueous film on the fuel surface that is capable of suppressing the evolution of fuel vapors. The foam produced with AFFF concentrate is suitable for combined use with dry chemicals.

Fluoroprotein-Foam Concentrate. Fluoroprotein-foam concentrate is very similar to protein-foam concentrate but has a synthetic fluorinated surfactant additive. In addition to an air-excluding foam blanket, it also can deposit a vaporization-preventing film on the surface of a liquid fuel. It is diluted with water to form 3 percent to 6 percent solutions, depending on the type. These concentrates are compatible with certain dry chemicals.

Film-Forming Fluoroprotein (FFFP) Foam Concentrate. FFFP concentrate uses fluorinated surfactants to produce a fluid aqueous film for suppressing hydrocarbon fuel vapors. This foam utilizes a protein base plus stabilizing additives and inhibitors to protect against freezing, corrosion, and bacterial decomposition. It also resists fuel pickup. The foam, which is usually diluted with water to a 3 percent or 6 percent solution, is dry chemical-compatible.

Alcohol-Resistant Foam Concentrate. Alcohol-resistant foam concentrate is used for fighting fires on water-soluble materials and other fuels destructive to regular, AFFF, or FFFP foams. It is also used for fires involving hydrocarbons. There are three general types of alcohol-resistant foam concentrates.

1. One is based on water-soluble natural polymers, such as protein or fluoroprotein concentrates. It also contains alcohol-insoluble materials that precipitate as an insoluble barrier in the bubble structure.
2. The second type is based on synthetic concentrates and contains a gelling agent that surrounds the foam bubbles and forms a protective raft on the surface of water-soluble fuels. These foams can also have film-forming characteristics on hydrocarbon fuels.
3. The third type is based on water-soluble natural polymers, such as fluoroprotein, and contains a gelling agent that protects the foam from water-soluble fuels. This foam can also have film-forming and fluoroprotein characteristics on hydrocarbon fuels.

Alcohol-resistant foam concentrate is generally used in 3 to 10 percent solutions, depending on the nature of the hazard to be protected and the type of concentrate.

Historically, foam concentrates were not permitted to be substituted. However, due to recent environmental issues concerning AFFF in particular, NFPA 11 permits different brands of foam concentrate to be mixed if the manufacturer provides data to the authority having jurisdiction (AHJ) demonstrating compatibility. It cannot be overstressed, however, that under no circumstances should foam concentrates of different concentration be substituted. For example, 3 percent protein should only be replaced with 3 percent protein.

11.1.2.2 This section shall not include systems detailed in NFPA 11, *Standard for Low-, Medium-, and High-Expansion Foam.*

Sometimes thought of as a refinery standard, NFPA 11 covers a very different type of foam system from those addressed in NFPA 25. They must be maintained in accordance with the requirements of Chapter 11 of NFPA 11.

11.1.3 Foam-Water System.

11.1.3.1 If during routine inspection and testing it is determined that the foam-water system has been altered or changed (e.g., equipment replaced, relocated, or foam concentrate replaced), it shall be determined whether the design intent has been altered and whether the system operates properly.

Alterations or changes, as described in 11.1.3.1, must be evaluated for compliance with NFPA 16.

11.1.3.1.1 Mechanical waterflow devices, including but not limited to water motor gongs, shall be tested quarterly.

11.1.3.1.2 Valve-type and pressure switch–type waterflow devices shall be tested semiannually.

11.1.3.1.3 Waterflow Devices. Waterflow devices shall be inspected quarterly to verify that they are free of physical damage.

11.1.3.2 The inspection shall verify that all components, including foam concentrate discharge devices and proportioning equipment, are installed in accordance with their listing.

11.1.4 Impairments.

The procedures outlined in Chapter 15 shall be followed where an impairment to protection occurs.

11.1.5 Notification to Supervisory Service.

To avoid false alarms where a supervisory service is provided, the alarm receiving facility shall be notified by the property owner or designated representative as follows:

(1) Before conducting any test or procedure that could result in the activation of an alarm
(2) After such tests or procedures are concluded

The notification of supervisory service required by 11.1.5 is part of the impairment procedure required by Chapter 15.

11.2 Inspection

Systems shall be inspected in accordance with the frequency specified in Table 11.1.

11.2.1 Deluge Valves.

Deluge valves shall be inspected in accordance with the provisions of Chapter 13.

11.2.2 Automatic Detection Equipment.

Automatic detection equipment shall be inspected, tested, and maintained in accordance with *NFPA 72, National Fire Alarm Code*, to ensure that the detectors are in place, securely fastened, and protected from corrosion, weather, and mechanical damage and that the communication wiring, control panels, or pneumatic tubing system is functional.

For more information on the automatic detection equipment addressed in 11.2.2, see the commentary following 10.2.3.

11.2.3 System Piping and Fittings.

System piping and fittings shall be inspected for the following:

(1) Mechanical damage (e.g., broken piping or cracked fittings)
(2) External conditions (e.g., missing or damaged paint or coatings, rust, and corrosion)
(3) Misalignment or trapped sections
(4) Low-point drains (automatic or manual)
(5) Location and condition of rubber-gasketed fittings

11.2.4 Hangers and Supports.

Hangers and supports shall be inspected for the following and repaired or replaced as necessary:

(1) Condition (e.g., missing or damaged paint or coating, rust, and corrosion)
(2) Secure attachment to structural supports and piping
(3) Damaged or missing hangers

11.2.5* Foam-Water Discharge Devices.

Discharge devices for foam-water spray systems can be either aspirating or non-aspirating. Discharge devices for foam-water sprinkler systems, such as wet, dry, or preaction systems, are automatic in operation and must be non-air aspirating. See Exhibit 11.1 for one type of air aspirating discharge device.

A.11.2.5 Directional-type foam-water discharge devices are quite often located in heavy traffic areas and are more apt to be dislocated compared to ordinary sprinkler locations. Of particular concern are low-level discharge devices in loading racks in and around low-level tankage and monitor-mounted devices that have been pushed out of the way for convenience. Inspection frequency might have to be increased accordingly.

11.2.5.1 Foam-water discharge devices shall be inspected visually and maintained to ensure that they are in place, continue to be aimed or pointed in the direction intended in the system design, and are free from external loading and corrosion.

11.2.5.2 Where caps or plugs are required, the inspection shall confirm they are in place and free to operate as intended.

11.2.5.3 Misaligned discharge devices shall be adjusted (aimed) by visual means, and the discharge patterns shall be checked at the next scheduled flow test.

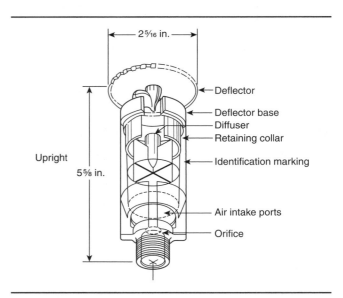

EXHIBIT 11.1 Air Aspirating Discharge Device. (Courtesy of "Automatic" Sprinkler/ Kidde Fire Fighting)

11.2.5.4* Inspection shall verify that unlisted combinations of discharge devices and foam concentrate have not been substituted.

A.11.2.5.4 Discharge devices are listed or approved for particular foam concentrates.

11.2.6 Water Supply.

11.2.6.1 The dependability of the water supply shall be ensured by regular inspection and maintenance, whether furnished by a municipal source, on-site storage tanks, a fire pump, or private underground piping systems.

11.2.6.2* Water supply piping shall be maintained free of internal obstructions.

A.11.2.6.2 Water supply piping should be free of internal obstructions that can be caused by debris (e.g., rocks, mud, tubercles) or by closed or partially closed control valves. See Chapter 5 for inspection and maintenance requirements.

11.2.7 Strainers.

11.2.7.1 Mainline and individual discharge device strainers (basket or screen) shall be inspected in accordance with the provisions of Chapter 10.

11.2.7.2 Foam concentrate strainers shall be inspected visually to ensure the blow-down valve is closed and plugged.

11.2.7.2.1 Baskets or screens shall be removed and inspected after each operation or flow test.

11.2.8 Drainage.

The area beneath and surrounding a foam-water spray system shall be inspected to ensure that drainage facilities, such as trap sumps and drainage trenches, are not blocked and retention embankments or dikes are in good repair.

11.2.9* Proportioning Systems.

The following description of various types of proportioning systems is based on NFPA 16, 2007 edition.

In *vacuum induction*, a venturi device, called an inductor, decreases the pressure in the foam concentrate proportioning inlet tube to the system flow passageway to below the atmospheric pressure in which the foam is pushed by atmospheric pressure through a calibrated orifice. This system is limited to deluge systems, and the friction loss of the discharge system to the discharge devices must be considered for proper performance. Proportioning percentage can vary based on the discharge and supply characteristics as well as the foam concentrate. These systems require full system discharge flow testing to determine proper installation and sizing.

A *foam concentrate pump* discharges foam through a metering orifice directly into the discharge system through a metering orifice specifically sized for the system performance that is being installed. This type of system is limited to deluge type systems and must be flow tested at the desired maximum flow rate of the system. Each installation must be full system flow tested due to specific design parameters for particular installation.

Balanced pressure proportioning with a foam pump or bladder tank uses a modified venturi proportioning device in the system water supply pipe. The waterflow to the system through the modified venturi causes a metered pressure drop in the foam concentrate inlet chamber. As the flow increases, the metered pressure loss increases, causing increased flow of foam concentrate through a calibrated orifice into the system water supply. With a foam pump system a balancing valve is used to measure the supply pressure at the inlet of the modified venturi proportioner and balance the foam concentrate pressure down to equal that pressure. When using a bladder tank, the foam concentrate is stored inside a bladder, within a pressure vessel where the supply water in the system supply is directed to the outside of the bladder, causing the foam concentrate to push out from the top or bottom of the tank and travel to the inlet of the modified venturi device. As the water pressure increases or decreases due to flow conditions, so does the foam concentrate pressure. In both the balanced pump system and the bladder tank system, when the inlet pressures of both foam concentrate and water supply match, the inlet calibrated orifice of the modified venturi device provides the proper mixture of foam to water as specified by the manufacturer. Also, these systems allow for system isolation test valves to be applied and divert test foam solution flow to a containment area that can be easily disposed of.

Positive pressure proportioning with a foam pump or pressure controlled bladder tank system uses a system that supplies foam concentrate to a modified venturi at a higher pressure than the supply water pressure. The modified venturi includes a calibrated orifice that is sized to match the system equipment and the foam concentrate being used.

Exhibit 11.2 illustrates a hydraulically operated concentrate valve. As the deluge valve on the system activates, water pressure is applied to the valve, allowing it to open. This valve should also be included in the inspection, testing, and maintenance program for foam-water systems.

A.11.2.9 Proportioning systems might or might not include foam concentrate pumps. If pumps are part of the proportioning system, the driver, pump, and gear reducer should be checked in accordance with the manufacturer's recommendations, and the check can include items such as lubrication, fuel, filters, oil levels, and clutches.

11.2.9.1 The components of the various proportioning systems described in 11.2.9 shall be inspected in accordance with the frequency specified in Table 11.1.

11.2.9.2 Valves specified to be checked shall be permitted to be open or closed, depending on specific functions within each foam-water system.

11.2.9.3 The position (open or closed) of valves shall be verified in accordance with specified operating conditions.

11.2.9.4* Inspection of the concentrate tank shall include verification that the quantity of foam concentrate satisfies the requirements of the original design.

A.11.2.9.4 In some cases, an adequate supply of foam liquid is available without a full tank. This is particularly true of foam liquid stored in nonmetallic tanks. If liquid is stored in metallic tanks, the proper liquid level should be one-half the distance into the expansion dome.

EXHIBIT 11.2 Hydraulically Operated Concentrate Valve.

As specified in NFPA 16, the quantity of foam concentrate is determined by the following hydraulic calculations:

% concentration × amount of flow × duration = foam concentrate quantity

This minimum quantity should be maintained in the storage tank at all times.

11.2.9.5 Additional inspection requirements shall be performed as detailed for the proportioning systems specified in 11.2.9.

11.2.9.5.1* Standard Pressure Proportioner. The pressure shall be removed before the inspection to prevent injury. The inspection shall verify the following:

(1) Ball drip valves (automatic drains) are free and opened.

(2) External corrosion on foam concentrate storage tanks is not present.

A.11.2.9.5.1 The standard pressure proportioner is a pressure vessel. Although under normal standby conditions this type of proportioning system should not be pressurized, some installations allow for inadvertent pressurization. Pressure should be removed before inspection.

Pressure can be removed, as prescribed by 11.2.9.5.1, by slowly opening the drain or vent valve.

11.2.9.5.2* Bladder Tank Proportioner. The pressure shall be removed before the inspection to prevent injury. The inspection shall include the following:

(1) Water control valves to foam concentrate tank
(2) A check for external corrosion on foam concentrate storage tanks
(3) A check for the presence of foam in the water surrounding the bladder (annual)

A.11.2.9.5.2 The bladder tank proportioner is a pressure vessel. Where inspecting for a full liquid tank, the manufacturer's instructions should be followed. If checked incorrectly, the tank sight gauges could indicate a full tank when the tank actually is empty of foam liquid. Some foam liquids, due to their viscosity, might not indicate true levels of foam liquid in the tank where checked via the sight glass.

> CAUTION: Depending on system configuration, this type of proportioner system might be pressurized or nonpressurized under normal conditions. Pressure should be removed before inspection.

Pressure can be removed, as required by 11.2.9.5.2, by slowly opening the drain or vent valve.

11.2.9.5.3 Line Proportioner. The inspection shall include the following:

(1)* Strainers

A.11.2.9.5.3(1) See 11.2.7.1.

(2)* Verification that pressure vacuum vent is operating freely

A.11.2.9.5.3(2) See Figure A.3.3.26.

(3) A check for external corrosion on foam concentrate storage tanks

11.2.9.5.4 Standard Balanced Pressure Proportioner. The inspection shall include the following:

(1)* Strainers

A.11.2.9.5.4(1) See 11.2.7.1.

(2)* Verification that pressure vacuum vent is operating freely

A.11.2.9.5.4(2) See Figure A.3.3.26.

(3) Verification that gauges are in good operating condition
(4) Verification that sensing line valves are open
(5) Verification that power is available to foam liquid pump

11.2.9.5.5 In-Line Balanced Pressure Proportioner. The inspection shall include the following:

(1)* Strainers

A.11.2.9.5.5(1) See 11.2.7.1.

(2)* Verification that pressure vacuum vent is operating freely

A.11.2.9.5.5(2) See Figure A.3.3.26.

(3) Verification that gauges are in good working condition
(4) Verification that sensing line valves at pump unit and individual proportioner stations are open
(5) Verification that power is available to foam liquid pump

11.2.9.5.6 Orifice Plate Proportioner. The inspection shall include the following:

(1)* Strainers

A.11.2.9.5.6(1) See 11.2.7.1.

(2)* Verification that pressure vacuum vent is operating freely

A.11.2.9.5.6(2) See Figure A.3.3.26.

(3) Verification that gauges are in good working condition
(4) Verification that power is available to foam liquid pump

Exhibit 11.3 illustrates an orifice plate proportioning arrangement. This method employs a second deluge valve supplied by a foam concentrate pump. The diameter of the orifice is determined by hydraulic calculations and controls the flow of foam concentrate.

11.2.10 Foam Concentrate Samples.

Samples shall be submitted in accordance with the manufacturer's recommended sampling procedures.

11.3* Operational Tests

Frequency of system tests shall be in accordance with Table 11.1.

A.11.3 Operational tests generally should be comprised of the following:

(1) A detection/actuation test with no flow to verify that all components such as automated valves, foam and water pumps, and alarms operate properly
(2) A water-only flow test to check piping continuity, discharge patterns, pressures, and line flushing
(3) A foam flow test to verify solution concentration
(4) Resetting of system to its normal standby condition, including draining of lines and filling of foam liquid tank

As required by 11.3.2, system tests must be conducted in accordance with Table 11.1. This includes annual testing of the foam concentrate, which helps ensure the reliability of the foam system. Properly stored foam concentrates have an excellent record of a long useful life. The expected shelf life of most foam concentrate solutions, when stored in a properly installed and maintained tank, is at least 10 years.

However, several problems can affect the efficiency of the foam blanket. These problems include dilution, evaporation, and contamination. Manufacturers of foam concentrates provide laboratory testing services to analyze foam concentrate samples. The samples are tested for such chemical characteristics as pH level and specific gravity. They are also tested for

EXHIBIT 11.3 Orifice Plate Proportioning Arrangement.

1. Strainer
2. Foam liquid storage tank
3. Flange with orifice plate (for calibration of foam liquid injection)
4. Flange with orifice plate (for pump test calibration)
5. Vent-vacuum breather
6. Sight glass
7. Foam liquid pump
8. Re-circulating and test line gate valve (N.C.)
9. Relief valve
10. Check valve
11. Foam liquid supply strainer
12. Deluge valve
13. Companion deluge valve
14. Deluge system control panel
15. Deluge valve solenoid
16. Water supply strainer
17. To system foam water solution
18. Pressure indicating gauge
19. Foam liquid control valve (N.O.)
20. Water supply control valve (N.O.)
21. Companion deluge valve solenoid

N.C. = Normally closed.
N.O. = Normally open.

contamination from water, oil, noncompatible foam concentrates, and possible microbiological organisms. In addition, changes in the concentrate due to storage temperature fluctuations and sediment are analyzed. All of these factors and conditions may reduce the effectiveness of the foam-water solution.

Samples for testing should be gathered according to the manufacturer's instructions. Each sample should represent the found condition of the concentrate in the storage tank.

A test connection is normally provided as a means of periodically checking the performance of the proportioners used in foam sprinkler systems. Typical test connections are illustrated in Figure A.11.3.3. If the proportioning controller is located after the main sprinkler valve, an additional supervised OS&Y valve is provided to isolate the sprinkler overhead during the proportioner test. This is done to eliminate the problems caused by air cushions in wet pipe sprinkler systems or the servicing delays caused where charging and draining preaction or deluge sprinkler systems. The manufacturer's test procedures should be followed closely.

11.3.1* Test Preparation.

Precautions shall be taken to prevent damage to property during the test.

The precautions required by 11.3.1 include the following:

- Check related facility systems for proper operation. Make sure the water supply to the deluge system has been inspected and tested.

◀ FAQ
What precautions must be taken to prevent property damage prior to a discharge test?

- Make sure sump tanks are empty and sump pumps are operating properly.
- Inspect any drainage provisions required to contain or collect the foam-water discharging from the system.
- Make sure the owner or the owner's authorized representative has protected all exposed equipment or property that may be damaged by the system when discharging.

A.11.3.1 The property owner's representative should take care to prevent damage to equipment or the structure during the test. Damage could be caused by the system discharge or by runoff from the test site. It should be verified that there is adequate and unobstructed drainage. Equipment should be removed or covered as necessary to prevent damage. Means such as curbing or sandbagging should be used to prevent entry of the foam-water solution.

In the execution of maintenance tests, only a small amount of foam concentrate should be discharged to verify the correct concentration of foam in the foam-water solution. Designated foam-water test ports can be designed into the piping system so that the discharge of foam-water solution can be directed to a controlled location, such as a container that would be transported to an approved disposal site by a licensed hazardous waste contractor. The remainder of the test should be conducted using water only.

11.3.1.1 Mechanical waterflow devices, including but not limited to water motor gongs, shall be tested quarterly.

11.3.1.2 Vane-type and pressure switch–type waterflow devices shall be tested semiannually.

11.3.1.3 Waterflow Devices. Waterflow devices shall be inspected quarterly to verify that they are free of physical damage.

11.3.2* Operational Test Performance.

A.11.3.2 An alternative method for achieving flow can be permitted to be an installation as shown in Figure A.11.3.2. This type of testing does not verify system pipe conditions or discharge device performance but only the water supply, foam concentrate supply, and proportioning accuracy.

11.3.2.1 Operational tests shall be conducted to ensure that the foam-water system(s) responds as designed, both automatically and manually.

Knowledge of the equipment and test apparatus is necessary to perform the tests required by 11.3.2.1. The tests should, therefore, be conducted by a contractor or a foam equipment supplier (or both).

11.3.2.2 The test procedures shall simulate anticipated emergency events so the response of the foam-water system(s) can be evaluated.

The following procedure is recommended, with variations based on the equipment installed, when conducting a full flow operational test of a foam-water deluge system, as prescribed by 11.3.2.2:

Step 1. Inspect the deluge valve.
    A. Check the deluge valve, trim piping and valves, and associated riser piping and valves, for proper arrangement.
    B. Position all system and trim valves properly, including any low point drains, in accordance with the system installation drawings or valve schematic.

Step 2. Isolate the foam supply from the deluge system.
    A. Shut tank foam liquid supply valve.
    B. Disconnect electrical service to foam concentrate pump(s) if applicable, or shut off water pressure supply to bladder tank.

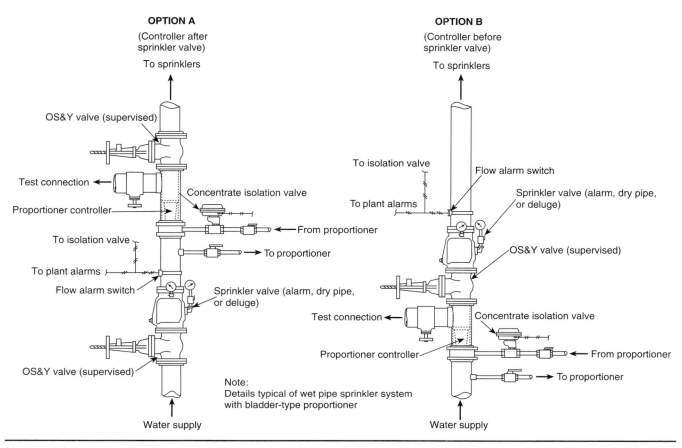

FIGURE A.11.3.2 *Foam System/Test Header Combination.*

Step 3. Check deluge system for proper operation.

    A. Run and test any fire pumps.
    B. Open test valve to trip the deluge valve.
    C. Allow water to flow through the system, flushing all piping until water runs clear.
    D. Check for obstructed nozzles.

Step 4. Prepare system for concentration test.

    A. Test detectors.
    B. Test manual actuation.
    C. Test alarms and auxiliary equipment shutdowns.

Step 5. Return foam tank and proportioning system to normal condition.

    A. If foam concentrate pump is used, check oil level and reenergize power supply to the pump.
    B. Inspect foam tank for proper set-up and check level of foam concentrate.
    C. If bladder tank is used, carefully open water pressure supply to the tank in accordance with the manufacturer's procedures. Care must be taken to avoid damage to the bladder.
    D. Set up remaining proportioning equipment, and open tank foam liquid supply valve.

Step 6. Make sample foam mixtures.

    A. Both the foam concentrate and the water used for the mixtures must be taken from the site supply.

B. Mix in 100 ml graduated cylinders. If a single cylinder and pipette is used for both samples, clean them thoroughly before each use.
C. For 3 percent mixture, add 3 parts foam concentrate to 97 parts water. For 6 percent mixture, add 6 parts foam concentrate to 94 parts water.
D. Using a pipette, stir each sample until thoroughly mixed.
E. Using the pipette, place a small amount of the 3 percent mixture on the refractometer.
F. Thoroughly clean the refractometer and repeat for the 6 percent mixture.
H. Graph should now include all information except the plot point from the actual system full flow discharge test sample and the end head residual pressure.

Step 7. Prepare the highest, most remote branch line for the end head residual pressure recording.
  A. Remove the end head (discharge device) from the highest, most remote branch line.
  B. Install $\frac{1}{2}$ in. to $\frac{1}{4}$ in. adapter, or size as needed.
  C. Attach $\frac{1}{4}$ in. hose to the adapter. A 20 ft hose is recommended unless a longer one is necessary.
  D. Attach a pressure gauge 0-250 psi (0-17 bar) and a valve to the end of the hose.
  E. The valve will be used to take a foam sample, and will then be shut to record the residual pressure.

Step 8. Prepare to perform the operational test of the system.
  A. From the system installation drawings and hydraulic calculations, determine the pressure required at the highest, most remote discharge outlet. Record this pressure on the test graph in the appropriate place.
  B. Check fire pump(s), foam tank, foam concentrate piping, and proportioning system, to make sure all components are set up in normal mode.
  C. Check all system valves to make sure they are in normally open or normally closed position, as appropriate for the system discharge test.
  D. Prepare the area for full flow operational test.
  E. Coordinate the efforts of the testing personnel before the test begins, to minimize the amount of time the system discharges.
  F. Notify all testing and witnessing personnel that the test is about to begin.

Step 9. Trip the system and record the results.
  A. Manually trip the system using the manual pull station at the deluge valve.
  B. Allow the system to flow long enough for the discharge from the nozzles to stabilize before checking spray patterns and taking foam-water samples.
  C. Visually inspect nozzles for proper discharge. Note any deficiencies.
  D. Take three foam-water samples: one from the valve attached to the highest, most remote discharge outlet (see Step 7E), and two from the foam blanket.
  E. Shut off the valve attached to the highest, most remote discharge outlet.
  F. Take residual pressure readings at the system deluge valve and at the gauge installed at the highest, most remote discharge outlet.

Step 10. Begin system shut-down procedures.
  A. Shut off foam supply. Allow only water to flow through the system until the system is flushed clear.
  B. Shut water supply valve to deluge system.
  C. Open system drains.

Step 11. Determine foam concentrate levels from samples taken during the test.
  A. Using the pipette, apply a drop of foam-water sample #1 to the refractometer and take reading #1.

B. Cleaning the pipette and refractometer after each reading, follow this procedure for foam-water samples #2 and #3.
C. Plot the single best result from the three foam-water samples on the graph.
D. Adjust the pressure recorded at the gauge connected to the highest, most remote discharge outlet by the head pressure caused by the elevation difference between the gauge and the branch line fitting. [Assuming the gauge is below the branch line fitting, multiply the elevation difference measured in feet between them and multiply that by 0.433 to determine the head pressure in psi (0.0978 bar/m). Subtracting the head pressure from the pressure recorded at the gauge will equal the actual pressure at the highest, most remote discharge outlet.] Record this pressure in the appropriate place on the graph.

Step 12. Analyze test results.

11.3.2.3 Where discharge from the system discharge devices would create a hazardous condition or conflict with local requirements, an approved alternate method to achieve full flow conditions shall be permitted.

When an alternate method of achieving full flow conditions is permitted by 11.3.2.3, full flow can be achieved from the discharge test connection at the system riser.

11.3.2.4 Response Time. Under test conditions, the automatic fire detection systems, when exposed to a test source, shall operate within the requirements of *NFPA 72, National Fire Alarm Code,* for the type of detector provided, and the response time shall be recorded.

11.3.2.5 Discharge Time. The time lapse between operation of detection systems and water delivery time to the protected area shall be recorded for open discharge devices.

11.3.2.6 Discharge Patterns.

11.3.2.6.1 The discharge patterns from all of the open spray devices shall be observed to ensure that patterns are not impeded by plugged discharge devices and to ensure that discharge devices are correctly positioned and that obstructions do not prevent discharge patterns from covering surfaces to be protected.

11.3.2.6.2 Where obstructions occur, the piping and discharge devices shall be cleaned and the system retested.

11.3.2.6.3 Discharge devices shall be permitted to be of different orifice sizes and types.

11.3.2.7* Pressure Readings.

A.11.3.2.7 Specific foam concentrates typically are listed or approved with specific sprinklers. Part of the approval and listing is a minimum sprinkler operating pressure. Sprinkler operating pressure affects foam quality, discharge patterns, and fire extinguishment (control) capabilities. Discharge pressures less than this specified minimum pressure should be corrected immediately; therefore, it is necessary to test under full flow conditions.

11.3.2.7.1 Pressure readings shall be recorded at the highest, most remote discharge device.

11.3.2.7.2 A second pressure reading shall be recorded at the main control valve.

11.3.2.7.3 Readings shall be compared to the hydraulic design pressures to ensure the original system design requirements are met.

11.3.3 Multiple Systems.

The maximum number of systems expected to operate in case of fire shall be tested simultaneously to check the adequacy of the water supply and concentrate pump.

11.3.4 Manual Actuation Devices.

Manual actuation devices shall be tested annually.

11.3.5 Concentration Testing.

11.3.5.1 During the full flow foam test, a foam sample shall be taken.

11.3.5.2 The foam sample shall be checked by refractometric or other methods to verify concentration of the solution.

Two methods are used to determine foam concentration as required by 11.3.5.2. They are the refractometric and the conductivity methods.

The refractometric method uses a prism that has a much greater refractive index than the sample solution to be measured. Measurements are made at the interface of the prism and the sample solution. For a weak (low-concentration) solution, the difference between the refractive index of the prism and the solution is high. For a strong (high-concentration) solution, the difference between the refractive index of the prism and the sample is smaller. The following is a sample test procedure for foam concentration determination.

The foam solution concentration determination test is used to determine the percent concentration of a foam concentrate in the water being used to generate foam. It is typically used as a means of determining the accuracy of a system's proportioning equipment. If the level of foam concentrate injection varies widely from that of the design, it may abnormally influence the expansion and drainage foam quality values, which may, in turn, affect the foam's fire performance.

There are two acceptable methods for measuring foam concentrate percentage in water: the refractive index method and the conductivity method. Both involve comparing foam solution test samples to premeasured solutions that are plotted on a baseline graph of percent concentration versus instrument reading.

Refractive Index Method. A handheld refractometer is used to measure the refractive index of the foam solution samples. This method is not particularly accurate for AFFF or alcohol-resistant AFFFs, since they typically exhibit very low refractive index readings. For this reason, the conductivity method may be preferred where these products are used.

Equipment. Prepare a base (calibration) curve using the following apparatus:

- Four 3.4 fl oz (100 ml) or larger plastic bottles with caps
- One measuring pipette [0.34 fl oz (10 ml)] or syringe [0.34 fl oz (10 ml)]
- One 3.4 fl oz (100 ml) or larger graduated cylinder
- Three plastic-coated magnetic stirring bars
- One handheld refractometer (American Optical Model 10400 or 10441, Atago NI, or equivalent)
- Standard graph paper
- Ruler or other straight edge

Procedure. Using water and foam concentrate from the system to be tested, make up three standard solutions using the 3.4 fl oz (100 ml) or larger graduated cylinders. These samples should include the nominal intended percentage of injection, the nominal percentage plus 1 percent, and the nominal percentage minus 1 percent.

Place the water in the graduated cylinders (leaving adequate space for the foam concentrate), and then carefully measure the foam concentrate samples into the water using the pipette or syringe. Use care not to pick up air in the foam concentrate samples.

Pour each measured foam solution from the graduated cylinder into a 3.4 fl oz (100 ml) or larger plastic bottle. Each bottle should be marked with the percent solution it contains. Add a plastic stirring bar to the bottle, cap it, and shake it thoroughly to mix the foam solution.

After thoroughly mixing the foam solution samples, take a refractive index reading of each sample. To do this, place a few drops of the solution on the refractometer prism, close the cover plate, and observe the scale reading at the dark field intersection. Since the refractometer is temperature compensated, it may take 10 to 20 seconds for the sample to be read properly. It is important to take all refractometer readings at ambient temperatures of 50°F (10°C) or above.

Using standard graph paper, plot the refractive index readings on one axis and the percent concentration readings on the other (see Exhibit 11.4). This plotted curve will serve as the known baseline for the test series. Set the solution samples aside in the event the measurements need to be checked.

Sampling and Analysis. Collect foam solution samples from the proportioning system, being careful to ensure that the samples are taken at an adequate distance downstream of the proportioner being tested. Take refractive index readings of the sample and compare them to the plotted curve to determine the percentage of the samples.

Conductivity Method. This method is based on changes in electrical conductivity as foam concentrate is added to water. A handheld conductivity meter (as shown in Exhibit 11.5) is used to measure the conductivity of foam solutions in microsiemens units.

Conductivity is a very accurate method, provided there are substantial changes in conductivity, as foam concentrate is added to the water in relatively low percentages. Because it is very conductive, the use of salt or brackish water may not be suitable for this method, due

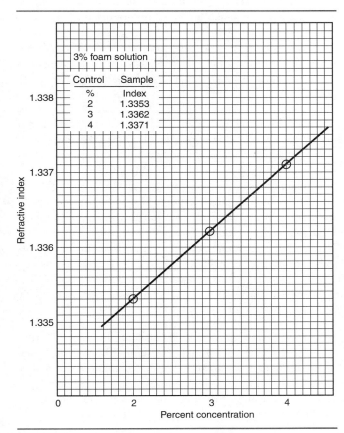

EXHIBIT 11.4 *Typical Graph of Refractive Index Versus Foam Concentration. (Source: NFPA 11, 2005, Figure D.2.1.1.2)*

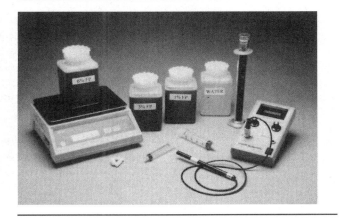

EXHIBIT 11.5 *Equipment Needed for Conductivity Method of Proportioning Measurement. (Source: NFPA 11, 2005, Figure D.2.1.2)*

to small conductivity changes as foam concentrate is added. It will be necessary to make foam and water solutions in advance to determine if adequate changes in conductivity can be detected if the water source is salty or brackish.

Equipment. Prepare a base (calibration) curve using the following apparatus:

- Four 3.4 fl oz (100 ml) or larger plastic bottles with caps
- One measuring pipette [0.34 fl oz (10 ml)] or syringe [0.34 fl oz (10 ml)]
- One 3.4 fl oz (100 ml) or larger graduated cylinder
- Three plastic-coated magnetic stirring bars
- A portable, temperature compensated conductivity meter (Omega Model CDH-70, VWR Scientific Model 23198-014, or equivalent)
- Standard graph paper
- Ruler or other straight edge

Procedure. Using water and foam concentrate from the system to be tested, make up three standard solutions using the 3.4 fl oz (100 ml) or larger graduated cylinders. These samples should include the nominal intended percentage of injection, the nominal percentage plus 1 percent, and the nominal percentage minus 1 percent.

Place the water in the graduated cylinders (leaving adequate space for the foam concentrate), and then carefully measure the foam concentrate samples into the water using the pipette or syringe. Use care not to pick up air in the foam concentrate samples.

Pour each measured foam solution from the graduated cylinder into a 3.4 fl oz (100 ml) or larger plastic bottle. Each bottle should be marked with the percent solution it contains. Add a plastic stirring bar to the bottle, cap it, and shake it thoroughly to mix the foam solution.

After making the three foam solutions in this manner, measure the conductivity of each. Refer to the instructions that came with the conductivity meter to determine proper procedures for taking readings. It will be necessary to switch the meter to the correct conductivity range setting to obtain a proper reading.

Most synthetic-based foams used with fresh water result in foam solution conductivity readings of less than 2000 microsiemens. Protein-based foams in fresh water solutions generally produce conductivity readings in excess of 2000 microsiemens. Due to the temperature compensation feature of the conductivity meter, it may take a short time to obtain a consistent reading.

Once the solution samples have been measured and recorded, set the bottles aside for control sample reference. The conductivity readings then should be plotted on the graph paper (see Exhibit 11.6). It is most convenient to place the foam solution percentage on the horizontal axis and the conductivity readings on the vertical axis.

Use a ruler or straight edge to draw a line that approximates connecting all three points. Although it may not be possible to hit all three points with a straight line, they should be very close. If not, repeat the conductivity measurements and, if necessary, make new control sample solutions until all three points plot in a nearly straight line. This plot will serve as the known base (calibration) curve to be used for the test series.

Sampling and Analysis. Collect foam solution samples from the proportioning system using care to take the samples at an adequate distance downstream of the proportioner being tested. Using foam solution samples that are allowed to drain from expanded foam is not recommended, as doing so may produce misleading conductivity readings.

Once one or more samples have been collected, read their conductivity and find the corresponding percentage from the base curve prepared from the control sample solutions.

11.3.5.3 Concentration shall be within 10 percent of the acceptance test results but in no case more than 10 percent below minimum design standards.

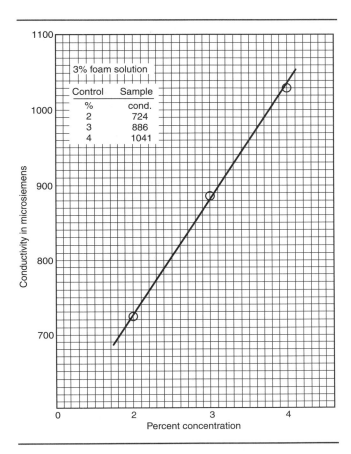

EXHIBIT 11.6 Typical Graph of Conductivity Versus Foam Concentration. (Source: NFPA 11, 2005, Figure 2.1.2.2)

11.3.6 Return to Service.

After the full flow test, the foam-water system shall be returned to service and the foam concentrate tank shall be replenished to design level.

11.4* Maintenance

A.11.4 The maintenance items specified in the body of this standard are in addition to the typical inspection and test procedures indicated. Foam-water systems are, as are all fire protection systems, designed to be basically maintenance free. There are, however, some areas that need special attention. Foam concentrate shelf life varies between liquids and is affected by factors such as heat, cold, dilution, contamination, and many others. As with all systems, common sense dictates those maintenance-sensitive areas that should be given attention. Routine testing and inspection generally dictate the need for additional maintenance items. Those maintenance items specified are key procedures that should be performed routinely.

11.4.1 Maintenance of foam-water systems shall be in accordance with the requirements of those chapters covering the specific component parts.

11.4.2 Maintenance of specific foam components shall be in accordance with 11.4.3 through 11.4.7.

11.4.3 Standard Pressure Proportioner.

11.4.3.1 The ball drip (automatic type) drain valves shall be disassembled, cleaned, and reassembled.

11.4.3.2* The foam liquid storage tank shall be drained of foam liquid and flushed. (Foam liquid shall be permitted to be salvaged and reused.)

A.11.4.3.2 Foam concentrates tend to settle out over time. Depending on the specific characteristics of the foam concentrate, sedimentation accumulates in the bottom of the storage vessel. This sediment can affect proportioning and foam concentrate integrity. Some concentrates tend to settle out more rapidly than others. If the annual samples indicate excessive sediment, flushing the tank could be required more frequently.

11.4.3.3 The foam liquid tank shall be inspected for internal and external corrosion and hydrostatically tested to the specified working pressure.

11.4.4 Bladder Tank Proportioner.

11.4.4.1 Sight glass, where provided, shall be removed and cleaned.

11.4.4.2* The foam concentrate tank shall be hydrostatically tested to the specified working pressure.

A.11.4.4.2 Where hydrostatically testing bladder tanks, the generation of a pressure differential across the diaphragm should not be allowed. The manufacturer should be consulted for specific procedures.

11.4.5 Line Proportioner.

11.4.5.1 The foam concentrate tank shall be inspected for internal corrosion. Pickup pipes inside the tank shall be inspected for corrosion, separation, or plugging.

11.4.5.2 The foam concentrate tank shall be drained and flushed. (Foam concentrate shall be permitted to be salvaged and reused.)

11.4.6 Standard Balanced Pressure Proportioner.

11.4.6.1 The foam concentrate pump shall be operated. Foam concentrate shall be circulated back to the tank.

11.4.6.2 Foam pumps, drive train, and drivers shall be serviced in accordance with the manufacturer's instructions and frequency, but not at intervals of more than 5 years.

11.4.6.3 The diaphragm balancing valve shall be flushed through the diaphragm section with water or foam concentrate until fluid appears clear or new.

11.4.6.4 The foam concentrate tank shall be inspected internally for corrosion and sediment. Excessive sediment shall require draining and flushing of the tank.

11.4.7 In-Line Balanced Pressure Proportioner.

11.4.7.1 The foam concentrate pump shall be operated. Foam concentrate shall be circulated back to the tank.

11.4.7.2 Foam pumps, drive train, and drivers shall be serviced in accordance with the manufacturer's instructions and frequency, but not at intervals of more than 5 years.

11.4.7.3 The diaphragm balancing valve shall be flushed through the diaphragm section with water or foam concentrate until fluid appears clear or new.

11.4.7.4 The foam concentrate tank shall be inspected internally for corrosion and sediment. Excessive sediment shall require draining and flushing of the tank.

11.4.8 Pressure Vacuum Vents.

The procedures specified in 11.4.8.1 through 11.4.8.8 shall be performed on pressure vacuum vents every 5 years.

11.4.8.1 The vent shall be removed from the expansion dome. While the vent is removed, it shall be ensured that the opening is not blocked and that dirt or other foreign objects do not enter the tank.

11.4.8.2 The vest bonnet shall be removed. The vacuum valve and pressure valve shall be lifted out.

11.4.8.3 The vent body shall be flushed internally, and the vacuum valve and the pressure valve shall be washed thoroughly. It shall be ensured that the screen is not clogged, and the use of any hard, pointed objects to clear the screen shall be avoided.

11.4.8.4 If the liquid has become excessively gummy or solidified, the vent body and parts shall be soaked in hot soapy water.

11.4.8.5 The vent body shall be turned upside down and drained thoroughly. Parts shall be dried by placing them in a warm and dry area or by using an air hose.

11.4.8.6 Parts shall be sprayed with a light Teflon® coating, and the vent shall be reassembled. The use of any type of oil for lubrication purposes shall be avoided, as oil is harmful to the foam liquid.

11.4.8.7 The vent bonnet shall be replaced, and the vent shall be turned upside down slowly a few times to ensure proper freedom of the movable parts.

11.4.8.8 The vent shall be attached to the liquid storage tank expansion dome.

11.5 Component Action Requirements

Component replacement tables were added in the 2008 edition to offer guidance to the user of the standard when system components are adjusted, repaired, rebuilt, or replaced. It is not necessary in each case to require a complete acceptance test for each component when maintenance is performed.

11.5.1 Whenever a component in a foam-water sprinkler system is adjusted, repaired, reconditioned, or replaced, the action required in Table 11.5.1 shall be performed.

11.5.2 Where the original installation standard is different from the cited standard, the use of the appropriate installing standard shall be permitted.

11.5.3 A main drain test shall be required if the system control or other upstream valve was operated in accordance with 13.3.3.4.

11.5.4 The actions of 11.5.1 through 11.5.3 shall not require a design review, which is outside the scope of this standard.

TABLE 11.5.1 Summary of Component Replacement Action Requirements

Component	Adjust	Repair/ Recondition	Replace	Required Action
Water Delivery Components				
Pipe and fittings on open head system	X	X	X	(1) Operational flow test
Pipe and fittings on closed head system	X	X	X	(1) Hydrostatic test in conformance with NFPA 16, *Standard for the Installation of Foam-Water Sprinkler and Foam-Water Spray Systems*
Discharge devices	X		X	(1) Check for leaks at system working pressure (2) Check for impairments at orifice
Fire department connections	X	X	X	See Chapter 13
Manual release	X	X	X	(1) Operational test (2) Check for leaks at system working pressure (3) Test all alarms
Valves	X	X	X	See Chapter 13
Fire pump	X	X	X	See Chapter 8
Foam Components				
Foam concentrate strainer(s)				See Chapter 13
Proportioning system(s)	X	X	X	Conduct flow test and check proportioning by refractometer test or equivalent
Water supply tank(s)				See Chapter 9
Foam concentrate	X		X	Submit 1 pint (473 mL) sample for laboratory analysis for conformance with manufacturer's specifications
Foam concentrate pump				See Chapter 8
Ball drip (automatic type) drain valves				See Chapter 13
Foam concentrate tank	X	X	X	Inspect for condition, repair as appropriate
Bladder tank	X	X	X	Check water jacket for presence of foam concentrate
Alarm and Supervisory Components				
Vane-type waterflow	X	X	X	(1) Operational test using inspector's test connection
Pressure switch–type waterflow	X	X	X	(1) Operational test using inspector's test connection
Water motor gong			X	(1) Operational test using inspector's test connection
Valve supervisory device			X	Test for conformance with NFPA 16 and/or *NFPA 72, National Fire Alarm Code*
Detection system	X	X	X	Operational test for conformance with NFPA 16 and/or *NFPA 72*
Status-Indicating Components				
Gauges			X	Verify at 0 psi (0 bar) and system working pressure
Testing and Maintenance Components				
Main drain	X	X	X	Full flow main drain test
Auxiliary drains	X	X	X	(1) Check for leaks at system working pressure
Inspector's test connection	X	X	X	(1) Check for leaks at system working pressure

TABLE 11.5.1 Continued

Component	Adjust	Repair/ Recondition	Replace	Required Action
Structural Components				
Hanger/seismic bracing	X	X	X	Check for conformance with NFPA 16 and /or NFPA 13, *Standard for the Installation of Sprinkler Systems*
Pipe stands	X	X	X	Check for conformance with NFPA 16 and /or NFPA 13
Informational Components				
Valve signs	X	X	X	Check for conformance with NFPA 16 and /or NFPA 13
Hydraulic placards	X	X	X	Check for conformance with NFPA 16 and /or NFPA 13

SUMMARY

Foam-water systems are highly specialized due to the complex nature of the equipment and chemicals they use. Inspection, testing, and maintenance of these systems, as well as the analysis of foam-water solution, require special equipment and skill. It should only be attempted by those possessing the training and knowledge of the functioning of these systems and equipment.

REVIEW QUESTIONS

1. Describe the methods permitted to determine foam concentration.
2. A foam-water spray system is designed to discharge 750 gpm (2839 L/min) with a 3 percent protein foam-water solution. What is the total quantity of foam concentrate needed for the auxiliary supply?
3. An analysis of a foam sample indicates a 3.6 percent concentration of foam in the solution. The system acceptance test indicates a 3.2 percent concentration. Is the test sample acceptable?

REFERENCES CITED IN COMMENTARY

National Fire Protection Association, 1 Batterymarch Park, Quincy, MA 02169-7471.

NFPA 11, *Standard for Low-, Medium-, and High-Expansion Foam*, 2005 edition.
NFPA 13, *Standard for the Installation of Sprinkler Systems*, 2007 edition.
NFPA 15, *Standard for Water Spray Fixed Systems for Fire Protection*, 2007 edition.
NFPA 16, *Standard for the Installation of Foam-Water Sprinkler and Foam-Water Spray Systems*, 2007 edition.

Water Mist Systems

CHAPTER 12

Chapter 12 is new to the 2008 edition of NFPA 25 and covers the inspection, testing, and maintenance of water mist systems. Water mist systems are now widespread in the marine sector as alternatives to marine sprinkler systems on passenger cruise ships. Water mist is used in heavy industry for protecting turbines, generators, and process equipment, including industrial deep fat frying equipment. Increasingly, water mist systems are being used in hotels, heritage buildings, and offices as alternatives to standard sprinkler systems or for special hazards such as computer room subfloors. As for any fire protection system, water mist systems require reliable, competent inspection, testing, and maintenance to ensure they will perform properly if a fire occurs.

Water mist systems utilize different types of hardware and control concepts not previously used in water-based fire protection systems. Some system configurations involve components borrowed from gaseous agent extinguishing systems, such as compressed gas cylinders, pressure regulators, and pneumatic control valves. Others utilize positive displacement pumps and high pressure piping or tubing not previously used in fire protection. Many inspectors familiar with inspection, testing, and maintenance of fire pumps, sprinkler, and water spray or foam-deluge systems will be unfamiliar with some of the hardware used in a water mist system. Similarly, inspectors familiar with gaseous clean agent systems may not be familiar with pumps, controllers, filters, and water storage equipment. Special training by the manufacturer's representative is often necessary, especially where proprietary differences exist between water mist systems developed by different manufacturers.

◄ **FAQ**
Can water mist systems use the same components as a sprinkler system?

The requirements for inspection, testing, and maintenance of water mist systems in NFPA 25 are taken from Chapter 13 of NFPA 750, *Standard on Water Mist Fire Protection Systems,* 2006 edition. They must be supplemented by the specific requirements contained in each manufacturer's design, installation, operation, and maintenance (DIOM) manual, which is required for all listed or approved water mist system equipment.

12.1 Responsibility of the Owner or Occupant. [750:13.1]

12.1.1 General. [750:13.1.1]

12.1.1.1 The responsibility for properly maintaining a water mist fire protection system shall be the obligation of the property owner. [**750:13.1.1.1**]

12.1.1.2 By means of periodic inspection, tests, and maintenance, in accordance with the standard and manufacturers' requirements, either this equipment shall be shown to be in good operating condition or that defects or impairments exist. [**750:13.1.1.2**]

Some of the components of water mist systems are similar to other water-based extinguishing systems, such as ancillary fire and smoke detection and activation systems. Detection and activation systems are addressed by *NFPA 72*®, *National Fire Alarm Code*®. All other components specific to water mist systems, such as special pump assemblies, gas cylinders, control valves, nozzles, pipe and stainless steel tubing, hangers, supports, strainers and filters, and manual releases, are addressed in specific paragraphs in this chapter.

FAQ ▶
What is involved in inspecting a water mist system?

"Inspecting" a component means that a person who is knowledgeable about its intended function examines it, usually visually, to ascertain whether it appears to be in its normal state, and verifies that there is no obvious condition that could immediately or at a future time prevent it from performing as intended. If anything is noted that is out of the ordinary, a closer examination is warranted, and the condition should be corrected if necessary.

"Testing" involves actual operation of devices and is more involved than inspecting components. There are different levels of testing. Some components can be tested without involving other components in the system. For example, a fire pump can be started weekly to confirm that electric power or diesel fuel is available and that starting switches work. There are other tests that cannot be performed without involving most elements of the system. Testing a pressure regulating valve or an automatic sectional control valve may require releasing compressed gas cylinders or flowing water into overhead piping. In that case, significant expertise is required to drain the system, replace gas cylinders, and put the system back in operating order.

12.1.1.3 Inspection, testing, and maintenance activities shall be implemented in accordance with procedures meeting or exceeding those established in this document and in accordance with the manufacturer's instructions. [**750:13.1.1.3**]

The manufacturer of a water mist system must produce a design, installation, operation, and maintenance (DIOM) manual as part of the listing or approval documentation for the water mist system. The DIOM manual contains specific inspection, testing, and maintenance instructions for specialized or proprietary components or system features, which must be followed in addition to the requirements of this chapter. Subsection 11.4.3 in NFPA 750 requires that a copy of the manufacturer's DIOM manual be provided to the user upon completion of the installation. The DIOM manual is required to provide enough information to allow the user or a third party to verify that the system has been designed and installed properly.

12.1.1.4 These tasks shall be performed by personnel who have developed competence through training and experience. [**750:13.1.1.4**]

FAQ ▶
What qualifications are needed to service a water mist system?

The owner or occupant may employ a specialized contractor with the necessary training and experience to carry out the inspections. The inspection, testing, and maintenance service contractor should provide certificates issued by the water mist system manufacturer confirming the training of the employees/inspectors who will inspect, test, or provide maintenance on a water mist system.

12.1.2 Notification. [750:13.1.2]

12.1.2.1 The owner or occupant shall notify the authority having jurisdiction, the fire department (if required), and the alarm receiving facility before shutting down a system or its supply. [**750:13.1.2.1**]

Where a water mist system is interlocked with other mechanical systems such as automatic door closers, emergency power shutdown, stopping or starting of ventilation fans, opening or closing of automatic fire dampers, stopping of production lines, or other ancillary actions integrated in a cause and effect protocol, the relevant plant operations personnel should be informed of the temporary shutdown in addition to those mentioned in 12.1.2.1. Where such interlocks exist, the owner should ensure that the service contractor is familiar with the cause and effect protocol before beginning work on the system, to ensure that the temporary shutdown of the water mist system will not inadvertently interrupt other important processes.

12.1.2.2 The notification shall include the purpose of the shutdown, the system or component involved, and the estimated time needed. [**750:13.1.2.2**]

12.1.2.3 The authority having jurisdiction, the fire department, and the alarm receiving facility shall be notified when the system, supply, or component is returned to service. [**750:13.1.2.3**]

All affected parties, including the relevant plant operations personnel, should be notified after the inspection, testing, or maintenance activity is completed that the water mist system and all interlocks have been returned to normal operational status (see Chapter 15).

12.1.3 Correction or Repair. [750:13.1.3]

12.1.3.1 The owner or occupant shall promptly correct or repair deficiencies, damaged parts, or impairments found while performing the inspection, test, and maintenance requirements of this standard. [**750:13.1.3.1**]

If the inspection and testing activity identifies a deficiency or impairment in the system, the inspection, testing, and maintenance provider must bring the problem to the owner's attention. This should be done both by (1) verbally informing the owner, or a representative of the owner who has some authority to act, and by (2) submitting a signed inspection form with the deficiency clearly marked. The owner should take immediate action to restore the system to its intended level of operation in as short a time as possible.

◀ **FAQ**
How should the inspector notify the owner of any deficiency or impairment found in the system?

12.1.3.2 Corrections and repairs shall be performed by qualified maintenance personnel or a qualified contractor. [**750:13.1.3.2**]

Minor impairments may be repaired by the owner or owner's representative if they have relevant experience. Repairs requiring specialized knowledge of the components should only be undertaken by the manufacturer's representative with the required level of knowledge and skill.

12.1.4 System Reevaluation. [750:13.1.4]

12.1.4.1 The owner or occupant shall give special attention to factors that might alter the requirements for a continued approved installation. [**750:13.1.4.1**]

12.1.4.2 Such factors shall include, but shall not be limited to, the following:

(1) Occupancy changes
(2) Process or material changes
(3) Structural revisions such as relocated walls, added horizontal or vertical obstructions, or ventilation changes
(4) Removal of heating systems in spaces with piping subject to freezing [**750:13.1.4.2**]

The inspection, testing, and maintenance services provider is not required to confirm that the original design for a particular installation was appropriate for the hazard. However, information on the original design is provided on the "system design information sign" that is required by 11.4.4 in NFPA 750. The sign provides a description of the hazard protected that the maintenance service provider can check against the observed conditions. The service provider should also review the as-built installation drawings, which are required to be provided to the owner at completion of the installation (see 11.4.2 of NFPA 750). It is very important to note changes in ventilation, compartment size, and obstructions, which may significantly alter the ability of the water mist system to achieve its intended performance.

12.1.5 Changes of Occupancy. [750:13.1.5]

12.1.5.1 Where changes in the occupancy, hazard, water supply, storage arrangement, structural modification, or other condition that affect the installation criteria of the system are identified, the owner or occupant shall promptly take steps to evaluate the adequacy of the installed system to protect the hazard in question, such as contacting a qualified contractor, consultant, or engineer. [**750:13.1.5.1**]

12.1.5.2 Where the evaluation reveals a deficiency, the owner shall notify the insurance underwriter, the authority having jurisdiction, and the local fire department. [**750:13.1.5.2**]

12.1.6 Return to Service. [750:13.1.6]

12.1.6.1 Where a water mist system is returned to service following an impairment, it shall be verified that it is working properly. [**750**:13.1.6.1]

12.1.6.2 Chapter 12 of NFPA 750 shall be referenced to provide guidance on the type of inspection or test, or both, that is required. [**750**:13.1.6.2]

The inspection service provider is directed to Chapter 12 of NFPA 750, 2006 edition, which provides details on functionality testing of water mist systems. Subsection 12.2.5 of NFPA 750 describes preliminary functional tests to be performed to confirm that detection devices, supervisory devices, automatic and manual releasing devices, and pneumatic and hydraulic components are functional. Auxiliary functions such as alarm functions, remote annunciators, air handling shutdown, and power shutdown are checked for operation in accordance with the system requirements. Depending on the nature of the repair that was performed to remediate the deficiency or impairment, some or all of the functional and operational tests of the system described in Chapter 12 should be performed.

12.2 Inspection and Testing. [750:13.2]

12.2.1 Components and Systems. [750:13.2.1]

12.2.1.1 All components and systems shall be inspected and tested to verify that they function as intended. [**750**:13.2.1.1]

12.2.1.2 Water mist systems that are equipped with an additive system shall be tested with the specific additive system engaged or used during the acceptance testing. [**750**:13.2.1.2]

Some marine water mist systems utilize AFFF additive for the bilge area nozzles. The AFFF foam solution storage and injection system is typically supplied by pressurized nitrogen gas, and special valves must open in order to propel the foam solution into the water stream. The requirement in 12.2.1.2 recognizes that the additive feed line is an essential part of the system and must be included in any functional or operational tests. The actual additive, not a substitute, must be used. The testing must confirm that the correct quantity of additive is injected at the correct rate. Details of the intended operation of all elements of the system, including the additive injection system, are described in the "written sequence of operations" specified in 11.4.5 of NFPA 750, which must be submitted with the system documentation.

12.2.2 Requirements.

The components of typical water mist systems to be inspected and tested are provided in Table 12.2.2. [**750**:13.2.2]

Table 12.2.2 is a detailed list of the types of components that may be included in water mist systems. Not all of the components listed are found with every water mist system. The table indicates the number of times per year each component should be inspected or tested to the degree indicated. The manufacturer's DIOM manual must be consulted to determine if more frequent inspection or testing is recommended.

In Table 12.2.2, each section or grouping of components is listed followed by additional explanatory information. Schematic diagrams in Exhibits 12.1, 12.2, and 12.3 are provided to illustrate different types of components and arrangements typically encountered in water mist systems.

TABLE 12.2.2 Maintenance of Water Mist Systems

Item	Task	Weekly	Monthly	Quarterly	Semi-annually	Annually	Other
Water supply (general)	Check source pressure.			X			
	Check source quality (*first year).				X*	X	
	Test source pressure, flow, quantity, duration.					X	

A reliable water source is essential for any water-based fire protection system. Chapter 10 of NFPA 750 requires that the water supply be capable of supplying the largest single hazard or group of hazards to be protected simultaneously for a specified duration. An annual water flow test is required to confirm flow rate and pressure. Where the water supply is taken from a reliable large water supply system, testing for "quantity" may not be required. Where the water supply is taken from a potentially limited source such as a tank, the quantity of water available should be checked. The quantity available must be sufficient to meet the design flow rate for the specified duration of protection. The basic requirement in NFPA 750 is for 30 minute duration of protection. That may be relaxed for pre-engineered systems. If the minimum duration intended for the system is less than 30 minutes, it should be identified on the "system design information sign" required in 11.4.4 of NFPA 750 or in other system documentation. The list shown in 11.4.4 in NFPA 750 does not include "minimum duration of protection." The intent of the committee was to ensure that all pertinent design information be displayed on the sign. If the design duration of protection is not indicated on the sign, the as-built design documents and acceptance test forms should be consulted.

The water supply must be taken from a source that is equivalent in quality to potable water with respect to bacteria, particulates, and dissolved solids or from a source of natural seawater. Ships may utilize natural seawater after the potable water supply has been depleted. As ships enter waters of different quality in different ports, however, bacteria-free and silt-free seawater is not always available. The high chloride content of seawater would be damaging to stainless steel components, so switching to seawater would shorten the service life of such piping systems.

Checking water quality requires sending at least one sample of water to a water quality testing laboratory. At a minimum, tests for turbidity, algae, *Legionella* bacteria, *E. coli* bacteria, and chloride content should be conducted during the first semi-annual inspection. For piping systems in stainless steel, the chloride content of water should not exceed 50 ppm. The need for laboratory testing may be reduced in future inspections if the quality of the source water is found to be acceptable over at least two consecutive semi-annual tests.

◀ **FAQ**
What types of water quality tests should be conducted during the first semi-annual inspection?

It is advised to check the quality of the source water semi-annually during the first year. If no adverse conditions are noted, it is acceptable to extend water quality testing to annual frequency. If adverse conditions are found, remedial measures should be taken and the frequency of quality testing increased to monthly until it is assured that the remedial measures are working.

Item	Task	Weekly	Monthly	Quarterly	Semi-annually	Annually	Other
Water storage tanks	Check water level (unsupervised).	X					
	Check water level (supervised).				X		
	Check sight glass valves are open.			X			
	Check tank gauges, pressure.				X		
	Check all valves, appurtenances.					X	
	Drain tank, inspect interior, and refill.					X	
	Inspect tank condition (corrosion).					X	
	Check water quality.					X	
	Check water temperature						Extreme weather

(continues)

TABLE 12.2.2 *Continued*

Item	Task	Weekly	Monthly	Quarterly	Semi-annually	Annually	Other
	The owner's primary maintenance concerns with the water storage tank should be to confirm that (1) it contains the right quantity of water, (2) the water quality is acceptable, and (3) the discharge valves are open. It is also important to confirm that the ambient temperature at the tank location is within the acceptable range of 40°F to 130°F (4°C to 54°C) and that there are no signs of leakage or corrosion.						
	For water storage tanks that are not supervised by means of a low water level switch connected to an alarm panel, the water level in the tank must be verified visually weekly. A sight glass or other means of quickly verifying water level in a tank is required. If a tank is supervised by low water level switches connected to an alarm panel, a quarterly inspection will suffice.						
	The quality of the source water used to fill a tank may be more than adequate, but the quality of stored water must be carefully monitored. The list above for storage tanks indicates that water quality (in tanks) should be checked "annually." As indicated previously for the water source, a semi-annual check should be made during the first year at least. It is not advisable to wait for a full year to evaluate whether the quality of the water in the storage tank is deteriorating. The two greatest concerns with respect to quality of stored water are bacteria and algae growth. Algae will plug screens and small nozzle orifices; bacteria such as *Legionella* may cause illness.						
	A visual check for algae growth and discoloration may be enough to identify the presence of fouling substances. However, where people could be exposed to the water mist taken from stored water, it is advisable to send water samples to a laboratory to test for *Legionella* and *E. coli* bacteria. *Legionella* bacteria have been identified in some water storage tanks for water mist systems, and ultraviolet (UV) antibacterial systems have been installed as a remedial measure. Efficient UV devices can be installed inside each storage tank, which eliminates the need for re-circulation piping. If present, ancillary devices associated with the tank must be included in the inspection, testing, and maintenance task list.						
	Paragraph 10.5.1.5.3 of NFPA 750 requires that a filter or strainer be supplied on the downstream (i.e., the system side) of any reservoir or break tank with an air-water interface greater than 11 ft² (1 m²). For a circular tank, that corresponds to a tank diameter of approximately 4 ft or 1.2 m. (See Exhibit 12.1 on p. 231.) The objective of the strainer was to remove any algae that might grow in the tank before it enters the water mist system piping. The NFPA 750 committee believed that this was a concern only in tanks with a "large" air-water interface. The listed water mist systems currently on the market, which utilize a water tank, ensure that water going into the tank is filtered. However, if algae grow in the water once inside the tank, fibrous material will clog valves and nozzles in the water mist system if discharged. Notwithstanding the limit in NFPA 750, the potential for algae growth in stored water may exist in tanks of diameters less than 4 ft (1.2 m).						
	An inspector should be cognizant of the need to check for the presence of algae in stored water. A sample of water [e.g., approximately 1 quart (1 liter)] should be removed via the tank clean-out access and examined under good lighting for the presence of algae. A water sample should be sent to a laboratory for bacterial analysis, turbidity, and chloride tests.						
	As indicated under preventive maintenance, once per year, the tank must be completely emptied, cleaned, inspected for corrosion, and refilled with filtered water.						
Water storage cylinder (high pressure)	Check water level (load cells).					X	
	Check water level (unsupervised).			X			
	Check support frame/restraints.					X	
	Check vent plugs at refilling.					X	
	Check cylinder pressure on discharge.					X	
	Inspect filters on refill connection					X	

TABLE 12.2.2 Continued

Item	Task	Weekly	Monthly	Quarterly	Semi-annually	Annually	Other

Water storage cylinders are modified compressed gas cylinders, with a special head assembly to allow water to enter and air to escape from the cylinder for refilling. (Exhibit 12.2 on p. 232 is a schematic diagram of a water mist system utilizing water storage cylinders.) The cylinders are typically lined with a corrosion-resistant coating. Refilling is done by connecting a fill line to an inlet and opening a vent port. The cylinder is full when water starts to flow out of the open vent port. Only an experienced technician should perform the filling operation. A permanent water supply connection with filter and hose valve should be mounted near the unit to facilitate refilling with filtered water. The NFPA 750 committee believes that concern about algae growth in water stored in cylinders is minimal because the air-water interface in a full cylinder is very small.

Some manufacturers provide a dipstick on each water-storage cylinder to allow an inspector to easily verify that the tanks are full. However, with some water mist systems, once the cylinders are filled and the vent ports sealed, it is not possible to visually confirm that water storage cylinders are full of water. A written record of who last filled the cylinders must be maintained at the cylinder station. If the person doing a monthly or quarterly inspection is not the person who last filled the system, he or she must rely on the written record as the only evidence that the cylinders were at least full of water at some time in the recent past.

◀ **FAQ**
How can it be determined whether the water storage cylinders are full?

Some end users require placing the cylinder rack on load cells to continuously monitor the weight of the cylinders. This is the best way to supervise water level in such closed cylinders.

Item	Task	Weekly	Monthly	Quarterly	Semi-annually	Annually	Other
Additive storage cylinders	Inspect general condition, corrosion.			X			
	Check quantity of additive agent.				X		
	Test quality of additive agent.					X	
	Test additive injection, full discharge test.					X	

AFFF is the most commonly used additive with water mist systems. It is used in some marine systems for bilge nozzles. Inspection, testing, and maintenance procedures for reservoirs of AFFF solution should comply with Chapter 11 of NFPA 25. Specialized water mist systems may utilize additives to prevent freezing or to enhance extinguishing effectiveness. For the most part, these are small pre-engineered systems for which the additive is pre-mixed with water in the water storage container. Inspection, testing, and maintenance for such systems should follow the manufacturer's recommendations.

Item	Task	Weekly	Monthly	Quarterly	Semi-annually	Annually	Other
Water recirculation tank	Check water level (unsupervised).		X				
	Check water level (supervised).				X		
	Inspect supports, attachments.					X	
	Test low water level alarm.					X	
	Check water quality, drain, flush and refill.					X	
	Test operation of float operated valve.					X	
	Test pressure at outlet during discharge.					X	
	Test backflow prevention device (if present).					X	
	Inspect and clean filters, strainers, cyclone separator.					X	

(continues)

TABLE 12.2.2 Continued

Item	Task	Weekly	Monthly	Quarterly	Semi-annually	Annually	Other
	Water re-circulation tanks are also referred to as "break tanks" and are commonly used on marine systems. The break tank is typically built into the frame of the pump assembly with direct connections to the suction inlets of the pumps. The water level is usually supervised in that if the water level drops to a certain level, the "float-operated valve" will open and allow water from a permanent source to flow into the tank. It is preferable to have an additional low-level alarm for supervision of water level so that one is alerted to declining water level without waiting for the refill switch to activate. The inspection and testing must include the water supply to the break tank, which may include a backflow prevention device and/or a seawater connection with cyclone separator. The water supply to the break tank must be tested for capacity — the minimum flow rate must equal or exceed the maximum output capacity of the pump assembly. The quality of the water in the break tank should be checked more frequently than indicated in Table 12.2.2. If algae grow in the water, there is no strainer on the outlet from the tank to the suction inlet of the pumps. Fibrous algae material may pass through the pumps and into the water mist system piping, where it will clog vital components. The possibility of algae growth may be reduced by frequent changeover of the water in the tank, such as occurs with weekly startup of the pump assembly. Otherwise, fixed UV antibacterial units can be installed directly in the break tank.						
Compressed gas cylinders	Inspect support frame and cylinder restraints.			X			
	Check cylinder pressure (unsupervised).	X					
	Check cylinder pressure (supervised).			X			
	Check cylinder control valve is open.	X					
	Check cylinder capacity and pressure rating.					X	
	Check cylinder compliance specification.					X	
	Confirm compressed gas meets specifications (moisture, cylinder pressure).					X	
	Hydrostatic test cylinders						5–12 years

Exhibits 12.1 and 12.2 illustrate two types of water mist systems that utilize compressed gas cylinders as the propellant or as the atomizing medium. Exhibit 12.1 shows a twin-fluid system, which includes a pressure-reducing valve to create a high pressure side and a low pressure side. The compressed air is reduced from typically 2175 psi (150 bar) to 116 psi (8 bar), which drives the water out of the water storage tank (acts as propellant), and also goes to the air-inlet of each of several twin-fluid nozzles (to act as the atomizing medium). The twin-fluid system is a low pressure water mist system; therefore, the water storage tank is not rated for high pressure. A similar arrangement is used for single-fluid low or intermediate pressure water mist systems. Typical applications include pre-engineered systems for turbine enclosures and machinery spaces.

Exhibit 12.2 represents a decaying pressure, high pressure water mist system. The compressed gas cylinder pressure is typically between 2175 and 2900 psi (150 and 200 bar), and the pressure is not regulated to a lower pressure. The compressed gas forces the water out of the high pressure water storage cylinders and into the water mist system piping.

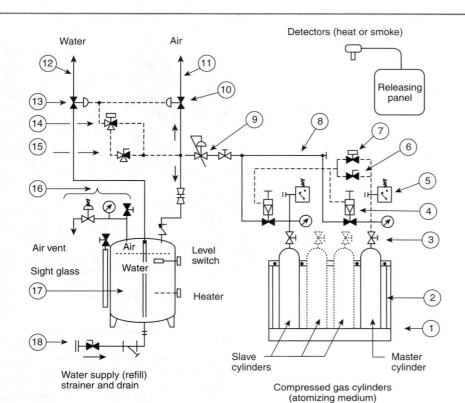

EXHIBIT 12.1 *Gas-Driven System with Water Stored in Tank with Air-Water Interface Greater than 10.8 ft² (1 m²).*

1. Steel base and frame
2. Compressed gas cylinders (atomizing medium): master and slave cylinders
3. Cylinder control valve
4. Pneumatic cylinder release valve
5. Pressure supervisory switch with burst disc
6. Manually operated master release valve
7. Solenoid operated master release valve
8. High pressure tubing manifold
9. Air pressure control valve (high to low pressure)
10. Air-actuated globe valve
11. Air line to twin-fluid nozzles (low pressure)
12. Water line to twin-fluid nozzles (low pressure)
13. Air-actuated globe valve (cycle water line)
14. Low pressure solenoid valves (for operating air-actuated globe valves)
15. Manual release valve (opens globe valves)
16. Pressure gauge, pressure relief valve, and vent valve
17. Low pressure rated water tank
18. Drain and refill connection with strainer

TABLE 12.2.2 Continued

The pressure in the system piping typically starts at about 1450 psi (100 bar) and declines over approximately 10 minutes to about 290 psi (20 bar). Typical applications for a high pressure water mist system include turbine enclosures and machinery spaces.

Gas-driven water mist systems involve components not previously used with water-based fire protection systems. There are "master" and "slave" pneumatic releasing valves that are interconnected by compressed gas tubing. If one accidentally activates a "slave" valve instead of a "master" valve, it is possible that not all cylinders will discharge. The testing of compressed gas cylinders and their interconnections should only be performed by technicians with specialized training and experience with the particular equipment involved.

(continues)

EXHIBIT 12.2 Pre-Engineered, Gas-Driven Single Fluid High Pressure Water Mist System with Water Stored in High Pressure Cylinders.

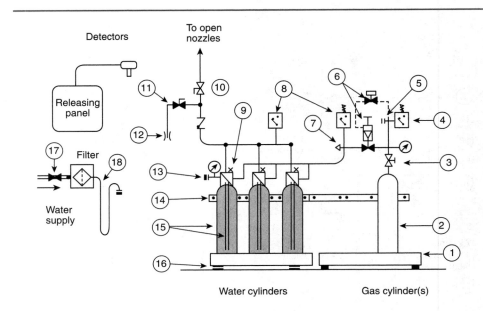

1. Steel base and frame
2. Compressed gas cylinders (driving medium)
3. Cylinder control valve
4. Pressure switch, supervise cylinder pressure
5. Burst disc
6. Solenoid operated master release valve
7. Micro-leakage valve
8. Pressure switches, alarm if system trips
9. Vent port, for filling water cylinders (fill until water discharges from open port)
10. Primary system or sectional control valve
11. Test connection and drain
12. Test orifice (alternative to full discharge)
13. Cylinder discharge header with filling port
14. Cylinder rack with restraints
15. Pressure-rated water = 150 cylinders with dip tube
16. Optional load cells to confirm water level in cylinders
17. Water supply valve, normally closed
18. Filter and hose with adaptor fitting for filling

TABLE 12.2.2 Continued

Item	Task	Weekly	Monthly	Quarterly	Semi-annually	Annually	Other
Plant air, compressors, and receivers	Check air pressure (unsupervised).	X					
	Check air pressure (supervised).		X				
	Start compressor.	X					
	Check compressor/receiver capacity, changes.				X		
	Check compressed air moisture content.					X	
	Clean filters, moisture traps.				X		
	Test full capacity, duration, and any changes in other demands.					X	

In the author's experience, systems that utilize plant air are small systems, such as a twin-fluid system for wet benches or fume hood protection. Generally, it is preferable to have a dedicated air supply, which can be better controlled than a compressed gas system serving other purposes. It is difficult for a fire protection inspection, testing, and maintenance service provider to have access to test plant air compressors. The full capacity of the plant supply will have been verified during the original system acceptance tests. The plant supply should be inspected annually to confirm that assumptions about the plant capacity are still valid (e.g., that

TABLE 12.2.2 Continued

Item	Task	Weekly	Monthly	Quarterly	Semi-annually	Annually	Other
there are no new demands or operational constraints). A means of confirming the capacity of the air supply at the point of connection to the water mist system should be devised.							
Pumps and drivers	Inspection, testing, and maintenance shall be in accordance with the requirements of NFPA 20, *Standard for the Installation of Stationary Pumps for Fire Protection*, and NFPA 25.	X	X	X	X	X	

Large water mist systems used to protect entire passenger ships, or as equivalent sprinkler systems in buildings, may utilize a pumped water supply. The inspection, testing, and maintenance procedures for standard fire pumps, as provided in Chapter 8 of NFPA 25, generally apply equally well to water mist systems involving pumps. Some additional requirements or cautions, however, should be observed for water mist systems that utilize positive displacement pumps.

The challenge in using positive displacement pumps for fire protection systems with widely varying demands is to provide a means of bypassing pump water flow that cannot be discharged from the number of nozzles open. The operation of flow-bypass valves, called "unloader" valves, is critically important to the operation of the system. If only a few thermally activated water mist nozzles operate, the pump assembly must deliver the proper flow to those nozzles, but bypass the unused portion of the flow. If many nozzles are opened by heat, such that the entire output from the pump assembly is required, the unloader valves should not allow any water to flow to the bypass. The operation of the pump assembly for both small and maximum demands will have been verified at the acceptance test. It is very important not to disturb any of the settings on the unloader valves in the course of the annual pump testing.

Exhibit 12.3 is a schematic of one type of high pressure pump assembly for an engineered water mist system. It consists of an assembly of two electric motors, each driving two small piston pumps. The total capacity of the pump assembly is the sum of the output of each pump. If there are not enough nozzles opened on the system to flow the full capacity of the assembly, the system pressure will rise to the point of causing one or more unloader valves to divert the pump flow to the bypass line. The pressure at which an unloader valve opens will be pre-set at the time of the system design. Typical applications for this pump assembly include passenger ships, multi-zone industrial systems, and sprinkler-equivalent systems in buildings.

As shown by the tag numbered 9 in Exhibit 12.3, a test connection should be provided on the discharge manifold. The test connection should include a pre-designed orifice device that will allow the full flow of the pump at the desired backpressure. Since the unloader valves respond to the system backpressure, it will not be possible to verify the proper operation of all contributing pumps without approximately matching the corresponding system pressure. A throttling valve can be installed on the discharge line to allow one to vary the backpressure in the system.

Some high pressure pump assemblies do not have unloader valves installed on each pump, but instead have an equivalent large volume pressure relief valve on the discharge manifold. See the valve identified as tag number 8 in Exhibit 12.3. Before conducting annual flow testing, the inspector must understand the intended operation of all pressure or flow control devices.

(continues)

EXHIBIT 12.3 *Engineered Water Mist System with Positive Displacement Pump Assembly That Has Unloader Valves on Individual Pumps and Sectional Control Valves for Separate Zones.*

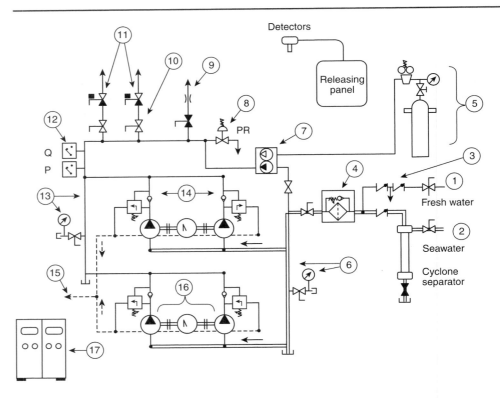

1. Fresh water supply
2. Seawater supply with cyclone separator (marine systems only)
3. Backflow device (optional)
4. Filters or screens with bypass
5. Air-supply and regulator for standby pressure maintenance pump
6. Suction manifold with NPSH gauge (+/−)
7. Pneumatic standby pressure pump
8. Pressure relief valve
9. Test connection with flowmeter
10. Isolation valve for sectional valves
11. Solenoid actuated sectional control valves
12. Control sensors, pressure (P) and flow (Q), connected to programmable controller
13. Discharge manifold with pressure gauge
14. Unloader valve (one per pump)
15. Unloader valve discharge bypass line (to drain)
16. Positive displacement pumps (2 per motor)
17. Programmable pump controllers in bank (one per motor)

TABLE 12.2.2 Continued

Note in Exhibit 12.3 that the fire pump controller is shown as a pair of controllers. The FM listing for the pump assembly requires one controller per electric motor (as per NFPA 20, *Standard for the Installation of Stationary Pumps for Fire Protection*). The controller is usually programmed to start or stop individual motors according to the demands on the system. Pressure and flow sensors are provided on the pump assembly, which are utilized in the system control logic program, to determine if another motor is needed or if one of several motors can be shut down. The programmable features are very different from a standard fire pump installation, which usually involves only one motor and one pump.

The waterflow test results from the annual pump test need to be compared to the flow test results obtained during the acceptance testing of the system. This is relatively straightforward for low and intermediate pressure systems that utilize centrifugal-type pumps. As is indicated in Chapter 8 of NFPA 25, the usual practice is to plot at least three points from the

TABLE 12.2.2 Continued

flow test on a pressure versus flow plot, subtract the water supply pressures from the test pressures, and draw a smooth curve through the resulting points. The smooth curve should lie on or above the "factory" pump curve. If it does not, one concludes that some impairment has occurred.

The situation is more complex for a positive displacement pump or pump assembly. One reason is because the unloader valves or pressure relief valve on the manifold are usually set at less than the maximum possible pump pressure and there may be at least a 5 percent variation in setting of individual unloader valves. The pump curve for the output of an assembly of positive displacement pumps, with all pumps contributing, will invariably be less than the presumed unloader setting, and the flow will be less than the theoretical sum of the individual pumps. Exhibit 12.4 illustrates the theoretical pump curve for a positive displacement pump assembly with eight pumps and eight unloader valves set nominally at 1597 psi (110 bar).

The key to determining whether the pump assembly is operating as it should is to ensure that the pressure achieved at maximum flow is within 5 percent of the setting of the pressure relief valve or highest unloader valve setting, *and* the total flow is within 10 percent of the theoretical maximum flow rate, as shown in Exhibit 12.4.

◀ **FAQ**
What is the key determining factor as to whether the pump assembly is operating properly?

Exhibit 12.4 illustrates one flow test comparison with the acceptance test result. During the acceptance test, a maximum flow/pressure result of 9 gpm at 1552 psi (360 L/min at 107 bar) was accepted. That was 10 percent less flow than the theoretical maximum of 10 gpm at 1552 psi (400 L/min at 107 bar), and 1552 psi (107 bar) was within ±5 bar precision of the

(continues)

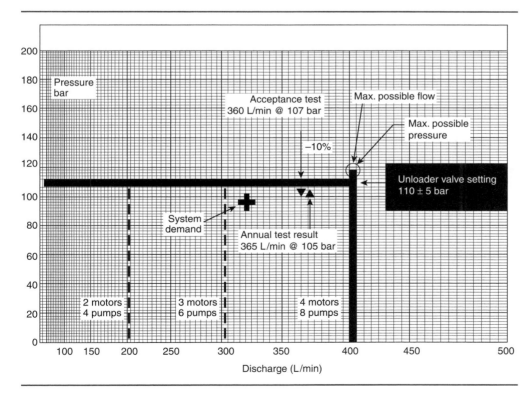

EXHIBIT 12.4 *Composite Pump Curve for a Four-Motor, Eight-Pump Assembly, with All Eight Unloader Valves Set at 1595 psi (±72.5 psi) [110 bar (±5 bar)].*

TABLE 12.2.2 Continued

Item	Task	Weekly	Monthly	Quarterly	Semi-annually	Annually	Other
	unloader valves. During the annual flow test, a maximum flow rate of 9 gpm (365 L/min) was recorded, at a pressure of 1524 psi (105 bar). This was deemed to be an acceptable condition, since the flow was slightly greater than the acceptance test, and within 10 percent of the maximum; and the pressure was within the ±5 bar precision of the unloader valve settings.						
	Note that with high pressure pumps operating in the vicinity of 1450 psi (100 bar), the supply water pressure [typically less than 30 psi (2 bar)] does not show up on the scale. The pressure at the suction inlet to the pumps is nevertheless important. Positive displacement pumps are more sensitive to net positive suction head (NPSH) problems than centrifugal pumps. The need to filter all water going to the system adds hydraulic resistance to the suction inlet line. If filters are present on the suction line, friction losses in the suction supply may become too high. A suction booster pump may in some cases be used to increase the suction line pressure to overcome friction losses through the filters. Starting of the positive displacement pumps must then be preceded by starting the suction booster pump.						
	It is important that filters be able to pass the maximum system demand flow rate for an extended period of time, without clogging. Unlike with the manual-fill systems suitable for pre-engineered systems (see Exhibit 12.2), which only need to pass the flow needed to refill a reservoir in a reasonable period of time, the filters on the suction supply to a pumped system have to be sized for the maximum system flow rate. In Exhibit 12.3, a spring-loaded check valve is placed on a bypass around the filters (tag number 4). If the filters become plugged, the bypass line should automatically open to allow flow of unfiltered water to the pumps. This must be tested during the annual discharge test.						
	Some high pressure pump unit suppliers install a gauge on the suction side of the pumps that is not suitable for reading in the range of suction line pressures. This must be corrected by installing a new gauge with the appropriate scale to read the NPSH. At the other end of the spectrum, another pump system manufacturer uses two banks of filters in parallel, with a pressure transducer connected to the pump controller. If the head loss across the first bank of filters becomes too high (as the filters become clogged), a solenoid valve opens and reroutes the flow through the redundant bank of filters.						
	The maintenance plan for a water mist pump system must be designed to evaluate the specific details of the suction supply to the pumps. It is recommended to install temporary supplemental instrumentation to measure performance factors during the annual discharge test.						
Standby pump	Inspect moisture trap, oil injection (pneumatic).				X		
	Check compressed gas supply, inlet air pressure.		X				
	Check outlet water (standby) pressure.		X				
	Test start/stop pressure settings for standby pressure.				X		
	Inspection, testing, and maintenance tasks for standby pumps on low or intermediate pressure water mist systems are covered in Chapter 8 of NFPA 25. High pressure water mist systems may utilize a pneumatic pump to maintain the system piping at an elevated pressure [e.g., 360 psi (25 bar)]. Inspection and testing of such pumps may not be covered in Chapter 8. Tag number 7 in Exhibit 12.3 shows the symbol for a pneumatic standby pump with dedicated cylinders of compressed gas. The pneumatic pump activates on a signal from the pump controller panel (tag number 17), which receives pressure and flow information from the switches						

Section 12.2 • Inspection and Testing

TABLE 12.2.2 Continued

Item	Task	Weekly	Monthly	Quarterly	Semi-annually	Annually	Other
	identified by tag number 12 in the exhibit. It is important to check the compressed gas supply and outlet pressure on a monthly basis. The owner's personnel can monitor the standby pump on a monthly basis, but the inspection, testing, and maintenance service provider should perform a test of the unit during the annual servicing.						
Pneumatic valves	Check cylinder valves, master release valves.		X				
	Inspect all tubing associated with release valves.				X		
	Test solenoid release of master release valve.					X	
	Test manual release of master release valve.					X	
	Test operation of slave valves.					X	
	Reset all pneumatic cylinder release valves.					X	
	Test on-off cycling of valves intended to cycle.					X	

Exhibits 12.1 and 12.2 show compressed gas cylinders used for two types of gas-driven water mist systems. Special pneumatic valves are released by a solenoid valve to activate the systems. On some systems, there is a master valve on the first cylinder, which is electrically released on a signal from the releasing panel. A pneumatic line then connects the first releasing valve to additional valves on the remaining cylinders; these are referred to as "slave" cylinders.

The integrity of pneumatic tubing connecting master and slave valves is seldom or never supervised. For that reason, pneumatic valves should be inspected carefully for possible problems with the control tubing. Generally, it can only be determined to function properly at the time of an annual trip test of the system.

Some water mist systems are designed to "cycle" flow on and off for specified periods of time. Cycling flow may be a means of passing the plate cooling test for the FM Global approval for water mist systems for turbine enclosures. For example, the system may be designed to discharge water for 60 seconds, stop for 40 seconds, and then resume discharging water mist for another 60 seconds. The cycle is repeated for the intended duration of protection, which is usually 10 minutes.

The total volume of water required is based on the total number of minutes of actual discharge, not the elapsed time since discharge.

◄ FAQ
How should the total volume of water required for 10 minutes of protection be calculated for a cycling system?

Under ideal conditions, cycling systems can extinguish hydrocarbon fires in enclosures more quickly and with less total water usage than continuous discharge systems. They rely, however, on the confinement of heat and water vapor inside the enclosure to assist in extinguishing fire. During each "off" cycle at the start of discharge, a hydrocarbon fire may regrow and increase the temperature in the enclosure and hence the amount of water vapor that can be held in the vapor state. By the third or fourth cycle, the enclosure is warm enough to sustain a high water vapor concentration, which reduces the oxygen concentration and results in extinguishing the fire. Therefore, all ventilation to an enclosure must be interlocked to stop on activation of the water mist system. It is also very important to verify that the enclosure integrity is maintained for such systems.

(continues)

TABLE 12.2.2 Continued

Item	Task	Weekly	Monthly	Quarterly	Semi-annually	Annually	Other
	The timing of the on and off cycles is controlled by a system control panel and specialized solenoid valves. The operation of the timing circuit should be tested during the annual inspection.						
System control valves	Inspection, testing, and maintenance shall be in accordance with the requirements of NFPA 25.	X	X	X	X	X	
	As on any water-based fire protection system, primary control valves should all be supervised. Valves identified by tag number 10 in Exhibit 12.3 are lever-handled ball valves. They can be visually determined to be "open," but a means of supervising them is usually devised. This may be done utilizing a standard tamper switch arranged to indicate a movement of the lever handle. Zone control valves operate in a manner similar to deluge valves on sprinkler and water spray systems. On high pressure systems, zone control valves are specially designed to be able to hold back high pressure water, but open the valve with a small solenoid valve typical of standard deluge systems (e.g., 1 amp maximum draw). See valves identified by tag number 11 in Exhibit 12.3. Special steps are required to ensure that such valves are properly set to be released upon a signal from the releasing panel. During the annual inspection and testing, all solenoid valves and zone control valves must be tested operationally. It is particularly important to test the sectional control valves used in large water mist systems. The valves are electrically activated. They may be non-indicating, and it may not be clear how to open them manually. Even during a trip test it is not possible to tell by visual inspection whether they are fully or partially open. On the high pressure systems, friction loss through sectional control valves may be very high. In order to keep friction losses within reasonable bounds, the total flow to a system may be divided among two or more sectional valves. This introduces the possibility that one of the valves in the group could fail to open, and the hydraulic performance of the system will be adversely affected. The only way to test sectional control valves is to open them under full pressure and flow water. In facilities where a full flow test can be conducted annually, the performance of sectional valves can be easily tested. If it is not possible to do a full system discharge, an alternative means of testing the functionality of the valves must be found. It may be necessary to install a test line on each individual zone in a water mist system, on the system (downstream) side of each sectional control valve so that the valve can be tested.						
Control equipment	Inspection, testing, and maintenance shall be in accordance with the requirements of *NFPA 72, National Fire Alarm Code*.						
	With the exception of water mist systems that are sprinkler equivalents, and use thermally activated (automatic) nozzles, water mist systems rely on a separate detection system to activate. Inspection and testing procedures, therefore, must include the detection system. The type of detection is often dictated by the listing for the system. For example, turbine enclosures and machinery spaces listed by FM are required to use thermal detectors. The listing also indicates that certain shutdown events must be synchronized with the release of the system. The ventilation system should be shut down, door holders released, dampers closed, and						

TABLE 12.2.2 Continued

Item	Task	Weekly	Monthly	Quarterly	Semi-annually	Annually	Other
lubrication or fuel lines shut off. Obviously, the annual testing of a water mist system must confirm the proper operation of all events associated with release of the water mist system. Where a cause and effect protocol has been developed to identify how the water mist system integrates with other process and safety operations, the protocol should be available to the water mist system inspector.							
Water mist system piping and nozzles	Inspection, testing, and maintenance shall be in accordance with NFPA 25. Inspect sample of nozzle screens and strainers (see 10.3.7).	X	X	X	X	X	After discharge
Enclosure features, interlocks	Inspect enclosure integrity.				X		
Ventilation	Test interlocked systems (e.g., ventilation shutdown). Test shutdown of fuel/lubrication systems.					X X	

[**750**: Table 13.2.2]

It may be necessary to coordinate the annual inspection, testing, and maintenance of the water mist system with other shutdown activity, so that access may be gained to protected spaces for inspection of the piping and nozzles. For example, it is not normally possible to enter a turbine enclosure while it is running.

12.2.3 Frequencies.

The frequency of inspections and tests shall be in accordance with Table 12.2.2 or as specified in the manufacturer's listing, whichever is more frequent. [**750**:13.2.3]

12.2.4* Restoration.

Following tests of components or portions of water mist systems that require valves to be opened or closed, the system shall be returned to service, with verification that all valves are restored to their normal operating position, that the water has been drained from all low points, that screens and filters have been checked and cleaned, and that plugs or caps for auxiliary drains or test valves have been replaced. [**750**:13.2.4]

A.12.2.4 If differences indicate a significant change or deterioration in performance, appropriate maintenance actions should be taken to restore the component or system to its original performance. [**750**:13.2.4]

The tasks involved in inspection, testing, and maintenance of the water mist system may consume some of the reserve water supply, compressed gas, or additive. Water filters may need to be replaced following inspection or water flow testing. Sufficient quantities of such consumables must be assembled prior to beginning work on the water mist system, so that the system can be returned to service without delay.

12.2.5 Specialized Equipment.

Specialized equipment required for testing shall be in accordance with the manufacturer's specifications. [**750**:13.2.5]

Specialized equipment may include flowmeters, orifice tips, and pressure gauges in order to conduct water flow testing of the pump system. Electrical instruments may be required to measure electrical conditions. Since water mist systems may be low pressure or high pressure, the equipment to be used on the system must be properly rated for the operating conditions. Even on a high pressure water mist system, gauges for the suction side of a pump inlet are typically low pressure gauges and should be graduated in appropriately fine increments to provide useful readings.

12.2.6 High Pressure Cylinders.

High pressure cylinders used in water mist systems shall not be recharged without a hydrostatic test (and remarking) if more than 5 years have elapsed from the date of the last test. Cylinders that have been in continuous service without discharging shall be permitted to be retained in service for a maximum of 12 years, after which they shall be discharged and retested before being returned to service. [**750**:13.2.6]

Safe procedures for handling compressed gas cylinders must be observed at all times. High pressure compressed gas cylinders present the opportunity for a serious accident to occur. Problems are particularly likely to occur, for example, when replacement cylinders are being maneuvered into position and reconnected after a trip test. Once in place, safety chains or other substantial restraints must be secured. In some facilities, racks of compressed gas cylinders have been placed too close to walls or other equipment, so that it is difficult to remove cylinders or move new ones into place. Ongoing maintenance is greatly facilitated if access to the cylinder rack is unrestricted.

As indicated in Table 12.2.2, high pressure compressed gas cylinders are permitted to be kept in service (if not discharged) for up to 12 years. Generally, cylinders that are discharged are replaced by new cylinders provided by a compressed gas supplier; therefore, it is seldom necessary to send cylinders out for hydrostatic testing. Some water mist systems, however, utilize customized compressed gas cylinders, or fixed banks of cylinders, that are intended to be refilled in place. A special fill connection should be provided on the manifold to permit refill in place. If the system is discharged after the first five years, the cylinders must be removed and sent for hydrostatic testing.

If in-situ recharge cannot be easily accomplished within 24 hours of a system release, replacement cylinders should be kept in stock. The spent cylinders should be removed and sent to a supplier to be refilled, and the stock replacement cylinders installed.

Exhibit 12.5 shows an assembly of high pressure nitrogen cylinders used as the driving medium for a water mist system. Each cylinder has its own low pressure alarm pressure switch. The manifold is designed to permit refill in place.

Different countries have different pressure ratings for cylinders. In Europe, Asia, and Australia, 2900 psi (200 bar) cylinders are standard. In North America, commercially available cylinders are typically 2250 psi (155 bar). It is important to check the actual cylinder volume and pressure whenever a new cylinder is installed. It is not uncommon to find a commercially supplied cylinder at, for example, 2175 psi (150 bar), about 3 percent less than the 2250 (155 bar) expected. A small difference (e.g., ~3 percent) will probably not significantly affect the overall performance of the system. The calculation to determine the number of cylinders needed for a given duration of discharge is based on the volume of the cylinders and the pressure. If the water mist system was designed on the basis of 50 liter cylinders at 2900 psi (200 bar) and similarly sized replacement cylinders were only 2250 (155 bar), the duration of protection will be reduced. Consult the system documentation to determine the minimum acceptable cylinder volume and pressure before replacing any cylinders.

EXHIBIT 12.5 Two Banks of Compressed Gas Cylinders for Use as Propellant for a Water Mist System.

12.3 Maintenance. [750:13.3]

Certain components of water mist systems may have relatively long delivery times for replacement. If a component must be removed for repair, such work should be scheduled to minimize the amount of time the system will be out of service.

12.3.1 Maintenance shall be performed to keep the system equipment operable or to make repairs. [**750:**13.3.1]

12.3.1.1 Mechanical waterflow devices including but not limited to water motor gongs shall be tested quarterly.

12.3.1.2 Vane-type and pressure switch–type waterflow devices shall be tested semiannually.

12.3.1.3 Waterflow devices shall be inspected quarterly to verify that they are free of physical damage.

12.3.2 As-built system installation drawings, original acceptance test records, and device manufacturer's maintenance bulletins shall be retained to assist in the proper care of the system and its components. [**750:**13.3.2]

Inspection, testing, and maintenance of a water mist system should not be undertaken without the as-built drawings and original acceptance test records. The inspection, testing, and maintenance service provider should also ensure that copies of the drawings and acceptance test records are kept with the new test record.

12.3.3 Preventive maintenance includes, but is not limited to, lubricating control valve stems, adjusting packing glands on valves and pumps, bleeding moisture and condensation from air compressors and air lines, and cleaning strainers. [**750:**13.3.3]

The purpose of preventive maintenance is to service a critical component before it breaks down or fails. The manufacturer's DIOM manual should identify equipment that merits preventive maintenance attention.

12.3.4 Scheduled maintenance shall be performed as outlined in Table 12.3.4. [**750:**13.3.4]

12.3.5 Corrective maintenance includes, but is not limited to, replacing loaded, corroded, or painted nozzles, replacing missing or loose pipe hangers, cleaning clogged fire pumps, replacing valve seats and gaskets, and restoring heat in areas subject to freezing temperatures where water-filled piping is installed. [**750:**13.3.5]

TABLE 12.3.4 Maintenance Frequencies

Item	Activity	Frequency
Water tank	Drain and refill	Annually
System	Flushing	Annually
Strainers and filters	Clean or replace as required	After system operation

[**750**:13.3.4]

12.3.6 Emergency maintenance includes, but is not limited to, repairs due to piping failures caused by freezing or impact damage, repairs to broken water mains, and replacement of frozen or fused nozzles, defective electric power, or alarm and detection system wiring. [**750**:13.3.6]

12.3.7 Specific maintenance activities, where applicable to the type of water mist system, shall be performed in accordance with the schedules in Table 12.3.4. [**750**:13.3.7]

Maintenance activities should be scheduled to coincide with the annual inspection and testing.

12.3.8 Replacement components shall be in accordance with the manufacturer's specifications and the original system design. [**750**:13.3.8]

12.3.9 Spare components shall be accessible and shall be stored in a manner to prevent damage or contamination. [**750**:13.3.9]

It is advisable to provide a locked cabinet for keeping spare components such as nozzles, filters, test orifices, extra gauges, and test records for the water mist system(s).

12.3.10* After each system operation, a representative sample of operated water mist nozzles in the activated zone shall be inspected. [**750**:13.3.10]

A.12.3.10 The representative sample should include 10 percent of the water mist nozzles in the activated zone. If contamination of filters or strainers is found on inspection, it is recommended that all nozzles within the activated zone be inspected. [**750**: A.13.3.10]

12.3.11 After each system operation due to fire, the system filters and strainers shall be cleaned or replaced. [**750**:13.3.11]

Annex D of NFPA 750 provides a "Water Mist System Questionnaire" that is intended to collect information on the reliability of water mist systems. If a water mist system operates for any reasons, accidentally or due to fire, it is likely that the inspection, testing, and maintenance provider will be involved in restoring the system. It would be very valuable if the service provider would consent to complete the questionnaire and send it to the staff person responsible for NFPA 750.

12.4 Training

12.4.1 All persons who might be expected to inspect, test, maintain, or operate water mist systems shall be trained thoroughly in the functions they are expected to perform. [**750**:13.4.1]

Most manufacturers of water mist systems equipment provide training on their specific equipment. Some provide certificates for confirmation that specific employees have completed the specialized training. Specialized training must also be accompanied by meaningful experience in the field.

In addition to familiarity about the various types of hardware encountered in water mist systems, it is important that the service provider be familiar with the features and limitations of the particular system. The service provider should obtain and study all of the system documentation provided before beginning testing or maintenance work on the system.

12.4.2 Refresher training shall be provided as recommended by the manufacturer or by the authority having jurisdiction. [**750:**13.4.2]

Personnel changes in an inspection, testing, and maintenance service company may result in the employee who received specialized training leaving the company. It is intended that all employees who work on a water mist system have received the specialized training provided by the manufacturer.

SUMMARY

The activities required in this chapter are intended to provide the preventive maintenance that can reveal potential problems with the water mist system before those problems compromise system performance. Unlike with other water-based systems where many of the activities can be performed by the property owner, only some of the activities required in this chapter can be performed by the owner because of the complexity of the water mist system. Most of the required activities in this chapter should be performed by those who have been trained by the manufacturer of the water mist equipment and who possess a certificate from the manufacturer verifying that training.

REVIEW QUESTIONS

1. Who is responsible for maintaining a water mist system?
2. If a deficiency is found during inspection, testing, or maintenance, what is the proper method for initiating corrective action?
3. When should a water mist system be re-evaluated?
4. What type of testing must be conducted following component replacement or system repair?
5. What precautions should be taken when shutting down a water mist system?

REFERENCES CITED IN COMMENTARY

National Fire Protection Association, 1 Batterymarch Park, Quincy, MA 02169-7471.

NFPA 20, *Standard for the Installation of Stationary Pumps for Fire Protection,* 2007 edition.
NFPA 72®, *National Fire Alarm Code*®, 2007 edition.
NFPA 750, *Standard on Water Mist Fire Protection Systems*, 2006 edition.

Valves, Valve Components, and Trim

CHAPTER 13

Chapter 13 of NFPA 25 is devoted to ensuring the operational status of valves. As indicated in Commentary Table 1.1, shut valves are the most common cause of sprinkler system failure. Because every type of fire protection system uses a valve of some sort, this chapter was written to consolidate the inspection, testing, and maintenance of all types of valves.

13.1* General

This chapter shall provide the minimum requirements for the routine inspection, testing, and maintenance of valves, valve components, and trim. Table 13.1 shall be used to determine the minimum required frequencies for inspection, testing, and maintenance.

TABLE 13.1 *Summary of Valves, Valve Components, and Trim Inspection, Testing, and Maintenance*

Item	*Frequency*	*Reference*
Inspection		
Control Valves		
Sealed	Weekly	13.3.2.1
Locked	Monthly	13.3.2.1.1
Tamper switches	Monthly	13.3.2.1.1
Alarm Valves		
Exterior	Monthly	13.4.1.1
Interior	5 years	13.4.1.2
Strainers, filters, orifices	5 years	13.4.1.2
Check Valves		
Interior	5 years	13.4.2.1
Preaction/Deluge Valves		
Enclosure (during cold weather)	Daily/weekly	13.4.3.1
Exterior	Monthly	13.4.3.1.6
Interior	Annually/5 years	13.4.3.1.7
Strainers, filters, orifices	5 years	13.4.3.1.8
Dry Pipe Valves/Quick-Opening Devices		
Enclosure (during cold weather)	Daily/weekly	13.4.4.1.1
Exterior	Monthly	13.4.4.1.4
Interior	Annually	13.4.4.1.5
Strainers, filters, orifices	5 years	13.4.4.1.6
Pressure Reducing and Relief Valves		
Sprinkler systems	Annually	13.5.1.1
Hose connections	Annually	13.5.2.1

(continues)

TABLE 13.1 Continued

Item	Frequency	Reference
Inspection *(continued)*		
Hose racks	Annually	13.5.3.1
Fire pumps		
Casing relief valves	Weekly	13.5.6.1, 13.5.6.1.1
Pressure relief valves	Weekly	13.5.6.2, 13.5.6.2.1
Backflow Prevention Assemblies		
Reduced pressure	Weekly/monthly	13.6.1
Reduced pressure detectors	Weekly/monthly	13.6.1
Fire Department Connections	Quarterly	13.7.1
Test		
Main Drains	Annually/quarterly	13.2.6, 13.2.6.1, 13.3.3.4
Waterflow Alarms	Quarterly/semiannually	13.2.6
Control Valves		
Position	Annually	13.3.3.1
Operation	Annually	13.3.3.1
Supervisory	Semiannually	13.3.3.5
Preaction/Deluge Valves		
Priming water	Quarterly	13.4.3.2.1
Low air pressure alarms	Quarterly	13.4.3.2.10
Full flow	Annually	13.4.3.2.2
Dry Pipe Valves/Quick-Opening Devices		
Priming water	Quarterly	13.4.4.2.1
Low air pressure alarm	Quarterly	13.4.4.2.6
Quick-opening devices	Quarterly	13.4.4.2.4
Trip test	Annually	13.4.4.2.2
Full flow trip test	3 years	13.4.4.2.2.2
Pressure Reducing and Relief Valves		
Sprinkler systems	5 years	13.5.1.2
Circulation relief	Annually	13.5.6.1.2
Pressure relief valves	Annually	13.5.6.2.2
Hose connections	5 years	13.5.2.2
Hose racks	5 years	13.5.3.2
Backflow Prevention Assemblies	Annually	13.6.2
Maintenance		
Control Valves	Annually	13.3.4
Preaction/Deluge Valves	Annually	13.4.3.3.2
Dry Pipe Valves/Quick-Opening Devices	Annually	13.4.4.3.2

The term *valve* as used in this standard includes control, operational, pressure regulating, and other types of valves. Alarm check, dry pipe, preaction, and deluge are types of operational valves. OS&Y (outside screw and yoke), IBV (indicating butterfly valve), PIV (post indicator valve), and WPIV (wall post indicator valve) are types of control valves. Pressure regulating valves include pressure reducing valves and pressure relief valves. This chapter also covers backflow prevention valves and fire department connections. The term *valve components* refers to the parts that make up an operational or control valve. *Trim* refers to the exterior parts of an operational valve, such as the basic alarm trim, retard trim, and the main drain connection.

A.13.1 *Alarm Valves.* Alarm valves are installed in water-based fire protection systems to sound a fire alarm when a flow of water from the system equals or exceeds the flow of a single discharge device. A retarding chamber, which minimizes false alarms due to surges and fluctuating water supply pressure, can be supplied with the alarm valve.

An alarm valve is a check valve with several trim options for alarm connections, gauges, a retarding chamber, and main drain attachment. The alarm trim option provides pipe to a pressure switch and/or water motor alarm gong for the purpose of generating a waterflow alarm. A retarding chamber is used for the purpose of absorbing a small amount of water due to pressure surges to avoid a momentary waterflow and subsequent false alarm. A retarding chamber is not needed when a fire pump is installed due to the constant pressure conditions created by the pressure maintenance (jockey) pump. Exhibit 13.1 shows a wet pipe system riser with alarm valve and trim.

Backflow Prevention Devices. Backflow prevention devices are used to prevent water in a fire protection system from entering the public water supply due to a reverse flow of water, thermal expansion, hydraulic shock, back pressure, or back siphonage. *[See Figure A.13.1(a).]*

Although not required for fire protection purposes, backflow prevention devices, also known as backflow preventers, are provided to isolate potable water supplies from backflow from fire protection systems. The backflow preventer should be installed between the fire department connection (FDC) and the domestic water supply because the greatest potential for backflow exists when the FDC is in use. A reduced pressure backflow preventer provides a level of protection higher than a double check valve assembly and is normally used when chemicals, such as antifreeze or foam concentrates, are involved in the system.

See Exhibit 3.1 for an illustration of a double check valve assembly. Exhibit 3.7 shows a reduced pressure backflow preventer.

Ball Valves. Ball valves are manually operated through their full range of open to closed positions with a one-quarter turn.

Ball valves are usually $\frac{1}{4}$ turn. They are used for trim for valves such as alarm valves.

EXHIBIT 13.1 Wet Pipe System Riser with Alarm Valve and Trim.

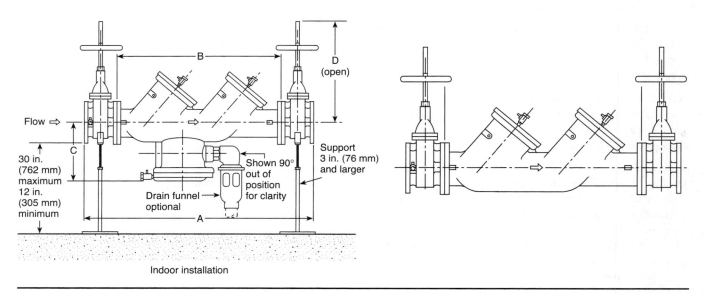

FIGURE A.13.1(a) *Reduced Pressure Backflow Preventers (left) and Double Check Valve Assemblies (right).*

FIGURE A.13.1(b) *Butterfly Post Indicator Valve. (Courtesy of Henry Pratt Co.)*

Butterfly Valves. Butterfly valves are water supply control valves with gear operators to assist in opening and closing. Butterfly valves can be of the wafer or grooved-end type. *[See Figure A.13.1(b).]*

Butterfly valves, which are lighter in weight than OS&Y valves, are currently used for control applications. They are indicating and are usually equipped with a factory-installed tamper switch. Exhibit 13.2 shows a butterfly valve.

Check Valves. Check valves allow water flow in one direction only. *[See Figure A.13.1(c).]*

Check valves may have connections for automatic drains such as ball drip valves. They must be used in FDCs. Exhibit 13.3 shows a typical check valve.

DCA. A double check assembly (DCA) consists of two independently operating spring-loaded check valves. The assembly includes two resilient-seated isolation valves and four test cocks required for testing.

EXHIBIT 13.2 Butterfly Valve.

EXHIBIT 13.3 *Detector Check Valve. (Courtesy of John Jensen)*

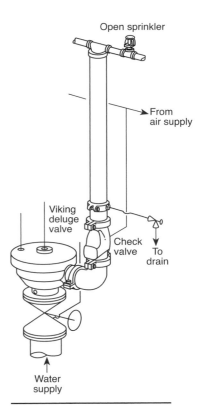

FIGURE A.13.1(d) *Deluge Valve.*

FIGURE A.13.1(c) *Detector Check Valve.*

DCDA. A double check detector assembly (DCDA) is hydraulically balanced to include a metered bypass assembly to detect system leakage. The main valve assembly and bypass assembly afford equal levels of backflow prevention and are each equipped with two resilient-seated isolation valves and four test cocks required for testing.

Deluge Valves. Deluge valves hold water at the valve until actuated by the operation of a detection system or manual release. *[See Figure A.13.1(d).]*

Drip Valves. Drip valves automatically drain condensation or small amounts of water that have leaked into system piping or valves. Drip valves close when exposed to system pressure.

Dry Pipe Valves. Dry pipe valves control the flow of water to areas that could be exposed to freezing conditions. Water is held at the valve by air pressure in the system piping. When the air pressure is reduced, the valve operates and floods the system. *[See Figure A.13.1(e) and Figure A.13.1(f).]*

Indicating Valves. Indicating valves provide a dependable, visible indication of the open position, even at a distance.

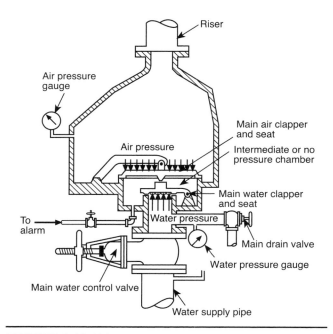

FIGURE A.13.1(e) *Dry Pipe Valve.*

FIGURE A.13.1(f) *Dry Pipe System Accelerator. (Courtesy of Reliable Automatic Sprinkler Co.)*

Exhibit 13.4 shows OS&Y indicating valves. Because the stem of the valve on the right is flush with the handwheel or yoke, it is clear that the valve is closed. The open status of the valve on the left is easy to detect because the stem is protruding.

Indicator Posts. Indicator posts include wall and underground types and are intended for use in operating inside screwed pattern gate valves and for indicating the position of the gates in the valves. [See Figure A.13.1(g).]

NRS Gate Valves, OS&Y Gate Valves. Nonrising stem (NRS) gate valves are used underground with indicator posts attached or as roadway box valves (curb-box installation). Outside screw and yoke (OS&Y) gate valves are used indoors and in pits outdoors. The valve stem moves out when the valve is open and moves in when it is closed. The stem indicates the position of the valve. [See Figure A.13.1(h) and Figure A.13.1(i).]

RPA. A reduced-pressure zone principle assembly (RPA) consists of two independently spring-loaded check valves separated by a differential-sensing valve. The differential-sensing valve includes a relief port to atmosphere that discharges excess water resulting from supply system fluctuations. The assembly includes two resilient-seated isolation valves and four test cocks required for testing.

EXHIBIT 13.4 *OS&Y Indicating Valves.*

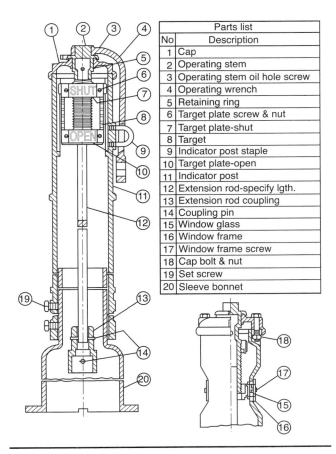

FIGURE A.13.1(g) *Vertical Indicator Post.*

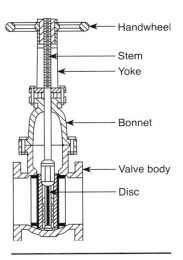

FIGURE A.13.1(h) *OS&Y Gate Valve.*

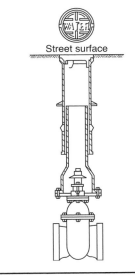

FIGURE A.13.1(i) *Nonindicating-Type Gate Valve.*

RPDA. A reduced-pressure detector assembly (RPDA) is hydraulically balanced to include a metered bypass assembly to detect system leakage. The main valve assembly and bypass assembly afford equal levels of backflow prevention, and each assembly is equipped with two resilient-seated isolation valves and four test cocks required for testing.

Strainers. Strainers are used for protection against clogging of water discharge openings.

Waterflow Detector Check Valves. Detector-type check valves allow flow in one direction only and have provisions for the connection of a bypass meter around the check valve. [See Figure A.13.1(c).]

13.2 General Provisions

13.2.1 The property owner shall have manufacturers' literature available to provide specific instructions for inspecting, testing, and maintaining the valves and associated equipment.

The term *owner* in 13.2.1 applies to whoever is responsible for the inspection, testing, and maintenance of the equipment (see 4.1.2.3). The literature should be that which was provided in the operation and maintenance manuals as required by the Contractor's Material and Test Certificate at the time of installation.

The operation and maintenance manuals are an excellent source of information to the inspector and maintenance contractor. The manual should include such information as the system narrative report, which describes all aspects of the initial construction project, and a complete list of equipment installed, including the contact information for the equipment suppliers. The manual also should include operation and maintenance instructions for all system components, any previous inspection/test reports, and recommended spare parts. If such a manual is not present, it may be appropriate for the maintenance contractor to produce one for future reference, complete with a telephone number for emergency service.

13.2.2 All pertinent personnel, departments, authorities having jurisdiction, or agencies shall be notified that testing or maintenance of the valve and associated alarms is to be conducted.

It is not necessary to notify the authority having jurisdiction (AHJ) of when inspection activities will take place because alarm initiating devices are not activated. However, when testing certain valves, alarms will be activated and authorities need to know not to respond but to acknowledge that the alarms were received. It is important to retain inspection records for review by the AHJ and/or insurance inspectors to verify that the work has been completed to demonstrate compliance with this standard. Such records should include notes about problems, solutions, and suggestions for correcting deficiencies. See Supplement 4 for sample inspection forms.

13.2.3* All system valves shall be protected from physical damage and shall be accessible.

A.13.2.3 The valves are not required to be exposed. Doors, removable panels, or valve pits can be permitted to satisfy this requirement. Such equipment should not be obstructed by features such as walls, ducts, columns, direct burial, or stock storage.

Most manufacturers will provide clearance dimensions for their valves in their product data. These dimensions should be maintained on all sides of the valve.

For example, operational valves, such as alarm check, dry pipe, and deluge valves, must have at least 3 ft (1 m) clearance on all sides to allow normal inspection, testing, and maintenance activities. If the valve is inside an enclosure for heating, that clearance should be increased to 4 ft (1.2 m).

In warm climates, the sprinkler riser and valve may be located outside of the building. In this situation, the clearance should be 3 ft (1 m) on all sides.

Exhibit 13.5 is a typical valve clearance dimension diagram. The dimensions shown should be used as a minimum to maintain access to the valve for operation and maintenance purposes.

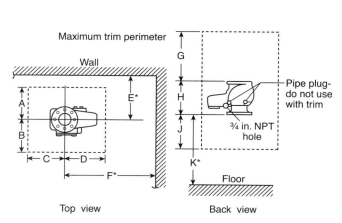

Dimension	2-½ in. (63.5 mm) valve	4 in. (101.6 mm) valve	6 in. (152.4 mm) valve
A	8-¾ in. (222.2 mm)	9-½ in. (241.3 mm)	9 in. (228.6 mm)
B	9 in. (228.6 mm)	10-¼ in. (260.3 mm)	11-¼ in. (285.7 mm)
C	9-¾ in. (247.6 mm)	9-¾ in. (247.6 mm)	11 in. (279.4 mm)
D	10-½ in. (266.7 mm)	15-¾ in. (400.0 mm)	13-½ in. (342.9 mm)
E*	12 in. (304.8 mm)	12-½ in. (317.5 mm)	15 in. (381.0 mm)
F*	24 in. (609.6 mm)	24 in. (609.6 mm)	24 in. (609.6 mm)
G	10-½ in. (266.7 mm)	7 in. (177.8 mm)	5 in. (127.0 mm)
H	9 in. (228.6 mm)	12-¼ in. (311.1 mm)	18 in. (457.2 mm)
J	9-¼ in. (234.9 mm)	10-¼ in. (260.3 mm)	5-¼ in. (133.3 mm)
K*	48 in. (1219.2 mm)	48 in. (1219.2 mm)	48 in. (1219.2 mm)

*These dimensions represent minimum recommended clearances.

EXHIBIT 13.5 *Typical Valve Clearance Dimensions.*

13.2.4 Before opening a test or drain valve, it shall be verified that adequate provisions have been made for drainage.

The best location for the sprinkler riser is along the building outside wall since the riser must be equipped with a main drain. The main drain discharge pipe can be installed directly to the building exterior, providing discharge directly outdoors. If the riser is located on the interior of a building, the drain should be piped to a floor or waste water drain having twice the diameter of the main drain [for example, a 2 in. (50 mm) main drain to a 4 in. (100 mm) floor drain]. A utility sink can be used if the flow capacity, which is approximately 200 gpm (757 L/min), is adequate for the main drain without overflowing onto the floor.

In high-rise buildings with the sprinkler riser in the stairway, a separate drain line is often installed that is one diameter larger than the largest main drain valve that connects to the drain line. This drain line should not be considered a main drain connection but should be viewed as an auxiliary drain for each floor. The auxiliary drain is not intended to be flow tested as is a main drain connection. The auxiliary drain line can be used to flow test pressure regulating valves that are installed in the fire sprinkler system and in the fire department standpipe system and to test the system flow switch.

13.2.5* Main Drain Test.

A main drain test shall be conducted annually at each water-based fire protection system riser to determine whether there has been a change in the condition of the water supply piping and control valves. *(See also 13.3.3.4.)*

A.13.2.5 Main drains are installed on system risers for one principal reason: to drain water from the overhead piping after the system is shut off. This allows the contractor or plant maintenance department to perform work on the system or to replace nozzles after a fire or other incident involving system operation.

The test for standpipe systems should be done at the low-point drain for each standpipe or the main drain test connection where the supply main enters the building.

These drains also are used to determine whether there is a major reduction in waterflow to the system, such as could be caused by a major obstruction, a dropped gate, a valve that is almost fully closed, or a check valve clapper stuck to the valve seat.

A large drop in the full flow pressure of the main drain (as compared to previous tests) normally is indicative of a dangerously reduced water supply caused by a valve in an almost fully closed position or other type of severe obstruction. After closing the drain, a slow return to normal static pressure is confirmation of the suspicion of a major obstruction in the waterway and should be considered sufficient reason to determine the cause of the variation.

A satisfactory drain test (i.e., one that reflects the results of previous tests) does not necessarily indicate an unobstructed passage, nor does it prove that all valves in the upstream flow of water are fully opened. The performance of drain tests is not a substitute for a valve check on 100 percent of the fire protection valving.

The main drain test is conducted in the following manner:

(1) Record the pressure indicated by the supply water gauge.
(2) Close the alarm control valve on alarm valves.
(3) Fully open the main drain valve.
(4) After the flow has stabilized, record the residual (flowing) pressure indicated by the water supply gauge.
(5) Close the main drain valve slowly.
(6) Record the time taken for the supply water pressure to return to the original static (nonflowing) pressure.
(7) Open the alarm control valve.

FAQ ▶
What is the purpose of the main drain test?

There are two major reasons to conduct a main drain test:

1. To verify that the water supply pressure is similar to that in previous years
2. To verify that the system control valve is fully open after having been closed and reopened during annual testing

Exhibit 13.6 shows the discharge from a 2 in. (50 mm) main drain test.

EXHIBIT 13.6 Discharge from a Main Drain Test.

The results of the main drain test should be compared to those of previous tests, including the original commissioning test, to identify any deterioration in water supply.

Main drain test results should not be solely relied on to determine whether valves are open. A main drain test may not reveal a partially closed valve in a large-diameter supply pipe with relatively high pressure. Only visual inspection can verify that a valve is in the open position.

13.2.5.1 Systems where the sole water supply is through a backflow preventer and/or pressure reducing valves, the main drain test of at least one system downstream of the device shall be conducted on a quarterly basis.

FAQ ▶
Why are backflow preventers tested on a quarterly basis?

The intent of 13.2.5.1 is to exercise the backflow preventers by producing periodic flow through them. Fire protection systems are relatively static, only flowing when a fire occurs. Because backflow preventers were originally designed for use in domestic systems, where a continual cycling of water exists, they must be exercised to prevent failure.

 13.2.5.2 When there is a 10 percent reduction in full flow pressure when compared to the original acceptance test or previously performed tests, the cause of the reduction shall be identified and corrected if necessary.

Note that 13.2.5.2 does not necessarily require any corrective action for a water supply that is exhibiting lower pressure readings during the main drain test. What is required is an investigation as to the cause of the depleted pressure. While a reduction in the available pressure from the attached water supply is a concern, it is critical that the water supply can still meet or exceed the system demand. The system should not be considered to be impaired unless it is determined that the water supply no longer meets the system demand. The 10 percent (or more) reduction in available pressure is considered to be sufficient to cause an investigation.

 13.2.6 Alarm Devices.

 13.2.6.1 Mechanical waterflow devices, including but not limited to water motor gongs, shall be tested quarterly.

As required in 5.3.3.1, water motor gongs must be tested quarterly. According to 5.3.3.2, however, vane-type waterflow devices and pressure switches need only be tested semi-annually. This recent revision to the standard for the 2002 edition was made as a result of the excellent performance of these devices. Exhibit 13.7 shows two vane-type waterflow switches. Exhibit 13.8 shows a pressure switch.

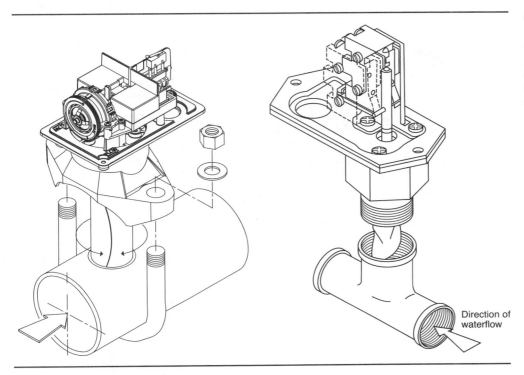

EXHIBIT 13.7 Vane-Type Waterflow Switch with Retard (left) and and Vane-Type Waterflow Alarm Switch — Small Pipe (right). (Courtesy of Potter Electric Signal Company)

EXHIBIT 13.8 Pressure Switch.

Exhibit 13.9 illustrates closed (set) and open (operating) positions for alarm valves, respectively. Exhibit 13.10 illustrates closed (set) and open (operating) positions for deluge valves, respectively.

The waterflow alarm test is conducted by opening the inspector's test valve or alarm bypass valve and recording the time required for the local alarm to activate. An alarm must sound within 5 minutes, as required by NFPA 13, *Standard for the Installation of Sprinkler*

EXHIBIT 13.9 Set Position for Alarm Valves (top) and Operating Position for Alarm Valves (bottom).

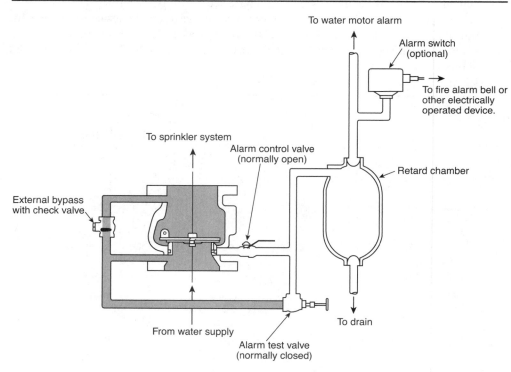

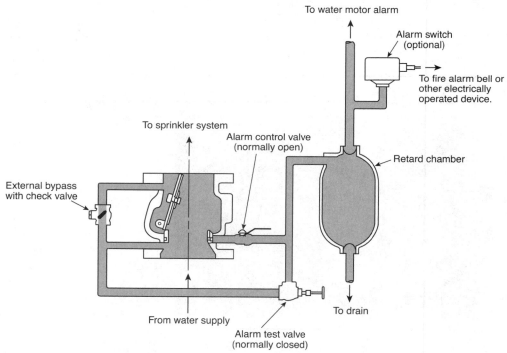

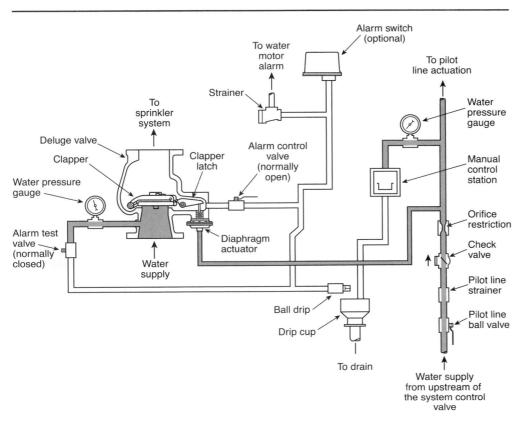
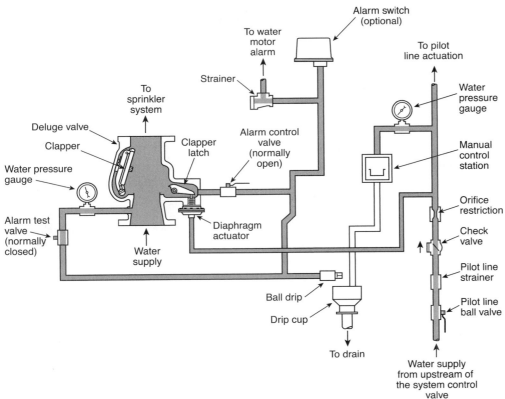

EXHIBIT 13.10 Set Position for Deluge Valves (top) and Operating Position for Deluge Valves (bottom).

Systems (or, for a remote alarm, 90 seconds, as required by *NFPA 72®*, *National Fire Alarm Code®*).

The valve is then closed. In most cases, the alarms should stop sounding. However, if the local alarm device is part of the building fire alarm system, the local alarm will sound continuously until the fire alarm system is reset. Exhibit 13.11 shows a pressure switch waterflow alarm.

EXHIBIT 13.11 *Pressure Switch Waterflow Alarm.*

The purpose of the waterflow alarm test is to verify that the local and any remote alarm signals operate correctly. If available, the inspector's test connection should be used to conduct this test, because the test is intended to simulate the flow of a single sprinkler, and the alarm must operate with a single sprinkler open.

A dry pipe system with a capacity of over 500 gal (1893 L) without a quick opening device, and dry pipe systems that protect dwelling unit portions of any occupancy, must deliver water to the inspector's test connection within 60 seconds. Dry pipe systems with a capacity of 500 gal or less, and dry systems with a capacity between 500 and 750 gal (1893 and 2839 L) with a quick opening device installed, do not have to meet the 60 second rule. In theory, there is no time limit for water delivery for these systems. However, all test results should be compared to previous tests, and significant degradation should be investigated. As of the 2007 edition of NFPA 13, system water delivery time is permitted to be calculated. These requirements are intended to limit the size of a dry pipe system and also the water delivery time.

Most preaction systems, with the exception of double interlock preaction systems, do not have an inspector's test connection that uses water. Testing the supervisory alarm involves lowering the supervisory air pressure, which can be done by opening the inspector's test connection to bleed off supervisory air. When the air pressure is lowered to the alarm set point, a signal will sound at the preaction control panel. This will activate the preaction valve, allowing the system to fill with water. When the system is filled with water, the waterflow pressure switch or water motor gong will sound, sending a signal to the preaction control panel and sounding the audible alarm signal.

Double interlock preaction systems with a capacity in excess of 750 gal (2839 L) have an inspector's test connection similar to that of a dry pipe system.

Preaction systems should not be equipped with a vane-type waterflow switch because the plastic vane can be dislodged, presenting a potential obstruction to the system piping.

13.2.6.2 Vane-type and pressure switch–type waterflow devices shall be tested semiannually.

13.2.7 Gauges.

13.2.7.1 Gauges shall be inspected monthly to verify that they are in good condition and that normal pressure is being maintained.

In 13.2.7.1, "good condition" means that the gauge pointer is responsive to changes in pressure. The "normal pressure" is the average of water pressure readings taken for several months. Large fluctuations from the normal pressure may indicate that something is wrong with the system or the water supply.

13.2.7.1.1 Where other sections of this standard have different frequency requirements for specific gauges, those requirements shall be used.

13.2.7.2 Gauges shall be replaced every 5 years or tested every 5 years by comparison with a calibrated gauge.

Paragraph 13.2.7.2 provides the option of replacing a gauge in lieu of testing. In some cases, replacement may be more economical than testing. If, after five years, an installed gauge is still operating within the original range and with the required accuracy, it need not be replaced.

13.2.7.3 Gauges not accurate to within 3 percent of the full scale shall be recalibrated or replaced.

Paragraph 13.2.7.3 requires that inaccurate gauges be recalibrated or replaced. Many gauges cannot be recalibrated due to their design. When such a gauge is found to be inaccurate, there is an alternative to replacement. The gauge's accuracy should be checked and a comparison pressure table developed to determine the actual readings at various gauge settings. This table should be used in lieu of the uncorrected readings.

13.2.8 Records.

Records shall be maintained in accordance with Section 4.4.

13.3 Control Valves in Water-Based Fire Protection Systems

13.3.1* Each control valve shall be identified and have a sign indicating the system or portion of the system it controls.

A.13.3.1 Signs identifying underground fire service main control valves in roadway boxes should indicate the direction of valve opening, the distance and direction of the valve from the sign location (if the valve is subject to being covered by snow or ice), and the location of the wrench if not located with the sign.

13.3.1.1 Systems that have more than one control valve that must be closed to work on a system shall have a sign on each affected valve referring to the existence and location of other valves.

13.3.1.2* When a normally open valve is closed, the procedures established in Chapter 15 shall be followed.

A.13.3.1.2 Valves that normally are closed during cold weather should be removed and replaced with devices that provide continuous fire protection service.

Paragraph 13.3.1.2 stipulates under what conditions the requirements in Chapter 15, which refer to fire protection impairments, must be followed.

If the annual testing of a valve requires the closing of that valve, the valve should be returned to its normal open position immediately after testing. The number of turns required to close the valve should be recorded. That information can then be used to make sure the valve is reopened all the way. The manufacturer's literature will also indicate the number of turns necessary to operate the valve through its full range of motion.

Any supervisory device should be returned to its normal position and no trouble signal should be observed at the fire alarm panel.

13.3.1.2.1 When the valve is returned to service, a drain test (either main or sectional drain, as appropriate) shall be conducted to determine that the valve is open.

Paragraph 13.3.1.2.1 does not apply to underground sectional control valves. These are verified by opening a hydrant connected to the underground main system.

13.3.1.3 Each normally open valve shall be secured by means of a seal or a lock or shall be electrically supervised in accordance with the applicable NFPA standards.

The "applicable NFPA standards" in 13.3.1.3 are those standards having to do with water extinguishing systems. They include the following:

NFPA 13, *Standard for the Installation of Sprinkler Systems*

NFPA 14, *Standard for the Installation of Standpipe and Hose Systems*

NFPA 15, *Standard for Water Spray Fixed Systems for Fire Protection*

NFPA 16, *Standard for the Installation of Foam-Water Sprinkler and Foam-Water Spray Systems*

NFPA 20, *Standard for the Installation of Stationary Pumps for Fire Protection*

NFPA 22, *Standard for Water Tanks for Private Fire Protection*

13.3.1.4 Normally closed valves shall be secured by means of a seal or shall be electrically supervised in accordance with the applicable NFPA standard.

If the valves are electrically supervised, the supervisory device must be located so that opening the valve will send a signal within two turns of the fully closed position. An example of a normally closed valve is the isolation valve on the fire pump test header piping.

13.3.1.4.1 Sealing or electrical supervision shall not be required for hose valves.

13.3.2 Inspection.

The valve inspection required by 13.3.2 is specifically for system control valves. The following observations must be made during inspection of these valves:

- Is the valve open? (check the target)
- Is the valve leaking water?
- Is the valve lubricated?
- Is the valve wrench available?
- If the valve is supervised, is the seal or chain and lock in position and secured? Where is the key for the lock?
- Is the electrical supervision in place and operational?

Exhibit 13.12 shows a wall post indicator valve (WPIV) that is inoperable due to a tree branch.

EXHIBIT 13.12 Inoperable WPIV. (Courtesy of John Jensen)

13.3.2.1 All valves shall be inspected weekly.

13.3.2.1.1 Valves secured with locks or supervised in accordance with applicable NFPA standards shall be permitted to be inspected monthly.

13.3.2.1.2 After any alterations or repairs, an inspection shall be made by the property owner to ensure that the system is in service and all valves are in the normal position and properly sealed, locked, or electrically supervised.

13.3.2.2* The valve inspection shall verify that the valves are in the following condition:

(1) In the normal open or closed position
(2)* Properly sealed, locked, or supervised

A.13.3.2.2(2) The purpose of the valve sealing program is as follows:

(1) The presence of a seal on a control valve is a deterrent to closing a valve indiscriminately without obtaining the proper authority.
(2) A broken or missing seal on a valve is cause for the plant inspector to verify that protection is not impaired and to notify superiors of the fact that a valve could have been closed without following procedures.
(3) Accessible
(4) Provided with appropriate wrenches
(5) Free from external leaks
(6) Provided with appropriate identification

A.13.3.2.2 Valves should be kept free of snow, ice, storage, or other obstructions so that access is ensured.

13.3.3 Testing.

13.3.3.1 Each control valve shall be operated annually through its full range and returned to its normal position.

The testing required by 13.3.3.1 should include counting the number of turns needed to fully close and open the valve. Also, the valve stem should be lubricated before closing and opening.

Note that 13.3.3.1 requires that all control valves be operated annually through their full range of motion.

How often do post indicator valves need to be cycled?

◀ FAQ

13.3.3.2* Post indicator valves shall be opened until spring or torsion is felt in the rod, indicating that the rod has not become detached from the valve.

A.13.3.3.2 These "spring tests" are made to verify that a post indicator valve is fully open. If an operator feels the valve is fully open, he or she should push in the "open" direction. The handle usually moves a short distance (approximately a one-quarter turn) and "springs" back toward the operator in a subtle move when released. This spring occurs when the valve gate pulls up tight against the top of its casting and the valve shaft (being fairly long) twists slightly. The spring indicates that the valve is fully opened and that the gate is attached to the handle. If the gate is jammed due to a foreign particle, the handle is not likely to spring back. If the gate is loose from the handle, the handle continues to turn in the "open" direction with little resistance.

13.3.3.2.1 This test shall be conducted every time the valve is closed.

13.3.3.3 Post indicator and outside screw and yoke valves shall be backed a one-quarter turn from the fully open position to prevent jamming.

 13.3.3.4 A main drain test shall be conducted any time the control valve is closed and re-opened at system riser.

The main drain test required by 13.3.3.4 is intended to verify that the control valve has been returned to the fully open position.

13.3.3.5* Supervisory Switches.

A.13.3.3.5 For further information, see *NFPA 72, National Fire Alarm Code*.

13.3.3.5.1 Valve supervisory switches shall be tested semiannually.

13.3.3.5.2 A distinctive signal shall indicate movement from the valve's normal position during either the first two revolutions of a hand wheel or when the stem of the valve has moved one-fifth of the distance from its normal position.

If the valve is normally closed, the valve supervisory switch must sound a distinctive signal when the valve is opened and return to normal when the valve is closed. Exhibit 13.13 illustrates a supervisory switch.

13.3.3.5.3 The signal shall not be restored at any valve position except the normal position.

EXHIBIT 13.13 *Supervisory Switch.*

13.3.4 Maintenance.

13.3.4.1 The operating stems of outside screw and yoke valves shall be lubricated annually.

13.3.4.2 The valve then shall be completely closed and reopened to test its operation and distribute the lubricant.

The lubricant in 13.3.4.2 should have been applied while exercising the valve as required by 13.3.3.1.

13.4 System Valves

13.4.1 Inspection of Alarm Valves.

Alarm valves shall be inspected as described in 13.4.1.1 and 13.4.1.2.

13.4.1.1* Alarm valves and system riser check valves shall be externally inspected monthly and shall verify the following:

(1) The gauges indicate normal supply water pressure is being maintained.
(2) The valve is free of physical damage.
(3) All valves are in the appropriate open or closed position.
(4) The retarding chamber or alarm drains are not leaking.

A.13.4.1.1 A higher pressure reading on the system gauge is normal in variable pressure water supplies. Pressure over 175 psi (12.1 bar) can be caused by fire pump tests or thermal expansion and should be investigated and corrected.

With minimal training, the property owner or in-house maintenance personnel can inspect an alarm valve. See Exhibit 13.14 for the required inspection points.

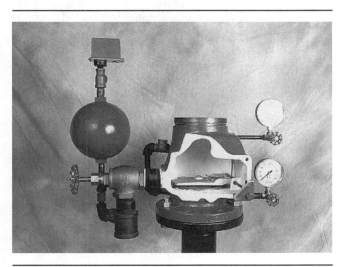

EXHIBIT 13.14 Alarm Valve Inspection Points.

13.4.1.2* Alarm valves and their associated strainers, filters, and restriction orifices shall be inspected internally every 5 years unless tests indicate a greater frequency is necessary.

A.13.4.1.2 The system should be drained for internal inspection of valve components as follows:

> **FAQ ▶**
> Is the requirement for the five-year internal inspection of an alarm check valve valid for alarm check valves that no longer have active retard chambers or other trim that depends on internal orifices?

(1) Close the control valve.
(2) Open the main drain valve.
(3) Open the inspector's test valve.
(4) Wait for the sound of draining water to cease and for all gauges to indicate 0 psi (0 bar) before removing the handhole cover or dismantling any component.

The internal inspection of alarm valves is not contingent upon retard chambers or orifices. The purpose of the inspection is to verify the operation of the clapper and the condition of the valve seats and to look for evidence of corrosion or obstructing material.

> **FAQ ▶**
> What test results would indicate that more frequent inspections are needed?

Test results that may indicate more frequent inspections are needed include the following:

- Water motor alarm gong does not operate.
- Retard chamber does not drain.
- Local bells continue to cycle on and off after the test is complete.
- Alarm bypasses do not operate the local alarm device.

Exhibit 13.15 illustrates the internal components of an alarm check valve.

EXHIBIT 13.15 Alarm Valve Internal Inspection.

13.4.1.3 Maintenance.

13.4.1.3.1 Internal components shall be cleaned/repaired as necessary in accordance with the manufacturer's instructions.

Before any major maintenance is started, a complete set of gaskets, filters, and strainers should be readily available to minimize the system downtime required to make any repairs. In addition, it is suggested that repair kits for each type of valve on the premises be kept on hand for emergency repairs as well as routine maintenance. Many manufacturers have repair kits available for their valves. Doing so will minimize downtime.

13.4.1.3.2 The system shall be returned to service in accordance with the manufacturer's instructions.

13.4.2 Check Valves.

13.4.2.1 Inspection. Valves shall be inspected internally every 5 years to verify that all components operate correctly, move freely, and are in good condition.

13.4.2.2 Maintenance. Internal components shall be cleaned, repaired, or replaced as necessary in accordance with the manufacturer's instructions.

Gaskets should be replaced at the time of the five-year inspection because they can dry out and harden, causing leaks.

13.4.3 Preaction Valves and Deluge Valves.

Preaction and deluge valves are basically identical but are trimmed differently, based on the type of operation needed. For example, a deluge valve activates and discharges water into a system of open nozzles or sprinklers, whereas a preaction system trips and floods a closed piping system (one with automatic sprinklers) that will not discharge water until a sprinkler activates. A preaction system has a check valve installed immediately above the deluge valve to isolate the deluge valve from the system air pressure. Otherwise, the two valve bodies (preaction and deluge) are basically the same. See Exhibit 13.16 for typical preaction and deluge valve arrangements.

◄ **FAQ**
What is the difference between a preaction valve and a deluge valve?

EXHIBIT 13.16 *Preaction Valve with Trim Piping and Accessories (left) and Deluge Valve with Trim Piping and Accessories (right).*

13.4.3.1 Inspection.

13.4.3.1.1 Valve enclosure heating equipment for preaction and deluge valves subject to freezing shall be inspected daily during cold weather for its ability to maintain a minimum temperature of at least 40°F (4.4°C).

13.4.3.1.1.1 Valve enclosures equipped with low temperature alarms shall be inspected weekly.

13.4.3.1.2 Low temperature alarms, if installed in valve enclosures, shall be inspected annually at the beginning of the heating season.

Low temperature supervisory signals must be tested on a six-month schedule, in accordance with *NFPA 72*. Each year, one of those tests should be performed just prior to the start of the heating season.

13.4.3.1.3 Gauges shall be inspected weekly.

13.4.3.1.3.1 The gauge on the supply side of the preaction or deluge valve shall indicate that the normal supply water pressure is being maintained.

13.4.3.1.4 The gauge monitoring the preaction system supervisory air pressure, if provided, shall be inspected monthly to verify that it indicates that normal pressure is being maintained.

Normally, the supervisory pressure in 13.4.3.1.4 is very low [1 to 2 psi (0.06 to 0.14 bar)] for non-interlocked preaction systems. The supervisory air or alarm air used in a double interlock preaction system will be higher [2 to 20 psi (0.14 to 1.4 bar)], unless an accelerator is installed. An accelerator is installed on systems having a water capacity of over 500 gal (1893 L).

The normal dry system air maintenance device maintains too high an air pressure for the system. The air used for supervisory purposes should be dried to minimize internal pipe corrosion.

13.4.3.1.5 The gauge monitoring the detection system pressure, if provided, shall be tested monthly to verify that it indicates that normal pressure is being maintained.

Paragraph 13.4.3.1.5 refers to a pilot sprinkler or rate-of-rise thermal detection system used to activate a preaction or deluge system. As in 13.4.3.1.4, the air should be dried and kept at the pressure required in the manufacturer's operations and maintenance manual.

An air dryer is filled with a chemical desiccant. The desiccant, which is usually white, turns purple when it is no longer capable of absorbing moisture. When that color change occurs, the desiccant should be replaced.

13.4.3.1.6 The preaction or deluge valve shall be externally inspected monthly to verify the following:

(1) The valve is free from physical damage.
(2) All trim valves are in the appropriate open or closed position.
(3) The valve seat is not leaking.
(4) Electrical components are in service.

The external inspection of the preaction or deluge valve required by 13.4.3.1.6 should also include inspection of pneumatic components.

13.4.3.1.7 The interior of the preaction or deluge valve and the condition of detection devices shall be inspected annually when the trip test is conducted.

13.4.3.1.7.1 Internal inspection of valves that can be reset without removal of a faceplate shall be permitted to be conducted every 5 years.

Paragraph 13.4.3.1.7.1 acknowledges the fact that some deluge valves do not have to be opened to be reset. See Exhibit 13.17 for an illustration of an externally reset deluge valve.

13.4.3.1.8 Strainers, filters, restricted orifices, and diaphragm chambers shall be inspected internally every 5 years unless tests indicate a greater frequency is necessary.

An example of a case when the greater inspection frequency requirement in 13.4.3.1.8 would apply would be where water supply conditions produce obstructing material or rapid corrosion.

13.4.3.2 Testing.

13.4.3.2.1* The priming water level in supervised preaction systems shall be tested quarterly for compliance with the manufacturer's instructions.

A.13.4.3.2.1 High priming water levels can adversely affect the operation of supervisory air. Test the water level as follows:

(1) Open the priming level test valve.
(2) If water flows, drain it.

EXHIBIT 13.17 Externally Reset Deluge Valve.

(3) Close the valve when water stops flowing and air discharges.
(4) If air discharges when the valve is opened, the priming water level could be too low. To add priming water, refer to the manufacturer's instructions.

13.4.3.2.2* Each deluge valve shall be trip tested annually at full flow in warm weather and in accordance with the manufacturer's instructions. Protection shall be provided for any devices or equipment subject to damage by system discharge during tests.

A.13.4.3.2.2 Preaction and deluge valves in areas subject to freezing should be trip tested in the spring to allow time before the onset of cold weather for all water that has entered the system or condensation to drain to low points or back to the valve.

13.4.3.2.2.1* Where the nature of the protected property is such that water cannot be discharged for test purposes, the trip test shall be conducted in a manner that does not necessitate discharge in the protected area.

One method for complying with 13.4.3.2.2.1 is to conduct a trip test by throttling the water supply control valve so that it is one or two turns open. When the deluge valve operates, the water supply control valve can then be closed quickly before water discharges into the protected area.

Another method for conducting a trip test without flowing water is to remove the head of the deluge valve solenoid. When a signal is received at the solenoid, the solenoid becomes magnetized. The magnetization confirms that the alarm system is functioning. This method is not preferred for performing a discharge test of a deluge valve as it does not verify that the valve is operating properly.

A.13.4.3.2.2.1 Full flow tests should incorporate full functionality of the system as a unit, including automatic detection and manual activation.

13.4.3.2.2.2 Where the nature of the protected property is such that water cannot be discharged unless protected equipment is shut down (e.g., energized electrical equipment), a full flow system test shall be conducted at the next scheduled shutdown. In all cases, the test frequency shall not exceed 3 years.

 13.4.3.2.2.3 The water discharge patterns from all of the open spray nozzles or sprinklers shall be observed to ensure that patterns are not impeded by plugged nozzles, that nozzles are correctly positioned, and that obstructions do not prevent discharge patterns from wetting surfaces to be protected.

(A) Where the nature of the protected property is such that water cannot be discharged, the nozzles or open sprinklers shall be inspected for proper orientation and the system tested with air to ensure that the nozzles are not obstructed.

(B) Where obstructions occur, the piping and sprinklers or nozzles shall be cleaned and the system retested.

Paragraph 13.4.3.2.2.3 addresses the discharge of water into the piping system for a system protecting a freezer since water in this type of system could cause the immediate formation of ice.

 13.4.3.2.3 Except for preaction systems covered by 13.4.3.2.5, every 3 years the preaction valve shall be trip tested with the control valve fully open.

 13.4.3.2.4 During those years when full flow testing in accordance with 13.4.3.2.3 is required, the preaction valve shall be trip tested with the control valve partially open.

13.4.3.2.5 Preaction or deluge valves protecting freezers shall be trip tested in a manner that does not introduce moisture into the piping in the freezer.

13.4.3.2.6 Pressure Readings.

13.4.3.2.6.1 Pressure readings shall be recorded at the hydraulically most remote nozzle or sprinkler.

Hydraulic calculations for deluge systems may be based on nozzle pressure rather than the area/density design approach. This measurement will confirm that adequate pressure exists at the discharge device.

13.4.3.2.6.2 A second pressure reading shall be recorded at the deluge valve.

Most hydraulic calculation programs calculate pressure at the base of the riser. A reading taken at the deluge valve, as required by 13.4.3.2.6.2, provides an adequate comparison.

13.4.3.2.6.3 These readings shall be compared to the hydraulic design pressures to ensure the original system design requirements are met by the water supply.

13.4.3.2.6.4 Where the hydraulically most remote nozzle or sprinkler is inaccessible, nozzles or sprinklers in other than foam-water systems shall be permitted to be checked visually without taking a pressure reading on the most remote nozzle or sprinkler.

13.4.3.2.6.5 Where the reading taken at the riser indicates that the water supply has deteriorated, a gauge shall be placed on the hydraulically most remote nozzle or sprinkler and the results compared with the required design pressure.

13.4.3.2.7 Multiple Systems. The maximum number of systems expected to operate in case of fire shall be tested simultaneously to check the adequacy of the water supply.

Paragraph 13.4.3.2.7 requires the simultaneous testing of multiple systems because the discharge of multiple systems will have an effect on the water supply. It is important to test under conditions that may prevail during a fire. It is not the intent to require testing of multiple systems wherever they exist, only when the discharge of multiple systems is needed based on the system design documents should multiple systems be discharged for this test. Exhibit 13.18 illustrates several different types of systems in a building, not all of which would require simultaneous discharge or testing.

EXHIBIT 13.18 Multiple System Manifold.

13.4.3.2.8 Manual Operation. Manual actuation devices shall be operated annually.

13.4.3.2.9 Return to Service. After the full flow test, the system shall be returned to service in accordance with the manufacturer's instructions.

13.4.3.2.10 Grease or other sealing materials shall not be applied to the seating surfaces of preaction or deluge valves.

13.4.3.2.11* Records indicating the date the preaction or deluge valve was last tripped and the tripping time, as well as the individual and organization conducting the test, shall be maintained at a location or in a manner readily available for review by the authority having jurisdiction.

A.13.4.3.2.11 Methods of recording maintenance include tags attached at each riser, records retained at each building, and records retained at one building in a complex.

To comply with the requirements of 13.4.3.2.11, a maintenance log including all inspection, testing, and maintenance reports may be kept near the valve in a valve room or mechanical equipment room.

See Supplement 4 for sample report forms. Note that 13.4.3.2.11 does not preclude the use of electronic reporting systems or other methods for documenting the service history of valves and other equipment.

13.4.3.2.12 Low air pressure alarms, if provided, shall be tested quarterly in accordance with the manufacturer's instructions.

13.4.3.2.13 Low temperature alarms, if installed in valve enclosures, shall be tested annually at the beginning of the heating season.

13.4.3.2.14 Automatic air pressure maintenance devices, if provided, shall be tested yearly at the time of the annual preaction or deluge valve trip test, in accordance with the manufacturer's instructions.

13.4.3.3 Maintenance.

13.4.3.3.1 Leaks causing drops in supervisory pressure sufficient to sound warning alarms, and electrical malfunctions causing alarms to sound, shall be located and repaired.

13.4.3.3.2 During the annual trip test, the interior of the preaction or deluge valve shall be cleaned thoroughly and the parts replaced or repaired as necessary.

13.4.3.3.2.1 Interior cleaning and parts replacement or repair shall be permitted every 5 years for valves that can be reset without removal of a faceplate.

13.4.3.3.3* Auxiliary drains in preaction or deluge systems shall be operated after each system operation and before the onset of freezing conditions.

A.13.4.3.3.3 Suitable facilities should be provided to dispose of drained water. Low points equipped with a single valve should be drained as follows:

(1) Open the low-point drain valve slowly.
(2) Close the drain valve as soon as water ceases to discharge, and allow time for additional accumulation above the valve.
(3) Repeat this procedure until water ceases to discharge.
(4) Replace plug or nipple and cap as necessary.

Low points equipped with dual valves should be drained as follows:

(1) Close the upper valve.
(2) Open the lower valve, and drain the accumulated water.
(3) Close the lower valve, open the upper valve, and allow time for additional water accumulation.
(4) Repeat this procedure until water ceases to discharge.
(5) Replace plug or nipple and cap in lower valve.

13.4.3.3.4 Additional maintenance as required by the manufacturer's instructions shall be provided.

13.4.4 Dry Pipe Valves/Quick-Opening Devices.

A dry pipe system valve is a differential type valve and is more sophisticated than an alarm check valve. Consequently, a dry pipe valve requires more testing and maintenance than an alarm check valve. See Exhibit 13.19 for a typical dry pipe valve.

EXHIBIT 13.19 Typical Dry Pipe Valve.

13.4.4.1 Inspection.

13.4.4.1.1 Valve enclosure heating equipment shall be inspected daily during cold weather for its ability to maintain a minimum temperature of at least 40°F (4°C).

13.4.4.1.1.1 Valve enclosures equipped with low temperature alarms shall be inspected weekly.

13.4.4.1.1.2 Low temperature alarms, if installed in valve enclosures, shall be inspected annually at the beginning of the heating season.

13.4.4.1.2 Gauges shall be inspected weekly.

13.4.4.1.2.1 The gauge on the supply side of the dry pipe valve shall indicate that the normal supply water pressure is being maintained.

13.4.4.1.2.2 The gauge on the system side of the dry pipe valve shall indicate that the proper ratio of air or nitrogen pressure to water supply pressure is being maintained in accordance with the manufacturer's instructions.

13.4.4.1.2.3* The gauge on the quick-opening device, if provided, shall indicate the same pressure as the gauge on the system side of the dry pipe valve.

A.13.4.4.1.2.3 A conflict in pressure readings could indicate an obstructed orifice or a leak in the isolated chamber of the quick-opening device, either of which could make the quick-opening device inoperative.

13.4.4.1.3 Systems with auxiliary drains shall require a sign at the dry or preaction valve indicating the number of auxiliary drains and location of each individual drain.

13.4.4.1.4 The dry pipe valve shall be externally inspected monthly to verify the following:

(1) The valve is free of physical damage.
(2) All trim valves are in the appropriate open or closed position.
(3) The intermediate chamber is not leaking.

13.4.4.1.5 The interior of the dry pipe valve shall be inspected annually when the trip test is conducted.

13.4.4.1.6 Strainers, filters, and restricted orifices shall be inspected internally every 5 years unless tests indicate a greater frequency is necessary.

13.4.4.2 Testing.

13.4.4.2.1* The priming water level shall be tested quarterly.

A.13.4.4.2.1 High priming water levels can affect the operation of supervisory air or nitrogen pressure maintenance devices. Test the water level as follows:

(1) Open the priming level test valve.
(2) If water flows, drain it.
(3) Close the valve when water stops flowing and air discharges.
(4) If air discharges when the valve is opened, the priming water level could be too low. To add priming water, refer to the manufacturer's instructions.

13.4.4.2.2* Each dry pipe valve shall be trip tested annually during warm weather.

A.13.4.4.2.2 Dry pipe valves should be trip tested in the spring to allow time before the onset of cold weather for all water that has entered the system or condensation to drain to low points or back to the valve.

13.4.4.2.2.1 Dry pipe valves protecting freezers shall be trip tested in a manner that does not introduce moisture into the piping in the freezers.

> **CASE STUDY**
>
> **Dry Pipe Valves Fail to Operate**
>
> Fire in a warehouse in Illinois caused $28 million in damage. The facility was used as a cold-storage warehouse for perishable foods. The single-story warehouse was of unprotected noncombustible construction with a ground-floor area of 89,000 ft^2. The facility was operating at the time of the fire.
>
> The warehouse was fully equipped with sprinklers. Coverage consisted of one wet pipe system for the office and mezzanine areas and two dry pipe preaction systems with heat actuating devices. The systems' water supply was supplemented by a 1000 gpm electric fire pump.
>
> The running lights of a tractor trailer truck that had been backed into a loading dock were accidentally left on, and the heat they generated eventually ignited the polyurethane truck bumpers on the dock. The fire was discovered when the driver pulled the vehicle from the dock. Workers tried to extinguish the fire before notifying the fire department, but the fire spread through an open docking bay into the polystyrene insulation surrounding the cold-storage areas. The fire department arrived within 5 minutes but was unable to contain the fire due to its rapid extension and the collapse of the mezzanine. The warehouse and its contents were a total loss.
>
> Although the wet pipe sprinkler system operated as designed, the dry pipe valves for the two preaction systems failed to operate. The reason for this failure was not reported. The failure of the preaction system resulted in rapid fire extension throughout the warehouse.

13.4.4.2.2.2* Every 3 years and whenever the system is altered, the dry pipe valve shall be trip tested with the control valve fully open and the quick-opening device, if provided, in service.

Traditionally, the purpose of the full flow trip test was to verify the presence of water at the inspector's test connection within 60 seconds. It was never the intent of NFPA 25 to reverify the 60 second requirement every three years. It is the intent of NFPA 25 to reveal substantial delays on the water delivery time of dry pipe systems since drastic differences from one test to another are an indication of operational problems with the system. Of utmost concern when full flow trip testing a dry pipe system is a drastic increase in the water delivery time, because such an increase is most likely due to potential corrosion of the system piping, unless physical changes have been made to the system such as a change in the layout, orifice size, or water supply characteristics. Any drastic change in water delivery time from one test to another is cause for investigation as to the source of the problem.

Recently, NFPA 13 was revised to permit an alternate method for determining water delivery times for dry pipe systems. It is important for the inspector to be aware of this change when testing newer systems. The current requirements establish no specific water delivery time for systems of 500 gal (1893 L) capacity or less and for systems up to 750 gal (2839 L) equipped with a quick opening device. All other systems must meet the 60 second water delivery time unless calculated using a listed calculation program. When the calculation method is used, the maximum time of water delivery must not exceed that shown in Commentary Table 13.1.

A.13.4.4.2.2.2 A full flow trip test generally requires at least two individuals, one of whom is situated at the dry pipe valve while the other is at the inspector's test. If possible, they should be in communication with each other. A full flow trip test is conducted as follows:

COMMENTARY TABLE 13.1 Dry System Water Delivery

Hazard	Number of Most Remote Sprinklers Initially Open	Maximum Time of Water Delivery
Residential	1	15 seconds
Light	1	60 seconds
Ordinary I	2	50 seconds
Ordinary II	2	50 seconds
Extra I	4	45 seconds
Extra II	4	45 seconds
High piled	4	40 seconds

(1) The main drain valve is fully opened to clean any accumulated scale or foreign material from the supply water piping. The main drain valve then is closed.
(2) The system air or nitrogen pressure and the supply water pressure are recorded.
(3) The system air or nitrogen pressure is relieved by opening the inspector's test valve completely. Concurrent with opening the valve, both testers start their stopwatches. If two-way communication is not available, the tester at the dry valve is to react to the start of downward movement on the air pressure gauge.
(4) Testers at the dry pipe valve note the air pressure at which the valve trips and note the tripping time.
(5) Testers at the inspector's test note the time at which water flows steadily from the test connection. This time is noted for comparison purposes to previous tests and is not meant to be a specific pass/fail criterion. Note that NFPA 13, *Standard for the Installation of Sprinkler Systems*, does not require water delivery in 60 seconds for all systems.
(6) When clean water flows, the test is terminated by closing the system control valve.
(7) The air or nitrogen pressure and the time elapsed are to be recorded as follows:
  (a) From the complete opening of the test valve to the tripping of the valve
  (b) From the complete opening of inspector's valve to the start of steady flow from the test connection
(8) All low-point drains are opened and then closed when water ceases to flow.
(9) The dry pipe valve and quick-opening device are reset, if installed, in accordance with the manufacturer's instructions, and the system is returned to service.

For dry pipe systems that were designed and installed using either a manual demonstration or a computer calculation to simulate multiple openings to predict water delivery time, a full flow trip test from a single inspector's test connection should have been conducted during the original system acceptance and a full flow trip test from the single inspector's test should continue to be conducted every 3 years. The system is not required to achieve water delivery to the inspector's test connection in 60 seconds, but comparison to the water delivery time during the original acceptance will determine if there is a problem with the system.

13.4.4.2.2.3* During those years when full flow testing in accordance with 13.4.4.2.2.2 is not required, each dry pipe valve shall be trip tested with the control valve partially open.

A.13.4.4.2.2.3 A partial flow trip test is conducted in the following manner:

(1) Fully open the main drain valve to clean any accumulated scale or foreign material from the supply water piping.
(2) Close the control valve to the point where additional closure cannot provide flow through the entire area of the drain outlet.
(3) Close the valve controlling flow to the device if a quick-opening device is installed.

(4) Record the system air or nitrogen pressure and the supply water pressure.
(5) Relieve system air or nitrogen pressure by opening the priming level test valve.
(6) Note and record the air or nitrogen pressure and supply water pressure when the dry pipe valve trips.
(7) Immediately close the system control valve, and open the main drain valve to minimize the amount of water entering the system piping.
(8) Trip test the quick-opening device, if installed, in accordance with the manufacturer's instructions.
(9) Open all low point drains; close when water ceases to flow.
(10) Reset the dry pipe valve and quick-opening device, if installed, in accordance with the manufacturer's instructions and return the system to service.

CAUTION: A partial flow trip test does not provide a high enough rate of flow to latch the clappers of some model dry pipe valves in the open position. When resetting such valves, check that the latching equipment is operative.

13.4.4.2.3 Grease or other sealing materials shall not be applied to the seating surfaces of dry pipe valves.

13.4.4.2.4* Quick-opening devices, if provided, shall be tested quarterly.

A.13.4.4.2.4 Except when a full flow trip test is conducted in accordance with A.13.4.4.2.2.2, a quick-opening device should be tested in the following manner:

(1) Close the system control valve.
(2) Open the main drain valve, and keep it in the open position.
(3) Verify that the quick-opening device control valve is open.
(4) Open the inspector's test valve. A burst of air from the device indicates that it has tripped.
(5) Close the device's control valve.
(6) Return the device to service in accordance with the manufacturer's instructions, and return the system to service.

13.4.4.2.5 A tag or card that shows the date on which the dry pipe valve was last tripped, and the name of the person and organization conducting the test, shall be attached to the valve.

See Supplement 4 for sample trip test report forms.

13.4.4.2.5.1 Separate records of initial air and water pressure, tripping air pressure, and dry pipe valve operating conditions shall be maintained on the premises for comparison with previous test results.

13.4.4.2.5.2 Records of tripping time shall be maintained for full flow trip tests.

13.4.4.2.6 Low air pressure alarms, if provided, shall be tested quarterly in accordance with the manufacturer's instructions.

13.4.4.2.7 Low temperature alarms, if installed in valve enclosures, shall be tested annually at the beginning of the heating season.

13.4.4.2.8 Automatic air pressure maintenance devices, if provided, shall be tested annually during the dry pipe valve trip test in accordance with the manufacturer's instructions.

13.4.4.2.9 Dry pipe systems shall be tested once every three years for air leakage, using one of the following test methods:

(1) A pressure test at 40 psi for two hours. The system shall be permitted to lose up to 3 psi (0.2 bar) during the duration of the test. Air leaks shall be addressed if the system loses more than 3 psi (0.2 bar) during this test.
(2) With the system at normal system pressure, shut off the air source (compressor or shop air) for 4 hours. If the low air pressure alarm goes off within this period, the air leaks shall be addressed.

The previous 10 psi (0.7 bar) pressure loss per week repair requirement was difficult to measure because an automatic air supply would generally maintain pressure by cycling on and off. The intent of 13.4.4.2.9 is to require an air pressure test to determine if a significant problem exists. The dry pipe system must be capable of staying closed without any trips of the dry pipe valve for short periods of time such as a power outage without the air compressor maintaining pressure. Either method listed in 13.4.4.2.9 can be used to demonstrate the tightness of the system.

EXAMPLE

If a dry pipe valve is designed to trip at 25 psi (1.7 bar), the air supply should be set to provide 20 psi (1.4 bar) in excess of this pressure or 45 psi (2.8 bar). The automatic air compressor could be set to turn on at 35 psi (2.4 bar) and turn off at 45 psi (3.1 bar). The low air pressure switch may be set at 30 psi (2.1 bar) to produce a low air alarm. In this case, assuming the air compressor is about to turn on (a worst case scenario), the system would only be permitted to lose 5 psi (0.3 bar). At the high end of the air compressor cycle, the system would be permitted to lose 15 psi (1 bar).

13.4.4.3 Maintenance.

13.4.4.3.1 During the annual trip test, the interior of the dry pipe valve shall be cleaned thoroughly, and parts replaced or repaired as necessary.

13.4.4.3.2* Auxiliary drains in dry pipe sprinkler systems shall be drained after each operation of the system, before the onset of freezing weather conditions, and thereafter as needed.

A.13.4.4.3.2 A quick-opening device, if installed, should be removed temporarily from service prior to draining low points.

The low point drain in a dry pipe system is commonly referred to as a condensate nipple. The condensate nipple consists of a 1 ft (0.3 m) length of 2 in. (50 mm) pipe with a 1 in. (25 mm) valve on each end. The upper valve is closed to isolate the condensate nipple from the dry pipe system. Next, the lower valve is opened to drain any moisture from within. Once draining is complete, the upper valve is opened to allow moisture to drain into the condensate nipple.

Exhibit 13.20 shows a low point drain on a dry pipe system. Exhibit 13.21 shows a corroded pipe from a dry pipe system that was not drained properly.

13.5 Pressure Reducing Valves and Relief Valves

13.5.1 Inspection and Testing of Sprinkler Pressure Reducing Valves.

Sprinkler pressure reducing valves shall be inspected and tested as described in 13.5.1.1 and 13.5.1.2.

13.5.1.1 All valves shall be inspected quarterly to verify that the valves are in the following condition:

(1) In the open position
(2) Not leaking
(3) Maintaining downstream pressures in accordance with the design criteria
(4) In good condition, with handwheels installed and unbroken

Pressure reducing and relief valves are installed to control pressures in the fire protection system. They are usually set to reduce pressures on the system side of the valve to pressures not exceeding 175 psi (12.1 bar). This is the maximum pressure that system components can

EXHIBIT 13.20 Low Point Drain.

withstand without failure. Exhibit 13.22 illustrates pressure reducing valves for both a sprinkler and a standpipe system, Exhibit 13.23 illustrates a master pressure reducing valve, and Exhibit 13.24 illustrates a pressure relief valve.

13.5.1.2* A full flow test shall be conducted on each valve at 5-year intervals and shall be compared to previous test results.

A.13.5.1.2 The sectional drain valve should be opened to compare the results with the original installation or acceptance tests.

As noted in A.13.5.1.2, the sectional drain valve is opened to produce flow for testing purposes. Pressures should be measured on each side of the valve to verify that high pressures on the supply side of the valve are reduced to not more than 175 psi (12.1 bar) on the system side of the valve. Discharge should be to a drain connection or receptacle that can accommodate the flow without spilling.

13.5.1.2.1 Adjustments shall be made in accordance with the manufacturer's instructions.

Not all valves are field adjustable. Valves that are adjustable should be adjusted in accordance with the design and installation documents and, as required by 13.5.1.2.1, the manufacturer's instructions. Exhibit 13.25 illustrates test apparatus for testing a pressure reducing valve.

13.5.1.3 A partial flow test adequate to move the valve from its seat shall be conducted annually.

The test required in 13.5.1.3 is intended to exercise the valve and ensure that it is not obstructed and will re-seat.

EXHIBIT 13.21 Corroded Pipe.

EXHIBIT 13.22 Pressure Reducing Valves for Both a Sprinkler and a Standpipe System.

EXHIBIT 13.23 Master Pressure Reducing Valve.

EXHIBIT 13.24 Pressure Relief Valve.

EXHIBIT 13.25 Pressure Reducing Valve Test Assembly. (Courtesy of Oliver Sprinkler Co., Inc.)

13.5.2 Hose Connection Pressure Reducing Valves.

Pressure reducing valves for hose connections are installed to reduce pressures to a safe level for the system user.

Hose connections intended for use by trained industrial fire brigades are provided in $1\frac{1}{2}$ in. (40 mm) diameters and must be restricted to not more than 100 psi (7 bar). Hose connections intended for fire department use must be restricted to not more than 175 psi (12.1 bar).

For more information regarding hose connection pressures, see Chapter 7 of NFPA 14, 2007 edition.

 13.5.2.1 All valves shall be inspected annually to verify the following:

(1) The handwheel is not broken or missing.
(2) The outlet hose threads are not damaged.
(3) No leaks are present.
(4) The reducer and the cap are not missing.

The hose connection cap, which is permitted to be plastic, is intended to protect the hose threads only. If the cap is removed, material such as trash could be placed in the valve, creating an obstruction.

13.5.2.2* A full flow test shall be conducted on each valve at 5-year intervals and shall be compared to previous test results.

A.13.5.2.2 PRV devices can be bench tested in accordance with the manufacturer's instructions or tested in place. To test in place, a gauge is connected on both the inlet side and the outlet side of the device, and flow readings are taken using a Pitot tube or a flowmeter. Water is discharged through a roof manifold, if available, or through hose to the exterior of the building. Another acceptable method for systems having at least two risers is to take one standpipe out of service and use it as a drain by removing PRV devices and attaching hoses at the outlets near the ground floor level. When testing in this manner, a flowmeter should be used and a hose line utilized to connect the riser being tested and the drain riser.

Readings are to be compared to the system's hydraulic demands at the test location. Field-adjustable valves are to be reset in accordance with manufacturer's instructions. Non-adjustable valves should be replaced. Extreme caution should be exercised because of the high pressure involved when testing.

Standpipe systems installed prior to 1993 must be tested as described in A.13.5.2.2. Since 1993, NFPA 14 has required the installation of a 3 in. (75 mm) drain riser whenever pressure reducing devices are installed in a standpipe system. The drain riser must provide a $2\frac{1}{2}$ in. (65 mm) swivel outlet on every other floor for connection of hose.

13.5.2.2.1 Adjustments shall be made in accordance with the manufacturer's instructions.

13.5.2.3 A partial flow test adequate to move the valve from its seat shall be conducted annually.

The test in 13.5.2.3 is intended to exercise the valve and ensure that it is not obstructed and will re-seat.

13.5.3 Hose Rack Assembly Pressure Reducing Valves.

Hose rack pressure reducing valves, covered in 13.5.3, may be either $1\frac{1}{2}$ in. (40 mm) or $2\frac{1}{2}$ in. (65 mm) in size and will be pre-connected to a hose rack complete with hose and nozzle. A hose rack may be connected directly to a sprinkler system or may be part of a standpipe system. In either case, the test requirements for the valve are identical to those specified in 13.5.2.

 13.5.3.1 All valves shall be inspected annually to verify the following:

(1) The handwheel is not missing or broken.
(2) No leaks are present.

13.5.3.2 A full flow test shall be conducted on each valve at 5-year intervals and compared to previous test results.

13.5.3.2.1 Adjustments shall be made in accordance with the manufacturer's instructions.

13.5.3.3 A partial flow test adequate to move the valve from its seat shall be conducted annually.

13.5.4 Master Pressure Reducing Valves.

NFPA 14, 2007 edition, now recognizes the use of a master pressure reducing valve (PRV) in standpipe systems. The installation of these devices is permitted downstream of the fire pump discharge piping and can be installed in such a way as to control the pressure to an entire system or an entire zone of a standpipe system. The master PRV must be installed with a bypass arrangement that will permit maintenance of the device without initiating an impairment procedure. Sufficient valving must also be installed to allow isolation of the device for maintenance and testing purposes. Some installations may include multiple master PRVs: one for the higher flow of standpipe systems and one for the lower flows of sprinkler systems. Other installations may include two PRVs installed in series to prevent overpressurization of a standpipe system should one valve fail. See Exhibit 13.26 for typical PRV installations.

13.5.4.1* Valves shall be inspected weekly to verify that the valves are in the following condition:

(1)*The downstream pressures are maintained in accordance with the design criteria.

A.13.5.4.1(1) Pressures downstream of the master PRV should not exceed the maximum pressure rating of the system components.

(2) The supply pressure is in accordance with the design criteria.
(3) The valves are not leaking.
(4) The valve and trim are in good condition.

A.13.5.4.1 When the PRV is located in or immediately downstream of the fire pump discharge, the weekly inspection of the master PRV can be performed during the weekly fire pump operating test.

13.5.4.2* A partial flow test adequate to move the valve from its seat shall be conducted quarterly.

A.13.5.4.2 The partial flow test of the master PRV can be performed during the quarterly main drain test. *(See 13.2.5.1.)*

13.5.4.3* A full flow test shall be conducted on each valve annually and shall be compared to previous test results.

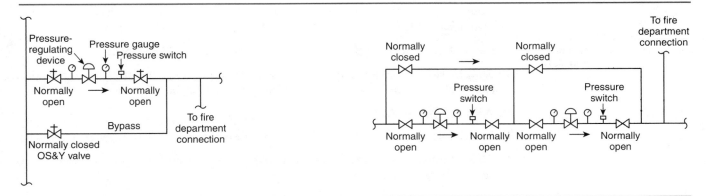

EXHIBIT 13.26 *Typical PRV Installations. Pressure regulating device arrangement (left) and dual pressure regulating device arrangement (right). [Source: NFPA 14, 2007, Figures A.7.2.2(a) and A.7.2.2(b)]*

A.13.5.4.3 When the PRV is located in the fire pump discharge, the full flow test of the master PRV can be performed during the annual fire pump flow test.

13.5.4.4 When valve adjustments are necessary, they shall be made in accordance with the manufacturer's instructions.

13.5.5 Pressure Reducing Valves.

13.5.5.1 All pressure reducing valves installed on fire protection systems not covered by 13.5.1, 13.5.2, 13.5.3, or 13.5.4 shall be inspected in accordance with 13.5.1.1.

13.5.5.2 All pressure reducing valves installed on fire protection systems not covered by 13.5.1, 13.5.2, 13.5.3, or 13.5.4 shall be tested in accordance with 13.5.1.2.

13.5.6 Hose Valves.

13.5.6.1 Inspection.

13.5.6.1.1 Hose valves shall be inspected quarterly.

13.5.6.1.2 Hose valves shall be inspected to ensure that hose caps are in place and not damaged.

13.5.6.1.3 Hose threads shall be inspected for damage.

13.5.6.1.4 Valve handles shall be present and not damaged.

13.5.6.1.5 Gaskets shall be inspected for damage or deterioration.

13.5.6.1.6 Hose valves shall be inspected for leaks.

13.5.6.1.7 Hose valves shall be inspected to ensure no obstructions are present.

13.5.6.1.8 Hose valves shall be inspected to ensure that restricting devices are present.

13.5.6.2 Testing.

The testing and maintenance requirements described in 13.5.6.2 should be performed by a qualified contractor.

13.5.6.2.1* Class I and Class III standpipe system hose valves shall be tested annually by opening and closing the valves.

A.13.5.6.2.1 Hose valves can be tested without a full flow if the cap is left on the hose threads. The purpose of this requirement is to exercise the valve so it can be operated easily.

FAQ ▶ Does the requirement in 13.5.6.2.1 mean the valves need to be flowed, or just opened and closed?

The purpose of this requirement is to exercise the valve so it can be operated easily. Basically, exercising the valve is the intent of this requirement.

13.5.6.2.1.1 Hose valves that are difficult to operate or leak shall be repaired or replaced.

13.5.6.2.2* Hose valves on hose stations attached to sprinkler systems and Class II standpipe systems shall be tested every 3 years by opening and closing the valves.

A.13.5.6.2.2 See A.13.5.6.2.1.

13.5.6.2.2.1 Hose valves that are difficult to operate or that leak shall be repaired or replaced.

13.5.6.3 Maintenance. Hose valves that do not operate smoothly or open fully shall be lubricated, repaired, or replaced.

13.5.7 Fire Pump Pressure Relief Valves.

13.5.7.1 All circulation relief valves shall be inspected weekly.

13.5.7.1.1 The inspection shall verify that water flows through the valve when the fire pump is operating at shutoff pressure (i.e., churn), to prevent the pump from overheating.

13.5.7.1.2 During the annual fire pump test, the closure of the circulation relief valve shall be verified to be in accordance with the manufacturer's specifications.

13.5.7.2 All pressure relief valves shall be inspected weekly.

13.5.7.2.1 The inspection shall verify that the pressure downstream of the relief valve fittings in the fire pump discharge piping does not exceed the pressure for which the system components are rated.

13.5.7.2.2 During the annual fire pump flow test, the pressure relief valve shall be verified to be correctly adjusted and set to relieve at the correct pressure and to close below that pressure setting.

Paragraph 13.5.7.2.2 applies only to those pump installations equipped with a pressure relief valve. Such installations are those that produce pressures in excess of the rating for system components. Other installations will not be so equipped (see Exhibit 13.27).

EXHIBIT 13.27 Pressure Relief Valve with Sight Glass. (Courtesy of John Jensen)

13.5.8 Maintenance.

All damaged or missing components noted during the inspections specified in 13.5.6.1 through 13.5.6.2.2 shall be repaired or replaced in accordance with the manufacturer's instructions.

See Chapter 8 for commentary regarding fire pumps and related equipment.

13.6 Backflow Prevention Assemblies

13.6.1 Inspection.

Inspection of backflow prevention assemblies shall be as described in 13.6.1.1 through 13.6.1.2.2.

13.6.1.1 The double check assembly (DCA) valves and double check detector assembly (DCDA) valves shall be inspected weekly to ensure that the OS&Y isolation valves are in the normal open position.

13.6.1.1.1 Valves secured with locks or electrically supervised in accordance with applicable NFPA standards shall be inspected monthly.

OS&Y valve inspections can be accomplished by the property owner.

CASE STUDY

Sprinkler System Inoperative

Seven people died in a fire in a New York clothing store. The two-story store of fire-resistive construction was in a multiple-occupancy business block and was operating at the time of the fire.

A wet pipe sprinkler system was located in the basement of the clothing store. The system was inoperative because an OS&Y valve, located in an adjacent store, was closed. The fire department siamese connection was covered by a construction shed at the front entrance and could not be used. There were no detection systems.

The fire began when a gunman entered the store, herded employees into a basement office, then poured flammable liquid over racks of clothing and ignited them. The gunman then went to the first floor and ignited additional clothing racks there, trapping three more employees and himself. Although some employees were able to escape, a total of seven fatalities occurred as a result of the fire and failure of the system.

This case study illustrates the frequently published statistics indicating that control valves are the leading cause of sprinkler system failure. NFPA 25 devotes an entire chapter to valves of all types in an effort to maintain the operating status of valves. Inspection of control valves cannot be overemphasized, because this is the best method to ensure that they are open.

13.6.1.2* Reduced pressure assemblies (RPA) and reduced pressure detector assemblies (RPDA) shall be inspected weekly to ensure that the differential-sensing valve relief port is not continuously discharging and the OS&Y isolation valves are in the normal open position.

A.13.6.1.2 Intermittent discharge from a differential-sensing valve relief port is normal. Continuous discharge is a sign of malfunction of either or both of the check valves, and maintenance is necessary.

The property owner or occupant can accomplish this inspection with minimal training.

13.6.1.2.1 Valves secured with locks or electrically supervised in accordance with applicable NFPA standards shall be inspected monthly.

13.6.1.2.2 After any testing or repair, an inspection by the property owner shall be made to ensure that the system is in service and all isolation valves are in the normal open position and properly locked or electrically supervised.

13.6.2 Testing.

13.6.2.1* All backflow preventers installed in fire protection system piping shall be tested annually in accordance with the following:

(1) A forward flow test shall be conducted at the designed flow rate, including hose stream demand, of the system, where hydrants or inside hose stations are located downstream of the backflow preventer.
(2) A backflow performance test, as required by the authority having jurisdiction, shall be conducted at the completion of the forward flow test.

A.13.6.2.1 The full flow test of the backflow prevention valve can be performed with a test header or other connections downstream of the valve. A bypass around the check valve in the fire department connection line with a control valve in the normally closed position can be an acceptable arrangement. When flow to a visible drain cannot be accomplished, closed loop flow can be acceptable if a flowmeter or sight glass is incorporated into the system to ensure flow.

The tests required by 13.6.2 typically test only for operation of the device under backflow conditions. Forward-flow test conditions are required by other portions of this standard.

The flow test and backflow test are best performed by a qualified contractor. Local health regulations may require a special testing license for the backflow performance test.

On systems installed under the requirements of the 2007 edition of NFPA 13, a test connection will be provided to accommodate the necessary flow for this test. On older systems, either a bypass or a method of holding the FDC clappers open will be needed.

13.6.2.1.1 For backflow preventers sized 2 in. (50 mm) and under, the forward flow test shall be acceptable to conduct without measuring flow, where the test outlet is of a size to flow the system demand.

13.6.2.1.2 Where water rationing is enforced during shortages lasting more than 1 year, an internal inspection of the backflow preventer to ensure the check valves will fully open shall be acceptable in lieu of conducting the annual forward flow test.

13.6.2.1.3 Where connections do not permit a full flow test, tests shall be completed at the maximum flow rate possible.

13.6.2.1.4 The forward flow test shall not be required where annual fire pump testing causes the system demand to flow through the backflow preventer device.

Many local health authorities require that a backflow preventer be the first device installed on a water supply pipe. In such cases, the annual fire pump test will produce sufficient flow through the backflow preventer to meet the full forward flow test requirement. In some cases, a backflow preventer can be installed on the system side of a fire pump (ordinarily between the FDC and the potable water supply), so the annual fire pump test does not produce any flow through the backflow preventer. When this is the case, a separate flow test is necessary.

13.6.2.2 Where connections do not permit a full flow test, tests shall be conducted at the maximum flow rate possible.

13.6.3 Maintenance.

13.6.3.1 Maintenance of all backflow prevention assemblies shall be conducted by a trained individual following the manufacturer's instructions in accordance with the procedure and policies of the authority having jurisdiction.

13.6.3.2 Rubber parts shall be replaced in accordance with the frequency required by the authority having jurisdiction and the manufacturer's instructions.

It is good practice to maintain a spare parts kit on-site with the installation of a backflow preventer. In some instances, local health regulations may require doing so.

13.7 Fire Department Connections

13.7.1 Fire department connections shall be inspected quarterly. The inspection shall verify the following:

(1) The fire department connections are visible and accessible.
(2) Couplings or swivels are not damaged and rotate smoothly.

(3) Plugs or caps are in place and undamaged.
(4) Gaskets are in place and in good condition.
(5) Identification signs are in place.
(6) The check valve is not leaking.
(7) The automatic drain valve is in place and operating properly.
(8) The fire department connection clapper(s) is in place and operating properly.

Exhibit 13.28 illustrates a typical FDC inspection where the caps are removed. The inspector examines inside the FDC for evidence of obstructing material and checks that the couplings are not damaged, the gaskets are in good condition, and the swivels rotate properly. Exhibit 13.29 illustrates an FDC that is obscured from view by vegetation. Visibility, particularly at night, is important so that the fire department can locate the device easily. Exhibit 13.30 illustrates an FDC that has been broken or vandalized. In this case, a cap is missing, which can allow obstructing material to enter the device. Exhibit 13.31 illustrates an FDC that is in good condition and has all caps intact.

EXHIBIT 13.28 FDC Inspection.

EXHIBIT 13.29 FDC Obstructed by Vegetation.

13.7.2 If fire department connection plugs or caps are not in place, the interior of the connection shall be inspected for obstructions, and it shall be verified that the fire department connection clapper is operational over its full range.

The property owner or occupant can accomplish this inspection with just minimal training.

13.7.3 Components shall be repaired or replaced as necessary in accordance with the manufacturer's instructions. Any obstructions that are present shall be removed.

SUMMARY

System and control valves form a critical part of any fire protection system. Water supply control valves, in particular, are the leading cause of system failure. A comprehensive inspection,

EXHIBIT 13.30 FDC Missing Caps.

EXHIBIT 13.31 FDC Showing Clear Access with Caps Intact.

testing, and maintenance program is necessary to ensure that a system will operate correctly. Further, because a fire protection system may have many water supply control valves, it is very important that they all be open at all times. When a water supply control valve must be closed for any reason, the requirements in Chapter 15 must be met.

Although system control valves are not necessarily complex, they do require specialized care by a qualified contractor to remain functional throughout their life cycle.

A property owner or his or her designee can arrange to have a qualified contractor perform all of the operations required to inspect, test, and maintain any type of fire protection system.

REVIEW QUESTIONS

1. When must a main drain test be performed?
2. When is an interior inspection required for a deluge/preaction valve?
3. Discuss the pass/fail criteria for the full flow trip test for a dry pipe system having a capacity in excess of 750 gal (2839 L).
4. Describe the test procedure and frequency for a forward flow test of a backflow preventer.
5. How often must pressure reducing valves for sprinkler and hose connections be flow tested?

REFERENCES CITED IN COMMENTARY

National Fire Protection Association, 1 Batterymarch Park, Quincy, MA 02169-7471.

NFPA 13, *Standard for the Installation of Sprinkler Systems,* 2007 edition.
NFPA 14, *Standard for the Installation of Standpipe and Hose Systems,* 2007 edition.
NFPA 15, *Standard for Water Spray Fixed Systems for Fire Protection,* 2007 edition.
NFPA 16, *Standard for the Installation of Foam-Water Sprinkler and Foam-Water Spray Systems,* 2007 edition.
NFPA 20, *Standard for the Installation of Stationary Pumps for Fire Protection,* 2007 edition.
NFPA 22, *Standard for Water Tanks for Private Fire Protection,* 2003 edition.
NFPA 72®, *National Fire Alarm Code®,* 2007 edition.

Obstruction Investigation

CHAPTER 14

Piping systems of any type can become inoperative when the pipe is filled with obstructing material. Sprinkler systems are no exception. Any maintenance program must include means for revealing potential obstructions and removing any obstructions that exist. Chapter 14 of NFPA 25 directly addresses this issue by listing potential causes of piping obstructions and requiring an obstruction investigation when any of those situations exist.

14.1 General

This chapter shall provide the minimum requirements for conducting investigations of fire protection system piping for possible sources of materials that could cause pipe blockage.

14.2* Obstruction Investigation and Prevention

A.14.2 For obstruction investigation and prevention, see Annex D.

14.2.1 An inspection of piping and branch line conditions shall be conducted every 5 years by opening a flushing connection at the end of one main and by removing a sprinkler toward the end of one branch line for the purpose of inspecting for the presence of foreign organic and inorganic material.

Subsection 14.2.1 applies to all four types of sprinkler systems (wet, dry, preaction, and deluge), as well as standpipes, foam water sprinklers, and private fire service mains.

Subsection 14.2.1 was added to the 2002 edition of NFPA 25 in an attempt to prevent the destruction of piping systems caused by microbiologically influenced corrosion (MIC). The 2008 edition modified this requirement to specifically indicate that 14.2.1 is intended to require an inspection, not a complete (and subsequently expensive) obstruction investigation. The investigation inspection required by 14.2.1 is intended to reveal the presence of MIC, zebra mussels, or inorganic material such as rust and scale, which are an increasingly common sources of piping obstruction. See Supplement 1 for more information on MIC.

The inspection required by 14.2.1 should be coordinated with the internal inspection requirement of system control valves, such as is required by 14.4.1.2 for alarm valves. Subsection 14.2.1 is not intended to place an additional burden on the property owner by requiring an additional inspection every five years.

Zebra mussels and quagga mussels can enter system piping when the fire protection system is supplied by a natural source of water, such as a river or lake. The mussels can enter the piping as larvae or young adults and then can grow into obstructions. The zebra mussel species entered the United States through the Great Lakes while attached to the hulls of foreign ships. The mussels then spread throughout the Great Lakes region and beyond. See Exhibit 14.1 for a photo of a typical zebra mussel, which is named for the stripes on the shell. The map in Exhibit 14.2 illustrates the extent of zebra mussel infestation.

◄ **FAQ**
Does 14.2.1 apply to only dry systems or to both dry and wet systems?

EXHIBIT 14.1 Zebra Mussel.

EXHIBIT 14.2 Zebra Mussel Map. (Source: USGS)

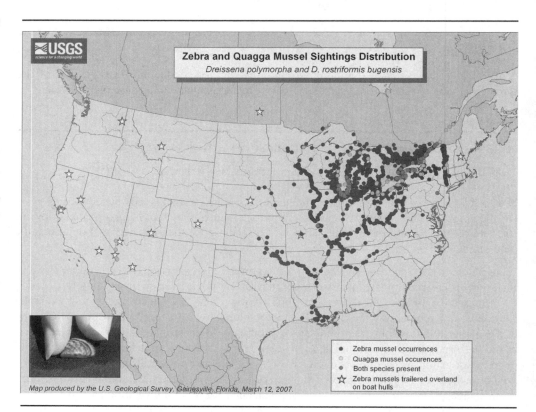

Other types of obstructing materials can take the form of work gloves, rust, scale, and other types of material. Exhibits 14.3 and 14.4 illustrate inorganic obstructions in system piping.

CASE STUDY

Sediment Blocks Sprinkler Heads

Fire in a furniture showroom in California caused $6 million in damage. This four-story building of protected, noncombustible construction covered a ground-floor

EXHIBIT 14.4 Weld Coupons Residing in System Piping.

EXHIBIT 14.3 Work Gloves Removed from System Piping. (Courtesy of John Jensen)

> area of approximately 44,000 ft². The showroom was closed but construction workers were in the building.
>
> The building had no automatic detection system but did have a partial-coverage sprinkler system. Sprinklers helped control fire spread on the second and third floors but weren't effective on the fourth floor because of sediment in the system.
>
> Fire fighters found sediment blocking several heads. The building also had portable extinguishers and a standpipe system. Investigators believe that workers used the extinguishers. Molten slag came in contact with furniture during welding operations and ignited a fire. The fire spread out the second-floor windows and into the third floor. Flames then breached a ceiling and entered the fourth floor where there was a flashover. No injuries were reported.

14.2.1.1 Alternative nondestructive examination methods shall be permitted.

The alternative nondestructive examination methods in 14.2.1.1 include the use of ultrasound, x-ray, and remote video techniques for the internal examination of piping systems. These methods are permitted in lieu of impairing the systems to conduct a visual examination. See also 14.2.3.3 and 14.3.1.

14.2.1.2 Tubercules or slime, if found, shall be tested for indications of microbiologically influenced corrosion (MIC).

◀ **FAQ**
Does NFPA 25 permit x-ray technology to "substitute" for the traditional flushing methodology to investigate for scale and debris within a dry pipe system?

FAQ ▶
What steps should be taken if MIC is found in the fire protection system?

With increasing awareness of the occurrence of MIC, 14.2.1.2 recognizes risk factors for its development. Test kits and services are available to analyze test samples for the presence of MIC. If MIC is found, special equipment and flushing services are available for treatment. For more information on MIC and other causes of obstruction, see Annex D and Supplement 1. See Exhibits 14.5 and 14.6 for examples of MIC in fire protection system piping.

EXHIBIT 14.5 *Pipe Severely Corroded by MIC.*

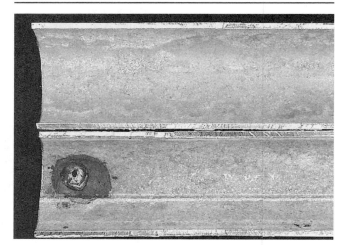

EXHIBIT 14.6 *Localized Corrosion Caused by MIC.*

14.2.2* An obstruction investigation shall be conducted for system or yard main piping wherever any of the following conditions exist:

(1) Defective intake for fire pumps taking suction from open bodies of water
(2) The discharge of obstructive material during routine water tests
(3) Foreign materials in fire pumps, in dry pipe valves, or in check valves

---

CASE STUDY

Corroded Valve Prevents Sprinkler Operation in Dairy Plant Fire

When a main sprinkler valve failed to operate because of corrosion, fire spread through the packaging area of a South Carolina dairy plant.

The single-story building had areas for receiving, processing, packaging, loading, cold storage, and offices. The large processing area was in the middle. There were two automatic sprinkler systems monitored by a central station. A preaction dry pipe sprinkler system actuated by rate-of-rise heat detectors protected the packaging, loading, and cold-storage areas. A wet pipe sprinkler system protected the processing area.

The fire began in the packaging and loading areas where two employees were using a cutting torch to remove a bolt on a stacker machine. The employees were not following any set procedure for safe torch operations and had exposed nearby combustibles. There were no fire extinguishers in the area.

Arriving five minutes after the initial call, fire fighters observed dense smoke issuing from the bay doors, and advanced a single $1\frac{3}{4}$ in. (43 mm) handline into the building. Because smoke was thick and reaching the seat of the fire was difficult, a second alarm was struck.

> A second engine company established a feeder line from a municipal hydrant to the first engine company and backed up the first line with a second 1¾ in. (43 mm) handline. With help from additional companies, positive-pressure ventilation, and more handlines, the fire was extinguished in about an hour.
>
> Officials found that all 40 preaction dry pipe sprinklers in the loading room were fused without any evidence of waterflow. When investigators opened the valve chamber, they discovered that corrosion — so extensive that a steel bar was needed to open the valve — had prevented proper operation of the main valve.
>
> A single sprinkler — located away from the immediate fire area and equipped with the wet pipe system — had operated and set off a waterflow alarm, which the central station received.
>
> Severe fire damage was limited to the packaging and loading areas. There was minor smoke and heat damage in the processing area. Thousands of gallons of milk and ice cream had to be disposed of because of smoke damage in the coolers. The combined loss for the structure and its contents was estimated at $375,000. No injuries were reported.

Aside from the lack of proper procedures for the use of cutting equipment, this case study presents a classic case of poor maintenance. Had the combined dry pipe–preaction system been maintained correctly, the quarterly waterflow alarm test required as part of a maintenance program would have revealed a problem with the system. Failure of the waterflow alarm during the test would have resulted in an internal examination of the valves and subsequent obstruction investigation of the entire system. Paragraph 14.2.2(3) requires that an obstruction investigation take place whenever obstructing material is found in pumps, dry pipe valves, and so on.

Had the operational status of the system been maintained, it likely would have substantially reduced the monetary loss.

(4) Foreign material in water during drain tests or plugging of inspector's test connection(s)
(5) Plugged sprinklers
(6) Plugged piping in sprinkler systems dismantled during building alterations
(7) Failure to flush yard piping or surrounding public mains following new installations or repairs
(8) A record of broken public mains in the vicinity
(9) Abnormally frequent false tripping of a dry pipe valve(s)
(10) A system that is returned to service after an extended shutdown (greater than 1 year)
(11) There is reason to believe that the sprinkler system contains sodium silicate or highly corrosive fluxes in copper systems
(12) A system has been supplied with raw water via the fire department connection
(13) Pinhole leaks
(14) A 50 percent increase in the time it takes water to travel to the inspector's test connection from the time the valve trips during a full flow trip test of a dry pipe sprinkler system when compared to the original system acceptance test

◀ FAQ
How can the presence of MIC be detected without an inspection?

The situations listed in 14.2.2 may not reveal the presence of MIC, unlike the inspection required by 14.2.1, which is intended to look specifically for MIC, zebra mussels, or other organic or inorganic obstructing material. In the absence of such an inspection, MIC is usually revealed by the presence of a pinhole leak. Unfortunately, by the time such a leak develops and is noticed, it is usually too late to save the affected pipe. It may even be too late to salvage much of the entire system.

An obstruction investigation that is poorly conducted or conducted too soon following the filling of a system may not reveal the presence of MIC.

For more information on MIC, see Annex D and Supplement 1.

A.14.2.2 For obstruction investigation procedures, see Section D.3. The type of obstruction investigation should be appropriately selected based on the observed condition. For instance, ordering an internal obstruction investigation would be inappropriate where the observed condition was broken public mains in the vicinity. On the other hand, such an investigation would be appropriate where foreign materials are observed in the dry pipe valve.

14.2.3* Systems shall be examined for internal obstructions where conditions exist that could cause obstructed piping.

A.14.2.3 For obstruction prevention program recommendations, see Section D.4.

14.2.3.1 If the condition has not been corrected or the condition is one that could result in obstruction of the piping despite any previous flushing procedures that have been performed, the system shall be examined for internal obstructions every 5 years.

It is important to note that, per 14.2.3.1, obstruction investigations are only required to be repeated on a five-year basis if the cause of the obstruction cannot be corrected (such as zebra mussel infestation). In all other cases, an obstruction investigation is required only when one of the conditions listed in 14.2.2 is found and corrected. An obstruction investigation is not intended to be a routine activity.

14.2.3.2 Internal inspections shall be accomplished by examining the interior of the following four points:

(1) System valve
(2) Riser
(3) Cross main
(4) Branch line

FAQ ▶
How many pipes have to be looked at if obstructions in the system are suspected?

The intent of 14.2.3.2 is that each of the main components be visually inspected. If an obstruction is found in one of these areas of the system, during an obstruction, it would be good practice to open a few other areas of the system to verify that the obstruction was an isolated incident. For branch lines, the intent is to open a representative number of branch lines as opposed to every branch line in the system.

It is also a good idea to run a flow test to make sure that the amount of water flowing through the system has not been compromised and that any obstructions have been cleared. However, in a case where the system runs through a freezer, water should not be in the piping unless there is a fire. For this type of circumstance, air flow through the system may be the only way to verify that the piping network is not blocked.

14.2.3.3 Alternative nondestructive examination methods shall be permitted.

The alternative nondestructive examination methods in 14.2.3.3 include the use of ultrasound, x-ray, and remote video camera technology.

14.2.4* If an obstruction investigation carried out in accordance with 14.2.1 indicates the presence of sufficient material to obstruct sprinklers, a complete flushing program shall be conducted by qualified personnel.

A.14.2.4 For obstruction investigation flushing procedures, see Section D.5.

The flushing procedures outlined in Section D.5 (Annex D) are intended to be carried out with water. In cases of severe MIC, chemical scouring may be necessary. This process is highly specialized and should only be attempted by those thoroughly experienced in the procedure. Chemicals should not be added to water used for flushing without a determination of the possible effects on sprinkler and valve seats.

14.3 Ice Obstruction

Dry pipe or preaction sprinkler system piping that protects or passes through freezers or cold storage rooms shall be inspected internally on an annual basis for ice obstructions at the point where the piping enters the refrigerated area.

The inspection described in Section 14.3 should be conducted regularly to ensure that piping is not obstructed with ice and that fittings are not fractured due to ice buildup. Commentary Table 1.1 indicated that frozen systems account for 1.4 percent of system failures.

14.3.1 Alternative nondestructive examinations shall be permitted.

The alternative nondestructive examination methods in 14.3.1 include the use of ultrasound, x-ray, and remote video camera technology.

14.3.2 All penetrations into the cold storage areas shall be inspected and, if an ice obstruction is found, additional pipe shall be examined to ensure no ice blockage exists.

SUMMARY

As indicated in Commentary Table 1.1, obstruction to water distribution accounts for 8.2 percent of system failures. This chapter provides inspection and obstruction investigation procedures and guidance on preventing the occurrence of obstructions in fire protection systems. Each system should be evaluated to determine if the potential sources of obstruction in 14.2.2 exist. If any of those obstruction sources do exist, preventive measures should be taken. Such measures may include treatment for MIC, the installation of strainers to prevent the introduction of obstructing material, or more frequent inspection and testing.

REVIEW QUESTIONS

1. How often must an obstruction investigation be conducted?
2. What four points of a system must be examined during an obstruction investigation?
3. What methods can be used to perform an obstruction investigation?
4. What should be done if tubercles or slime are found during an obstruction investigation?
5. How often is an inspection for foreign material required to be performed?

Impairments

CHAPTER 15

Rarely is a facility more exposed to a potentially catastrophic fire loss than when a fire protection system, or portion of a system, is impaired and out of service. Loss experience has shown time and again that many devastating fires could have been mitigated had a proper impairment program been in place and adhered to.

Whether an impairment is preplanned or occurs in an emergency situation seems to have little bearing on the propensity for catastrophe. What has been identified as a major factor leading to large fire losses is failure to do the following:

- Recognize the exposures created by the impairment
- Establish a fire watch and provide backup fire protection
- Control ignition sources and/or shut down hazardous processes
- Expedite repairs
- Ensure that fire protection is properly placed back into service

The requirements in Chapter 15 of NFPA 25 are straightforward and cover all important aspects of a suitable impairment program. However, the success of any program will be determined by the support it receives from those responsible for implementing it. Often, the best way to ensure that an impairment program is properly set up is to review it with the authority having jurisdiction (AHJ), whether that be the fire department, the insurance company, or other officials. If any situations arise that are not covered by the impairment program, seek the advice of the AHJ.

15.1 General

This chapter shall provide the minimum requirements for a water-based fire protection system impairment program. Measures shall be taken during the impairment to ensure that increased risks are minimized and the duration of the impairment is limited.

An impairment is the shutdown of a water-based fire protection system or a portion thereof. This chapter deals primarily with preplanned impairments — those impairments in which the system or a portion thereof is out of service due to work, such as modifications to the water supply or sprinkler system piping, that has been planned in advance.

◄ **FAQ**
What is an impairment?

15.2 Impairment Coordinator

15.2.1 The property owner shall assign an impairment coordinator to comply with the requirements of this chapter.

The impairment coordinator in 15.2.1 is usually someone from the building maintenance department. In larger facilities, it might be a member of the safety department.

The impairment coordinator should be on-site and familiar with the building layout, operations, and systems. In addition, he or she should have the ability to initiate proper notification when a system impairment is planned. This notification should include, but not be limited to, the fire department, insurance carrier, alarm company, property owner, property supervisors, and tenants.

Prior to removing a system from service, the impairment coordinator should have the necessary tools, equipment, and replacement parts on hand to minimize the duration of the impairment. For example, a stock of spare parts for each type of valve should be kept on the premises to avoid delays due to ordering and shipment of such equipment.

15.2.2 In the absence of a specific designee, the property owner shall be considered the impairment coordinator.

15.2.3 Where the lease, written use agreement, or management contract specifically grants the authority for inspection, testing, and maintenance of the fire protection system(s) to the tenant, management firm, or managing individual, the tenant, management firm, or managing individual shall assign a person as impairment coordinator.

15.3 Tag Impairment System

15.3.1* A tag shall be used to indicate that a system, or part thereof, has been removed from service.

A.15.3.1 A clearly visible tag alerts building occupants and the fire department that all or part of the water-based fire protection system is out of service. The tag should be weather resistant, plainly visible, and of sufficient size [typically 4 in. × 6 in. (100 mm × 150 mm)]. The tag should identify which system is impaired, the date and time impairment began, and the person responsible. Figure A.15.3.1 illustrates a typical impairment tag.

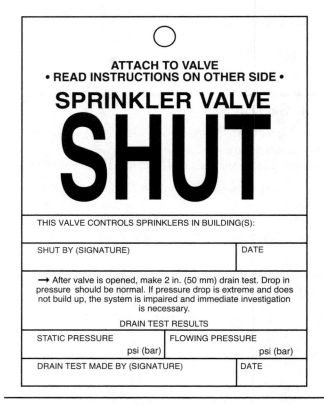

FIGURE A.15.3.1 Sample Impairment Tag.

A tag, once retrieved, can also be used for verification that a valve or system has been restored to service. Exhibit 15.1 illustrates a valve that has been closed for testing purposes and has been tagged indicating that an impairment program is in place.

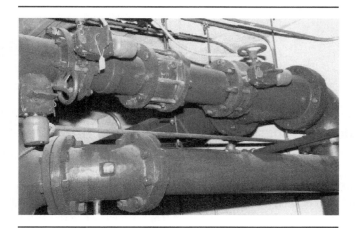

EXHIBIT 15.1 Closed and Tagged Valve.

15.3.2* The tag shall be posted at each fire department connection and system control valve, indicating which system, or part thereof, has been removed from service.

A.15.3.2 An impairment tag should be placed on the fire department connection to alert responding fire fighters of an abnormal condition. An impairment tag that is located on the system riser only could go unnoticed for an extended period if fire fighters encounter difficulty in gaining access to the building or sprinkler control room.

15.3.3 The authority having jurisdiction shall specify where the tag is to be placed.

CASE STUDY

Shut Valve

A gunman entered a clothing store in New York and herded six employees into a basement office, shooting one in the stomach. He then poured flammable liquid over racks of clothing and ignited a fire in the basement. The gunman then went to the first floor and ignited additional clothing racks, trapping three employees and himself. Two occupants, one wounded, escaped from the basement through a hole in a rear wall. Four of the seven fatalities were found in the cellar, three others were on the first floor. The gunman was found on the first floor with a fatal gunshot wound to the chest.

A wet pipe sprinkler system was located in the basement of the clothing store. The system was inoperative because an OS&Y valve, located in an adjacent store, was closed. The fire department siamese connection was covered by a construction shed at the front entrance and couldn't be used.

15.4 Impaired Equipment

15.4.1 The impaired equipment shall be considered to be the water-based fire protection system, or part thereof, that is removed from service.

15.4.2 The impaired equipment shall include, but shall not be limited to, the following:

(1) Sprinkler systems
(2) Standpipe systems
(3) Fire hose systems
(4) Underground fire service mains
(5) Fire pumps
(6) Water storage tanks
(7) Water spray fixed systems
(8) Foam-water systems
(9) Fire service control valves

15.5* Preplanned Impairment Programs

A.15.5 The need for temporary fire protection, termination of all hazardous operations, and frequency of inspections in the areas involved should be determined. All work possible should be done in advance to minimize the length of the impairment. Where possible, temporary feedlines should be used to maintain portions of systems while work is completed.

Water-based fire protection systems should not be removed from service when the building is not in use. Where a system that has been out of service for a prolonged period, such as in the case of idle or vacant properties, is returned to service, qualified personnel should be retained to inspect and test the systems.

15.5.1 All preplanned impairments shall be authorized by the impairment coordinator.

15.5.2 Before authorization is given, the impairment coordinator shall be responsible for verifying that the following procedures have been implemented:

(1) The extent and expected duration of the impairment have been determined.
(2) The areas or buildings involved have been inspected and the increased risks determined.
(3) Recommendations have been submitted to management or the property owner/manager. Where a required fire protection system is out of service for more than 10 hours in a 24-hour period, the impairment coordinator shall arrange for one of the following:

   (a) Evacuation of the building or portion of the building affected by the system out of service
   (b)*An approved fire watch

A.15.5.2(3)(b) A fire watch should consist of trained personnel who continuously patrol the affected area. Ready access to fire extinguishers and the ability to promptly notify the fire department are important items to consider. During the patrol of the area, the person should not only be looking for fire, but making sure that the other fire protection features of the building such as egress routes and alarm systems are available and functioning properly.

FAQ ▶ What are the qualifications and responsibilities of the fire watch?

Fire watch personnel should be trained to understand the hazards involved in the job site and the nature of hot work. They should ensure that safe conditions are maintained during all hot work operations and should have the authority to stop work if unsafe conditions develop. Fire-extinguishing equipment should be readily available, and fire watch personnel should be trained in its use. The fire watch personnel should be familiar with the facilities and procedures for sounding an alarm in the event of a fire.

Fire watch personnel should attempt to extinguish fires only when they are obviously within the capacity of the equipment available. If it is determined that the fire is not within the capacity of the equipment, an alarm should be sounded immediately.

Fire watch personnel may be allowed to perform additional tasks, but those tasks should not distract them from their watch responsibilities. For additional guidance on fire watch operations, see NFPA 51B, *Standard for Fire Prevention During Welding, Cutting, and Other Hot Work.*

A fire watch must be posted for the duration of the work and for 60 minutes thereafter for torch-applied roofing operations.

Areas where smoking should be prohibited include, but are not limited to, temporary holding areas for combustible construction materials, storage areas, and areas where oil, gasoline, propane, or flammable material is stored or used.

(c)*Establishment of a temporary water supply

A.15.5.2(3)(c) Temporary water supplies are possible from a number of sources including use of a large-diameter hose from a fire hydrant to a fire department connection, use of a portable tank and a portable pump, or use of a standby fire department pumper and/or tanker.

Temporary water supplies can also be established through the use of existing systems or portions of systems that may not be involved in the maintenance or repair activity. Some examples are as follows.

Temporary Fire Protection. A temporary water supply for fire protection should be available as soon as existing fire protection systems are shut down for maintenance or repair. At this time, if underground water mains and hydrants are to be provided, they should be in service prior to the commencement of work.

Sprinkler Protection. If automatic sprinkler protection is installed, the system should be returned to service as soon as possible. Where sprinklers are required for safety to life, the building should not be occupied until the sprinkler repair or maintenance procedure has been entirely completed and tested so that the protection is not susceptible to frequent impairment caused by testing and correction.

The operation of sprinkler control valves should be limited to properly authorized personnel only. The valve should be properly tagged in accordance with the impairment program, and the impairment coordinator should be notified. Where the sprinkler protection is regularly turned off and on to facilitate connection of newly repaired segments, the sprinkler control valves should be checked at the end of each work shift and the impairment program implemented to verify that protection is in service.

Standpipe Systems. If at all possible, sprinkler systems should be isolated from standpipes in order to keep the standpipe system in service and readily available for manual fire fighting. For Class II and III standpipe systems, hose and nozzles should remain available for use. In systems where occupant hose is not provided, temporary hose and nozzles should be provided during construction.

(d)*Establishment and implementation of an approved program to eliminate potential ignition sources and limit the amount of fuel available to the fire

An approved fire prevention program should include all essential items, such as the following:

1. Good housekeeping
2. On-site security
3. Activation of new fire protection systems as construction and repair progresses
4. Preservation of existing systems during construction
5. Organization and training of an on-site fire brigade
6. Development of a prefire plan with the local fire department
7. Rapid communication
8. Consideration of special hazards resulting from construction operations
9. Protection of existing structures and equipment from exposure fires resulting from construction, alteration, and demolition operations

Temporary fire extinguishers should be provided in accordance with NFPA 10, *Standard for Portable Fire Extinguishers*. At least one approved fire extinguisher should also be provided in plain sight on each floor at each usable stairway, and access to permanent, temporary, or portable first aid fire equipment should be maintained at all times.

CASE STUDY

Lack of Impairment Program Cited in Motel Fire

Fire caused over $5 million in damage to a motel in Maine. The motel was closed for the season, but areas of it were being renovated. It was four stories high, had a ground floor area of 12,700 ft^2, and was of unprotected ordinary construction. No one was on the site at the time of the fire.

There were smoke detectors throughout the structure, and the alarms were sounding in the building when fire fighters arrived. A dry pipe sprinkler system was not operable — the main drains had been left open and the water supply turned off, due to the renovations.

An accidental fire started when a space heater was left too close to combustible structural components. The fire spread undetected for several hours in numerous voids throughout the building until a worker arrived at the motel. Fire fighters made an interior attack, but abandoned it several hours later.

Contributing factors in this fire included the following:

- The sprinkler system was inoperable.
- The automatic detection system wasn't heard outside the building, delaying notification of the fire department and giving the fire time to spread undetected.
- Many years of renovations had left voids throughout the building.
- There were no attic separations.

In addition, the lack of an impairment program was obviously a significant issue. Paragraph 15.5.2(3) requires that several precautions be taken when a fire protection system is out of service for more than 4 hours in a 24-hour period: evacuation of the building, a fire watch, a temporary water supply, and a fire prevention program to eliminate sources of ignition. None of these procedures was implemented in this case, resulting in a significant loss.

A.15.5.2(3)(d) Depending on the use and occupancy of the building, it could be enough in some circumstances to stop certain processes in the building or to cut off the flow of fuel to some machines. It is also helpful to implement "No Smoking" and "No Hot Work" (cutting, grinding, or welding) policies while the system is out of service because these activities are responsible for many fire ignitions.

According to *Introduction to Employee Fire and Life Safety*, hot work, such as that referred to in A.15.5.2(3)(d), is responsible for 5 percent of all nonresidential fires, accounting for 8 percent of property damage. However, should welding or cutting be necessary during system repair, these activities should be carried out in accordance with NFPA 51B. Because OSHA 29 CFR 1910.252, "Welding, Cutting, and Brazing, General," references NFPA 51B and ANSI Z49.1, *Safety in Welding, Cutting and Allied Processes*, the requirements of those two documents must be followed for OSHA compliance.

FAQ ▶
Is welding of pipe and pipe supports permitted?

Although shop welding of pipe is preferred, welding of pipe and pipe supports is permitted by NFPA 13, *Standard for the Installation of Sprinkler Systems*, provided that welding operations are conducted in accordance with NFPA 51B. In some areas, a permit issued by the local AHJ may be required.

(4) The fire department has been notified.

Although many fire departments may not require a permit for maintenance activities, notification that a system is out of service serves two purposes:

1. Knowledge of an impaired system assists fire department personnel should an emergency response to the affected property be necessary.
2. An extended impairment may be avoided if a fire department representative, notified when a system was impaired but having received no word that the system has been returned to service, inquires as to the status of the impairment.

(5) The insurance carrier, the alarm company, property owner/manager, and other authorities having jurisdiction have been notified.

Notifying the insurance carrier, the alarm company, the property owner/manager, and the AHJs is important for a number of reasons. Among them are the facts that the shutting of valves can transmit a trouble signal to alarm companies, insurance companies may provide suggestions for temporary protection during impairments, and water or health departments may have special requirements when systems are shut off.

(6) The supervisors in the areas to be affected have been notified.
(7) A tag impairment system has been implemented. *(See Section 15.3.)*
(8) All necessary tools and materials have been assembled on the impairment site.

Prior to any scheduled inspection or testing, the service company should consult with the property owner or impairment coordinator. Issues of advance notification in certain occupancies, including advance notification time, building posting, systems interruption and restoration of protection, evacuation procedures, accommodation for evacuees, and other related issues should be agreed upon by all parties prior to any inspection or testing.

Everyone who may be affected by fire protection system testing in a protected premises must be notified that testing will take place. This notification should include, but not be limited to, the property owner, property manager, switchboard operator, building engineer, building or floor fire wardens, and building maintenance personnel. In addition, the staff of the public fire service communication center, alarm supervising station, and building occupants should also be notified.

Effective methods of notification include bulletin board postings, electronic mail, public address announcements, and lobby signs. A fire emergency plan that makes provision for notifying occupants, the public fire service communication center, and the supervising station of the alarm company, in the event of a fire while testing is being conducted, should also be established for each protected premises.

15.6 Emergency Impairments

15.6.1 Emergency impairments include but are not limited to system leakage, interruption of water supply, frozen or ruptured piping, and equipment failure.

An emergency impairment is a condition in which the water-based fire protection system or a portion thereof is out of service due to an unexpected occurrence. Such an impairment may also be necessary following a successful system operation during a fire.

◄ FAQ
What is an emergency impairment?

Following a fire, sprinklers that have operated must be replaced and the system returned to service. The replacement of sprinklers illustrates the importance of the sprinkler cabinet inspections required by Chapter 5, since the impairment time can be substantially reduced if replacement sprinklers are on site.

15.6.2 When emergency impairments occur, emergency action shall be taken to minimize potential injury and damage.

15.6.3 The coordinator shall implement the steps outlined in Section 15.5.

The impairment coordinator can use an emergency impairment notice (see Exhibit 15.2) to coordinate emergency repairs of a fire protection system.

EXHIBIT 15.2 Impairment Notice. (Courtesy of Tyco/Fire & Security Simplex-Grinnell)

IMPAIRMENT NOTICE

DURING A RECENT INSPECTION OF YOUR FIRE PROTECTION SYSTEM(S), AN **EMERGENCY IMPAIRMENT** WAS DISCOVERED AND INDICATED ON THE INSPECTION REPORT. AS DEFINED BY NFPA 25, AN **EMERGENCY IMPAIRMENT** IS "A CONDITION WHERE A WATER-BASED FIRE PROTECTION SYSTEM OR PORTION THEREOF IS OUT OF ORDER DUE TO AN UNEXPECTED OCCURRENCE, SUCH AS A RUPTURED PIPE, OPERATED SPRINKLER, OR AN INTERRUPTION OF WATER SUPPLY TO THE SYSTEM." NFPA 25 FURTHER STATES, "EMERGENCY IMPAIRMENTS INCLUDE BUT ARE NOT LIMITED TO SYSTEM LEAKAGE, INTERRUPTION OF WATER SUPPLY, FROZEN OR RUPTURED PIPING, AND EQUIPMENT FAILURE."

WE RECOMMEND THAT IMMEDIATE STEPS BE TAKEN, AS DESCRIBED IN THE ATTACHED COPY OF CHAPTER 11 OF NFPA 25, TO CORRECT THE FOLLOWING IMPAIRMENT(S) TO YOUR FIRE PROTECTION SYSTEM(S):

[ ] CONTROL VALVE SHUT. SYSTEM OUT OF SERVICE.
[ ] LOW WATER PRESSURE DURING FLOW TEST. POSSIBLE OBSTRUCTION IN WATER SUPPLY OR PARTIALLY SHUT VALVE.
[ ] PIPE(S) FROZEN.
[ ] PIPE(S) LEAKING.
[ ] PIPE(S) ARE OBSTRUCTED.
[ ] SYSTEM PIPING OR PORTIONS OF SYSTEM PIPING ARE DISCONNECTED.
[ ] FIRE DEPT. CONNECTION MISSING OR DAMAGED OR OBSTRUCTED.
[ ] DRY PIPE VALVE IS OBSOLETE AND WAS NOT TRIP TESTED.
[ ] DRY PIPE VALVE CANNOT BE RESET.
[ ] DRY PIPE SYSTEM QUICK OPENING DEVICE IS OUT OF SERVICE.
[ ] SPRINKLERS ARE PAINTED, CORRODED, DAMAGED, OR LOADED.
[ ] FIRE PUMP IS OUT OF SERVICE.
[ ] JOCKEY PUMP IS OUT OF SERVICE.
[ ] DETECTION/ACTUATION SYSTEM IS OUT OF SERVICE.
[ ] OTHER: _____

15.7 Restoring Systems to Service

When all impaired equipment is restored to normal working order, the impairment coordinator shall verify that the following procedures have been implemented:

(1) Any necessary inspections and tests have been conducted to verify that affected systems are operational. The appropriate chapter of this standard shall be consulted for guidance on the type of inspection and test required.

The most common tests required following a system repair are the hydrostatic pressure test to verify that the piping system will not leak and the main drain test to verify that all water supply control valves have been returned to their full open position. Other tests, such as a running test for fire pumps and flow or alarm tests to check the functioning of replaced components, are required when these components are repaired or replaced.

(2) Supervisors have been advised that protection is restored.
(3) The fire department has been advised that protection is restored.
(4) The property owner/manager, insurance carrier, alarm company, and other authorities having jurisdiction have been advised that protection is restored.

All parties involved in the original impairment notification must be advised when protection is restored.

(5) The impairment tag has been removed.

Impairment tags should be retrieved and accounted for as further verification that all valves have been returned to their proper operating position.

SUMMARY

As discussed in Chapter 1, the leading reason for unsatisfactory sprinkler performance is that the water supply is shut off (see Commentary Table 1.1). With a formal impairment program, procedures for maintaining system control valves in their correct position can be established. Further, implementation of an organized impairment program can minimize the length of time that a system is out of service. This chapter provides the basic requirements for such an impairment program.

REVIEW QUESTIONS

1. What procedures should be used for an impairment that will last for more than one working shift?
2. What are the basic elements of an impairment program?
3. What are some common reasons for emergency impairment?
4. Why should a closed valve be tagged?

REFERENCES CITED IN COMMENTARY

National Fire Protection Association, 1 Batterymarch Park, Quincy, MA 02169-7471.

Colonna, G., ed., *Introduction to Employee Fire and Life Safety*, 2001.
NFPA 10, *Standard for Portable Fire Extinguishers*, 2007 edition.
NFPA 13, *Standard for the Installation of Sprinkler Systems,* 2007 edition.
NFPA 51B, *Standard for Fire Prevention During Welding, Cutting, and Other Hot Work,* 2003 edition.

American National Standards Institute, Inc., 25 West 43rd Street, 4th Floor, New York, NY 10036.

ANSI Z49.1, *Safety in Welding, Cutting and Allied Processes,* 2005 edition.

U.S. Government Printing Office, Washington, DC 20402.

Title 29, Code of Federal Regulations, Part 1910.252, "Welding, Cutting, and Brazing, General."

Explanatory Material

ANNEX A

The material contained in Annex A of NFPA 25 is included within the text of this handbook and, therefore, is not repeated here.

Forms for Inspection, Testing, and Maintenance

ANNEX B

This annex is not a part of the requirements of this NFPA document but is included for informational purposes only.

B.1

Forms need to be complete with respect to the requirements of NFPA 25 for the system being inspected, tested, or maintained, or any combination thereof. Because water-based fire protection systems are comprised of many components, it could be necessary to complete more than one form for each system.

Authorities having jurisdiction are legitimately concerned that the forms used are comprehensive. Therefore, they could develop their own forms or utilize those already developed and reviewed by their jurisdiction.

At least five formats can be used and are described as follows:

(1) One form in which all requirements for NFPA 25 are specified and large sections of information do not apply to most systems
(2) Individual forms that provide requirements corresponding to each chapter of NFPA 25 and address the following:
   (a) Sprinkler systems
   (b) Standpipe systems
   (c) Private fire service mains
   (d) Fire pumps
   (e) Storage tanks
   (f) Water spray systems
   (g) Foam-water sprinkler systems
(3) Forms that include information from the specific system chapter: Chapter 1, Chapter 13, and Chapter 14
(4) A series of forms similar to option (2) but with a more detailed breakdown of system types. For example, fire sprinkler systems are divided into five separate forms such as:
   (a) Wet pipe fire sprinkler systems
   (b) Dry pipe fire sprinkler systems
   (c) Preaction fire sprinkler systems
   (d) Deluge fire sprinkler systems
   (e) Foam-water sprinkler systems
(5) Separate forms for each individual component of each fire protection system

Regardless of the format used, it is of utmost importance to comply with the reporting requirements of NFPA 25. The general formats referenced in B.1(1) through (5) can be used to create custom forms.

Supplement 4 contains NFPA forms that are also appropriate for this application. These forms are provided in a full size format that may be easily reproduced.

Several electronic reporting systems are also available. Any system that meets the minimum reporting requirements of this standard can be used.

B.2

Sample forms are available for downloading at www.nfpa.org, www.nfsa.org, and www.sprinklernet.org.

Possible Causes of Pump Troubles

ANNEX C

This annex is not a part of the requirements of this NFPA document but is included for informational purposes only.

This annex is extracted from NFPA 20, Standard for the Installation of Stationary Pumps for Fire Protection.

The information provided in Annex C is intended to assist the user in troubleshooting fire pump problems. Figure C.1 supplements the text of Annex C, identifying potential problems with fire pumps and their likely causes in an easy-to-read, concise format. Figure C.1 makes it clear that a given problem can be attributed to a variety of causes.

C.1 Causes of Pump Troubles

This annex contains a partial guide for locating pump troubles and their possible causes *(see Figure C.1)*. It also contains a partial list of suggested remedies. *(For other information on this subject, see Hydraulic Institute Standard for Centrifugal, Rotary and Reciprocating Pumps.)* The causes listed here are in addition to possible mechanical breakage that would be obvious on visual inspection. In case of trouble, it is suggested that those troubles that can be checked easily should be corrected first or eliminated as possibilities.

C.1.1 Air Drawn into Suction Connection Through Leak(s).

Air drawn into suction line through leaks causes a pump to lose suction or fail to maintain its discharge pressure. Uncover suction pipe and locate and repair leak(s).

C.1.2 Suction Connection Obstructed.

Examine suction intake, screen, and suction pipe and remove obstruction. Repair or provide screens to prevent recurrence.

C.1.3 Air Pocket in Suction Pipe.

Air pockets cause a reduction in delivery and pressure similar to an obstructed pipe. Uncover suction pipe and rearrange to eliminate pocket.

C.1.4 Well Collapsed or Serious Misalignment.

Consult a reliable well drilling company and the pump manufacturer regarding recommended repairs.

C.1.5 Stuffing Box Too Tight or Packing Improperly Installed, Worn, Defective, Too Tight, or of Incorrect Type.

Loosen gland swing bolts and remove stuffing box gland halves. Replace packing.

Annex C • Possible Causes of Pump Troubles

Fire pump troubles	Suction				Pump																Driver and/or Pump					Driver						
	Air drawn into suction connection through leak(s)	Suction connection obstructed	Air pocket in suction pipe	Well collapsed or serious misalignment	Stuffing box too tight or packing improperly installed, worn, defective, too tight, or incorrect type	Water seal or pipe to seal obstructed	Air leak into pump through stuffing boxes	Impeller obstructed	Wearing rings worn	Impeller damaged	Wrong diameter impeller	Actual net head lower than rated	Casing gasket defective, permitting internal leakage (single-stage and multistage pumps)	Pressure gauge is on top of pump casing	Incorrect impeller adjustment (vertical shaft turbine-type pump only)	Impellers locked	Pump is frozen	Pump shaft or shaft sleeve scored, bent, or worn	Pump not primed	Seal ring improperly located in stuffing box, preventing water from entering space to form seal	Excess bearing friction due to lack of lubrication, wear, dirt, rusting, failure, or improper installation	Rotating element binds against stationary element	Pump and driver misaligned	Foundation not rigid	Engine-cooling system obstructed	Faulty driver	Lack of lubrication	Speed too low	Wrong direction of rotation	Speed too high	Rated motor voltage different from line voltage	Faulty electrical circuit, obstructed fuel system obstructed steam pipe, or dead battery
	1	2	3	4	5	6	7	8	9	10	11	12	13	14	15	16	17	18	19	20	21	22	23	24	25	26	27	28	29	30	31	32
Excessive leakage at stuffing box					X													X					X									
Pump or driver overheats				X	X	X		X		X			X					X	X	X	X	X	X	X	X	X			X	X	X	
Pump unit will not start				X	X								X	X	X						X					X	X					X
No water discharge	X	X	X					X											X													
Pump is noisy or vibrates				X	X			X	X									X			X	X	X	X		X						
Too much power required				X	X			X	X		X		X		X			X			X	X	X	X		X			X	X	X	
Discharge pressure not constant for same gpm	X				X	X	X																									
Pump loses suction after starting	X	X	X		X	X														X												
Insufficient water discharge	X	X	X		X	X	X	X	X	X	X				X													X	X		X	
Discharge pressure too low for gpm discharge	X	X	X		X	X	X	X	X	X	X	X	X															X	X		X	

FIGURE C.1 *Possible Causes of Fire Pump Troubles.*

C.1.6 Water Seal or Pipe to Seal Obstructed.

Loosen gland swing bolt and remove stuffing box gland halves along with the water seal ring and packing. Clean the water passage to and in the water seal ring. Replace water seal ring, packing gland, and packing in accordance with manufacturer's instructions.

C.1.7 Air Leak into Pump Through Stuffing Boxes.

Same as possible cause in C.1.6.

C.1.8 Impeller Obstructed.

Does not show on any one instrument, but pressures fall off rapidly when an attempt is made to draw a large amount of water.

For horizontal split-case pumps, remove upper case of pump and remove obstruction from impeller. Repair or provide screens on suction intake to prevent recurrence.

For vertical shaft turbine-type pumps, lift out column pipe and pump bowls from wet pit or well and disassemble pump bowl to remove obstruction from impeller.

For close-coupled, vertical in-line pumps, lift motor on top pull-out design and remove obstruction from impeller.

C.1.9 Wearing Rings Worn.

Remove upper case and insert feeler gauge between case wearing ring and impeller wearing ring. Clearance when new is 0.0075 in. (0.19 mm). Clearances of more than 0.015 in. (0.38 mm) are excessive.

C.1.10 Impeller Damaged.

Make minor repairs or return to manufacturer for replacement. If defect is not too serious, order new impeller and use damaged one until replacement arrives.

C.1.11 Wrong Diameter Impeller.

Replace with impeller of proper diameter.

C.1.12 Actual Net Head Lower Than Rated.

Check impeller diameter and number and pump model number to make sure correct head curve is being used.

C.1.13 Casing Gasket Defective, Permitting Internal Leakage (Single-Stage and Multistage Pumps).

Replace defective gasket. Check manufacturer's drawing to see whether gasket is required.

C.1.14 Pressure Gauge Is on Top of Pump Casing.

Place gauges in correct location.

C.1.15 Incorrect Impeller Adjustment (Vertical Shaft Turbine-Type Pump Only).

Adjust impellers according to manufacturer's instructions.

C.1.16 Impellers Locked.

For vertical shaft turbine-type pumps, raise and lower impellers by the top shaft adjusting nut. If this adjustment is not successful, follow the manufacturer's instructions.

For horizontal split-case pumps, remove upper case and locate and eliminate obstruction.

C.1.17 Pump Is Frozen.

Provide heat in the pump room. Disassemble pump and remove ice as necessary. Examine parts carefully for damage.

C.1.18 Pump Shaft or Shaft Sleeve Scored, Bent, or Worn.

Replace shaft or shaft sleeve.

C.1.19 Pump Not Primed.

If a pump is operated without water in its casing, the wearing rings are likely to seize. The first warning is a change in pitch of the sound of the driver. Shut down the pump.

For vertical shaft turbine-type pumps, check water level to determine whether pump bowls have proper submergence.

C.1.20 Seal Ring Improperly Located in Stuffing Box, Preventing Water from Entering Space to Form Seal.

Loosen gland swing bolt and remove stuffing box gland halves along with the water-seal ring and packing. Replace, putting seal ring in proper location.

C.1.21 Excess Bearing Friction Due to Lack of Lubrication, Wear, Dirt, Rusting, Failure, or Improper Installation.

Remove bearings and clean, lubricate, or replace as necessary.

C.1.22 Rotating Element Binds Against Stationary Element.

Check clearances and lubrication and replace or repair the defective part.

C.1.23 Pump and Driver Misaligned.

Shaft running off center because of worn bearings or misalignment. Align pump and driver according to manufacturer's instructions. Replace bearings according to manufacturer's instructions.

C.1.24 Foundation Not Rigid.

Tighten foundation bolts or replace foundation if necessary.

C.1.25 Engine-Cooling System Obstructed.

Heat exchanger or cooling water systems too small. Cooling pump faulty. Remove thermostats. Open bypass around regulator valve and strainer. Check regulator valve operation. Check strainer. Clean and repair if necessary. Disconnect sections of cooling system to locate and remove possible obstruction. Adjust engine-cooling water-circulating pump belt to obtain proper speed without binding. Lubricate bearings of this pump.

If overheating still occurs at loads up to 150 percent of rated capacity, contact pump or engine manufacturer so that necessary steps can be taken to eliminate overheating.

C.1.26 Faulty Driver.

Check electric motor, internal combustion engine, or steam turbine, in accordance with manufacturer's instructions, to locate reason for failure to start.

C.1.27 Lack of Lubrication.

If parts have seized, replace damaged parts and provide proper lubrication. If not, stop pump and provide proper lubrication.

C.1.28 Speed Too Low.

For electric motor drive, check that rated motor speed corresponds to rated speed of pump, voltage is correct, and starting equipment is operating properly.

Low frequency and low voltage in the electric power supply prevent a motor from running at rated speed. Low voltage can be due to excessive loads and inadequate feeder capacity or (with private generating plants) low generator voltage. The generator voltage of private generating plants can be corrected by changing the field excitation. When low voltage is from the other causes mentioned, it can be necessary to change transformer taps or increase feeder capacity.

Low frequency usually occurs with a private generating plant and should be corrected at the source. Low speed can result in older type squirrel-cage-type motors if fastenings of copper bars to end rings become loose. The remedy is to weld or braze these joints.

For steam turbine drive, check that valves in steam supply pipe are wide open; boiler steam pressure is adequate; steam pressure is adequate at the turbine; strainer in the steam supply pipe is not plugged; steam supply pipe is of adequate size; condensate is removed from steam supply pipe, trap, and turbine; turbine nozzles are not plugged; and setting of speed and emergency governor is correct.

For internal combustion engine drive, check that setting of speed governor is correct; hand throttle is opened wide; and there are no mechanical defects such as sticking valves, timing off, or spark plugs fouled, and so forth. The latter can require the services of a trained mechanic.

C.1.29 Wrong Direction of Rotation.

Instances of an impeller turning backward are rare but are clearly recognizable by the extreme deficiency of pump delivery. Wrong direction of rotation can be determined by comparing the direction in which the flexible coupling is turning with the directional arrow on the pump casing.

With polyphase electric motor drive, two wires must be reversed; with dc driver, the armature connections must be reversed with respect to the field connections. Where two sources of electrical current are available, the direction of rotation produced by each should be checked.

C.1.30 Speed Too High.

See that pump- and driver-rated speed correspond. Replace electric motor with one of correct rated speed. Set governors of variable-speed drivers for correct speed. Frequency at private generating stations can be too high.

C.1.31 Rated Motor Voltage Different from Line Voltage.

For example, a 220 or 440 V motor on 208 or 416 V line. Obtain motor of correct rated voltage or a larger size motor.

C.1.32 Faulty Electric Circuit, Obstructed Fuel System, Obstructed Steam Pipe, or Dead Battery.

Check for break in wiring open switch, open circuit breaker, or dead battery. If circuit breaker in controller trips for no apparent reason, make sure oil is in dash pots in accordance with manufacturer's specifications. Make sure fuel pipe is clear, strainers are clean, and control valves open in fuel system to internal combustion engine. Make sure all valves are open and strainer is clean in steam line to turbine.

C.2 Warning

Chapters 6 and 7 of NFPA 20, *Standard for the Installation of Stationary Pumps for Fire Protection*, include electrical requirements that discourage the installation of disconnect means in the power supply to electric motor–driven fire pumps. This requirement is intended to ensure the availability of power to the fire pumps. When equipment connected to those circuits is serviced or maintained, the employee can have unusual exposure to electrical and other hazards. It can be necessary to require special safe work practices and special safeguards, personal protective clothing, or both.

C.3 Maintenance of Fire Pump Controllers After a Fault Condition

C.3.1 Introduction.

In a fire pump motor circuit that has been properly installed, coordinated, and in service prior to the fault, tripping of the circuit breaker or the isolating switch indicates a fault condition in excess of operating overload.

It is recommended that the following general procedures be observed by qualified personnel in the inspection and repair of the controller involved in the fault. These procedures are not intended to cover other elements of the circuit, such as wiring and motor, which can also require attention.

C.3.2 Caution.

All inspections and tests are to be made on controllers that are de-energized at the line terminal, disconnected, locked out, and tagged so that accidental contact cannot be made with live parts and so that all plant safety procedures will be observed.

C.3.2.1 Enclosure. Where substantial damage to the enclosure, such as deformation, displacement of parts, or burning has occurred, replace the entire controller.

C.3.2.2 Circuit Breaker and Isolating Switch. Examine the enclosure interior, circuit breaker, and isolating switch for evidence of possible damage. If evidence of damage is not apparent, the circuit breaker and isolating switch can continue to be used after the door is closed.

If there is any indication that the circuit breaker has opened several short-circuit faults, or if signs of possible deterioration appear within either the enclosure, circuit breaker, or isolating switch (e.g., deposits on surface, surface discoloration, insulation cracking, or unusual toggle operation), replace the components. Verify that the external operating handle is capable of opening and closing the circuit breaker and isolating switch. If the handle fails to operate the device, this would also indicate the need for adjustment or replacement.

C.3.2.3 Terminals and Internal Conductors. Where there are indications of arcing damage, overheating, or both, such as discoloration and melting of insulation, replace the damaged parts.

C.3.2.4 Contactor. Replace contacts showing heat damage, displacement of metal, or loss of adequate wear allowance of the contacts. Replace the contact springs where applicable. If deterioration extends beyond the contacts, such as binding in the guides or evidence of insulation damage, replace the damaged parts or the entire contactor.

C.3.2.5 Return to Service. Before returning the controller to service, check for the tightness of electrical connections and for the absence of short circuits, ground faults, and leakage current.

Close and secure the enclosure before the controller circuit breaker and isolating switch are energized. Follow operating procedures on the controller to bring it into standby condition.

Obstruction Investigation

ANNEX D

Annex D provides additional information relating to obstruction of piping and discharge outlets for fire protection systems. It outlines sources of obstruction, treatment strategies, and flushing procedures. Chapter 14 of NFPA 25 provides the enforceable requirements for conducting an obstruction investigation. This annex provides a recommended systematic approach to correcting obstruction problems.

In addition to the methods discussed in this annex, new technologies employing ultrasonic examination and remote video are now available for interior piping investigations. Methods for cleaning obstructed pipe other than those mentioned here (e.g., flushing with chemical solutions) are also available.

This annex is not a part of the requirements of this NFPA document but is included for informational purposes only.

D.1

For effective control and extinguishment of fire, automatic sprinklers should receive an unobstructed flow of water. Although the overall performance record of automatic sprinklers has been very satisfactory, there have been numerous instances of impaired efficiency because sprinkler piping or sprinklers were plugged with pipe scale, corrosion products, including those produced by microbiologically influenced corrosion, mud, stones, or other foreign material. If the first sprinklers to open in a fire are plugged, the fire in that area cannot be extinguished or controlled by prewetting of adjacent combustibles. In such a situation, the fire can grow to an uncontrollable size, resulting in greater fire damage and excessive sprinkler operation and even threatening the structural integrity of the building, depending on the number of plugged sprinklers and fire severity.

Keeping the inside of sprinkler system piping free of scale, silt, or other obstructing material is an integral part of an effective loss prevention program.

D.2 Obstruction Sources

D.2.1 Pipe Scale.

Loss studies have shown that dry pipe sprinkler systems are involved in the majority of obstructed sprinkler fire losses. Pipe scale was found to be the most frequent obstructing material (it is likely that some of the scale was composed of corrosion products, including those produced by microbiologically influenced corrosion). Dry pipe systems that have been maintained wet and then dry alternately over a period of years are particularly susceptible to the accumulation of scale. Also, in systems that are continuously dry, condensation of moisture in the air supply can result in the formation of a hard scale, microbiological materials, and corrosion products along the bottom of the piping. When sprinklers open, the scale is broken loose and carried along the pipe, plugging some of the sprinklers or forming obstructions at the fittings.

D.2.2 Careless Installation or Repair.

Many obstructions are caused by careless workers during installation or repair of yard or public mains and sprinkler systems. Wood, paint brushes, buckets, gravel, sand, and gloves have been found as obstructions. In some instances, with welded sprinkler systems and systems with holes for quick-connect fittings, the cutout discs or coupons have been left within the piping, obstructing flow to sprinklers.

Paragraph 6.5.2.4.6(2) of NFPA 13, *Standard for the Installation of Sprinkler Systems,* 2007 edition, requires that discs removed prior to attachment of a welded outlet be retrieved. This procedure can be verified by attaching the disc to the system at the point from which it was removed. Exhibit D.1 shows discs attached to the system from which they were cut. Exhibit D.2 shows weld coupons that were not properly retrieved and presented an obstruction to the system piping. Exhibit D.3 shows work gloves and other obstructing material removed from an underground pipe in a system water supply.

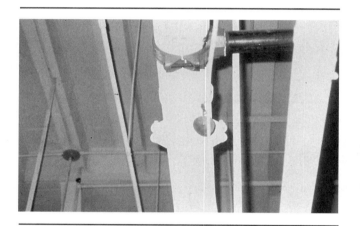

EXHIBIT D.1 Discs Attached to Piping from Which They Were Cut. (Source: Automatic Sprinkler Systems Handbook, 2007, Exhibit 6.15)

EXHIBIT D.2 Weld Coupons Found in ESFR Sprinkler Main During Obstruction Investigation.

D.2.3 Raw Water Sources.

Materials can be sucked up from the bottoms of rivers, ponds, or open reservoirs by fire pumps with poorly arranged or inadequately screened intakes and then forced into the system. Sometimes floods damage intakes. Obstructions include fine, compacted materials such as rust, mud, and sand. Coarse materials, such as stones, cinders, cast-iron tubercles, chips of wood, and sticks, also are common.

D.2.4 Biological Growth.

Biological growth has been known to cause obstructions in sprinkler piping. The Asiatic clam has been found in fire protection systems supplied by raw river or lake water. With an available food supply and sunlight, these clams grow to approximately $3/8$ in. to $7/16$ in. (9 mm to 11 mm) across the shell in 1 year and up to $2 1/8$ in. (54 mm) and larger by the sixth year. However, once in fire mains and sprinkler piping, the growth rate is much slower. The clams get into the fire protection systems in the larval stage or while still small clams. They then attach themselves to the pipe and feed on bacteria or algae that pass through.

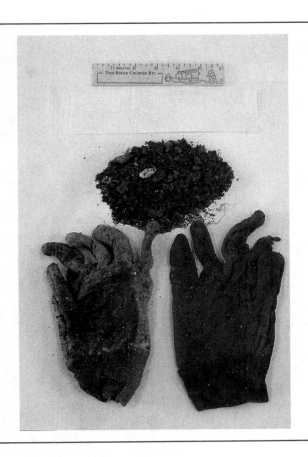

EXHIBIT D.3 Work Gloves and Other Obstructing Material. (Courtesy of John Jensen)

Originally brought to Washington state from Asia in the 1930s, the clams have spread throughout at least 33 states and possibly are present in every state. River areas reported to be highly infested include the Ohio River, Tennessee River Valley, Savannah River (S. Carolina), Altamaha River (Georgia), Columbia River (Washington), and Delta-Mendota Canal (California).

In addition to the Asian clam, the zebra mussel can create obstructions to fire protection piping systems. Since their introduction to the Great Lakes in the mid-1980s, zebra mussels have spread throughout the Great Lakes into many inland waterways, including the Mississippi Delta to the south and the Hudson River to the northeast. Like the Asian clam, the zebra mussels can enter the fire protection system in the larval stage or as young adults.

The mussel attaches to the inner surface of a pipe and feeds on algae or bacteria. Exhibit D.4 shows a 2 in. (50 mm) diameter pipe infested with zebra mussels. Exhibit D.5 shows the relative size of a zebra mussel. Exhibit D.6 indicates the extent of zebra mussel infestation in the United States and Canada.

The presence of zebra mussels can be revealed by reduced flow or discharge of the mussels or shells from the inspector's test or main drain test connections. If reduced flow or the discharge of obstructing material is observed, a complete obstruction investigation is required as per 14.2.2(2).

A strainer listed for use in a fire protection system may be of some value to combat the problem of clams and mussels obstructing piping. However, in the larval stage, these creatures can certainly pass through the $\frac{1}{8}$ in. perforations typically found in such devices. If clams or mussels are known to be present, the strainer must be cleaned more often than once a year, as is required by Table 10.1.

EXHIBIT D.4 Pipe Infested with Zebra Mussels. (Source: Michigan Sea Grant)

EXHIBIT D.5 Relative Size of Adult Zebra Mussel. (Source: Michigan Sea Grant)

EXHIBIT D.6 Extent of Zebra Mussel Infestation.

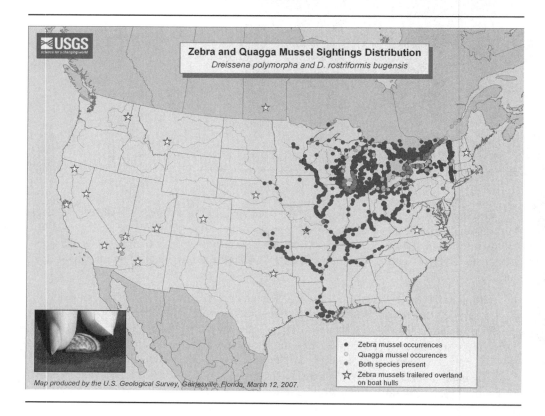

The use of potassium and heat has led to high mortality rates in zebra mussel populations. However, the use of more common chemicals (e.g., chlorine) should be considered for treatment of system water supplies. To prevent repopulation of systems with mussels or clams, flushing and testing in excess of the minimum requirements of this standard should be avoided.

D.2.5 Sprinkler Calcium Carbonate Deposits.

Natural freshwaters contain dissolved calcium and magnesium salts in varying concentrations, depending on the sources and location of the water. If the concentration of these salts is high, the water is considered hard. A thin film composed largely of calcium carbonate, $CaCO_3$, affords some protection against corrosion where hard water flows through the pipes. However, hardness is not the only factor to determine whether a film forms. The ability of $CaCO_3$ to precipitate on the metal pipe surface also depends on the water's total acidity or alkalinity, the concentration of dissolved solids in the water, and its pH. In soft water, no such film can form.

In automatic sprinkler systems, the calcium carbonate scale formation tends to occur on the more noble metal in the electrochemical series, which is copper, just as corrosion affects the less noble metal, iron. Consequently, scale formation naturally forms on sprinklers, often plugging the orifice. The piping itself could be relatively clear. This type of sprinkler obstruction cannot be detected or corrected by normal flushing procedures. It can be found only by inspection of sprinklers in suspected areas and then removed.

Most public water utilities in very hard water areas soften their water to reduce consumer complaints of scale buildup in water heaters. Thus, the most likely locations for deposits in sprinkler systems are where sprinklers are not connected to public water but supplied without treatment directly from wells or surface water in areas that have very hard water. These areas generally include the Mississippi basin west of the Mississippi River and north of the Ohio River, the rivers of Texas and the Colorado basin, and other white areas in Figure D.2.5(a). (The water of the Great Lakes is only moderately hard.)

Within individual plants, the sprinklers most likely to have deposits are located as follows:

(1) In wet systems only.
(2) In high temperature areas, except where water has unusually high pH *[see Figure D.2.5(b)]*. High temperature areas include those near dryers, ovens, and skylights or at roof peaks.
(3) In old sprinkler systems that are frequently drained and refilled.
(4) In pendent sprinklers that are located away from air pockets and near convection currents.

This annex section describes where the requirements to perform a main drain test (see Chapter 13), flow test (see Chapter 7), and records retention (see Chapter 4) come into play. Although the main drain test is primarily intended to reveal the presence of a closed or partially closed valve, comparison of results from one year to the next or to the results obtained from

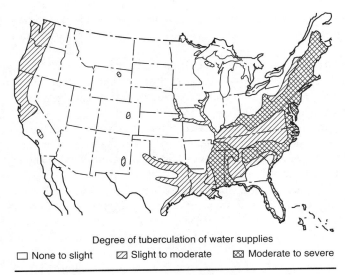

FIGURE D.2.5(a) *Map of Hard Water Areas. (Courtesy of Cast Iron Pipe Research Association)*

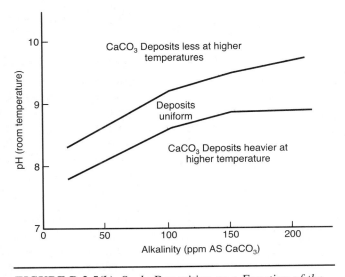

FIGURE D.2.5(b) *Scale Deposition as a Function of the Alkalinity/pH Ratio.*

the acceptance test can be of value. A periodic flow test can also be used for comparison purposes. A gradual decrease in available residual pressures from either test may indicate a buildup of tuberculation from $CaCO_3$ deposits. Although a slight deterioration can be expected over time, deterioration to the point where the water supply can no longer adequately supply the fire protection system must result in corrective action. Exhibit D.7 illustrates a pipe severely affected by $CaCO_3$ deposits. The effective diameter of this pipe has been substantially reduced to the point where the pipe was removed from service.

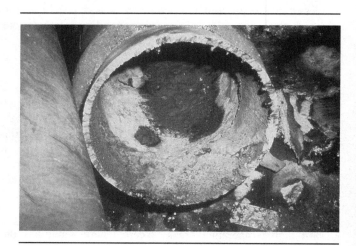

EXHIBIT D.7 Pipe Affected by $CaCO_3$ Deposits.

D.2.6 Forms of Corrosion.

Corrosion is defined as the deterioration of a material, usually a metal, resulting from a chemical or electrochemical reaction. The eight main forms of corrosion include: (1) uniform corrosion, (2) pitting, (3) galvanic corrosion, (4) crevice corrosion, (5) selective leaching (parting), (6) erosion corrosion, (7) environmental cracking, and (8) intergranular corrosion. Microbiologically influenced corrosion (MIC) is included herein as a ninth form of corrosion, although it is usually a secondary factor that accelerates or exacerbates the rate of another form of corrosion. Definitions of the different forms of corrosion are discussed next.

(1) *Uniform (or general) corrosion:* A regular loss of a small quantity of metal over the entire area or over a large section of the total area, which is evenly distributed within a pipe(s).
(2) *Pitting:* A localized form of corrosion that results in holes or cavities in the metal. Pitting is considered to be one of the more destructive forms of corrosion and is often difficult to detect. Pits may be covered or open and normally grow in the direction of gravity — for example, at the bottom of a horizontal surface.
(3) *Galvanic corrosion:* An electric potential exists between dissimilar metals in a conductive (corrosive) solution. The contact between the two materials allows electrons to transfer from one metal to the other. One metal acts as a cathode and the other as an anode. Corrosion usually occurs at anodic metal only.
(4) *Crevice corrosion:* A localized form of corrosion that occurs within crevices and other shielded areas on metal surfaces exposed to a stagnant corrosive solution. This form of corrosion usually occurs beneath gaskets, in holes, surface deposits, in thread and groove joints. Crevice corrosion is also referred to as gasket corrosion, deposit corrosion, and under-deposit corrosion.
(5) *Selective leaching:* The selective removal of one element from an alloy by corrosion. A common example is dezincification (selective removal of zinc) of unstabilized brass, resulting in a porous copper structure.

(6) *Erosion corrosion:* Corrosion resulting from the cumulative damage of electrochemical reactions and mechanical effects. Erosion corrosion is the acceleration or increase in the rate of corrosion created by the relative movement of a corrosive fluid and a metal surface. Erosion corrosion is observed as grooves, gullies, waves, rounded holes, or valleys in a metal surface.

(7) *Environmental cracking:* An acute form of localized corrosion caused by mechanical stresses, embrittlement, or fatigue.

(8) *Integranular corrosion:* Corrosion caused by impurities at grain boundaries, enrichment of one alloying element, or depletion of one of the elements in the grain boundary areas.

(9) *Microbiologically influenced corrosion (MIC):* Corrosion initiated or accelerated by the presence and activities of microorganisms, including bacteria and fungi. Colonies (also called bio-films and slimes) are formed in the surface of pipes among a variety of types of microbes. Microbes deposit iron, manganese, and various salts into the pipe surfaces, forming nodules, tubercles, and carbuncles. The formation of these deposits can cause obstruction to flow and dislodge, causing blockage (plugging) of system piping, valves, and sprinklers.

MIC can be found in many types of piping systems, not just water-based fire protection systems. MIC is thought to be responsible for 10 to 30 percent of corrosion in all piping systems in the United States.

Generally, two types of bacteria are responsible for MIC: sulfate-reducing bacteria (SRB) and acid-producing bacteria (APB). SRB are anaerobic (not requiring oxygen) and are typically found in carbon steel systems. APB are also anaerobic but can survive in aerated or anaerobic environments. APB will produce acids that stimulate SRB growth.

MIC can survive in a variety of environments from 0 to 100 percent oxygen-saturated water and water with a pH from 1 to 10. MIC can be recognized by the presence of a gray or black mud-like slime (typical of anaerobic bacteria) or a brown or rusty color (typical of aerobic bacteria).

Once established, MIC will create nodules on the inner surface of pipe. Left untreated, this will result in pitting of the pipe wall and eventual pinhole leaks (although pinhole leaks can be caused by other factors). Nodules can also increase in size, causing obstruction to waterflow. Exhibits D.8, D.9, and D.10 illustrate pipe with nodules from MIC.

If the presence of MIC is suspected, water samples should be taken and analyzed before treatment is initiated. In cases of extreme pitting, the affected pipe should be replaced. In addition, the water supplying the system should be treated with a biocide, such as chlorine, to prevent recurrence of the problem.

Exhibits D.11, D.12, and D.13 illustrate pipe with moderate to severe pitting and copper pipe with pitting and a pinhole leak. Where pitting is not extensive, flushing the entire system with a chemical cleaning agent to remove nodules is recommended. After flushing, the system should be refilled with treated water to prevent recurrence of MIC.

D.2.7 Microbiologically Influenced Corrosion (MIC).

The most common biological growths in sprinkler system piping are those formed by microorganisms, including bacteria and fungi. These microbes produce colonies (also called biofilms, slimes) containing a variety of types of microbes. Colonies form on the surface of wetted pipe in both wet and dry systems. Microbes also deposit iron, manganese, and various salts onto the pipe surface, forming discrete deposits (also termed nodules, tubercles, and carbuncles). These deposits can cause obstruction to flow and dislodge, causing plugging of fire sprinkler components. Subsequent under-deposit pitting can also result in pinhole leaks.

Microbiologically influenced corrosion (MIC) is corrosion influenced by the presence and activities of microorganisms. MIC almost always occurs with other forms of corrosion (oxygen corrosion, crevice corrosion, and under-deposit corrosion). MIC starts as microbial communities (also called biofilms, slimes) growing on the interior surface of the wetted

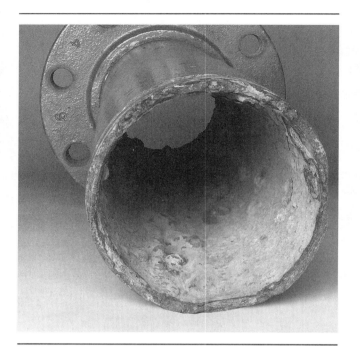

EXHIBIT D.8 *Pipe Sample Showing Effect of Nodule Formation. (Courtesy of Bioindustrial Technologies, Inc.)*

EXHIBIT D.9 *Pipe Sample Showing Effect of Severe Nodule Formation and Corrosion. (Courtesy of Bioindustrial Technologies, Inc.)*

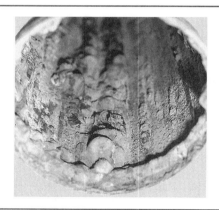

EXHIBIT D.10 *Close-Up View of Nodules. (Courtesy of Bioindustrial Technologies, Inc.)*

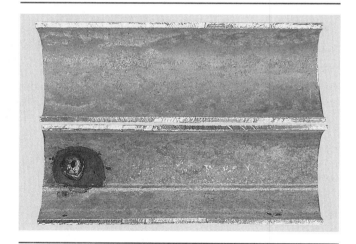

EXHIBIT D.11 *Pipe Sample with Pinhole Leak. (Courtesy of Bioindustrial Technologies, Inc.)*

sprinkler piping components in both wet and dry systems. The microbial communities contain many types of microbes, including slime formers, acid-producing bacteria, iron-depositing bacteria, and sulfate-reducing bacteria, and are most often introduced into the sprinkler system from the water source. The microbes deposit iron, manganese, and various salts onto the pipe surface, forming discrete deposits (also termed nodules, tubercles, and carbuncles). These deposits can cause obstruction to flow and dislodge, causing plugging of fire sprinkler components. MIC is most often seen as severe pitting corrosion occurring under deposits. Pitting is due to microbial activities such as acid production, oxygen consumption, and accumulation of salts. Oxygen and salts, especially chloride, can greatly increase the severity of MIC and other forms of corrosion.

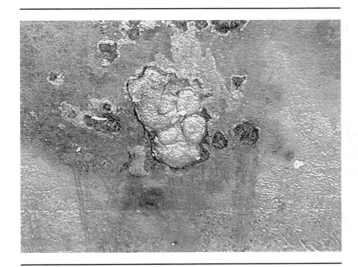

EXHIBIT D.12 *Pipe Sample with Severe Pitting (Pits Within Pits). (Courtesy of Bioindustrial Technologies, Inc.)*

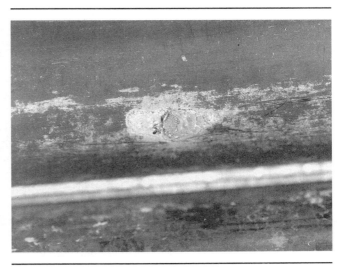

EXHIBIT D.13 *Copper Pipe with Localized Pitting. (Courtesy of Bioindustrial Technologies, Inc.)*

In steel pipe, MIC is most often seen as deposits on the interior surface of the pipes. The deposits can be orange, red, brown, black, and white (or combinations thereof), depending on local conditions and water chemistry. The brown, orange, and red forms are most common in oxygenated portions of the system and often contain oxidized forms of iron and other materials on the outside, with reduced (blacker) corrosion products on the inside. Black deposits are most often in smaller diameter piping farther from the water source and contain reduced forms (those with less oxygen) of corrosion products. White deposits often contain carbonate scales.

MIC of copper and copper alloys occurs as discrete deposits of smaller size, which are green to blue in color. Blue slimes can also be produced in copper piping or copper components (e.g., brass heads).

MIC is often first noticed as a result of pinhole leaks after only months to a few years of service. Initial tests for the presence of MIC should involve on-site testing for microbes and chemical species (iron, pH, oxygen) important in MIC. This information is also very important in choosing treatment methods. These tests can be done on water samples from source waters and various locations in the sprinkler system (e.g., main drain, inspector's test valve). Confirmation of MIC can be made by examination of interior of pipes for deposits and under-deposit corrosion with pit morphology consistent with MIC (cup-like pits within pits and striations).

The occurrence and severity of MIC is enhanced by the following:

(1) Using untreated water to test and fill sprinkler piping. This is made worse by leaving the water in the system for long periods of time.
(2) Introduction of new and untreated water containing oxygen, microbes, salts, and nutrients into the system on a frequent basis (during repair, renovation, and/or frequent flow tests).
(3) Leaving dirt, debris, and especially oils, pipe joint compound, and so forth in the piping. These provide nutrients and protection for the microbes, often preventing biocides and corrosion inhibitors from reaching the microbes and corrosion sites.

Once the presence of MIC has been confirmed, the system should be assessed to determine the extent and severity of MIC. Severely affected portions should be replaced or cleaned to remove obstructions and pipe not meeting minimal mechanical specifications.

D.3 Investigation Procedures

If unsatisfactory conditions are observed as outlined in Section 14.2, investigations should be made to determine the extent and severity of the obstructing material. From the fire protection system plan, determine the water supply sources, age of underground mains and sprinkler systems, types of systems, and general piping arrangement. Consider the possible sources of obstruction material.

Examine the fire pump suction supply and screening arrangements. If necessary, have the suction cleaned before using the pump in tests and flushing operations. Gravity tanks should be inspected internally, with the exception of steel tanks that have been recently cleaned and painted. If possible, have the tank drained and determine whether loose scale is on the shell or if sludge or other obstructions are on the tank bottom. Cleaning and repainting could be in order, particularly if it has not been done within the previous 5 years.

Investigate yard mains first, then sprinkler systems.

Where fire protection control valves are closed during investigation procedures, the fire protection impairment precautions outlined in Chapter 15 should be followed.

Large quantities of water are needed for investigation and for flushing. It is important to plan the safest means of disposal in advance. Cover stock and machinery susceptible to water damage and keep equipment on hand for mopping up any accidental discharge of water.

D.3.1 Investigating Yard Mains.

Flow water through yard hydrants, preferably near the extremes of selected mains, to determine whether mains contain obstructive material. It is preferable to connect two lengths of $2\frac{1}{2}$ in. (65 mm) hose to the hydrant. Attach burlap bags to the free ends of the hose from which the nozzles have been removed to collect any material flushed out, and flow water long enough to determine the condition of the main being investigated. If there are several water supply sources, investigate each independently, avoiding any unnecessary interruptions to sprinkler protection. In extensive yard layouts, repeat the tests at several locations, if necessary, to determine general conditions.

If obstructive material is found, all mains should be flushed thoroughly before investigating the sprinkler systems. *(See Section D.5.)*

D.3.2 Investigating Sprinkler Systems.

Investigate dry systems first. Tests on several carefully selected, representative systems usually are sufficient to indicate general conditions throughout the plant. If, however, preliminary investigations indicate the presence of obstructing material, this justifies investigating all systems (both wet and dry) before outlining needed flushing operations. Generally, the system can be considered reasonably free of obstructing material, provided the following conditions apply:

(1) Less than $\frac{1}{2}$ cup of scale is washed from the cross mains.
(2) Scale fragments are not large enough to plug a sprinkler orifice.
(3) A full, unobstructed flow is obtained from each branch line checked.

Where other types of foreign material are found, judgment should be used before considering the system unobstructed. Obstruction potential is based on the physical characteristics and source of the foreign material.

In selecting specific systems or branch lines for investigation, the following should be considered:

(1) Lines found obstructed during a fire or during maintenance work
(2) Systems adjacent to points of recent repair to yard mains, particularly if hydrant flow shows material in the main

Tests should include flows through 2½ in. (65 mm) fire hose directly from cross mains *[see Figure D.3.2(a) and Figure D.3.2(b)]* and flows through 1½ in. (40 mm) hose from representative branch lines. Two or three branch lines per system is a representative number of branch lines where investigating for scale accumulation. If significant scale is found, investigation of additional branch lines is warranted. Where investigating for foreign material (other than scale), the number of branch lines needed for representative sampling is dependent on the source and characteristic of the foreign material.

If provided, fire pumps should be operated for the large line flows, since maximum flow is desirable. Burlap bags should be used to collect dislodged material as is done in the investigation of yard mains. Each flow should be continued until the water clears (i.e., a minimum of 2 to 3 minutes at full flow for sprinkler mains). This is likely to be sufficient to indicate the condition of the piping interior.

D.3.3 Investigating Dry Pipe Systems.

Flood dry pipe systems one or two days before obstruction investigations to soften pipe scale and deposits. After selecting the test points of a dry pipe system, close the main control valve and drain the system. Check the piping visually with a flashlight while it is being dismantled. Attach hose valves and 1½ in. (40 mm) hose to the ends of the lines to be tested, shut the valves, have air pressure restored on the system, and reopen the control valve. Open the hose valve on the end branch line, allowing the system to trip in simulation of normal action. Any obstructions should be cleared from the branch line before proceeding with further tests.

After flowing the small end line, shut its hose valve and test the feed or cross main by discharging water through a 2½ in. (65 mm) fire hose, collecting any foreign material in a burlap bag.

After the test, the dry pipe valve should be cleaned internally and reset. Its control valve should be locked open and a drain test performed.

D.3.4 Investigating Wet Pipe Systems.

Testing of wet systems is similar to that of dry systems, except that the system should be drained after closing the control valve to permit installation of hose valves for the test. Slowly reopen the control valve and make a small hose flow as specified for the branch line, followed by the 2½ in. (65 mm) hose flow for the cross main.

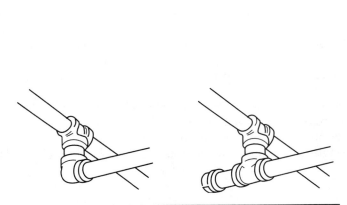

FIGURE D.3.2(a) *Replacement of Elbow at End of Cross Main with a Flushing Connection Consisting of a 50 mm (2 in.) Nipple and Cap.*

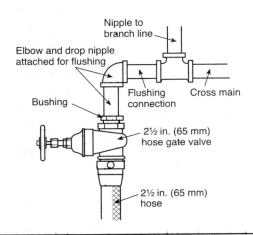

FIGURE D.3.2(b) *Connection of 65 mm (2½ in.) Hose Gate Valve with a 50 mm (2 in.) Bushing and Nipple and Elbow to 50 mm (2 in.) Cross Main.*

In any case, if lines become plugged during the tests, piping should be dismantled and cleaned, the extent of plugging noted, and a clear flow obtained from the branch line before proceeding further.

Perform similar tests on representative systems to indicate the general condition of the wet systems throughout the plant, keeping a detailed record of the procedures performed.

D.3.5 Other Obstruction Investigation Methods.

Other obstruction investigation methods, such as technically proven ultrasonic and X-ray examination, have been evaluated and if applied correctly, are successful at detecting obstructions.

The sources of the obstructing material should be determined and steps taken to prevent further entrance of such material. This entails work such as inspection and cleaning of pump suction screening facilities or cleaning of private reservoirs. If recently laid public mains appear to be the source of the obstructing material, waterworks authorities should be requested to flush their system.

D.4 Obstruction Prevention Program

D.4.1 Dry Pipe and Preaction Systems — Scale.

(1) Dry pipe and preaction systems using noncoated ferrous piping should be thoroughly investigated for obstruction from corrosion after they have been in service for 15 years, for 25 years, and every 5 years thereafter.

(2) Dry pipe systems with noncoated ferrous piping should be kept on air year-round, rather than on air and water alternately, to inhibit formation of rust and scale.

(3) Piping that has been galvanized internally for new dry pipe and preaction sprinkler system installations should be used. Fittings, couplings, hangers, and other appurtenances are not required to be galvanized. Copper or stainless steel piping also is permitted.

D.4.2 Flushing Connections.

Sprinkler systems installed in accordance with recent editions of NFPA 13, *Standard for the Installation of Sprinkler Systems*, should have provisions for flushing each cross main. Similarly, branch lines on gridded systems should be capable of being readily "broken" at a simple union or flexible joint. Property owners of systems installed without these provisions should be encouraged to provide them when replacement or repair work is being done.

D.4.3 Suction Supplies.

(1) Screen pump suction supplies and screens should be maintained. Connections from penstocks should be equipped with strainers or grids, unless the penstock inlets themselves are so equipped. Pump suction screens of copper or brass wire tend to promote less aquatic growth.

(2) Extreme care should be used to prevent material from entering the suction piping when cleaning tanks and open reservoirs. Materials removed from the interior of gravity tanks during cleaning should not be allowed to enter the discharge pipe.

(3) Small mill ponds could need periodic dredging where weeds and other aquatic growth are inherent.

D.4.4 Asian Clams.

Effective screening of larvae and small-size, juvenile Asian clams from fire protection systems is very difficult. To date, no effective method of total control has been found. Such controls can be difficult to achieve in fire protection systems.

D.4.5 Calcium Carbonate.

For localities suspected of having hard water, sample sprinklers should be removed and inspected yearly. Section D.2.5 outlines sprinkler locations prone to the accumulation of deposits where hard water is a problem. Sprinklers found with deposits should be replaced, and adjacent sprinklers should be checked.

D.4.6 Zebra Mussels.

Several means of controlling the zebra mussel are being studied, including molluscicides, chlorines, ozone, shell strainers, manual removal, robotic cleaning, water jetting, line pigging, sonic pulses, high-voltage electrical fields, and thermal backwashing. It is believed that these controls might need to be applied only during spawning periods when water temperatures are 57°F to 61°F (14°C to 16°C) and veligers are present. Several silicon grease-based coatings also are being investigated for use within piping systems.

While it appears that the use of molluscicides could provide the most effective means of controlling the mussel, these chemicals are costly. It is believed that chlorination is the best available short-term treatment, but there are problems associated with the use of chlorine, including strict Environmental Protection Agency regulations on the release of chlorine into lakes and streams. The use of nonselective poison, such as chlorine, in the amounts necessary to kill the mussels in large bodies of water could be devastating to entire ecosystems.

To provide an effective means of control against zebra mussels in fire protection systems, control measures should be applied at the water source, instead of within the piping system. Effective controls for growth of the zebra mussel within fire protection systems include the following:

(1) Selecting a water source that is not subject to infestation. This could include well water or potable or pretreated water.
(2) Implementing a water treatment program that includes biocides or elevated pH, or both.
(3) Implementing a water treatment program to remove oxygen, to ensure control of biological growth within piping.
(4) Relying on a tight system approach to deny oxygen and nutrients that are necessary to support growth.

D.5 Flushing Procedures

D.5.1 Yard Mains.

Yard mains should be flushed thoroughly before flushing any interior piping. Flush yard piping through hydrants at dead ends of the system or through blow-off valves, allowing the water to run until clear. If the water is supplied from more than one direction or from a looped system, close divisional valves to produce a high-velocity flow through each single line. A velocity of at least 10 ft/sec (3 m/sec) is necessary for scouring the pipe and for lifting foreign material to an aboveground flushing outlet. Use the flow specified in Table D.5.1 or the maximum flow available for the size of the yard main being flushed.

Connections from the yard piping to the sprinkler riser should be flushed. These are usually 6 in. (150 mm) mains. Although flow through a short, open-ended 2 in. (50 mm) drain can create sufficient velocity in a 6 in. (150 mm) main to move small obstructing material, the restricted waterway of the globe valve usually found on a sprinkler drain might not allow stones and other large objects to pass. If the presence of large size material is suspected, a larger outlet is needed to pass such material and to create the flow necessary to move it. Fire department connections on sprinkler risers can be used as flushing outlets by removing the clappers. Yard mains also can be flushed through a temporary Siamese fitting attached to the riser connection before the sprinkler system is installed. *[See Figure D.5.1.]*

TABLE D.5.1 Flushing Rates to Accomplish Flow of 10 ft/sec (3 m/sec)

Pipe Size	Steel			Copper			CPVC (gpm)	Polybutylene	
	SCH 10 (gpm)	SCH 40 (gpm)	XL (gpm)	K (gpm)	L (gpm)	M (gpm)		CTS (gpm)	IPS (gpm)
¾	—	—	—	14	15	16	19	12	17
1	29	24	30	24	26	27	30	20	27
1¼	51	47	52	38	39	41	48	30	43
1½	69	63	70	54	55	57	63	42	57
2	114	105	114	94	96	99	98	72	90
2½	170	149	163	145	149	152	144	—	—
3	260	230	251	207	212	217	213	—	—
4	449	396	—	364	373	379	—	—	—
5	686	623	—	565	582	589	—	—	—
6	989	880	—	807	836	846	—	—	—
8	1665	1560	—	1407	1460	1483	—	—	—
10	2632	2440	—	2185	2267	2303	—	—	—
12	—	3520	—	—	—	—	—	—	—

For SI units: 1 gpm = 3.785 L/min.

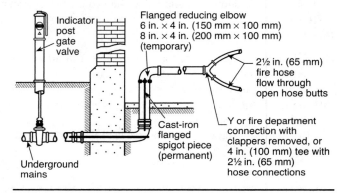

FIGURE D.5.1 Arrangement for Flushing Branches from Underground Mains to Sprinkler Risers.

D.5.2 Sprinkler Piping.

Two methods commonly are used for flushing sprinkler piping:

(1) The hydraulic method
(2) The hydropneumatic method

The hydraulic method consists of flowing water from the yard mains, sprinkler risers, feed mains, cross mains, and branch lines, respectively, in the same direction in which water would flow during a fire.

The hydropneumatic method uses special equipment and compressed air to blow a charge of about 30 gal (114 dm^3) of water from the ends of branch lines back into feed mains and down the riser, washing the foreign material out of an opening at the base of the riser.

The choice of method depends on conditions at the individual plant and the type of material installed. If examination indicates the presence of loose sand, mud, or moderate amounts of pipe scale, the piping generally can be flushed satisfactorily by the hydraulic method. Where the material is more difficult to remove and available water pressures are too low for effective scouring action, the hydropneumatic method generally is more satisfactory. The hydropneumatic method should not be used with listed CPVC sprinkler piping.

In some cases, where obstructive material is solidly packed or adheres tightly to the walls of the piping, the pipe needs to be dismantled and cleaned by rodding or other means.

Dry pipe systems should be flooded one or two days before flushing to soften pipe scale and deposits.

Successful flushing by either the hydraulic or hydropneumatic method is dependent on establishing sufficient velocity of flow in the pipes to remove silt, scale, and other obstructive material. With the hydraulic method, water should be moved through the pipe at least at the rate of flow indicated in Table D.5.1.

Where flushing a branch line through the end pipe, sufficient water should be discharged to scour the largest pipe in the branch line. Lower rates of flow can reduce the efficiency of the flushing operation. To establish the recommended flow, remove the small end piping and connect the hose to a larger section, if necessary.

Where pipe conditions indicate internal or external corrosion, a section of the pipe affected should be cleaned thoroughly to determine whether the walls of the pipe have seriously weakened. Hydrostatic testing should be performed as outlined in NFPA 13, *Standard for the Installation of Sprinkler Systems*.

Pendent sprinklers should be removed and inspected until it is reasonably certain that all are free of obstruction material.

Painting the ends of branch lines and cross mains is a convenient method for keeping a record of those pipes that have been flushed.

D.5.3 Hydraulic Method.

After the yard mains have been thoroughly cleaned, flush risers, feed mains, cross mains, and finally the branch lines. In multistory buildings, systems should be flushed by starting at the lowest story and working up. Branch line flushing in any story can immediately follow the flushing of feed and cross mains in that story, allowing one story to be completed at a time. Following this sequence prevents drawing obstructing material into the interior piping.

To flush risers, feed mains, and cross mains, attach $2\frac{1}{2}$ in. (65 mm) hose gate valves to the extreme ends of these lines *[see Figure D.5.3]*. Such valves usually can be procured from the manifold of fire pumps or hose standpipes. As an alternative, an adapter with $2\frac{1}{2}$ in. (65 mm) hose thread and standard pipe thread can be used with a regular gate valve. A length of fire hose without a nozzle should be attached to the flushing connection. To prevent kinking of the hose and to obtain maximum flow, an elbow usually should be installed between the

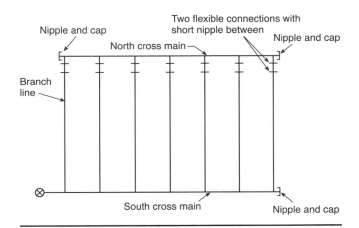

FIGURE D.5.3 Gridded Sprinkler System Piping.

end of the sprinkler pipe and the hose gate valve. Attach the valve and hose so that no excessive strain will be placed on the threaded pipe and fittings. Support hose lines properly.

Where feed and cross mains and risers contain pipe 4 in., 5 in., and 6 in. (100 mm, 125 mm, and 150 mm) in diameter, it could be necessary to use a Siamese with two hose connections to obtain sufficient flow to scour this larger pipe.

Flush branch lines after feed and cross mains have been thoroughly cleared. Equip the ends of several branch lines with gate valves, and flush individual lines of the group consecutively. This eliminates the need for shutting off and draining the sprinkler system to change a single hose line. The hose should be $1\frac{1}{2}$ in. (40 mm) in diameter and as short as practicable. Branch lines can be permitted to be flushed in any order that expedites the work.

Branch lines also may be permitted to be flushed through pipe $1\frac{1}{2}$ in. (40 mm) in diameter or larger while extended through a convenient window. If pipe is used, 45 degree fittings should be provided at the ends of branch lines. Where flushing branch lines, hammering the pipes is an effective method of moving obstructions.

Figure D.5.3 shows a typical gridded piping arrangement prior to flushing. The flushing procedure is as follows:

(1) Disconnect all branch lines and cap all open ends.
(2) Remove the cap from the east end of the south cross main, flush the main, and replace the cap.
(3) Remove the cap from branch line 1, flush the line, and replace the cap.
(4) Repeat step (3) for the remaining branch lines.
(5) Reconnect enough branch lines at the west end of the system so that the aggregate cross-sectional area of the branch lines approximately equals the area of the north cross main. For example, three $1\frac{1}{4}$ in. (32 mm) branch lines approximately equal a $2\frac{1}{2}$ in. (65 mm) cross main. Remove the cap from the east end of the north cross main, flush the main, and replace the cap.
(6) Disconnect and recap the branch lines. Repeat step (5), but reconnect branch lines at the east end of the system and flush the north cross main through its west end.
(7) Reconnect all branch lines and recap the cross main. Verify that the sprinkler control valve is left in the open and locked position.

D.5.4 Hydropneumatic Method.

The apparatus used for hydropneumatic flushing consists of a hydropneumatic machine, a source of water, a source of compressed air, 1 in. (25 mm) rubber hose for connecting to branch lines, and $2\frac{1}{2}$ in. (65 mm) hose for connecting to cross mains.

The hydropneumatic machine *[see Figure D.5.4(a)]* consists of a 30 gal (114 dm^3) (4 ft^3) water tank mounted over a 185 gal (700 dm^3) (4 ft^3) compressed air tank. The compressed air tank is connected to the top of the water tank through a 2 in. (50 mm) lubricated plug cock. The bottom of the water tank is connected through hose to a suitable water supply. The compressed air tank is connected through suitable air hose to either the plant air system or a separate air compressor.

To flush the sprinkler piping, the water tank is filled with water, the pressure is raised to 100 psi (6.9 bar) in the compressed air tank, and the plug cock between tanks is opened to put air pressure on the water. The water tank is connected by hose to the sprinkler pipe to be flushed. The lubricated plug cock on the discharge outlet at the bottom of the water tank then is snapped open, allowing the water to be "blown" through the hose and sprinkler pipe by the compressed air. The water tank and air tank should be recharged after each blow.

Outlets for discharging water and obstructing material from the sprinkler system should be arranged. With the clappers of dry pipe valves and alarm check valves on their seats and cover plates removed, sheet metal fittings can be used for connection to $2\frac{1}{2}$ in. (65 mm) hose lines or for discharge into a drum [maximum capacity per blow is approximately 30 gal (114 dm^3)]. If the 2 in. (50 mm) riser drain is to be used, the drain valve should be removed and a

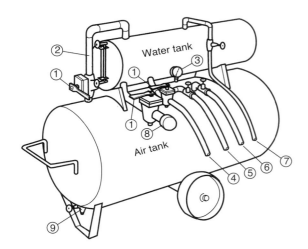

1. Lubricated plug cocks
2. Pipe connection between air and water tanks (This connection is open when flushing sprinkler system.)
3. Air pressure gauge
4. 1 in. (25 mm) rubber hose (air type) (Used to flush sprinkler branch lines.)
5. Hose connected to source of water (Used to fill water tank.)
6. Hose connected to ample source of compressed air (Used to supply air tank.)
7. Water tank overflow hose
8. 2½ in. (65 mm) pipe connection [Where flushing large interior piping, connect woven jacket fire hose here and close 1 in. (25 mm) plug cock hose connection (4) used for flushing sprinkler branch lines.]
9. Air tank drain valve

FIGURE D.5.4(a) *Hydropneumatic Machine.*

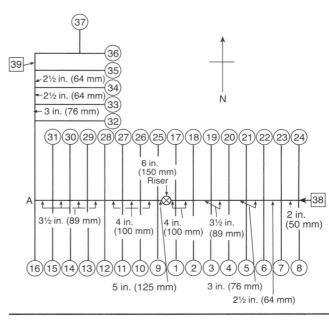

FIGURE D.5.4(b) *Schematic Diagram of Sprinkler System Showing Sequence To Be Followed Where Hydropneumatic Method Is To Be Utilized.*

direct hose connection made. For wet pipe systems with no alarm check valves, the riser should be taken apart just below the drain opening and a plate inserted to prevent foreign material from dropping to the base of the riser. Where dismantling of a section of the riser for this purpose is impractical, the hydropneumatic method should not be used.

Before starting a flushing job, each sprinkler system to be cleaned should be studied and a schematic plan prepared showing the order of the blows.

To determine that the piping is clear after it has been flushed, representative branch lines and cross mains should be investigated, using both visual examination and sample flushings.

(1) *Branch Lines.* With the yard mains already flushed or known to be clear, the sprinkler branch lines should be flushed next. The order of cleaning individual branch lines should be laid out carefully if an effective job is to be done. In general, the branch lines should be flushed, starting with the branch closest to the riser and working toward the dead end of the cross main. *[See Figure D.5.4(b).]* The order for flushing the branch lines is shown by the circled numerals. In this example, the southeast quadrant is flushed first, then the southwest, followed by the northeast, and, finally, the northwest. Air hose 1 in. (25 mm) in diameter is used to connect the machine with the end of the branch line being flushed. This hose air pressure should be allowed to drop to 85 psi (5.9 bar) before the valve is closed. The resulting short slug of water experiences less friction loss and a higher velocity and, therefore, cleans more effectively than if the full 30 gal (114 dm^3) of water were to be used. One blow is made for each branch line.

(2) *Large Piping.* Where flushing cross mains, fill the water tank completely and raise the pressure in the air receiver to 100 psi (6.9 bar) (690 kPa). Connect the machine to the end

of the cross main to be flushed with no more than 50 ft (15.2 m) of 2½ in. (65 mm) hose. After opening the valve, allow air pressure in the machine to drop to zero (0). Two to six blows are necessary at each location, depending on the size and length of the main. In Figure D.5.4(b), the numerals in squares indicate the location and order of the cross main blows. Because the last branch line blows performed were located west of the riser, clean the cross main located east of the riser first. Where large cross mains are to be cleaned, it is best, if practical, to make one blow at 38, one at 39, the next again at 38, then again at 39, alternating in this manner until the required number of blows has been made at each location.

(3) Where flushing cross mains and feed mains, arrange the work so that the water passes through a minimum of right-angle bends. In Figure D.5.4(b), blows at 38 should be adequate to flush the cross mains back to the riser. Do not attempt to clean the cross main from location A to the riser by backing out branch line 16 and connecting the hose to the open side of the tee. If this were to be done, a considerable portion of the blow would pass northward up the 3 in. (76 mm) line supplying branches 34 to 37, and the portion passing eastward to the riser could be ineffective. Where the size, length, and condition of cross mains necessitate blowing from a location corresponding to location A, the connection should be made directly to the cross main corresponding to the 3½ in. (90 mm) pipe so that the entire flow travels to the riser. Where flushing through a tee, always flush the run of the tee after flushing the branch. Note the location of blows 35, 36, and 37 in Figure D.5.4(b). Gridded systems can be flushed in a similar fashion. With branch lines disconnected and capped, begin flushing the branch line closest to the riser (branch line 1 in Figure D.5.3), working toward the most remote line. Then flush the south cross main in Figure D.5.3 by connecting the hose to the east end. Flushing the north cross main involves connecting the hose to one end while discharging to a safe location from the other end.

REFERENCES CITED IN COMMENTARY

National Fire Protection Association, 1 Batterymarch Park, Quincy, MA 02169-7471.

Dubay, C., ed., *Automatic Sprinkler Systems Handbook*, 2007 edition.
NFPA 13, *Standard for the Installation of Sprinkler Systems,* 2007 edition.

Informational References

ANNEX E

E.1 Referenced Publications

The documents or portions thereof listed in this annex are referenced within the informational sections of this standard and are not part of the requirements of this document unless also listed in Chapter 2 for other reasons.

E.1.1 NFPA Publications.

National Fire Protection Association, 1 Batterymarch Park, Quincy, MA 02169-7471.

NFPA 13, *Standard for the Installation of Sprinkler Systems*, 2007 edition.
NFPA 13R, *Standard for the Installation of Sprinkler Systems in Residential Occupancies up to and Including Four Stories in Height*, 2007 edition.
NFPA 14, *Standard for the Installation of Standpipe and Hose Systems*, 2007 edition.
NFPA 15, *Standard for Water Spray Fixed Systems for Fire Protection*, 2007 edition.
NFPA 16, *Standard for the Installation of Foam-Water Sprinkler and Foam-Water Spray Systems*, 2007 edition.
NFPA 20, *Standard for the Installation of Stationary Pumps for Fire Protection*, 2007 edition.
NFPA 22, *Standard for Water Tanks for Private Fire Protection*, 2003 edition.
NFPA 24, *Standard for the Installation of Private Fire Service Mains and Their Appurtenances*, 2007 edition.
NFPA 72®, National Fire Alarm Code®, 2007 edition.
NFPA 750, *Standard on Water Mist Fire Protection Systems*, 2006 edition.
NFPA 780, *Standard for the Installation of Lightning Protection Systems*, 2008 edition.

E.1.2 Other Publications.

E.1.2.1 ASTM Publications. ASTM International, 100 Barr Harbor Drive, P.O. Box C700, West Conshohocken, PA 19428-2959.

IEEE/ASTM-SI-10, *American National Standard for Use of the International System of Units (SI): The Modern Metric System*, 2002.

E.1.2.2 AWWA Publications. American Water Works Association, 6666 West Quincy Avenue, Denver, CO 80235.

AWWA, *Manual of Water Supply Practices — M42 Steel Water-Storage Tanks*, 1998.

E.1.2.3 Hydraulic Institute Publications. Hydraulic Institute, 9 Sylvan Way, Parsippany, NJ 07054.

Hydraulic Institute Standards for Centrifugal, Rotary and Reciprocating Pumps, 14th edition, 1983.

E.2 Informational References

The following documents or portions thereof are listed here as informational resources only. They are not a part of the requirements of this document.

E.2.1 NFPA Publications. National Fire Protection Association, 1 Batterymarch Park, Quincy, MA 02169-7471.

NFPA 1, *Uniform Fire Code*™, 2006 edition.

E.2.2 Other Publications.

E.2.2.1 AWWA Publications. American Water Works Association, 6666 West Quincy Avenue, Denver, CO 80235.

AWWA D101, *Inspecting and Repairing Steel Water Tanks, Standpipes, Reservoirs, and Elevated Tanks, for Water Storage,* 1986.

E.2.2.2 SSPC Publications. Society of Protective Coatings, 40 24th Street, 6th Floor, Pittsburgh, PA 15222.

SSPC Chapter 3, "Special Pre-Paint Treatments," 1993.
SSPC-PA 1, *Shop, Field, and Maintenance Painting,* 1991.
SSPC Paint 8, *Aluminum Vinyl Paint,* 1991.
SSPC Paint 9, *White (or Colored) Vinyl Paint,* 1995.
SSPC-SP 6, *Commercial Blast Cleaning,* 1994.
SSPC-SP 8, *Pickling,* 1991.
SSPC-SP 10, *Near-White Blast Cleaning,* 1994.

E.2.2.3 U.S. Government Publications. U.S. Government Printing Office, Washington, DC 20402.

Bureau of Reclamation Specification VR-3.
Federal Specification TT- P-86, *Specifications for Vinyl Resin Paint,* M-54, 1995.

E.2.2.4 Other Publications. Edward K. Budnick, P.E., "Automatic Sprinkler System Reliability," *Fire Protection Engineering*, Society of Fire Protection Engineers, Winter 2001.

Fire Protection Equipment Surveillance Optimization and Maintenance Guide, Electric Power Research Institute, July 2003.

William E. Koffel, P.E., *Reliability of Automatic Sprinkler Systems*, Alliance for Fire Safety.

NFPA's Future in Performance Based Codes and Standards, July 1995.

NFPA Performance Based Codes and Standards Primer, December 1999.

E.3 References for Extracts in Informational Sections

NFPA 20, *Standard for the Installation of Stationary Pumps for Fire Protection*, 2007 edition.
NFPA 24, *Standard for the Installation of Private Fire Service Mains and Their Appurtenances*, 2007 edition.
NFPA 750, *Standard on Water Mist Fire Protection Systems*, 2006 edition.

PART TWO

Supplements

The five supplements included in Part Two of the *Water-Based Fire Protection Systems Handbook* provide additional information as well as supporting reference material to assist users and enforcers of NFPA 25, *Standard for the Inspection, Testing, and Maintenance of Water-Based Fire Protection Systems*. The supplements are not part of NFPA 25 or the commentary presented in the previous part of this book.

1. Microbiologically Influenced Corrosion in Fire Sprinkler Systems
2. Foam Environmental Issues
3. The Role of the Inspector
4. Sample Inspection, Testing, and Maintenance Program
5. Technical/Substantive Changes from the 2002 Edition to the 2008 Edition of NFPA 25

SUPPLEMENT 1

Microbiologically Influenced Corrosion in Fire Sprinkler Systems

Bruce H. Clarke

Anthony M. Aguilera

Editor's Note: Supplement 1 has been included to provide the user with background information related to microbiologically influenced corrosion or MIC. While MIC has been in existence for some time, it has recently become a maintenance problem, through recognition of the source of corrosion, for those involved in the maintenance of fire protection systems. The supplement discusses the microbial corrosion process and recognition and treatment of MIC for fire protection systems.

Beginning in the early 1990s, concerns began to increase about microbiologically influenced corrosion (MIC) affecting fire sprinkler systems due to multiple cases involving the abnormally rapid development of pinhole-sized leaks and highly obstructive interior biological pipe growths. Most of these occurred in systems well before the end of the system's life expectancy, after 5 to 20 years of service. However, some systems began to show signs of critical obstruction and began developing leaks in less than one year. Research into the cause of these leaks, which have greatly increased in the last two decades, has led to a growing awareness of the problem. In the past 20 years MIC has grown from a relatively unknown topic of regional discussions to one now generating widespread concern throughout several countries. MIC in fire protection systems also has become the subject of a wide array of speculation, debate, and, in some cases, gross inaccuracies.

At the time this is being written, there are few time-proven solutions that are also universally accepted" best practices" in the fire protection industry. Although MIC and biological growth control have been extensively researched in many allied engineering fields for decades, treatment in fire sprinkler systems is relatively new. And although there are several detection and treatment systems that appear to be effective, long-term data to support overall effectiveness claims are still limited. This chapter will provide an overview of the issues related to microbiologically influenced corrosion in the fire protection industry.

For related topics, see the following chapters in Section 16 of the *Fire Protection Handbook* [1]: Chapter 1,

---

Bruce H. Clarke is Eastern Regional Manager of Field Services and a fire protection/loss prevention trainer with GE Global Asset Protection Services.

Anthony M. Aguilera is Director of Loss Prevention and Risk Management for Honeywell Aerospace.

"Principles of Automatic Sprinkler System Performance," Chapter 2, "Automatic Sprinklers," Chapter 3, "Automatic Sprinkler Systems," and Chapter 11, "Care and Maintenance of Water-Based Extinguishing Systems. "

DEFINING CORROSION

General Corrosion

Generating a universal "working definition" for MIC in fire protection is complex. And to understand the relationship of MIC and sprinkler components, general corrosion and its various causes must first be understood. One edition of *Webster's* defines *corrosion* as "the wearing away of materials by chemical action(s)." Another edition simply defines it as "the wearing away of material gradually." The on-line encyclopedia *Encarta* describes corrosion as "specifically being related to the gradual action of natural agents such as air or salt water on metals." All these definitions are true generally but are also very confusing to those in the fire protection industry who are looking for answers. Even definitions presented by the corrosion engineering communities, though more detailed, can at times appear somewhat contradictory and very confusing.

Corrosion can occur from many biophysical reactions and be described from a multitude of scientific viewpoints. "General corrosion" typically refers to uniform corrosion that occurs on most unprotected metallic systems. This can be associated with the uniform "rust" layer seen on many steel structures. Besides rust from oxidation, several other types of general corrosion include stray current, uniform biological, galvanic, molten salt, dealloying, chemical, high-temperature, and general carburization. And some specific nonbiological corrosion processes having adverse effects on fire protection sprinkler systems, which must always be considered with MIC, include oxygen, acid, and oxygen-acid corrosion. [2]

Microbiologically Influenced Corrosion

In contrast to general, or uniform, corrosion, MIC is a form of localized corrosion. Material is lost at discrete points instead of universally across an entire surface. There are several types of localized corrosion, including pitting, crevice, cratering, and filiform. (See Exhibit S1.1.)

As with the various other types of corrosion, MIC can take many forms and affect different systems in unique ways. But, for the fire protection profession, an industry-specific definition can be developed. This definition captures both the causes and effects of this problem with several continuous generalities to avoid some of the current inaccuracies. Thus, microbiologically influenced corrosion in fire protection systems can be described as "an

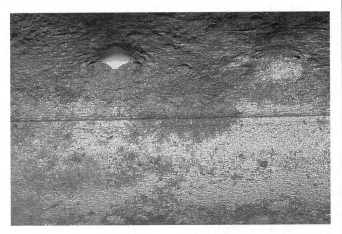

EXHIBIT S1.1 Interior Pinhole Leak.

electrochemical corrosion process that is concentrated and accelerated by the activity of specific bacteria within a fire sprinkler system resulting in the premature failure of metallic system components." [3]

THE CORROSION PROBLEM

Concentrated and Accelerated Corrosion

All metallic systems normally begin to corrode to various degrees from the instant moisture contacts metal. Typically, this appears as general corrosion. By definition, in comparison, the MIC process is both *concentrated* and *accelerated*. With general corrosion, when moisture is introduced into a system, a thin layer of oxidation occurs relatively evenly throughout the pipe wall surface. This type of corrosion is typically not a significant concern in fire sprinkler systems and does not require treatment because it does not change a pipe's interior surface roughness (i.e., C-factor) unevenly. The rate of decay is also typically slow. A typical corrosion rate in sprinkler pipe is highly dependent on water quality, but with MIC, this relatively slow corrosion rate is abnormally accelerated and not evenly dispersed. The result changes a relatively smooth pipe to one with pits and valleys.

$$p = \frac{4.52 Q^{1.85}}{C^{1.85} d^{4.87}}$$

where

p = frictional resistance in psi/ft
Q = flow in gpm
C = frictional loss coefficient
d = internal diameter of pipe

While there has not been a detailed study on exactly what amount of pipe change (or changes in pipe wall roughness) is unacceptable, the Hazen-Williams formula makes several points clear:

1. Pipe surface texture is critical in sprinkler system effectiveness. With "C" based on pipe wall smoothness, any increase in roughness decreases the value of "C" and thereby increases the pressure loss in each foot of sprinkler pipe. Even a small amount of internal corrosion, especially in main feed areas, could potentially make a system ineffective in fire control.
2. Pipe diameter is a significant contributor to friction loss. Thus, the smaller the pipe, the more dramatically corrosion affects friction loss and overall performance.

While this formula is not intended to be used for non-uniform pipe surfaces, it is clear that random pockets of biological growth have the potential to affect performance. This is especially true if a sprinkler system has a minimum design buffer, which is regularly the case. (See Exhibit S1.2 and Exhibit S1.3.)

EXHIBIT S1.3 *Interior View of Pipe Shown in Exhibit S1.2 with Approximate 65 percent Obstruction to 6 in. Tee Connection.*

EXHIBIT S1.2 *Exterior View of Pipe with No Signs of Internal Corrosion.*

MIC is the result of specific bacteria (see section on defining MIC). A multitude of bacteria are always omnipresent in all ecosystems, including the interior of sprinkler systems. Just as only a small number of bacteria on earth have the potential to cause human sickness, only a relatively small number of bacteria have the potential to cause the rapid system destruction currently linked to MIC. Thus, as defined, only a few specific bacteria concentrate and accelerate the general or uniform corrosion process.

Premature Failure

The ultimate effect of MIC in fire protection systems is the premature failure of metallic components.

Premature Failure Defined. It is clear that MIC and other related forms of corrosion cause sprinkler systems to fail prematurely. What constitutes "premature" with regard to the integrity of specific system components has not officially been defined in the industry. Long-term warranties are not typical with system premature failure of metallic components.by the industry to last over 50 years before major repairs are required. In most cases of MIC, it has appeared that after treatment of the affected areas, healthy systems can be expected to exceed this normal life expectancy.

However, as discussed, failure is a function of both integrity and function. And a system without leaks is never considered acceptable if it fails functionally (i.e., in fire control). Any time a system is in service and fails to operate as designed, it has experienced an unacceptable premature failure.

Forms of Premature Failure. Premature failure can take two forms, both requiring individual consideration. First is the failure of a system to hold water — in other words, the presence of pinhole leaks that require component replacement (see Exhibit S1.4). Leaks can cause significant damage and require immediate action. Leaks can also lead to excessive direct and indirect costs, as well as inherent risks from repeated system impairments.

In one known case, MIC-related leaks resulted in the shutdown of an aerospace manufacturer's global computing center. [4] A pinhole leak developed in a wet pipe

EXHIBIT S1.4 *Exterior Pinhole.*

sprinkler branch line located over the mainframe equipment of the computing center. Water from the leak not only damaged computing equipment but also resulted in over 5000 hours of lost operations time. Property damage from similar leaks at other sites has also been documented in the millions of dollars. [4,5]

The second and more worrisome effect is the failure of a system to operate as designed to achieve fire control. This failure not only affects property loss but also could threaten lives. Several systems with MIC have been found with obstructed sprinkler drops, the result of debris generated as a by-product of microbial activity (called biofilm or biosludge). Many in the industry have found systems with feed mains over 60 percent obstructed from biological growth. [5,6] An analysis of corrosion buildup indicates that thousands of pounds of debris can accumulate in medium-sized piping. [7] This buildup of debris presents an obvious hydraulic concern, as many affected sprinkler systems today will not provide fire control as the required discharge criteria, in terms of flow and pressure, are not available due to obstruction and associated frictional pressure loss.

Extent of MIC in the Sprinkler Industry

Currently only a limited number of credible national studies on the extent of the MIC problem in the fire protection industry exist. Most data are primarily anecdotal. No detailed comparative engineering-based study on effectiveness of the various treatment options currently being marketed has been completed. Complicating this lack of data, reported MIC cases rarely have secondary analysis to differentiate MIC as a primary contributing cause over other types of corrosion involving similar symptoms. Equally important, follow-up engineering data on treatment are rarely published.

The results of MIC in other industries are well documented. The Energy and Power Research Institute has estimated that corrosion in the U.S. electric power industry costs 5 to 10 billion dollars each year. Corrosion is said to be the culprit in half the forced outages each year in steam-generating plants. [8] And data collected from before the 2001 "U.S. power shortage" showed corrosion to be the primary factor in more than 10 percent of U.S. power generation costs. [8] In the gas and nuclear industries, MIC is specifically said to account for 15 to 30 percent of corrosion-related failures. [9] And, reports indicate that to prevent such problems, North American companies spend in excess of $1.5 billion per year (over $7 billion globally) on treatment chemicals to prevent microbial corrosion and fouling. Similar information, as mentioned, is not available for the fire protection industry. Although several industry groups have been attempting to compile data on this problem for several years, only two national studies have been published for the fire protection industry. [10,11] Thus, the true extent and cost of the problem are still not fully known.

In 1996, the National Fire Sprinkler Association sent out a questionnaire in its quarterly membership magazine requesting member information on sprinkler system failures that have been experienced. Results yielded approximately 40 responses from across the United States and Canada, which appeared to indicate that MIC was a "widespread problem." However, due to the reasons previously stated, the quality of analytical techniques used by respondents to confirm MIC as a cause, and the lack of investigation after detection and treatment bring the results of this study into question.

The only other related studies/data compilations completed have been conducted by FM Global. In their studies, in losses involving sprinkler systems, corrosion is the fifth leading cause of system failures. This is based on a review of data from 1988 to 1997. In another study by the FM Metallurgical Laboratory between 1994 and 2000 reviewing 155 cases of sprinkler system leaks, MIC was found to be present in approximately 40 percent of cases. Details on whether the presence of MIC directly caused leaks were not indicated in all these cases. [11] And, in another FM study of piping field samples between 1991 and 2002, over 60 percent of these failures are said to be attributed to MIC. [12]

Pinning the Blame

Part of the problem in obtaining detailed conclusive evidence is the complexity and cost of such investigations.

This is also a major point of frustration with owners experiencing these types of problems. Due to the nature of electrochemical corrosion, most corrosion engineers agree that the degree to which MIC specifically increases or contributes to general corrosion can never be conclusively determined. In fact, the complex biological interactions between bacteria and host materials are still not fully understood in many cases. There are simply too many variables and uncertainties that affect all corrosion reactions, especially those involving bacterial interactions. Thus, a percentage of blame or rate of corrosion from MIC likely cannot be numerically defined in any study. And, any database generated at this point will likely list only reported cases, not confirmed cases. Thus, the best that can be achieved in the future is a regularly updated database indicating where reported cases are occurring, method of failure detection, interior condition of pipe, methods of testing, and treatment with future corrosion monitoring.

Materials Affected by MIC

Although steel sprinkler pipe is the typically observed first point of failure, sprinkler orifice caps, control valves, fittings, and supply tanks may also be subject to damage from MIC. Numerous cases of obstructive growth and pinhole leaks associated with MIC have been found within 20 feet of the discharge side of site fire pumps — an interesting phenomenon due to the velocity of waterflow at this point in the system piping. While evidence suggests that only metallic components are susceptible to MIC, it is clear that some grades and alloys of steel are definitely more susceptible than others. As an example, certain grades of stainless steel are more susceptible than regular steel pipe, while others appear to show signs of resistivity. And although plastic components, such as underground water mains, are not subject to direct MIC, they are subject to the effects of biofouling or bacterial debris blockage from upstream corrosion activity.

DEFINING MIC

Oxygen Tolerance

MIC-related bacteria are classified primarily by oxygen tolerance; that is, they are *aerobic* or *anaerobic*. Aerobic bacteria require oxygen to flourish and reproduce. Anaerobic bacteria do not require oxygen to flourish and reproduce. [10] The most damaging MIC appears to take place within a highly complex community with multiple species of bacteria. This community includes not only aerobic and anaerobic bacteria, but also facultative bacteria, those MIC bacteria that function in both aerobic and anaerobic environments. All three types of bacteria can play a role in the somewhat random interactions that can occur in microbiologically influenced corrosion. [13]

Metabolism

In defining MIC bacteria further than previously done, classification is not absolute and can become confusing. The most commonly used method of categorizing bacteria associated with MIC is by metabolism. These categories are basically definitions of what each bacteria type eats (or metabolizes) and excretes as a by-product. As these terms imply, where plants use photosynthesis (i.e., light) to develop energy, bacteria use chemosynthesis (i.e., eating/breathing various chemicals or minerals) to sustain life.

Metabolic classifications are not universally replicated among scientists and can be somewhat confusing. A single bacteria type may fall under more than one metabolic definition. Some of the commonly referenced categories include sulfur-reducing bacteria, metal-reducing bacteria, acid-producing bacteria, iron-depositing bacteria, low-nutrient bacteria, iron-related bacteria, iron-reducing bacteria, iron-oxidizing bacteria, sulfate-oxidizing bacteria, slime-forming bacteria, and iron bacteria. [10,13,14,15]

Scientific Nomenclature

Finally, all bacteria (i.e., all plants and animals) can be classified by their scientific name under phylum, class, order, family, genus, and species. For example, one type of sulfate-reducing bacteria is anaerobic and metabolizes sulfate to sulfide. The sulfate-reducing bacteria group includes the genera *Desulfovibrio*, *Desulfobacter*, and *Desulfomaculum*. [3] All are of the phylum Thiopneutes, which interestingly translates from Greek to "sulfur-breathers."

SOURCES OF MIC INFECTION

Although there are no conclusive relational studies in the fire protection industry, as noted there are growing beliefs that a sprinkler system's water supply is not the only (and possibly primary) source of bacterial infection. Bacteria capable of causing MIC are potentially present in soil, air, and cutting oils, as well as in water. Thus, the manufacture, shipping, storage, and flushing of system materials should be addressed in all MIC investigations. MIC does not occur only in water-filled systems. Dry pipe systems are also susceptible. Dry systems may even be more susceptible to damage than are wet systems, due to the humidified atmosphere that is created after trip testing. A trip test and subsequent drain can create the right atmospheric moisture content for some bacterial colonies to thrive. Complete drainage and the subsequent use of truly dried air or nitrogen gas appear to mitigate this problem. With regard to tubercle growth alone, dissolved oxygen content, not bacteria, may be the only considerable factor in prevention.

MIC CORROSION PROCESS

As with other forms of corrosion, MIC removes material through a series of electrochemical interactions. As such, both an electrical and a chemical component occur with MIC. The electrical component occurs through electron transfer. Electron transfer removes pipe wall material one electron at a time. With MIC, this exact interaction is highly dependent on the specific bacteria involved. Within a sprinkler system, metallic parts become anodic in relation to the cathodic corrosion cell and surrounding water. Basic cathodic depolarization occurs as electrons are stripped away through various forms of oxidation and are pulled to an atom with another electrical potential. Although many of the complex cellular interactions of bacteria are still unclear and can vary by system, there appear to be several somewhat universal steps in the MIC process. They are as follows: [3]

1. Bacteria enter the system, attach to metallic components, and begin to rapidly colonize and reproduce.
2. Aerobic colonies metabolize nutrients from the water and/or the metal surfaces they are attached to, and subsequently excrete a polymer film by-product that bonds together to form crustaceous nodules called tubercles.
3. Tubercles and associated biofilms create microenvironments on the metallic material surface (under the tubercles). Tubercles are hard protective shells formed by biological activity. Tubercles typically have an open interior fluid cavity over the corrosion floor area with an approximate pH of 3 to 4. (See Exhibit S1.5.)
4. The underdeposit area (i.e., under the tubercles) becomes oxygen depleted (i.e., anaerobic and anodic) in relation to the surrounding system water or air, which remains aerobic and cathodic. Thus, electrons in the anodic metal flow to the cathode through a reduction reaction.
5. Underdeposit anaerobic bacteria metabolize pipe wall materials and excrete acids (such as acetic acid), as by-products, which are very aggressive to the carbon steels used in sprinkler piping. Relative acidity and alkalinity levels within the tubercle shells are reduced to an approximate pH of 2 to 4, which chemically attacks the metallic component surface.

On painted sprinkler piping, it is common to observe blisters where through-wall penetration has occurred. Test-

EXHIBIT S1.5 Interior Tubercle.

ing of the fluid within these blisters has shown a pH of less than 3.

The described corrosion process can continue indefinitely until the aerobic and anaerobic bacteria in the system are killed. The tubercles created from bacterial colonization must also be broken down to destroy the underdeposit microenvironment. Even without bacteria in the underdeposit of a corrosion cell, the process can continue indefinitely, as the corrosion chain in its final phases is no longer reliant on their activity.

PREVENTION AND TREATMENT

References from Allied Fields

Currently, the fire protection industry has a very limited number of directly usable references supported by scientific data. However, there is excellent information on data from other industries. The National Association of Corrosion Engineers (NACE) has published multiple studies about MIC detection and treatment for many years. ASTM (American Society for Testing and Materials) offers several publications on proper bacterial testing practices.

Also, the American Water Works Association (AWWA) offers standards describing the proper management of the somewhat hazardous chemicals typically used in injection devices attached to sprinkler systems for microbial control.

Fire Protection Codes and Standards

The National Fire Protection Association (NFPA) also addresses MIC. NFPA 25, *Standard for the Inspection, Testing, and Maintenance of Water-Based Fire Protection Systems*, discusses MIC treatment, inspection, and detection in some detail. NFPA 13, *Standard for the Installation of Sprinkler Systems*, contains a requirement that provides more guidance. In covering water supply treatment, NFPA 13 states:

> Water supplies and environmental conditions shall be evaluated for the existence of microbes and conditions that contribute to microbiologically influenced corrosion (MIC). Where conditions are found that contribute to MIC, the owner(s) shall notify the sprinkler system installer and a plan shall be developed to treat the system.

Although this requirement has generated curiosity, the resulting questions about effective treatment remain. NPFA 13 indicates that water supplies "shall be evaluated for the existence of microbes and conditions that contribute to microbiologically influenced corrosion (MIC)" and if present or suspected "a plan shall be developed to treat the system." The who, how, and when are still in debate by those addressing this issue. Who is best qualified to make the determination of when a failure is the result of MIC and if a biocidal treatment program will prevent all future failures? And how is a system best tested (i.e., with the most technical accuracy and cost effectiveness) to confirm MIC? Almost anything requiring laboratory work can be overtested, and undertesting can lead to a false sense of security. Answers to these questions are still evolving.

Operational Considerations

As the industry continues to develop methods to treat and prevent MIC, building owners are faced with the challenge of managing associated risks to their assets and business operations. At-risk businesses should review the interdependence of various operations to identify critical locations.

Proper Diagnosis

The analysis required to properly select a course of action to address MIC is typically outside of the scope of most sprinkler contractors and engineers. Thus, until treatment methods become universally proven and standardized, the most critical step in proper mitigation begins with the selection of a qualified corrosion control consultant.

With the wrong choice, a large amount of money can be spent on a problem that may not be correctable. A poor treatment choice could actually accelerate the corrosion rate and increase the affected area beyond that experienced before treatment. The company chosen to determine treatment must have detailed knowledge not only of microbial corrosion control but also of metallurgy and sprinkler system dynamics. Fire sprinkler systems have flow characteristics and concerns that are much different from other industrial processes and systems where MIC is typically addressed.

Specific Considerations for Fire Sprinkler Systems

Other industries deal with MIC in systems containing fluids that are either always static or always flowing, such as in cooling towers. Unlike sprinkler systems, dynamic systems have flow rates that are relatively constant, requiring that prescribed chemical dose rates be constant. A constant flow rate does not occur in sprinkler systems. Variable differences are seen with system drains and refills, inspector testing, and main drain tests. The dose rate for each of these flows must be considered to ensure that the chemical injection rate is always effective. Other industrial systems also have multiple points where biocidal chemicals can be injected. In contrast, sprinkler system water can realistically only be treated at system risers, back flow apparatus, or suction tanks.

Finally, it must again be stated that it is critical to understand that premature system failure can be a function of both bacterial infection and water quality that is incompatible with components. In the majority of premature system failure cases, water chemistry or poor design may be the only likely factors requiring consideration. A high bacterial count does not always indicate that MIC will occur, and conversely, a low bacterial count does not discount that MIC has occurred in the past, in a given system, and will not occur again in the near future.

Detection in Existing Systems

In existing systems suspected of being infected, the first step is to have all possible water supply sources (tank, city mains, ponds, rivers, etc.) and the interior of each system tested for bacterial levels and activity. Although current technology makes this detection easy, analysis of the results is somewhat complex. And, as previously stated, in determining treatment, bacterial detection is worthless without factoring in water quality. The laboratory used for analysis should be capable of giving conclusive details of water supply mineral and chemical levels, pipe wall deposit compositions, and type-specific bacterial counts. Multiple tests are used in these analyses, from simple bacterial incubation with visual inspection to sulfur print or DNA testing. Obviously, not all tests are required or are necessarily needed. Current preferred analysis methods run the spectrum, depending on the consultant chosen. Costs for such testing can also vary widely.

Mitigation in Affected Systems

When MIC is confirmed in operational systems, the building owner is first faced with a fundamental question. Can the system be salvaged (i.e., cleaned), or does it have to be replaced? Currently, this cost/benefit decision requires further study and is not supported one way or the other by documented best practices in the fire protection industry.

Pipe cleaning is typically an option when corrosion (i.e., pitting) is not excessive. However, "excessive" is a relative term. The after-cleaning quality of the pipe must be considered, for both future longevity and system hydraulics. The resulting frictional loss from numerous pits after cleaning could affect system performance. This, of course, is typically outside of the scope of work of most corrosion control consultants. When replacement materials are chosen that are different from those of the original system, they must also be accounted for in hydraulic analysis of the posttreated system. Case studies suggest that pipe cleaning may remove corrosion by-products that are, in effect, stopping the flow of water through existing pipe penetrations, subsequently resulting in leaks. [13] Prudence dictates that complete mitigation must include some form of treatment once pipes are cleaned.

Prevention in New Systems

In new systems, it is critical that susceptibility be determined before any systems are filled or tested with any water. If water tests are positive for MIC-related bacteria, a chemical injection system must be installed and used from the first fill. This includes hydrostatic testing and preliminary fills. The treatment methodology must also consider that water delivery of treatment chemicals might not effectively treat high spots, remote areas, or areas that trap air within the system.

If MIC is anticipated, one form of risk reduction may be to specify thicker piping in the design. Although this option only tends to buy time, it may serve as a prudent measure until universally accepted treatment methods are developed. Thicker walls buy time, not because they are less susceptible to MIC — a point on which evidence to date is inconclusive — but simply because a thicker wall has more material to corrode before a through-wall leak develops.

Another possibility in design is to remove the risk of leaks before MIC treatment is even evaluated. Areas such as those occupied by critical energized equipment can often be addressed by removing overhead sprinkler piping and limiting leak exposure through use of sidewall sprinkler coverage. Where it is not feasible to relocate large feed mains, piping can be sleeved so as not to permit potential leaks to contact energized equipment. Although these efforts provide some measure of immediate relief, they do not address the root cause or present a long-term solution to the problem.

Finally, the frequency of sprinkler and waterflow device testing must be addressed. Most agree that repeated draining and refilling of sprinkler systems can increase both biological and nonbiological forms of corrosion. Draining and refilling provide nutrients and oxygen to bacterial colonies and oxygen for general tuberculation. Some facilities have noted substantial reductions in the frequency of leaks by reducing the frequency of drain and alarm testing.

It is critical to note that once a system is filled with infected water, treatment can become exponentially more complex because any future treatment from a chemical injection system must now be effective in remote and stagnant system legs. In bacteria-positive areas, several additional water quality tests should be completed throughout the first year of service to ensure that contamination has not occurred from any other sources.

Chemical Injection

Once system components have been cleaned and sterilized or replaced, a chemical injection system must be installed to prevent recurrence. Once installed, this system will be required to be operational continuously. As with any other mechanical system, this will require continuous system preventive maintenance.

Several commercially available chemical injection systems have been specifically designed for installation on fire protection systems. Some simply use existing hardware and chemicals modified from MIC treatment in other utilities, such as cooling towers. At the time of this writing, none of the systems currently available was believed to be specifically listed or approved for use as a sprinkler system component. Although most systems appear to be effective when properly installed and maintained, reliability and effectiveness have not been time-proven when compared with other industrial system benchmarks. Past references should always be investigated with any choice.

Most injection systems currently available are designed to work with specific chemicals. These selected chemicals and dose rates are critical. Some bacteria can develop chemical resistance over time if doses are not strong enough and bacteria are not quickly killed. A small number of bacteria believed to be related to MIC (such as the genera *Bacillus* and *Clostridium*) are known to have the ability to convert to a spore state when they encounter adverse conditions that are not lethal. [3, 16] Spores are impervious to penetration by most chemicals and can thus survive biocide treatments indefinitely. Although subsequent treatments may slow or stop their activity, spores will reappear if and when treatments are stopped, and resume colonization. With a weak chemical attack, bacteria may also become resistant to the chemicals chosen.

As with other factors involved in treatment, the choice of chemical depends on the consultant. These chemicals generally include penetrants and biodispersants to break up the tubercles that protect underdeposit colonies, a biocide to kill the bacteria in the colonies, and a corrosion inhibitor to protect the interior surfaces of the system.

When such a system is chosen, the applicable authority having jurisdiction (AHJ) should be consulted. In addition to frictional loss concerns mentioned from changes in pipe surface roughness, increased back flow prevention hardware may be required. This could mean a pressure drop of 10 psi (0.7 bar) or more to sprinkler systems in addition to that created by pitting if cleaning is chosen. In new system designs, added alarm system contacts should also be planned for to monitor injection system chemical levels, operational status, and trouble conditions such as loss of power. Many pre-engineered systems provide contact points for these signals. As with fire detection, the perceived "best choice" is at the discretion of the person choosing, and opinions on this subject are highly variable.

Unlike other industrial systems treated for MIC, several unique interactions must be considered. First, sprinkler systems are typically located directly over people. The chemicals used must therefore be nontoxic in contemplation of accidental discharge or exposure to fire fighters under fire conditions. The effect of chosen chemicals on fire fighting (i.e., heat absorption) and chemical reaction with fire (i.e., heat) also needs consideration. Second, system designs typically place water discharge (such as from an inspector's test ports) into foliage or biologically sensitive drains and dry wells. Most municipal wastewater treatment plants (to which typical drains ultimately flow) require bacterial activity to decompose waste. Too large a quantity of biocides in municipal drains could be a problem.

In conclusion, a complete toxicity review with the highest possible biocidal chemical concentration must be completed. As much as possible, these chemicals should be noncombustible, colorless, odorless, and nontoxic. These chemicals must also be nondeteriorating to rubbers and polymers such as those used on pipe couplings, sprinkler o-rings, and valves. Chemical storage should also be reviewed, as several chemicals currently in use degrade rapidly with heat and may create relatively toxic vapors.

Opinions of which chemicals are believed to be "most effective" in control vary. Choices are available from those currently used for treatment in cooling towers and boilers to specifically patented compounds for fire protection systems. Some of the more common chemicals currently in use specifically for microbial control in sprinkler systems include quaternary ammonium compounds, organosulfur compounds, bromines, carbamates, isothiazalone, phosphates, and chlorines. Sodium silicate is effectively used in bulk quantities by several municipalities as an inhibitor, but this should be avoided for individual systems due to the potential for sprinkler plugging that overdosing can cause. At the writing of this article, no chemical can conclusively be said to be proven as the "most effective." The chemical choice may greatly depend on the bacteria present and system water quality.

SUMMARY

Current testing and treatment options can be confusing. Treatment is slowly evolving and research is continuing. Several industry groups, allied groups, and insurance companies are looking at the problem and applicable solutions. Many universities, governments, and private industry groups also continue to research microbial control in general industry, as they have for the past several decades. These efforts will continue to provide improved treatment options in our industry.

As the need to address this problem draws on the industry's creativity for resolution, prudent thinking requires us to evaluate each solution's impact on overall sprinkler system integrity and performance. In light of this problem, the overwhelming value of sprinkler systems should not be regarded as tarnished or tainted. Although some with repeated MIC problems may view sprinklers as a risk to property, the reduction of risk to life and property from fire these systems provide should never be overshadowed.

BIBLIOGRAPHY

References Cited

1. Cote, Arthur E., ed., *Fire Protection Handbook*®, 20th edition, National Fire Protection Association, Quincy, MA, 2008.
2. Christ, Bruce W., Ph.D., "Corrosion Process Inside Steel Fire Sprinkler Piping," *Fire Protection Engineering*, Summer 2005.
3. Clarke, B., "Microbiologically Influenced Corrosion in Fire Protection Systems," *Fire Protection Engineering*, Society of Fire Protection Engineers, No. 9, Winter 2001, pp. 14–16.
4. Cappers, M. A., "Investigation of Microbiological Influenced Corrosion in Sprinkler Systems," *Proceedings of Fire Suppression and Detection Research Application Symposium, Research and Practice:*

Bridging the Gap, February 12–14, 1997, Orlando, FL, National Fire Protection Research Foundation, Quincy, MA, 1997, pp. 69–81.
5. Kammen, J., "Bacteria Spell Doom for Fire Sprinklers," *The Arizona Republic*, October 24, 1999.
6. Shenkiryk, M., "Pipe-Klean Project—McCarran Airport," *HERC Products*, July 25, 1996.
7. Duncan, Bill, "Pipe Corrosion And Its Growing Threat to Office Building and Plant Operations," *Corrosioneering Journal*, July 2002.
8. Hoffman, S., *Bugging Water Systems for Corrosion Control*, Hoffman Publications, Inc., 1999.
9. The Mitigation and Detection of Microbial Corrosion," Argonne National Laboratory, Programs and Capabilities Database No. 526-002, 1999.
10. Bsharat, T. K., "Detection, Treatment, and Prevention of Microbiologically Influenced Corrosion in Water-Based Fire Protection Systems," National Fire Sprinkler Association, Inc., June 1998.
11. *Data Sheet 2-1: Internal Corrosion in Automatic Sprinkler Systems*, FM Global Property Loss Prevention Data Sheet, 2001.
12. Yee, Geary G., Ph.D., "Detection and Diagnostic Studies of MIC in Fire Protection Systems." Presented at the 2005 NFPA World Safety Conference and Exposition, June 9, 2005.
13. Little, B. J., Ray, R. I., and Wagner, P. A., "Tame Microbiologically Influenced Corrosion," *Chemical Engineering Progress*, September 1998.
14. Borenstein, S.W., *Microbiologically Influenced Corrosion Handbook*, Woodhead Publications Ltd., 1994.
15. Pope, D. H., Duquette, D. J., Johannes, A. H., and Wayner, P. C., "Microbiologically Influenced Corrosion of Industrial Alloys," *Materials Performance*, July 1984.
16. Hero, H. M., *The Nalco Guide to Cooling Water System Failure Analysis*, McGraw-Hill, Inc., 1993.

NFPA Codes, Standards, and Recommended Practices

Reference to the following NFPA codes, standards, and recommended practices will provide further information on microbiologically induced corrosion in fire sprinkler systems discussed in this chapter. (See the latest version of the NFPA Catalog for availability of current editions of the following documents.)

NFPA 13, *Standard for the Installation of Sprinkler Systems*

NFPA 25, *Standard for the Inspection, Testing, and Maintenance of Water-Based Fire Protection Systems*

SUPPLEMENT 2

Foam Environmental Issues

Editor's Note: Supplement 2 extracts Annex F from the 2005 edition of NFPA 11, Standard for Low-, Medium-, and High-Expansion Foam. It is included in this handbook to assist the user in managing the discharge of foam during foam flow tests in an environmentally responsible manner.

1 Overview

Fire-fighting foams as addressed in this standard serve a vital role in fire protection throughout the world. Their use has proven to be essential for the control of flammable liquid fire threats inherent in airport operations, fuel farms and petroleum processing, highway and rail transportation, marine applications, and industrial facilities. The ability of foam to rapidly extinguish flammable liquid spill fires has undoubtedly saved lives, reduced property loss, and helped minimize the global pollution that can result from the uncontrolled burning of flammable fuels, solvents, and industrial liquids.

However, with the ever-increasing environmental awareness, recent concern has focused on the potential adverse environmental impact of foam solution discharges. The primary concerns are fish toxicity, biodegradability, treatability in wastewater treatment plants, and nutrient loading. All of these are of concern when the end-use foam solutions reach natural or domestic water systems. Additionally, the U.S. Environmental Protection Agency (EPA) has highlighted a potential problem with some foam concentrates by placing glycol ethers and ethylene glycol, common solvent constituents in some foam concentrates, on the list of hazardous air pollutants under the 1990 Clean Air Act Amendments.

The purpose of this annex is to address the following:

(1) Provide foam users with summary information on foam environmental issues
(2) Highlight applicable regulatory status
(3) Offer guidelines for coping with regulations, and provide suggested sources for additional information
(4) Encourage planning for foam discharge scenarios (including prior contact with local wastewater treatment plant operators)

It should be emphasized that it is not the intent of this annex to limit or restrict the use of fire-fighting foams. The foam committee believes that the fire safety advantages of using foam are greater than the risks of potential environmental problems. The ultimate goal of this section is to foster use of foam in an environmentally responsible manner so as to minimize risk from its use.

2 Scope

The information provided in this section covers foams for Class B combustible and flammable liquid fuel fires. Foams for this purpose include protein foam, fluoroprotein foam, film-forming fluoroprotein foam (FFFP), and synthetic foams such as aqueous film-forming foam (AFFF).

Some foams contain solvent constituents that can require reporting under federal, state, or local environmental regulations. In general, synthetic foams, such as AFFF, biodegrade more slowly than protein-based foams. Protein-based foams can be more prone to nutrient loading and treatment facility "shock loading" due to their high ammonia nitrogen content and rapid biodegradation, respectively.

This section is primarily concerned with the discharge of foam solutions to wastewater treatment facilities and to the environment. The discharge of foam concentrates, while a related subject, is a much less common occurrence. All manufacturers of foam concentrate deal with clean-up and disposal of spilled concentrate in their MSDS sheets and product literature.

3 Discharge Scenarios

A discharge of foam water solution is most likely to be the result of one of four scenarios:
(1) Manual fire-fighting or fuel-blanketing operations
(2) Training

(3) Foam equipment system tests
(4) Fixed system releases

These four scenarios include events occurring at such places as aircraft facilities, fire fighter training facilities, and special hazards facilities (such as flammable/hazardous warehouses, bulk flammable liquid storage facilities, and hazardous waste storage facilities). Each scenario is considered separately in 3.1 through 3.4.

3.1 Fire-Fighting Operations. Fires occur in many types of locations and under many different circumstances. In some cases, it is possible to collect the foam solution used; and in others, such as in marine fire fighting, it is not. These types of incidents include aircraft rescue and fire-fighting operations, vehicular fires (i.e., cars, boats, train cars), structural fires with hazardous materials, and flammable liquid fires. Foam water solution that has been used in fire-fighting operations will probably be heavily contaminated with the fuel or fuels involved in the fire. It is also likely to have been diluted with water discharged for cooling purposes.

In some cases, the foam solution used during fire department operations can be collected. However, it is not always possible to control or contain the foam. This can be a consequence of the location of the incident or the circumstances surrounding it.

Event-initiated manual containment measures are the operations usually executed by the responding fire department to contain the flow of foam water solution when conditions and manpower permit. Those operations include the following measures:

(1) Blocking sewer drains: this is a common practice used to prevent contaminated foam water solution from entering the sewer system unchecked. It is then diverted to an area suitable for containment.
(2) Portable dikes: these are generally used for land-based operations. They can be set up by the fire department personnel during or after extinguishment to collect run-off.
(3) Portable booms: these are used for marine-based operations, which are set up to contain foam in a defined area. These generally involve the use of floating booms within a natural body of water.

3.2 Training. Training is normally conducted under circumstances conducive to the collection of spent foam. Some fire training facilities have had elaborate systems designed and constructed to collect foam solution, separate it from the fuel, treat it, and — in some cases — re-use the treated water. At a minimum, most fire training facilities collect the foam solution for discharge to a wastewater treatment facility. Training can include the use of special training foams or actual fire-fighting foams. Training facility design should include a containment system. The wastewater treatment facility should first be notified and should give permission for the agent to be released at a prescribed rate.

3.3 System Tests. Testing primarily involves engineered, fixed foam fire-extinguishing systems. Two types of tests are conducted on foam systems: acceptance tests, conducted pursuant to installation of the system; and maintenance tests, usually conducted annually to ensure the operability of the system. These tests can be arranged to pose no hazard to the environment. It is possible to test some systems using water or other nonfoaming, environmentally acceptable liquids in the place of foam concentrates if the AHJ permits such substitutions.

In the execution of both acceptance and maintenance tests, only a small amount of foam concentrate should be discharged to verify the correct concentration of foam in the foam water solution. Designated foam water test ports can be designed into the piping system so that the discharge of foam water solution can be directed to a controlled location. The controlled location can consist of a portable tank that would be transported to an approved disposal site by a licensed contractor. The remainder of the acceptance test and maintenance test should be conducted using only water.

3.4 Fixed System Releases. This type of release is generally uncontrolled, whether the result of a fire incident or a malfunction in the system. The foam solution discharge in this type of scenario can be dealt with by event-initiated operations or by engineered containment systems. Event-initiated operations encompass the same temporary measures that would be taken during fire department operations: portable dikes, floating booms, and so forth. Engineered containment would be based mainly on the location and type of facility, and would consist of holding tanks or areas where the contaminated foam water solution would be collected, treated, and sent to a wastewater treatment facility at a prescribed rate.

4 Fixed Systems

Facilities can be divided into those without an engineered containment system and those with an engineered containment system.

4.1 Facilities without Engineered Containment. Given the absence of any past requirements to provide containment, many existing facilities simply allow the foam water solution to flow out of the building and evaporate into the atmosphere or percolate into the ground. The choices for containment of foam water solution at such facilities fall into two categories: event-initiated manual containment measures and installation of engineered containment systems.

Selection of the appropriate choice is dependent on the location of the facility, the risk to the environment, the risk of an automatic system discharge, the frequency of automatic system discharges, and any applicable rules or regulations.

"Event-initiated manual containment measures" will be the most likely course of action for existing facilities without engineered containment systems. This can fall under the responsibility of the responding fire department and include such measures as blocking storm sewers, constructing temporary dikes, and deploying floating booms. The degree of such measures will primarily be dictated by location as well as available resources and manpower.

The "installation of engineered containment systems" is a possible choice for existing facilities. Retrofitting an engineered containment system is costly and can adversely affect facility operations. There are special cases, however, that can warrant the design and installation of such systems. Such action is a consideration where an existing facility is immediately adjacent to a natural body of water and has a high frequency of activation.

4.2 Facilities with Engineered Containment. Any engineered containment system will usually incorporate an oil/water separator. During normal drainage conditions (i.e., no foam solution runoff), the separator functions to remove any fuel particles from drainage water. However, when foam water solution is flowing, the oil/water separator must be bypassed so that the solution is diverted directly to storage tanks. This can be accomplished automatically by the installation of motorized valves set to open the bypass line upon activation of the fixed fire-extinguishing systems at the protected property.

The size of the containment system is dependent on the duration of the foam water flow, the flow rate, and the maximum anticipated rainfall in a 24-hour period. Most new containment systems will probably only accommodate individual buildings. However, some containment systems can be designed to accommodate multiple buildings, depending upon the topography of the land and early identification in the overall site planning process.

The specific type of containment system selected will also depend upon location, desired capacity, and function of facilities in question. The systems include earthen retention systems, belowground tanks, open-top inground tanks, and sump and pump designs (i.e., lift stations) piped to aboveground or inground tanks.

The earthen retention designs consist of open-top earthen berms, which usually rely upon gravity-fed drainage piping from the protected building. They can simply allow the foam water solution to percolate into the ground or can include an impermeable liner. Those containing an impermeable liner can be connected to a wastewater treatment facility or can be suction pumped out by a licensed contractor.

Closed-top, belowground storage tanks can be the least environmentally acceptable design approach. They usually consist of a gravity-fed piping arrangement and can be suction pumped out or piped to a wastewater treatment facility. A potential and frequent problem associated with this design is the leakage of groundwater or unknown liquids into the storage tank.

Open-top, belowground storage tanks are usually lined concrete tanks that can rely on gravity-fed drainage piping or a sump and pump arrangement. These can accommodate individual or multiple buildings. They must also accommodate the maximum anticipated rainfall in a 24-hour period. These are usually piped to a wastewater treatment facility.

Aboveground tanks incorporate a sump and pump arrangement to closed, aboveground tanks. Such designs usually incorporate the use of one or more submersible or vertical shaft, large-capacity pumps. These can accommodate individual or multiple buildings.

4.3 New Facilities. The decision to design and install a fixed foam water solution containment system is dependent on the location of the facility, the risk to the environment, possible impairment of facility operations, the design of the fixed foam system (i.e., automatically or manually activated), the ability of the responding fire department to execute event-initiated containment measures, and any pertinent regulations.

New facilities might not warrant the expense and problems associated with containment systems. Where the location of a facility does not endanger groundwater or any natural bodies of water, this can be an acceptable choice, provided the fire department has planned emergency manual containment measures.

Where conditions warrant the installation of engineered containment systems, there are a number of considerations. They include size of containment, design and type of containment system, and the capability of the containment system to handle individual or multiple buildings. Engineered containment systems can be a recommended protective measure where foam extinguishing systems are installed in facilities that are immediately adjacent to a natural body of water. These systems can also be prudent at new facilities, where site conditions permit, to avoid impairment of facility operations.

5 Disposal Alternatives

The uncontrolled release of foam solutions to the environment should be avoided. Alternative disposal options are as follows:

(1) Discharge to a wastewater treatment plant with or without pretreatment
(2) Discharge to the environment after pretreatment
(3) Solar evaporation
(4) Transportation to a wastewater treatment plant or hazardous waste facility

Foam users, as part of their planning process, should make provisions to take the actions necessary to utilize whichever of these alternatives is appropriate for their situation. Section 6 describes the actions that can be taken, depending on the disposal alternative that is chosen.

6 Collection and Pretreatment of Foam Solutions Prior to Disposal

6.1 Collection and Containment. The essential first step in employing any of these alternatives is collection of the foam solution. As noted above, facilities that are protected by foam systems normally have systems to collect and hold fuel spills. These systems can also be used to collect and hold foam solution. Training facilities are, in general, designed so that foam solution can be collected and held. Fire fighters responding to fires that are at other locations should attempt, insofar as is practical, to collect foam solution run-off with temporary dikes or other means.

6.2 Fuel Separation. Foam solution that has been discharged on a fire and subsequently collected will usually be heavily contaminated with fuel. Since most fuels present their own environmental hazards and will interfere with foam solution pretreatment, an attempt should be made to separate as much fuel as possible from the foam solution. As noted in 4.2, the tendency of foam solutions to form emulsions with hydrocarbon fuels will interfere with the operation of conventional fuel-water separators. An alternative is to hold the collected foam solution in a pond or lagoon until the emulsion breaks and the fuel can be separated by skimming. This can take from several hours to several days. During this time, agitation should be avoided to prevent the emulsion from reforming.

6.3 Pretreatment Prior to Discharge.

6.3.1 Dilution. Foam manufacturers and foam users recommend dilution of foam solution before it enters a wastewater treatment plant. There is a range of opinion on the optimum degree of dilution. It is generally considered that the concentration of foam solution in the plant influent should not exceed 1700 ppm (588 gal of plant influent per gallon of foam solution). This degree of dilution is normally sufficient to prevent shock loading and foaming in the plant. However, each wastewater treatment plant must be considered as a special case, and those planning a discharge of foam solution to a wastewater treatment facility should discuss this subject with the operator of the facility in advance.

Diluting waste foam solution 588:1 with water is an impractical task for most facilities, especially when large quantities of foam solution are involved. The recommended procedure is to dilute the foam solution to the maximum amount practical and then meter the diluted solution into the sewer at a rate which, based on the total volume of plant influent, will produce a foam solution concentration of 1700 ppm or less.

For example, if the discharge is to be made to a 6 million gal/day treatment plant, foam solution could be discharged at the rate of 7 gpm (6,000,000 gal/day divided by 1440 minutes/day divided by 588 equals 7 gpm). The difficulties of metering such a low rate of discharge can be overcome by first diluting the foam solution by 10:1 or 20:1, permitting discharge rates of 70 gpm or 140 gpm respectively. Dilution should also be considered if the foam solution is to be discharged to the environment in order to minimize its impact.

6.3.2 Defoamers. The use of defoamers will decrease, but not eliminate, foaming of the foam solution during pumping, dilution, and treatment. The foam manufacturer should be consulted for recommendations as to the choice of effective defoamers for use with a particular foam concentrate.

6.3.3 Method for Determining the Effective Amount of Antifoam Apparatus. The effective amount of antifoam is determined by using the following apparatus:

(1) Balance — 1600 g capacity minimum — readability 0.2 g maximum
(2) One 2 L beaker or similar container
(3) One 1 gal plastic or glass jug with cap
(4) Eyedropper
(5) Optional — 10 mL pipette

6.3.3.1 Procedure. Proceed with the following instructions to determine the effective amount of antifoam:

(1) In the 2 L beaker, weigh out 1 g (1 mL) of antifoam using an eyedropper or the pipette.
(2) Add 999 g of water.
(3) Mix well.
(4) Weigh out 1000 g of the solution to be defoamed and place it in the gallon jug.

(5) Add 10 g (10 mL) of the diluted antifoam to the gallon jug using the eyedropper or pipette, cap it, and shake vigorously.
(6) If the solution in the jug foams, go back to step 5 and repeat this step until little or no foam is generated by shaking the jug; keep a record of the number of grams (mL) that are required to eliminate the foaming.
(7) The number of grams (mL) of diluted antifoam required to eliminate foaming is equal to the number of parts per million (ppm) of the antifoam as supplied that must be added to the solution to be defoamed.
(8) Calculate the amount of neat antifoam to be added as follows:

$$W = 8.32 \; V \times D \div 1{,}000{,}000$$

where:

V = Volume of solution to be defoamed in U.S. gal

D = ppm of antifoam required

W = lb of antifoam required

Example

10,000 gal of foam solution require defoaming. The procedure above has determined that 150 ppm of antifoam are needed to defoam this solution: $8.32 \times 10{,}000 \times 150 \div 1{,}000{,}000 = 12.48$ lb.

(9) The amount of antifoam to be added will normally be quite small compared to volume of the solution to be defoamed. The antifoam must be uniformly mixed with the solution to be defoamed. It will aid in the achievement of this objective if the antifoam is diluted as much as is practical with water or the solution to be defoamed prior to addition to the solution containment area. The solution in the containment area must then be agitated to disperse the antifoam uniformly. One method of doing this is to use a fire pump to draft out of the containment area and discharge back into it using a water nozzle set on straight stream. Alternatively, if suitable metering equipment is available, antifoam as supplied or diluted antifoam can be metered into the solution discharge line at the proper concentration.

7 Discharge of Foam Solution to Wastewater Treatment Facilities

Biological treatment of foam solution in a wastewater treatment facility is an acceptable method of disposal. However, foam solutions have the potential to cause plant upsets and other problems if not carefully handled. The reasons for this are explained in 7.1 through 7.4.

7.1 Fuel Contamination. Foam solutions have a tendency to emulsify hydrocarbon fuels and some polar fuels that are only slightly soluble in water. Water-soluble polar fuels will mix with foam solutions. The formation of emulsions will upset the operation of fuel/water separators and potentially cause the carryover of fuel into the waste stream. Many fuels are toxic to the bacteria in wastewater treatment plants.

7.2 Foaming. The active ingredients in foam solutions will cause copious foaming in aeration ponds, even at very low concentrations. Aside from the nuisance value of this foaming, the foaming process tends to suspend activated sludge solids in the foam. These solids can be carried over to the outfall of the plant. Loss of activated sludge solids can also reduce the effectiveness of the wastewater treatment. This could cause water quality problems such as nutrient loading in the waterway to which the outfall is discharged. Because some surfactants in foam solutions are highly resistant to biodegradation, nuisance foaming can occur in the outfall waterway.

7.3 BOD (Biological Oxygen Demand). Foam solutions have high BODs compared to the normal influent of a wastewater treatment plant. If large quantities of foam solution are discharged to a wastewater treatment plant, shock loading can occur, causing a plant upset.

Before discharging foam solutions to a wastewater treatment plant, the plant operator should be contacted. This should be done as part of the emergency planning process. The plant operator will require, at a minimum, a Material Safety Data Sheet (MSDS) on the foam concentrate, an estimate of the five-day BOD content of the foam solution, an estimate of the total volume of foam solution to be discharged, the time period over which it will be discharged, and, if the foam concentrate is protein-based, an estimate of the ammonia nitrogen content of the foam solution.

The foam manufacturer will be able to provide BOD and ammonia nitrogen data for the foam concentrate, from which the values for foam solution can be calculated. The other required information is site-specific and should be developed by the operator of the facility from which the discharge will occur.

7.4 Treatment Facilities. Foam concentrates or solutions can have an adverse effect on microbiologically based oily water treatment facilities. The end user should take due account of this before discharging foam systems during testing or training.

8 Foam Product Use Reporting.

Federal (U.S.), state, and local environmental jurisdictions have certain chemical reporting requirements that apply to chemical constituents within foam concentrates. In addition, there are also requirements that apply to the flammable liquids to which the foams are being applied. For example, according to the U.S. Environmental Protection Agency (EPA), the guidelines in 8.1 through 8.4 must be adhered to.

8.1 Releases of ethylene glycol in excess of 5000 lb are reportable under Sections 102(b) and 103(a) of U.S. EPA Comprehensive Environmental Response Compensation & Liability Act (CERCLA). Ethylene glycol is generally used as a freeze-point suppressant in foam concentrates.

8.2 As of June 12, 1995, the EPA issued a final rule 60 CFR 30926 on several broad categories of chemicals, including the glycol ethers. The EPA has no reportable quantity for any of the glycol ethers. Thus foams containing glycol ethers (butyl carbitol) are not subject to EPA reporting. Consult the foam manufacturer's MSDS to determine if glycol ethers are contained in a particular foam concentrate.

8.3 The EPA does state that CERCLA liability continues to apply to releases of all compounds within the glycol ether category, even if reporting is not required. Parties responsible for releases of glycol ethers are liable for the costs associated with cleanup and any natural resource damages resulting from the release.

8.4 The end user should contact the relevant local regulating authority regarding specific current regulations.

9 Environmental Properties of Hydrocarbon Surfactants and Fluorochemical Surfactants

Fire-fighting foam agents contain surfactants. Surfactants or surface active agents are compounds that reduce the surface tension of water. They have both a strongly "water-loving" portion and a strongly "water-avoiding" portion.

Dish soaps, laundry detergents, and personal health care products such as shampoos are common household products that contain hydrocarbon surfactants.

Fluorochemical surfactants are similar in composition to hydrocarbon surfactants; however, a portion of the hydrogen atoms have been replaced by fluorine atoms. Unlike chlorofluorocarbons (CFCs) and some other volatile fluorocarbons, fluorochemical surfactants are not ozone depleting and are not restricted by the Montreal Protocol or related regulations. Fluorochemical surfactants also have no effect on global warming or climate change. AFFF, fluoroprotein foam, and FFFP are foam liquid concentrates that contain fluorochemical surfactants.

There are environmental concerns with use of surfactants that should be kept in mind when these products are used for extinguishing fires or for fire training. These concerns are as follows:

(1) All surfactants have a certain level of toxicity.
(2) Surfactants used in fire-fighting foams cause foaming.
(3) Surfactants used in fire-fighting foams can be persistent. (This is especially true of the fluorine-containing portion of fluorochemical surfactants.)
(4) Surfactants can be mobile in the environment. They can move with water in aquatic ecosystems and leach through soil in terrestrial ecosystems.

9.1 through 9.5 explain what each of these properties mean and what the properties mean in terms of how these compounds should be handled.

9.1 Toxicity of Surfactants. Fire-fighting agents, used responsibly and following Material Safety Data Sheet instructions, pose little toxicity risk to people. However, some toxicity does exist. The toxicity of the surfactants in fire-fighting foams, including the fluorochemical surfactants, is a reason to prevent unnecessary exposure to people and to the environment. It is a reason to contain and treat all fire-fighting foam wastes whenever feasible. One should always make plans to contain wastes from training exercises and to treat them following the suppliers' disposal recommendations as well as the requirements of local authorities.

Water that foams when shaken due to contamination from fire-fighting foam should not be ingested. Even when foaming is not present, it is prudent to evaluate the likelihood of drinking water supply contamination and to use alternate water sources until one is certain that surfactant concentrations of concern no longer exist. Suppliers of fire-fighting foams should be able to assist in evaluating the hazard and in recommending laboratories that can do appropriate analysis when necessary.

9.2 Surfactants and Foaming. Many surfactants can cause foaming at very low concentrations. This can cause aesthetic problems in rivers and streams, and both aesthetic and operational problems in sewers and wastewater treatment systems. When too much fire-fighting foam is discharged at one time to a wastewater treatment system, serious foaming can occur. The bubbles of foam that form in the treatment system can trap and bring flocks of the activated sludge that treat the water in the treatment system to the surface. If the foam blows off the surface of the treatment system, it leaves a black or brown sludge residue where the foam lands and breaks down.

If too much of the activated sludge is physically removed from the treatment system in foam, the operation of the treatment system can be impaired. Other waste passing through the system will then be incompletely treated until the activated sludge concentration again accumulates. For this reason, the rate of fire-fighting foam solution discharged to a treatment system has to be controlled. Somewhat higher discharge rates can be possible when antifoaming or defoaming agents are used. Foam concentrate suppliers can be contacted for guidance on discharge rates and effective antifoaming or defoaming agents.

9.3 Persistence of Surfactants. Surfactants can biodegrade slowly and/or only partially biodegrade. The fluorochemical surfactants are known to be very resistant to chemical and biochemical degradation. This means that, while the non-fluorochemical portion of these surfactants can break down, the fluorine-containing portion can likely remain. This means that after fire-fighting foam wastes are fully treated, the waste residual could still form some foam when shaken. It could also still have some toxicity to aquatic organisms if not sufficiently diluted.

9.4 Mobility of Surfactants. Tests and experience have shown that some surfactants or their residues can leach through at least some soil types. The resistance of some surfactants to biodegradation makes the mobility of such surfactants a potential concern. While a readily degradable compound is likely to degrade as it leaches through soil, this won't happen to all surfactants. Thus, if allowed to soak into the ground, surfactants that don't become bound to soil components can eventually reach groundwater or flow out of the ground into surface water. If adequate dilution has not occurred, surfactants can cause foaming or concerns about toxicity. Therefore, it is inappropriate to allow training waste to continually seep into soil, especially in areas where water resources could be contaminated.

9.5 Fluorochemical Surfactants and Living Systems. Some fluorochemical surfactants or their persistent degradation products have been found in living organisms.

SUPPLEMENT 3

The Role of the Inspector

Terry L. Victor

Editor's Note: In an effort to assist those involved in training inspectors and those who are in the process of being trained, this supplement has been included to outline the training needs, safety equipment, and basic duties and approach to the performance of work for the professional inspector. It includes information not found elsewhere for the training and education of the professional water-based fire protection systems inspector. This supplement is organized and written so that it can be used for formal training classes or for individual study.

Although NFPA 25 does require some training and experience for personnel involved in performing the tasks required by the standard, it cannot dictate the exact training and experience for every situation in which the standard may be adopted and used. For example, some jurisdictions may require licensing for individuals involved in providing professional inspection, testing, and maintenance services, whereas other jurisdictions do not.

It is important to note that "The Role of the Inspector" is not part of NFPA 25 and is not intended to be an enforceable part of the standard.

The inspection and testing of a water-based fire protection system requires expertise that is not usually associated with building engineers or building maintenance personnel. Paragraph 4.1.2.2 of NFPA 25 states: "These tasks shall be performed by personnel who have developed competence through training and experience." The inspection and testing of a fire protection system requires specialized knowledge of system components, including valves, piping, hangers, test connections, and sprinklers. This specialized knowledge includes recognition of different types of components, their intended use, manufacturer's testing requirements, and Underwriters Laboratories listing criteria. For this reason, many property owners and occupants rely on and contract with fire protection contractors who offer inspection and testing services.

Many activities in NFPA 25 require only a visual inspection to determine the status and/or the condition of a component or system. These activities can be taught to on-site personnel and require a minimal amount of training. Most of the weekly and monthly inspection requirements fall into this category, and training on-site personnel may prove to be the most cost-effective way to comply with the standard. This supplement deals primarily with the activities of the inspectors employed by fire protection contractors and the methods used to conduct the inspection and/or tests required by NFPA 25. Methods used to compile and record results are also addressed. For those who feel they are qualified to perform this work, proceed with caution and have the telephone number of a reputable contractor in your area available, in case you are unable to silence the fire alarms, reset the dry pipe valve, or restore the fire pump to normal service.

Terry L. Victor is the National Manager of Sprinkler Systems at Tyco/Fire & Security SimplexGrinnell and a member of the NFPA Technical Committee on Inspection, Testing, and Maintenance of Water-Based Systems.

INSPECTOR RESPONSIBILITIES VERSUS OWNER EXPECTATIONS

It must be understood that the inspection and testing of a fire protection system in accordance with NFPA 25 is *not* equivalent to any of the following engineering reviews, engineering evaluations, authority having jurisdiction (AHJ) acceptance inspections, or AHJ acceptance tests:

1. An engineering or AHJ review or evaluation of the adequacy of the system to control or extinguish a fire in the protected occupancy
2. An engineering or AHJ review or evaluation of the adequacy of the water supply, including water tanks and fire pumps, to provide the necessary waterflow and pressure needed to meet the fire protection system demand as designed
3. An engineering or AHJ review or evaluation of the hazards present in the facility to determine the required system criteria for minimum protection levels
4. An engineering or AHJ review, evaluation, acceptance, inspection, or test of the fire protection system to determine whether the system was designed and installed in accordance with the applicable installation standard, including NFPA 13, 13R, 14, 15, 16, 20, 22, and 214
5. An engineering or AHJ review or evaluation of all areas of the facilities to determine whether all areas of the facility are protected in accordance with the original applicable installation standard

The NFPA Technical Committee on Inspection, Testing, and Maintenance of Water-Based Systems has made it clear that the role of the inspector performing inspection and testing activities in accordance with NFPA 25 is not that of a fire protection engineer or an AHJ. The inspector is not required to be a fire protection engineer, nor is the inspector required to have detailed knowledge of the fire protection system installation requirements in the applicable NFPA standards. Therefore, when property owners contract with a fire protection contractor, they must not expect the inspector to provide an engineering analysis detailing the adequacy of the system design when compared to the hazard, or evaluate the accuracy and completeness of the installation.

The purpose of periodic inspection and testing of fire protection systems is to provide a reasonable amount of assurance at that moment in time that the system will operate as intended during a fire emergency. As stated in A.1.2 of NFPA 25, "History has shown that the performance reliability of a water-based fire protection system under fire-related conditions increases where comprehensive inspection, testing, and maintenance procedures are enforced." The inspector is not required to perform the following functions:

- Review the design criteria for the system to determine whether it is adequate for the commodity or hazard
- Review the installation shop drawing to make sure all pipes are correctly sized and the sprinklers are the correct type, orifice size, and temperature rating
- Perform a water supply analysis to determine whether the water supply is sufficient to provide the required gallons per minute at the required pounds per square inch for the system
- Perform a complete building inspection to determine whether closets, blind spaces, attics, crawl spaces, and all other special building features are adequately protected in accordance with the version of the installation standard in force at the time of the installation
- Recognize that storage commodities and/or arrangements are different from those anticipated when the system was designed and installed
- Research the installation contract files to determine whether special requirements were required by the AHJ

The bottom line is that the contract with a fire protection contractor to perform the inspection, testing, and maintenance of a water-based fire protection system is not an insurance policy or an engineering evaluation. If the property owner is ever uncertain whether the fire protection system as installed will provide the minimum protection required by the applicable codes, the owner should contact a fire protection engineer or qualified contractor to perform an analysis of the building, the building occupancy, the commodity classification of any storage present, the system design criteria, and the water supply.

INSPECTOR TRAINING AND QUALIFICATIONS

The inspector of a water-based fire protection system, whether an employee of the property owner, the owner's authorized representative, or an employee of a fire protection contractor, has a tremendous responsibility. Water-based fire protection systems protect both life and property. These systems have a history of reliability that exceeds most other mechanical systems. In those instances when a water-based fire protection system fails to extinguish or control a fire, poor maintenance and shut water control valves are the most frequent reasons for the failures. Although a properly conducted inspection, testing, and maintenance program cannot guarantee the performance of a water-based fire protection system, the chances of timely system operation and fire control during a fire emergency are increased dramatically. The inspector is responsible for conducting inspections and tests in accordance with NFPA 25, industry-accepted practices, and manufacturer's recommendations.

The choice of whom to hire and train as an inspector should be given careful consideration. A person hired as an inspector who has little or no knowledge of fire protection systems will obviously have a more difficult time during training than someone who has worked in or around the fire protection sprinkler industry and has knowledge of these systems. The inspector-in-training with little or no knowledge of fire protection systems will have to learn the systems and components, as well as the requirements of NFPA 25, and the proper ways to conduct an inspection and/or test.

Whatever the background of the individual hired as a water-based fire protection systems inspector or trainee, several personal traits must be considered as well. The role of the inspector is to properly perform the activities required by NFPA 25, which are meant to provide the highest level of assurance that systems will respond as expected during a fire emergency. Without a certain work ethic and sense of responsibility, the performance of these activities will not be carried out as accurately or thoroughly as expected, or in some cases not carried out at all. The following sections describe the desired personal traits of the person hired to perform inspection and testing activities and allow the reader to fully understand the total scope of competency and training required of the NFPA 25 inspector.

Inspector Qualifications

There are some basic traits that an inspector should possess to do the job completely, accurately, efficiently, and professionally. Aside from related work experience, the person being hired to be an inspector should meet a set of minimum qualifications.

An inspector must be conscientious and aware of the importance of the job being performed. The person who wants to be an inspector, but sees it as just another "8 to 5" job, doesn't understand the importance of inspecting and testing water-based fire protection systems. An inspector must consider the condition and operability of every system being inspected as if the lives of people are at stake, because, obviously, they are. The need to perform every inspection completely and thoroughly must be the number one priority of the inspector.

An inspector must possess a good work ethic, be self-motivated, and have the ability to self-supervise. Some contractors schedule their inspectors and even set up appointments to have systems inspected, whereas others rely on the inspector to establish their own schedule and make their own appointments. Whichever the case, the inspector needs to be at the property when scheduled.

The owner, the owner's authorized representative, and/or the occupants expect the inspection to take place as scheduled. The monitoring station and the local authorities may expect alarms to be tested at certain times. An inspector who ignores schedules and appointments has no regard for the other individuals involved in supervising and maintaining water-based fire protection systems.

An inspector should be mechanically inclined. Some of the activities that must be performed by an inspector include resetting or restoring to service dry valves, preaction valves, deluge valves, accelerators, exhausters, air compressors, fire pumps, controllers, alarm switches, and numerous other system components. The inspector must have the ability to read the manufacturer's maintenance information and follow the instructions. The inspector must also be able to interpret test results, perform basic troubleshooting, and make recommendations to correct deficiencies or impairments found.

An inspector must be educated and have legible writing. Local codes as well as NFPA 25 must be read, understood, and applied as inspections are performed. Inspection forms must be filled out so that those responsible for system maintenance can read them and correct any deficiencies found. In recent years, computer programs have been developed to aid in the inspection process. These programs include question sets that correspond to the activities required by NFPA 25. In addition, the programs provide standardized descriptions of, and describe corrective actions for, impairments and deficiencies. If computers are used as part of the inspection process, then computer skills are also necessary.

An inspector must be able to communicate well, both orally and in writing. Once an inspection is complete, the inspector must be able to translate the results of the inspection and/or test onto the inspection report. These results must then be explained to the owner or the owner's authorized representative. The inspector working for a fire protection contractor must have good customer relationship skills in addition to the ability to communicate well. For most fire protection contractors the inspector works face to face with the customer more than anyone else in the company; therefore, the inspector reflects the attitude and professionalism of the contractor.

An inspector must be organized and/or have an organized office behind the scenes. Keeping track of the names and addresses of the properties to be inspected, the testing frequencies required for the types of systems and components in each building or structure, and the previous inspection reports, requires organized tracking methods.

Some contractors rely entirely on the inspector to keep these records organized, whereas others provide office staff for this purpose.

Finally, an inspector should be neat in appearance and have good personal hygiene. The inspector has to interact with owners, owners' authorized representatives, building occupants, building engineers, approving authorities, and other inspectors. Having a neat appearance is especially important for the inspector who works for a fire protection contractor, as he or she represents the professionalism of the contractor. Having good personal hygiene is a common courtesy to those who have to work with the inspector. The pride or lack of pride that an inspector has in himself or herself is usually reflected in the quality of the work they do.

The candidate who does not have the complete set of minimum qualifications just described will struggle in the performance of the responsibilities of an inspector. Some of the traits described can be learned, such as communication skills, appearance, and education level. The others, however, are inherent for the most part, and either cannot be taught or are taught with much difficulty and effort.

Personal Background. Any inspector, whether working for a fire protection contractor or directly employed by the property owner, must also possess the personal characteristics necessary to perform work without being influenced by those who may not want the inspections or tests done properly. Every inspector should approach the inspection and testing of a water-based fire protection system as the most important job he or she has ever had. Again, lives and property are at stake. Therefore, the inspector should be public safety–oriented, self-motivated, responsible, a good communicator, thorough, organized, ethical, and honest (i.e., would not falsify a report for any reason). Many facilities being inspected require security clearance and background checks of their own employees as well as contractor employees entering their sites. Most fire protection contractors in turn perform background checks on candidates before hiring them as an inspector or inspector trainee.

Experience. Many inspectors working for fire protection contractors have previous experience with water-based fire protection systems either as a sprinkler fitter, an engineering technician, a fire service inspector, or an insurance underwriter. Inspectors with previous experience in these fields have a varied range of knowledge of water-based fire protection systems and how these systems should operate during a fire emergency.

The sprinkler fitter who installed fire protection systems knows how to read installation shop drawings, knows different system configurations, and knows all of the standard systems components. An inspector with this background also knows the methods of concealing piping systems, which is useful in locating pipes when inspecting a system for the first time. Many valves and system components that need to be inspected or tested are concealed behind access panels in walls or above ceilings. The sprinkler fitter who serviced and repaired water-based fire protection systems has a more thorough knowledge of how these systems work and what the possible reasons are for bad test results or failures. The service sprinkler fitter has replaced the gaskets in dry pipe valves, rebuilt exhausters and accelerators, and cleared obstructions from all parts of a system, from mains to branch lines to trim piping.

The engineering technician has a good theoretical knowledge of systems and their components. The layout of a fire protection system involves the ability to visualize how the pipe, fittings, and hangers fit in the building or structure. All of the various valves and system components are detailed on the shop drawings. With a background in performing and translating hydraulic calculations for various types of systems, the engineering technician should be able to better interpret main drain test and fire pump test results.

The fire service inspector has experience and skills that closely parallel those needed by the inspector of water-based fire protection systems. Knowledge of applicable codes and the ability to interpret them is required in both positions. The fire service inspector has performed a visual walkthrough of entire systems to ensure that they were installed in accordance with the approved shop drawings and has witnessed the testing of systems conducted by the installing contractor.

The insurance underwriter has performed the same inspections and tests that the fire service inspector has, and has seen firsthand how the systems perform. The insurance underwriter has been on the scene after a fire emergency and has documented how problems in a water-based fire protection system, such as a painted sprinkler or a clogged branch line, have impaired the system's ability to control fires. The insurance underwriter has also been involved in the rating of properties with and without fire systems and therefore knows the financial benefit of properly maintaining these systems.

Certifications. The National Institute for Certification in Engineering Technologies (NICET) sponsors certification programs for inspectors of water-based fire protection systems, in addition to other fire protection subfields. NICET certification can be used to determine the competency of inspectors in the application of NFPA 25 and other standards and practices used in the field. Three levels of certification are available, based on written verification of the individual's experience and ability to pass a written exam:

- Level I — less than 2 years of experience
- Level II — 2 to 5 years of experience
- Level III — over 5 years of experience

Once certified, inspectors must maintain their certification by accumulating professional development points from on-the-job experience, by attending training seminars or by participating in specific industry-related activities. Inspectors must accumulate 90 professional development points every three years to maintain an active certification. For more information regarding NICET certification, see www.nicet.org. Many states and local jurisdictions now require NICET certification for the licensing of fire protection contractors and/or inspectors. Contact the state or local jurisdiction for their licensing requirements.

Training the Inspector

Inspector training is a process of teaching skills to individuals over a period of time until a minimum competency level is obtained that will allow the inspector to perform basic inspection and test activities. However, training does not stop at this point, unless the inspector is limited to performing basic activities such as inspecting and testing wet and dry systems for the remainder of his or her career. In reality, the training process never stops. As the requirements of NFPA 25 change, as new types of systems and components are developed for the industry, and as more sophisticated inspection and testing equipment are introduced, additional training will be necessary.

A program developed for the training of inspectors has to address the various levels of experience of the individuals being trained. If the individual is new to the fire protection industry, training must start at a rudimentary level. Terminology within the fire protection industry is somewhat specialized and the knowledge of the components that make up a fire protection system must first be taught. Training seminars are available from several organizations, such as NFPA, National Fire Sprinkler Association (NFSA), and American Fire Sprinkler Association (AFSA). Internet-based training is also available. Some colleges and universities have developed curriculums specific to training inspectors of water-based fire protection systems.

If the individual has a background in fire protection, knows the industry terminology, and can identify system types and components, training can concentrate on the actual inspection and test activities.

Maintaining High Standards

In today's world, any incident that involves a loss of life or property becomes a target for lawsuits. One way to defend against lawsuits that attempt to prove negligence is to submit training records and personnel records that prove the competency of both the contractor and the individual inspectors. Maintaining high standards in hiring practices and providing continuous training is necessary if legal problems are to be avoided.

Code Enforcement

Laws enacted by local, city, and state legislatures allow for the enforcement of building codes as well as NFPA codes, including NFPA 25. Many of these codes are modified by local ordinances, with some mandating more stringent requirements, and some relaxing requirements. Each jurisdiction adopts those sections of a code that meet its needs and the needs of the community. Noncompliance with these laws can result in fines, loss of business license, and in some cases, businesses being shut down. (See the Case Study in Chapter 5.) In the case of NFPA 25, the inspection and testing requirements may be adopted without exception, but some frequencies may be amended to require fewer inspections and/or tests per year. If requirements are relaxed by local ordinance, the fire protection contractor may decide to still comply with NFPA 25 frequencies. NFPA 25 is a minimum standard and therefore many fire protection contractors make the business decision to fully comply with NFPA 25 requirements even though local ordinances don't require them to.

CONDUCTING AN INSPECTION AND/OR TEST

The process of conducting an inspection and/or test for the first time in a building will require more time than subsequent inspections and/or tests. The initial inspection will require that the inspector meet the owner or the owner's authorized representative for the first time. The inspector must then become familiar with the building, the systems installed in the building, the system components, and the records of the previous inspections and tests performed. The inspector must also determine the necessary inspection and test tools, equipment, and instruments needed. The inspector must plan the inspection and test sequence and prepare the inspection reports. Subsequent inspections and tests take less time because the building and systems are known and the inspector can focus on inspecting and testing the systems and components.

Agreements and Specifications

Paragraph 4.1.2.3 of NFPA 25 states: "Where the property owner is not the occupant, the property owner shall be permitted to pass on the authority for inspecting, testing, and maintaining the fire protection systems to the occupant, management firm, or managing individual through specific provisions in the lease, written use agreement, or management contract." The inspection agreement and the specifications should be read carefully by both parties. Each party believes they know what they are buying or providing. However, as with any contract for services, a properly executed document removes any ambiguities, misunderstandings, and/or confusion. More importantly, the inspection agreement defines the financial responsibilities and liabilities of the buyer and the fire protection contractor.

The Agreement. An inspection/test agreement describes the activities provided by a fire protection contractor for the owner, occupant, management firm, or managing individual. These agreements normally include the information that follows.

Building Address. Referencing the street address of the building being inspected is necessary to avoid any misunderstandings by the property owner, the fire protection contractor, and the inspector(s). Also, a building's name, owner, or management company may change, making it difficult to maintain a historical record of inspections and tests of the fire protection systems. The building's street address generally remains a consistent point of reference.

Description and Quantity of the System(s) Being Inspected/Tested. As in any contract, the services being provided for the price indicated must be clearly defined. Many buildings have several fire protection systems that are required to be inspected and tested by NFPA 25. For instance, a typical high-rise office building will have a backflow prevention device, one or more fire pumps, a standpipe system, wet pipe sprinkler systems on each floor, and dry pipe systems in the parking garage and mechanical penthouse. There may even be one or more preaction systems in computer rooms and a fixed water spray system in a cooling tower. Does the contract include all water-based fire protection systems in the building? Even if the intent of the contract is to include all systems in the building at the time the contract is executed, it is still advisable to include a list of all systems in the building. Systems can be added or changed, which will affect the contract price.

New to the 2008 edition of NFPA 25 is a series of tables that describe what action must be performed on systems and/or components when corrective action is taken to address a deficiency or impairment. The applicable text prior to each table reads: "Whenever a component in a [type of] system is adjusted, repaired, reconditioned, or replaced, the action required in Table x.x.1, Summary of Component Replacement Action Requirements, shall be performed." Many of the actions required include re-inspecting and/or retesting of systems or components after they are adjusted, repaired, reconditioned, or replaced. The fire protection contractor may be contacted to reinspect and/or retest the system or component, and the inspector may be called to return to the facility to perform the inspection or test. These inspections and tests are performed separately from those conducted on a periodic basis per NFPA 25, and the agreement should specify if these additional inspections and/or tests are included in the contract price. Because the types of reinspections or retests required or the frequency of these inspections and tests cannot accurately be determined ahead of time, most contracts will not include them. The fire protection contractor can price and provide these inspections and tests on an as-needed basis.

Frequency of the Inspection/Test Activities. NFPA 25 requires that certain activities be performed at prescribed frequencies. However, rarely does an inspection and test agreement assign all of these activities and frequencies to a fire protection contractor. To comply fully, an implied partnership is formed between the owner or the owner's representative and the fire protection contractor. All of the requirements of NFPA 25 are the responsibility of the owner, and the fire protection contractor relies on the owner to perform all inspections and tests not specified in the agreement.

A common agreement for a wet sprinkler system will specify that the fire protection contractor enter the building on a quarterly or semi-annual basis to inspect the gauges, control valves, alarm devices, hydraulic nameplate, and the fire department connection, as well as to test some of the alarm devices. In addition to these activities, flow switches and valve tamper switches will be tested semi-annually; the building, pipe hangers, pipes, fittings, and spare sprinklers will be inspected annually; and the main drain will be tested annually. Such an agreement would specify a quarterly or semi-annual inspection and test in accordance with NFPA 25. All other weekly and monthly activities are expected to be performed by the owner, occupant, management firm, or managing individual.

Unless specified in the agreement, all other less-frequent activities are not included, such as testing gauges every five years; testing sprinklers every five, ten, twenty, or fifty years; and internally inspecting pipes, alarm valve, check valves, strainers, filters, and orifices every five years. These activities are normally performed by the fire protection contractor but are invoiced as additional periodic services.

Reports. Records of inspection, testing, and maintenance activities are to be maintained by the property owner. The fire protection contractor generally provides reports of the inspection and test activities performed each visit. Many local jurisdictions require that such reports be kept at the site. Local ordinances may also require a copy of any reports be submitted directly to the AHJ.

Reports come in as many versions as there are contractors providing services. Supplement 4 has examples of forms that can be downloaded from NFPA (www.nfpa.org). Many forms are still filled out by hand, although computerized programs with printed reports are becoming more and more popular. The electronic reports are more professional and legible. When preprinted report forms are filled out by hand, multiple copies can be provided on the spot using multisheet forms or the occupant's copy machine. Some companies provide printers for technicians using computer-based inspection programs, but the more common practice is to input the test results off-site, create the report in the office, and e-mail it to the customer.

The inspector or the office administrator should be able to reference the agreement to determine who has been authorized by the owner or the owner's representative to receive the inspection report forms.

Annual Cost of the Agreement. Of course, the most important part of any agreement is the price. A typical agreement is priced for a one-year term. The price may be broken down into quarterly payments and in some cases per system. However, most agreements state a lump sum that covers all described activities and frequencies.

Payment Terms. Payment terms are extremely important to both the fire protection contractor and the owner. The owner needs to know when to expect billing for budgeting purposes, and the fire protection contractor needs to meet payroll and expense needs. Typically, one of two types of payment terms is used in the majority of agreements.

The first type stipulates that the full annual amount is invoiced up front and paid before any inspection or test activities begin. This ensures that the fire protection contractor will have the necessary funds available to meet payroll and expense needs throughout the term of the agreement and creates less paperwork for both parties. One invoice, one accounts payable transaction, one check posting, and no collection effort.

The second type stipulates that invoicing is done after each inspection in an amount equal to the work being performed. Many agreements with quarterly inspection frequencies invoice one-fourth of the annual amount after each inspection/test. This payment arrangement usually forces the fire protection contractor to wait until the last inspection of the agreement term to perform any time-intensive activities, such as tripping dry pipe valves and flow testing fire pumps. The owner may be able to budget more favorably with this payment schedule, but it requires four times the office work: four invoices, four accounts payable transactions, and four check postings. Another downside to invoicing quarterly after each inspection is that any delay in the payment of one of the invoices may affect the schedule for the subsequent inspections. Good business practices dictate that a fire protection contractor does not perform additional work if payment is late for the previous work performed.

Many other payment term options, such as one-twelfth invoiced per month, are commonly used by local governments on large agreements, but this option is the exception to the norm.

Renewal Terms. Every agreement should spell out the renewal terms. Typically, an annual agreement renews automatically each year with a cost of living increase tied to the federal index, or to pre-described increase percentages per year. Most agreements can be cancelled by either party with 30 days' written notice as long as all payments have been paid for services rendered.

Some agreements require special action on the part of the owner and/or fire protection contractor. The owner may require a new agreement each year, including any revised pricing, or a separate purchase order may have to be issued each year. These provisions should be spelled out in the agreement.

Limits of Liability. Every agreement should state the limits of liability for the fire protection contractor. This legal terminology protects the fire protection contractor for losses that are beyond the fire protection contractor's control and responsibility. No one can guarantee that a water-based fire protection system will work as designed and installed at any given time. As previously stated, fire protection systems are more reliable than just about any other mechanical system, and the fire protection industry has an excellent track record. However, unexpected and improbable occurrences do happen. A clogged sprinkler cannot be detected during an inspection or test, nor can an inspector predict when a leak in a pipe is going to develop. Despite the difficulty of predicting every potential problem or failure of a system, litigation is very common for a loss event that is associated with a fire protection system, in an attempt to assign blame to everyone from the architect, to the engineer, to the general contractor, to the installing sprinkler contractor, to the fire protection contractor performing inspections and tests. The unanticipated loss could cost any of these parties hundreds of thousands of dollars to defend, unless contractual language exists in the agreement limiting this exposure.

Signature Space Including Title, Signature, and Date. The agreement is a contractual document and therefore must be signed to be legally binding. The agreement should include designated lines at the end of the document for the fire protection contractor and the owner or the owner's authorized representative to sign, print their name and title, and insert the date. An unsigned agreement is not legally considered an agreement.

Specifications. Specifications are a binding part of an agreement when referenced as an attachment in the body of the agreement. Specifications normally include the more detailed requirements of a contract.

Criteria. Criteria indicate which specific NFPA 25 requirements are included.

Notification. Notification describes who is responsible for notifying occupants of an inspection or test.

Working Hours. Working hours are the hours during which the inspection/test activities may be performed.

Cancellation. Cancellation refers to a cancellation clause in the agreement.

Access. Access refers to the buyer's responsibility to provide access to the inspector.

Communication with the Property Owner

The fire protection contractor is required by NFPA 25 to communicate with the property owner; the owner's authorized representative; or the occupant, management firm, or managing individual as described in the agreement, to obtain certain information concerning occupancy, use, or building changes.

Communication with the owner or owner's authorized representative or the occupant, management firm, or managing individual is also necessary to schedule an inspection or test. Throughout the inspection and testing process, communication will be both verbal and written. Once an executed contract exists, communication then involves the actual performance of the inspection and/or test.

In some cases, arrangements are made so the inspector can arrive at the property without an appointment and perform the inspection and/or test. However, the owner or the owner's authorized representative or the occupant, management firm, or managing individual must be available either in person or by phone to answer certain questions on the inspection report. The inspector is provided with the necessary keys and access codes that will allow him to inspect throughout the protected areas of the property.

Safety Laws and Regulations

Generally, it is the responsibility of the property owner to advise anyone working in the building of specific hazards present. The inspector should, however, request copies of such items as a material safety data sheet (MSDS) and should determine whether any confined spaces or other hazardous conditions exist.

Tools and Equipment

To properly inspect and/or test a fire protection system, certain tools and equipment are necessary. This section will be divided into three parts. The first will include a list of safety equipment needed to perform an inspection or test. The second will describe standard inspection tools and equipment. The third will describe standard test equipment.

Safety Equipment. As described in the previous section, safety must be a matter of primary importance and be second nature to the inspector. Federal and state regulations require compliance with safety laws. Common sense dictates that care and caution must be taken to ensure the safety of the inspector, and those working around the inspector, from certain hazards inherent with performing inspection and test procedures. As a minimum, federal Occupational Safety and Health Administration (OSHA) regulations must be adhered to. If not, employee safety is at risk. OSHA levies stiff fines on companies whose employees do not follow these national regulations. If violations are repeated, the fines increase exponentially.

Safety equipment includes the following:

- Hard hat: Hard hats are required when head injury is a possibility from climbing, crawling through confined spaces, or falling objects.
- Safety shoes: Safety shoes are required in most industrial and manufacturing plants, and any location where foot injury is possible from stepping on sharp objects, or from heavy objects falling on the feet.
- Proper clothing; Loose clothing is not to be worn around machinery. In some plants, fire and/or chemical protective clothing must be worn.
- Eye protection: Safety glasses should always be worn when inspecting or testing fire protection systems.
- Hearing protection: Hearing protection should be a part of the standard safety equipment worn by inspectors. Many areas in industrial occupancies post signs where hearing protection must be worn.
- Safety harness: See OSHA regulations for the requirements for safety harnesses. Safety harnesses are required when inspectors must be on ladders or lifts above a certain height.
- Confined space equipment: See OSHA regulations for the requirements for confined space entry. Confined space equipment requirements are not limited to pits

or cellars, but can also extend to more common spaces, such as unventilated attics, crawl spaces, concealed spaces, and vaults.

Inspection Tools and Equipment. Inspectors commonly use the following inspection tools and equipment:

- Ladders: Ladders are needed to accomplish many of the inspection and test activities described in NFPA 25. Ladders are also one of the most hazardous pieces of equipment when working overhead. The inspector should use the right height ladder at all times, to avoid overextending his or her reach. When using extension ladders, always follow the 1 to 4 rule, setting the base of the ladder 1 foot away from the wall for every 4 feet in height. The inspector must follow all other safety rules concerning ladders as detailed in OSHA and local requirements, including safety harnesses and fall protection
- Flashlight
- Binoculars: Binoculars are typically used for a close-up examination of sprinklers from the floor level. The inspector is not expected to climb a ladder to inspect each sprinkler, but binoculars can be helpful when inspecting sprinklers for leakage, corrosion, foreign materials, loading, paint, physical damage, or improper orientation.
- Tape measure or folding rule: The size of some pipes or components may not be evident, and the inspector may need to measure the size to provide an accurate report.
- Assorted handtools: To perform certain internal inspections, the inspector will have to remove face plates from valves or remove sprinklers or flushing connections. Screwdrivers, a socket set, crescent wrench, hammer, and a set of open-end wrenches are included in the inspector's toolkit.

Test Equipment. Standard test equipment for the inspection and testing of water-based fire protection equipment includes the following:

- 12-in. pipe wrench
- Crescent wrench
- Calibrated test gauges
- Volt meter: Primarily used during the annual fire pump test. Proper safety equipment should also be worn while using this equipment.
- Amp meter tachometer
- Hydrometer or refractometer: Used to measure the concentration of antifreeze.
- Pitot tube: Used to measure velocity pressure during flow and fire pump tests. (See Chapters 6 and 8.)
- Stopwatch: A watch that reads in minutes and seconds. Used to measure trip times for dry pipe systems.
- Heat lamp: Used to activate heat detection systems on water spray fixed systems.
- Digital gauges
- Water analysis kit: Used when performing the five-year obstruction inspection if tubercules or slime are found in the piping.

INSPECTION PROCEDURE

NFPA 25 does not prescribe how to perform an inspection or test. The systems and components that are to be inspected and tested are listed, the acceptance criteria are spelled out, and the corrective action requirements are clear.

The user must realize that the actual inspection process is unique to each facility, building, and system. The NFPA 25 report forms available from NFPA's website, if followed line by line, lead the inspector through the process in one way. However, if these report forms or any other report forms are followed line by line, duplication of effort and doubling back over the same parts of the same systems will occur.

This section describes one method of performing an inspection. However, as each system is unique, so is each inspector. What works for one will be inefficient for another. Inspectors must develop their own process. However, certain bare essentials are universal. Before the experienced inspector can put together a game plan to perform the activities in the most efficient and comprehensive way possible, certain information must be reviewed. The inspector needs to know the entire scope of the agreement and the types of systems being inspected. He or she must become familiar with the layout of the building and identify where the system controls and valves are.

Familiarity with the Contract

Scope. The inspector must know how many systems are to be inspected and tested, what type of systems they are, and the frequency of the inspections to be performed. With this information the inspector can plan the time it will take to perform the inspection or test and the equipment needed to perform these activities. The terms and conditions of the contract must be reviewed at this time to determine whether the inspection and test activities are to be performed during off-hours, whether there are special payment terms, and whether there are any other special requirements that will affect the inspection and test process.

Property to Be Inspected. The inspector should review any as-built drawings available or should, as a minimum, review previous inspection reports or sketches that are retained in the owner's inspection file.

Contact Information. Usually, the inspection contract will have contact information. A scheduled appointment may be necessary prior to performing the inspection; contact with the property owner or representative or the occupant, management firm, or managing individual should be made at this time.

Planning. Based on the inspection contract and type of systems involved, the NFPA 25 inspection and test requirements should be reviewed prior to entering the building. Preplanning will save time and may reveal components or systems that require special attention.

Tools. Preplanning will reveal the inspection and test tasks to be performed, which in turn will dictate the tools and equipment needed to perform them.

Reports. Preplanning will also highlight the needed report forms.

Performing the Inspection/Test

The following procedures represent normal practices used when performing a sprinkler inspection and/or test.

Walk the Perimeter. Upon arrival at the location, the inspector will drive or walk the exterior perimeter of the building to become familiar with the building layout. This helps to determine some of the features of the fire protection systems and the location of many of the components being inspected and tested. The perimeter walk will locate the fire department connection, motor water gong, main valve room, inspector's tests, main drains, any outside valves [i.e., post indicator valve (PIV), wall post indicator valve (WPIV)], fire pump header — which also indicates the approximate size of the pump — and any specific conditions of which the inspector needs to be aware before flowing any water.

Meet the Owner or the Owner's Authorized Representative. Upon entering the building the inspector should report to the owner, owner's representative, or the occupant, management firm, or managing individual; provide company identification, which includes the inspector's name; and explain the purpose of the visit.

Ask Questions. Once the inspector meets the owner, owner's authorized representative, or the occupant, management firm, or managing individual, the steps involved to perform the inspection should be explained. The inspector then asks a series of questions that deal with occupancy changes, content changes, building changes, and system(s) status, filling out the inspection report forms as responses are provided. The inspector should then ask to be directed to the room housing the interior valves controlling the fire protection system(s). Generally, a conscientious inspector will observe the areas being passed through on the way to the valve room and make note of anything that is pertinent to the inspection.

Locate Systems and Equipment. At the system valve room the following steps are taken before starting any actual testing:

- Note the type, manufacturer, size, and number of systems to be inspected and/or tested.
- Note the location, type, manufacturer, size, and quantity of any flow switches or tamper switches.
- Alert the building occupants. Determine whether the alarms are monitored through the building fire alarm system and, if so, warn the building occupants of a possible alarm.
- Locate the fire alarm panel. Verify accessibility to the fire alarm equipment and determine whether the system is monitored by a central station.
- Notify the monitoring company. If it is monitored, call the central station and the appropriate local fire response authorities.
- Notify AHJ/local fire department. Every inspector's worst nightmare is to be conducting a flow test of a fire protection system and hearing sirens wailing as the fire trucks pull up in front of the building. Not only is this embarrassing, but a false alarm can be costly to all parties involved. The fire department potentially can be diverted from a real emergency. The owner, or the fire protection contractor, or both, can be fined by the fire department for causing a false alarm. The business at the property loses productivity because of the disruption. The inspector loses valuable time while meeting with the fire officials to assure them it was a mistake and not intentional. Finally, after the fire engines leave, the building occupants return to work, and the alarm company is properly notified, the test can continue. Do not depend on the central station, the property owner, or anyone else to notify local fire response authorities. Write down the names of individuals that were notified and the times they were notified.
- Verify that an authorized building engineer or in-house user is on site to reset any alarm systems.

Inspect/Test. The actual inspection and testing can now begin. The activities described in this section are for an annual inspection and test, with the exception of full flow tripping dry, preaction, or deluge systems. The requirements in NFPA 25 should be followed for these specific activities, and for inspections and tests during other frequencies.

Perform a Visual Inspection. The purpose of a walkthrough is to visually inspect all areas in the building visible from the floor level. The inspector is looking for areas of the building such as new additions, new rooms, or new mezzanines that are not protected by the fire protection system. It is the owner's responsibility to declare any such areas when answering the general questions at the beginning of the inspection. However, the inspector should look during the walkthrough to determine whether there are added areas, rooms, or additions that are not protected by the fire protection system. These should be brought to the owner's attention, noted, and recorded, as a courtesy to the owner, not as an obligation of the agreement or as a requirement of NFPA 25.

Special Note: The inspector is not required by NFPA 25 to determine whether the fire protection system as installed provides adequate coverage in the building, nor to determine whether all areas of the building are protected in accordance with the original applicable installation standard. The installation codes change every three to five years, and what is required in the most recent code may not have been required in earlier versions, and vice versa. The inspector is not required to examine concealed spaces, closets, attic spaces, or any other building feature to determine whether the fire protection system extends into those areas. If the property owner is ever uncertain whether the fire protection system as installed will provide the minimum protection required by the applicable codes, the owner should contact a fire protection engineer or qualified contractor to perform an analysis of the building, the building occupancy, the commodity classification of any storage present, the system design criteria, and the water supply.

- Perform a visual inspection of the alarm, dry, preaction and/or deluge system valve, and all associated trim. Look for any signs of leakage, rust, and any unusual problems or conditions.
- Determine where test water will discharge. Verify that the water from the main drain will discharge in a safe location. Note and record the static pressure on the supply side of the system valve. Conduct the main drain test by opening the main drain valve fully. Note and record the residual pressure with the valve wide open. Observe the water discharging from the main drain test fitting and note any debris or restrictions of any kind. Flow water through the main drain until the water stream is clear. Close the main drain valve and record the gauge readings. Note and record the static pressure after valve is fully closed.
- Verify the operation of the water motor alarm (if applicable). On a wet system it will ring during the main drain test. On dry, preaction, and deluge systems, open the test valve in the system valve trim.
- Close and fully open all control valves to verify proper operation. Test the tamper devices if applicable.
- Locate and flow all inspectors test valves for the required 90 seconds and verify alarm activation. Reset any alarms and continue with walkthrough.
- Look for proper coverage and orientation of sprinklers. Also observe sprinklers for evidence of loading, paint, corrosion, damage, and any physical obstructions to discharge patterns.
- Observe any exposed system piping and hangers for corrosion, leakage, misalignment, and physical damage.
- Record all conditions observed during the walkthrough. When this has been done, the inspection is complete.

Compile Results. Many inspectors fill out inspection and test reports as they perform the activities. Others compile results, observations, and deficiencies on a separate piece of paper and transfer the information to the actual report at the end of the inspection. Either method is acceptable. Using the inspection form as the inspection and test is performed helps to eliminate missed activities and aids in establishing a systematic process.

Complete Reports. All pressures, trip times, flows and other recorded results are written in the report. All deficiencies are to be clearly detailed, and corrective action recommended.

Special Note: Because many inspectors employed by fire protection contractors started their fire protection careers in other areas of the industry, they have specific knowledge of codes and standards other than NFPA 25. Because of this specific knowledge, these inspectors may observe potential problems with the water-based fire protection systems or problems related to other fire protection systems or problems related to general safety requirements that are unrelated to the NFPA 25 inspection being performed. Examples of some problems typically observed include:

- Sprinklers appear to be too far from walls or appear inadequately spaced.
- Sprinklers appear too close to building construction features that could obstruct spray patterns.
- Emergency exits are chained and locked.
- Flammable liquids are present and not stored properly.
- New kitchen cooking equipment has been added and is not protected by the hood suppression system.
- Fire extinguishers are discharged or are missing.
- Fire alarm zones are out of service.
- Fire doors are propped open.

Because inspectors are conscientious, they are compelled to alert the owner, or owner's representative, or the occupant, management firm, or managing individual of their observations. *The inspector should not note these observations on the NFPA 25 inspection form.* Instead, the inspector should provide a separate attachment describing the observations, and advise the owner or owner's representative to have them checked out by the appropriate fire protection system or building code professional.

Complete the Inspection and/or Test. To complete the inspection or test, many of the processes required to start these activities must be reversed. It is just as important to go through these steps upon the completion of the inspection or test as it was before starting. All systems are to be returned to normal service. Test and drain valves are to be closed tightly until no waterflow or dripping is observed. All flow and tamper switches are to be returned to normal settings. All alarm panels are to be returned to their normal condition. To do this, confirm that all alarm, trouble, or supervisor signals have been verified at the panel and then reset the panel. Fill out any tags provided and/or required. Some jurisdictions have a visual tag procedure that the inspector must follow. A green tag indicates the system did not have any deficiencies or impairments. A yellow tag indicates one or more deficiencies were found and therefore the application of one or more components is not within its designed limits or specifications, but there were no impairments found. A red tag indicates one or more impairments were found and the system may not function as intended during a fire emergency. If one or more impairments are found, the owner, the owner's representative, or the occupant, management firm, or managing individual must be notified immediately so emergency impairment procedures can be initiated.

Advise Occupants. If occupants were notified of the test, notify them that the test is complete.

Notify AHJ/Local Fire Department. Notify the proper authorities that the test is complete and the system has been returned to normal service. Write down the name of the individual who received the notification and the time of notification.

Notify the Monitoring Company. Notify the monitoring company, if applicable, that the test is complete. Have the individual perform a system check to verify they are receiving the proper signals. Write down the name of the individual who performed the system check and the time of notification.

Review Reports. After all systems are back in service and all notifications have been made, meet with the owner, the owner's authorized representative, or the occupant, management firm, or managing individual and review the results of the inspection and test. The report should be explained line by line, and any unusual readings or findings discussed. Deficiencies should be clearly explained, and if necessary show them the deficiencies. Explain the requirements of NFPA 25 pertaining to acceptable results, and corrective action that must be taken to address deficiencies.

Follow Impairment Procedures, if Applicable. If impairments are found, impairment procedures must be implemented by the owner. It is important for the owner to tag systems and/or components that are not working properly, especially those impaired. The inspector should advise the owner of the impairments and the steps required by NFPA 25 to correct those impairments.

Sign Reports. When the inspector has completely explained the inspection and test results with the owner or the owner's authorized representative, and is satisfied that the results and deficiencies have been clearly explained, the report is to be signed and dated by both parties. The owner or the owner's representative should be given a copy of the signed report.

After the Inspection

Inspection Follow-up. An inspection and/or test performed without proper follow-up is useless. Protection of lives and property is dependent on properly operating water-based fire protection systems. The owner has made a commitment to the employees, their insurance company, stockholders, and the community at large to have their system(s) in proper working order, by contracting with a reputable fire protection contractor. The fire protection contractor has an obligation to follow up with the inspection in a timely fashion.

Notification Process. Although not required by NFPA 25, a good practice is directly notifying the property owner upon completion of an inspection and/or test. In most cases, the owner is not the individual who meets with the inspector on site and who reviews the report after the inspection. NFPA 25 does require that the owner maintain records of all inspections and tests for a period of time. Therefore, the fire protection contractor should send a copy of each report to the owner. In addition, it is a good practice, as well as a preferred legal procedure, to attach a cover letter to any report that contains deficiencies or impairments. The cover letter should clearly explain all deficiencies and/or impairments and should detail the requirements of NFPA 25 to have them corrected. If the fire protection contractor is also in the repair and installation business, an offer to make the corrections and the price for those corrections can be included in the letter. For liability reasons, it is a common practice to send cover letters that describe impairments by certified mail and to call the owner to notify them of the impairment. Should a fire emergency occur between the time the impairment is discovered and the owner is notified of the impairment, lawsuits might be filed as insurance companies assess blame for the system failure.

Corrective Action. When an owner performs repairs to the system, or contracts the repair to a fire protection contractor, the date the repairs were made and the corrective action taken should be recorded. This information is then passed on to the inspector at the next inspection and noted on the new inspection report.

Component Replacement Action Requirements. New to the 2008 edition of NFPA 25 is a series of tables that describe what action must be performed on systems and/or components when corrective action is taken to address a deficiency or impairment. The applicable text prior to each table reads: "Whenever a component in a [type of] system is adjusted, repaired, reconditioned, or replaced, the action required in Table x.x.1, Summary of Component Replacement Action Requirements, shall be performed." Many of the actions required include reinspecting and/or retesting of systems or components after they are adjusted, repaired, reconditioned, or replaced. The fire protection contractor may be contacted to reinspect and/or retest the system or component, and the inspector may be called to return to the facility to perform the inspection or test. These inspections and tests are performed separately from those conducted on a periodic basis per NFPA 25, and the results should be recorded on a separate inspection/test report. These additional reports should be retained by both the owner and the fire protection contractor for future reference.

Maintaining Inspection Files. Proper filing of reports by the fire protection contractor is necessary to provide ongoing inspection and test services to the customer. The reports must be accessible for reference should the property owner call with questions about the reports, or if an additional copy is needed for the AHJ or insurance underwriter. The inspector will also need copies of previous reports when the next inspection is performed. Comparisons between current test results and previous test results must be made to determine whether degradation of water supplies, system piping, fire pump performance, and so on, is taking place.

Summary

This supplement is intended to offer examples illustrating the role and duties of the inspector. It is not intended to be an all-inclusive list of responsibilities, but rather an outline of the background and qualifications necessary to perform the work. The discussion of pre-planning, tools, and equipment, and the approach to an inspection is intended to provide the inspector and property owner with a basic plan for inspecting any building or system.

REVIEW QUESTIONS

1. Discuss the inspection, testing, and maintenance activities that can be performed by the property owner.
2. In an inspection contract, what items are usually the responsibility of the property owner?
3. What are some of the basic tools an inspector needs?
4. What safety equipment should every inspector use?

REFERENCES CITED

National Fire Protection Association, 1 Batterymarch Park, Quincy, MA 02169-7471.

NFPA 13, *Standard for the Installation of Sprinkler Systems*, 2007 edition.

NFPA 13R, *Standard for the Installation of Sprinkler Systems in Residential Occupancies up to and Including Four Stories in Height*, 2007 edition.

NFPA 14, *Standard for the Installation of Standpipe and Hose Systems*, 2007 edition.

NFPA 15, *Standard for Water Spray Fixed Systems for Fire Protection*, 2007 edition.

NFPA 16, *Standard for the Installation of Foam-Water Sprinkler and Foam-Water Spray Systems*, 2003 edition.

NFPA 20, *Standard for the Installation of Stationary Pumps for Fire Protection*, 2007 edition.

NFPA 22, *Standard for Water Tanks for Private Fire Protection*, 2003 edition.

NFPA 214, *Standard on Water-Cooling Towers*, 2005 edition.

SUPPLEMENT 4

Sample Inspection, Testing, and Maintenance Program

Editor's Note: *This supplement is intended to illustrate the use of reporting forms to document the required activities in NFPA 25. A sample dry pipe sprinkler system complete with a fire pump is used as an example of the systems being maintained.*

Exhibit S4.1 is an example of a typical dry pipe system and is used in this supplement for reference for each of the inspection, testing, and maintenance reports included in this section.

The Contractor's Material and Test Certificate for Aboveground Piping (Form S4.1) and the Contractor's Material and Test Certificate for Underground Piping (Form S4.2) must be kept on file by the property owner for

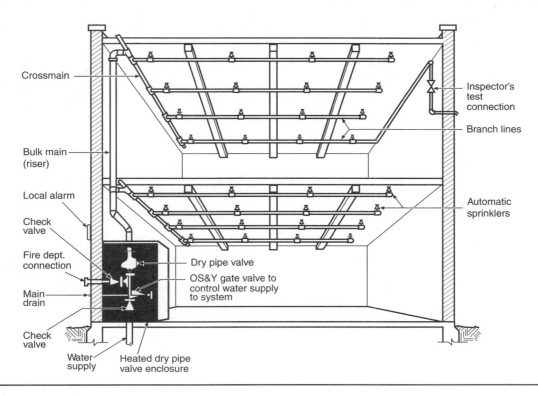

EXHIBIT S4.1 Sample Typical Dry-Type Sprinkler System in Storage Building.

the life of the system. The material and test certificates provide a basis for many of the subsequent inspections and tests required for their respective systems. Subsequent inspection, testing, and maintenance forms must be kept on file by the property owner for a period of one year after the activity is repeated.

Form S4.3 illustrates the reporting requirements necessary to verify compliance with the inspection, testing, and maintenance requirements for sprinkler systems found in Chapter 5 of NFPA 25. Forms S4.4 and S4.5 illustrate the reporting requirements necessary to demonstrate compliance with Chapter 8 for fire pumps.

The annual activities or activities that require specialized training and equipment should be conducted by a qualified fire protection contractor. Some jurisdictions require licensing of inspectors, and such licensing laws are commonly addressed to those involved in the business of providing such services. In most cases, licensing of an employee of the property owner is not required for "in-house" work. This should be verified with the local jurisdiction if an employee of the property owner is qualified to perform this type of work.

A file of all reports must be kept on the premises as stated in Chapter 4 of NFPA 25. This file provides direct evidence of compliance with NFPA 25 and a review of this file may be requested by the authority having jurisdiction (AHJ) during an inspection. An accurate and complete file will lead to a brief and successful review.

This supplement contains the following forms:

- Form S4.1 Sample Contractor's Material and Test Certificate for Aboveground Piping
- Form S4.2 Sample Contractor's Material and Test Certificate for Underground Piping
- Form S4.3 Sample Inspection, Testing, and Maintenance of Dry Pipe Sprinkler Systems
- Form S4.4 Sample Annual Performance Tests for Fire Pumps
- Form S4.5 Sample Annual Test Summary Page for Fire Pumps

Contractor's Material and Test Certificate for Aboveground Piping

PROCEDURE

Upon completion of work, inspection and tests shall be made by the contractor's representative and witnessed by an owner's representative. All defects shall be corrected and system left in service before contractor's personnel finally leave the job.

A certificate shall be filled out and signed by both representatives. Copies shall be prepared for approving authorities, owners, and contractor. It is understood the owner's representative's signature in no way prejudices any claim against contractor for faulty material, poor workmanship, or failure to comply with approving authority's requirements or local ordinances.

Property Name: XYZ Corporation   **Date:** 6-11-07

Property Address: 102 Main St., Taylorsville, USA

Plans	Accepted by approving authorities Taylorsville FD		
	Address 102 Main St.		
	Installation conforms to accepted plans	☑ Yes	☐ No
	Equipment used is approved. If no, explain deviations	☑ Yes	☐ No
Instructions	Has person in charge of fire equipment been instructed as to location of control valves and care and maintenance of this new equipment?	☑ Yes	☐ No
	If no, explain		
	Have copies of the following been left on the premises?	☑ Yes	☐ No
	1. System components instructions	☑ Yes	☐ No
	2. Care and maintenance instructions	☑ Yes	☐ No
	3. NFPA 25	☑ Yes	☐ No
Location of system	Supplies buildings		

	Make	Model	Year of manufacture	Orifice size	Quantity	Temperature rating
Sprinklers	Brass Upright	F	2001	1/2"	32	165° F

Pipe and fittings	Type of pipe ASTM A-53
	Type of fittings ANSI B16.1

1 of 4

FORM S4.1 *Sample Contractor's Certificate for Aboveground Sprinkler System Piping.*

Alarm valve or flow indicator	Alarm device			Maximum time to operate through test connection	
	Type	Make	Model	Minutes	Seconds

	Dry valve				Q.O.D		
Dry pipe operating test	Make	Model	Serial No.	Make	Model		Serial No.
	Dry	F	3377574				
	Time to trip through test connection[1,2]		Water pressure	Air Pressure	Trip point air pressure	Time water reached test outlet[1,2]		Alarm operated properly	
	Minutes	Seconds	psi	psi	psi	min.	sec.	Yes	No
Without Q.O.D	1	0	50 psi	40 psi	30 psi	0	55	☑	☐
With Q.O.D								☐	☐

If no explain: _____

Deluge and preaction valves

Operation ☐ Pneumatic ☐ Electric ☐ Hydraulic

Piping supervised ☐ Yes ☐ No   Detecting media supervised ☐ Yes ☐ No

Does valve operate from the manual trip, remote, or both control ☐ Yes ☐ No

Is there an accessible facility in each circuit for testing? ☐ Yes ☐ No

If no, explain _____

Make _____   Model _____

Does each circuit operate supervision loss alarm? ☐ Yes ☐ No

Does each circuit operate valve release? ☐ Yes ☐ No

Maximum time to operate release _____ Minutes _____ Seconds

Pressure reducing valve test	Location and floor _____	Static pressure	Residual pressure (flowing)		Flow rate
		Inlet (psi)	Outlet (psi)	Inlet (psi)	Outlet (psi)	Flow (gpm)
	Make and model _____					
	Setting _____					

Test description

Hydrostatic: Hydrostatic tests shall be made at not less than 200 psi (13.6 bar) for 2 hours or 50 psi (3.4 bar) above static pressure excess of 150 psi (10.2 bar) for 2 hours. Differential dry-pipe clappers shall be left open during the test to prevent damage. All aboveground piping leakage shall be stopped.

Pneumatic: Establish 40 psi (2.7 bar) air pressure and measure drop, which shall not exceed 1½ psi (0.1 bar) in 24 hours. Test pressure tanks at normal water level and air pressure and measure air pressure drop, which shall not exceed 1½ psi (0.1 bar) in 24 hours.

[1] Measured from time inspector's test connection is opened
[2] NFPA 13 only requires the 60-second limitation in specific sections

FORM S4.1 Continued

Tests	All piping hydrostatically tested at **200 psi** (_____ bar) for **2** hours If no, state _____ Dry piping pneumatically ☐ Yes ☑ No Equipment operates ☑ Yes ☐ No	
	Do you certify as the sprinkler contractor that additives and corrosive chemicals, sodium silicate or derivatives of sodium silicate, brine, or other corrosive chemicals were not used for testing systems or stopping leaks?	☑ Yes ☐ No
	Drain test: Reading of cutoff gauge located near water supply test connection: **60** psi (_____ bar) Residual pressure with valve in test connection open wide: **50** psi (_____ bar)	
	Underground mains and lead-in connections to system risers flushed before connection made to sprinkler piping	
	Verified by copy of the Contractor's Material and test Certificate for Underground Piping Explain _____	☑ Yes ☐ No ☐ Other
	Flushed by installer of underground sprinkler piping	☑ Yes ☐ No
	If powder-driven fasteners are used in concrete, has representative sample testing been satisfactorily completed? If no, explain _____	☑ Yes ☐ No
Blank testing gaskets	Number used: 1 Locations: Floor Flange Number removed: 1	
Welding	Welding piping ☑ Yes ☐ No If yes...	
	Do you certify as the sprinkler contractor that welding procedures used complied with the minimum requirements of AWS B2.1, ASME Section IX *Welding and Brazing Qualifications*, or other applicable qualification standard as required by the AHJ?	☐ Yes ☑ No
	Do you certify that the welding was performed by welders or welding operators qualified in accordance with the minimum requirements of AWS B2.1, ASME section IX *Welding and Brazing Qualifications*, or other applicable qualification standard as required by the AHJ?	☐ Yes ☐ No
	Do you certify that the welding was conducted in compliance with a documented quality control procedure to ensure that (1) all discs are retrieved; (2) that openings in piping are smooth, that slag and other welding residue are removed; (3) the internal diameters of piping are not penetrated; (4) completed welds are free from cracks, incomplete fusion, surface porosity greater than $1/16$ in. diameter, undercut deeper than the lesser of 25% of the wall thickness or $1/32$ in.; and (5) completed circumferential butt weld reinforcement does not exceed $3/32$ in.?	☐ Yes ☐ No
Cutouts (discs)	Do you certify that you have a control feature to ensure that all cutouts (discs) are retrieved?	☐ Yes ☑ No
Hydraulic data nameplate	Nameplate provided If no, explain	☑ Yes ☐ No
Remarks	Date left in service with all control valves open **6-11-07**	

3 of 4

FORM S4.1 *Continued*

Signatures	Name of sprinkler contractor WS Sprinkler		
	Tests witnessed by		
	The property owner or their authorized agent (signed) *John Doe*	Title Safety Manager	Date 6-11-07
	For sprinkler contractor (signed) *Jack Smith*	Title Foreman	Date 6-11-07

Additional explanations and notes:

_____
_____
_____
_____
_____
_____
_____
_____
_____
_____
_____
_____
_____
_____
_____
_____
_____
_____

FORM S4.1 *Continued*

Contractor's Material and Test Certificate for Underground Piping

PROCEDURE

Upon completion of work, inspection and tests shall be made by the contractor's representative and witnessed by an owner's representative. All defects shall be corrected and system left in service before contractor's personnel finally leave the job.
A certificate shall be filled out and signed by both representatives. Copies shall be prepared for approving authorities, owners, and contractor. It is understood the owner's representative's signature in no way prejudices any claim against contractor for faulty material, poor workmanship, or failure to comply with approving authority's requirements or local ordinances.

Property Name: XYZ Corp. **Date:** 4-15-07
Property Address: 100 Main St., Taylorsville, USA

Plans	Accepted by approving authorities (names)	Taylorsville FD
	Address	102 Main St.
	Installation conforms to accepted plans	☑ Yes ☐ No
	Equipment used is approved	☑ Yes ☐ No
	If no, state deviations	
Instructions	Has person in charge of fire equipment been instructed as to location of control valves and care and maintenance of this new equipment?	☑ Yes ☐ No
	If no, explain	
	Have copies of appropriate instructions and care and maintenance charts been left on the premises?	☑ Yes ☐ No
	If no, explain	
Location	Supplies buildings	Storage Building #100
Underground pipes and joints	Pipe types and class	Type joint MJ
	Pipe conforms to AWWA C151 standard	☑ Yes ☐ No
	Fittings conform to AWWA C110 standard	☑ Yes ☐ No
	If no, explain	
	Joints needed anchorage clamped, strapped, or blocked in in accordance with NFPA 24 standard	☑ Yes ☐ No
	If no, explain	

1 of 3

FORM S4.2 Sample Contractor's Certificate for Underground Sprinkler System Piping.

Test description	Flushing: Flow the required rate until water is clear as indicated by no collection of foreign material in burlap bags at outlets such as hydrants and blow-offs. Flush at flows not less than 390 gpm (1476 L/min) for 4 in. pipe, 880 gpm (3331 L/min) for 6 in. pipe, 1560 gpm (5905 L/min) for 8 in. pipe, 2440 gpm (9235 L/min) for 10 in. pipe, and 3520 gpm (13,323 L/min) for 12 in. pipe. When supply cannot produce stipulated flow rates, obtain maximum available.
	Hydrostatic: All piping and attached appurtenances subjected to system working pressure shall be hydrostatically tested at 200 psi (13.8 bar) or 50 psi (3.4 bar) in excess of the system working pressure, whichever is greater and shall maintain that pressure ± 5 psi for 2 hours.
	Hydrostatic Testing Allowance: Where additional water is added to the system to maintain the test pressures required by 10.10.2.2.1, the amount of water shall be measured and shall not exceed the limits of the following equation (for metric equation, see 10.10.2.2.4):
	
	$L = \dfrac{SD\sqrt{P}}{148{,}000}$ L = testing allowance (makeup water), in gallons per hour
	                S = length of pipe tested, feet
	                D = nominal diameter of the pipe, in inches
	                P = Average test pressure during the hydrostatic test, in pounds per square inch

Flushing tests	New underground piping flushed according to <u>**NFPA 24**</u> standard by (company) <u>**WS Sprinkler**</u> ☑ Yes ☐ No If no, explain _____ | How flushing was obtained	Through what type opening | |---|---| | ☑ Public water ☐ Tank or reservoir ☐ Fire pump | ☐ Hydrant butt ☑ Open pipe | Lead-ins flushed according to <u>**NFPA 24**</u> standard by <u>**WS Sprinkler**</u> ☑ Yes ☐ No If no, explain _____ | How flushing was obtained	Through what type opening | |---|---| | ☑ Public water ☐ Tank or reservoir ☐ Fire pump | ☐ Y connection to flange and spigot ☑ Open pipe |
Hydrostatic test	All underground piping hydrostatically tested at Joints covered <u>**200**</u> psi For <u>**2**</u> hours ☐ Yes ☑ No

FORM S4.2 *Continued*

Leakage test	Total amount of leakage measured		
	30 oz	2 hours	
	Allowable leakage		
	42 oz	2 hours	

Hydrants	Number installed	Type and make	All operate satisfactorily
	0		☐ Yes ☐ No

Control valves	Water control valves left wide open	☑ Yes ☐ No
	If no, state reason	
	Hose threads of fire department connections and hydrants interchangeable with those of fire department answering alarm	☑ Yes ☐ No

Remarks	Date left in service
	4-15-07

Signatures	Name of installing contractor		
	WS Sprinkler		
	Tests witnessed by		
	For property owner (signed)	Title	Date
	John Doe	Safety Manager	4-15-07
	For installing contractor (signed)	Title	Date
	Jack Smith	Foreman	4-15-07

Additional explanation and notes: _____

FORM S4.2 *Continued*

Inspection, Testing, and Maintenance of Dry Pipe Sprinkler Systems

Property Name: XYZ Corporation **Inspector:** John Doe
Property Address: 100 Main St., Taylorsville, USA **Contract No.:**
Phone Number: (000) 555-0000 **Date:** 4-15-07
This Report Covers: ☐ Monthly ☐ Quarterly ☑ Annual
☐ Three-Year ☐ Five-Year

Inspections
Monthly

☐ Yes ☐ No ☐ N/A Gauges—normal air and water pressure maintained

Control Valves

☐ Yes ☐ No ☐ N/A In the correct (open or closed) position
☐ Yes ☐ No ☐ N/A Sealed, locked, or supervised
☐ Yes ☐ No ☐ N/A Accessible
☐ Yes ☐ No ☐ N/A Free from damage or leaks
☐ Yes ☐ No ☐ N/A Proper signage

Dry Pipe Valves

☐ Yes ☐ No ☐ N/A Exterior is free of damage, trim valves are in correct open or closed position, and intermediate chamber is not leaking

Quarterly

☐ Yes ☐ No ☐ N/A Alarm devices—free of damage
☐ Yes ☐ No ☐ N/A Hydraulic data nameplate—securely attached to riser/legible

Fire Department Connections

☐ Yes ☐ No ☐ N/A Visible and accessible
☐ Yes ☐ No ☐ N/A Coupling/swivels operate correctly
☐ Yes ☐ No ☐ N/A Plugs/caps are in place
☐ Yes ☐ No ☐ N/A Gaskets are not damaged
☐ Yes ☐ No ☐ N/A Identification signs are in place
☐ Yes ☐ No ☐ N/A Ball drip is functional
☐ Yes ☐ No ☐ N/A FDC clapper is functional

Pressure Reducing Valve

☐ Yes ☐ No ☐ N/A In the open position/not leaking
☐ Yes ☐ No ☐ N/A Maintaining downstream pressure
☐ Yes ☐ No ☐ N/A In good condition

FORM S4.3 *Sample Inspection, Testing, and Maintenance of Dry Pipe Sprinkler Systems.*

Annual   **Sprinklers**

- [x] Yes   [ ] No   [ ] N/A   No damage or leaks
- [x] Yes   [ ] No   [ ] N/A   Free of corrosion, foreign material, or paint
- [x] Yes   [ ] No   [ ] N/A   Installed in proper orientation
- [x] Yes   [ ] No   [ ] N/A   Fluid in glass bulbs
- [x] Yes   [ ] No   [ ] N/A   Spare sprinklers—proper number and type. Complete with wrench?
- [x] Yes   [ ] No   [ ] N/A   Hangers and seismic bracing—not damaged or loose

Pipes and Fittings

- [x] Yes   [ ] No   [ ] N/A   In good condition/no external corrosion
- [x] Yes   [ ] No   [ ] N/A   No leaks or mechanical damage
- [x] Yes   [ ] No   [ ] N/A   Correct alignment—no external loads
- [x] Yes   [ ] No   [ ] N/A   Dry pipe valve interior—following trip test
- [x] Yes   [ ] No   [ ] N/A   Building—prior to onset of freezing weather—all openings are closed, no water-filled pipe is exposed to freezing temps

Five-Year

- [ ] Yes   [ ] No   [ ] N/A   Obstruction inspection—no foreign or obstructing material found
- [ ] Yes   [ ] No   [ ] N/A   Check valve—internal moves freely, in good condition
- [ ] Yes   [ ] No   [ ] N/A   Dry pipe valve strainers, filters, and orifices internal inspection

Test
Quarterly

- [ ] Yes   [ ] No   [ ] N/A   Alarm devices—water motor gong
- [ ] Yes   [ ] No   [ ] N/A   Main drain test—if the sole supply is through a backflow preventer or pressure reducing valve

    Static psi _____   Residual psi _____

- [ ] Yes   [ ] No   [ ] N/A   Do results differ by more than 10% from previous test?
- [ ] Yes   [ ] No   [ ] N/A   Priming water—test level
- [ ] Yes   [ ] No   [ ] N/A   Low air alarm—test per manufacturer's instructions
- [ ] Yes   [ ] No   [ ] N/A   Quick opening device tested

Semi-annual

- [ ] Yes   [ ] No   [ ] N/A   Supervisory switch functions
- [ ] Yes   [ ] No   [ ] N/A   Alarm devices—inspectors test or bypass opened/obstructed waterflow

Annual

- [x] Yes   [ ] No   [ ] N/A   Main drain test   Static psi: **59**   Residual psi: **50**
- [ ] Yes   [x] No   [ ] N/A   Do results differ by more than 10% from previous test?
- [x] Yes   [ ] No   [ ] N/A   All control valves operated through full range of motion and returned to normal position.

FORM S4.3 *Continued*

Dry pipe valve trip test (partial flow)

Water pressure **59 psi**   Air pressure **45 psi**

Tripping air pressure **30 psi**   Trip time **25** (sec)

- [x] Yes [ ] No [ ] N/A Results comparable to previous tests
- [x] Yes [ ] No [ ] N/A Backflow preventer—backflow test
- [ ] Yes [x] No [ ] N/A Backflow preventer—flow test

Three-Year

- [ ] Yes [ ] No [ ] N/A Dry pipe valve—full flow trip test

Water pressure _____   Air pressure _____

Tripping air pressure _____   Trip time _____ (sec)

Water delivery time _____ (min.) _____ (sec)

- [ ] Yes [ ] No [ ] N/A Results comparable to previous years

Five-Year

- [ ] Yes [ ] No [ ] N/A Gauges tested or replaced
- [ ] Yes [ ] No [ ] N/A Pressure reducing valve—flow test and comparable to previous results

Routine Maintenance

- [ ] Yes [ ] No [x] N/A Sprinklers tested or replaced per appropriate testing schedule

Comments: _____

_____

_____

_____

Signature: *John Doe*   Date: **4-15-07**

License/Certification No.: **Spr100500**

FORM S4.3 Continued

Annual Performance Tests for Fire Pumps

Date: 6-11-07   **Inspector:** Jack Smith   **System:** Main Fire Pump
Location: Pump house #1

Y = Satisfactory   **N** = Unsatisfactory (explain below)

Pump manufacturer and model: _____
Type: ☑ Centrifugal   ☐ Turbine
Controller manufacturer and model: ACME Model #100
Rated capacity: 1000 gpm (L/min)
Water supply source: City Water Supply
Rated pressure: 100 psi (bar)   Rated speed: 1770 rpm
Power: ☑ Electric   ☐ Diesel   ☐ Steam

Item	Y/N	Item	Y/N
Automatic starts performed 6 times	Y	Timer indicates total run time: 10 min.	Y
Automatic start functions properly	Y	Timer reset and graph paper changed?	Y
Automatic stop functions properly	Y	Test data and flow charts completed (attach all waterflow charts, electrical power charts, performance curves, etc.)	Y
Automatic start: 150 psi (bar)			
Automatic stop: 160 psi (bar)		Fire pump electrical readings recorded at each flow condition?	Y
Manual starts performed 6 times	Y	Fire pump motor speed: 1770 rpm	Y
Manual start functions properly	Y	Fire pump discharge flow: 1000 gpm (L/min)	Y
Manual start: 150 psi (bar)		Jockey pump operational	Y
Manual stop: 160 psi (bar)		Jockey pump appears properly aligned	Y
Remote start functions properly	Y	Jockey pump valves open	Y
Remote stop functions properly	Y	Jockey pump "turn-on": 150 psi (bar)	Y
Remote start: 150 psi (bar)			
Remote stop: 160 psi (bar)		Jockey pump "turn-off": 165 psi (bar)	Y

Notes: _____

FORM S4.4 Sample Annual Performance Tests for Fire Pumps.

Annual Test Summary Page for Fire Pumps

Date: 6-11-07 **Inspector:** Jack Smith **System:** Main Fire Pump
Location: Pump house #1

	Test 1	Test 2	Test 3
Approximate percent of rated pump discharge (gpm)(L/min)	0	100%	150%
Nozzle size in inches (mm)	No flow	1 1/8" (4)	1 3/4" (4)
Pitot pressure in psi (bar)	None	45/47 44/48	18/17 19/18
Flow in gpm (L/min)	None	246/251 243/254	376/365 383/376
Pump suction in psi (bar)	8	7	5
Pump discharge in psi (bar)	141	107	70
Net pump head (discharge pressure minus suction pressure)	133	100	65
Pump speed (rpm)	1775	1770	1770
Operate electric circuit breaker	Y	Y	Y
Test emergency power supply	Y	Y	Y
Check for excessive back pressure in exhaust system	NA	NA	NA

Notes:

FORM S4.5 *Sample Annual Test Summary Page for Fire Pumps*

SUPPLEMENT 5

Technical/Substantive Changes from the 2002 Edition to the 2008 Edition of NFPA 25

Editor's Note: Supplement 5 contains a useful table of major code changes from the 2002 to the 2008 edition of NFPA 25, Standard for the Inspection, Testing, and Maintenance of Water-Based Fire Protection Systems.

Subject/2008 Edition Text	Notes
Chapter 3 Definitions Selected definitions.	Revised to coordinate with terminology used in the NFPA Glossary of Terms.
Chapter 4 General Requirements **4.1.1* Responsibility for Inspection, Testing, and Maintenance.** The property owner or occupant shall provide ready accessibility to components of water-based fire protection systems that require inspection, testing, or maintenance.	The word "property" replaced "building" because legal arguments are being made that the owner of the contracting firm is responsible for inspection, testing, and maintenance. It is the intent of the NFPA 25 technical committee to place responsibility for inspection, testing, and maintenance on the property owner. Additionally, NFPA 13 accepted a similar proposal identifying "property owner" instead of just "owner."
4.1.2.4 Where an occupant, management firm, or managing individual has received the authority for inspection, testing, and maintenance, the occupant, management firm, or managing individual shall comply with the requirements identified for the owner or occupant throughout this standard.	Previously, this text was repeated throughout Chapter 4 and now is stated just once.
4.1.3 Notification of System Shutdown. The property owner or occupant shall notify the authority having jurisdiction, the fire department, if required, and the alarm-receiving facility before testing or shutting down a system or its supply.	
4.1.6 Addressing Changes in Hazard. Where changes in the occupancy, hazard, water supply, storage commodity, storage arrangement, building modification, or other condition that affects the installation criteria of the system are identified, the	The AHJ was added because he/she needs to be notified to ensure that proper protection is provided in the

Subject/2008 Edition Text	Notes
property owner or occupant shall promptly take steps, such as contacting a qualified contractor, consultant, or engineer, and the authority having jurisdiction, to evaluate the adequacy of the installed system in order to protect the building or hazard in question.	facility. The AHJ should be included from the beginning, not just after others have made an evaluation.
4.1.7 Valve Location. The property owner shall ensure that responsible occupants are made aware of the location of the shutoff valves and the procedures for shutting down the system.	Owners and occupants should know where the control valves are located and how to shut the system down. Time is often wasted after a system starts to discharge as occupants search for the shutoff valves. This wasted time during non-emergency discharges leads only to additional water damage.
4.1.8 Information Sign. A permanently marked metal or rigid plastic information sign shall be placed at the system control riser supplying an antifreeze loop, dry system, preaction system, or auxiliary system control valve. Each sign shall be secured with a corrosion-resistant wire, chain, or other acceptable means and shall indicate the following information: (1) Location of the area served by the system (2) Location of auxiliary drains and low-point drains (3) The presence and location of antifreeze or other auxiliary systems	The committee believes that the retroactive installation of these signs is necessary for the proper maintenance of systems. The language is consistent with the language in NFPA 13.
4.6.1.1.1* As an alternative means of compliance, subject to the authority having jurisdiction, components and systems shall be permitted to be inspected, tested and maintained under a performance-based program.	Performance-based testing is currently permitted in NFPA 25, but on a limited basis via the "equivalency" option. The guidance necessary to support this process is now provided in this paragraph. The implementation of this paragraph requires a significant undertaking by both the AHJ and the user in order to optimize the benefits of implementing a performance-based program, if desired.
Chapter 5 Sprinkler Systems **Table 5.1 Summary of Sprinkler System Inspection, Testing, and Maintenance** **5.1.4** Hose connections shall be inspected, tested, and maintained in accordance with Chapters 6 and 13.	This wording provides consistency with 5.1.1 Valves and Connections on how other chapters are referenced for criteria on components.
5.2.1.1.2 Any sprinkler shall be replaced that has signs of leakage; is painted, other than by the sprinkler manufacturer, corroded, damaged, or loaded; or in the improper orientation.	As previously written, any sprinkler that is painted, even by the manufacturer, must be replaced. This change clarifies the intent to require replacement of sprinklers that are painted *except* those painted by the manufacturer.
5.2.1.1.6 Sprinklers that are subject to recall shall be replaced per the manufacturer's requirements.	Sprinklers and devices that were recalled by the manufacturer must be replaced. This paragraph allows for the planned replacement of sprinklers and devices that are part of an ongoing replacement program, but is intended to require the immediate replacement of sprinklers or devices for which a replacement program has expired.

Subject/2008 Edition Text	Notes
5.2.1.2* The minimum clearance required by the installation standard shall be maintained below all sprinklers. Stock, furnishings, or equipment closer to the sprinkler than the clearance rules allow shall be corrected.	Because obstruction rules have changed in almost every edition of NFPA 13 since 1989, it is impossible for every inspector to know whether or not every sprinkler is clear of all obstructions. The inspector's job is not to point out all of the potential obstructions. The inspector should verify that the owner has not piled up anything that would prevent the sprinkler spray pattern from developing.
5.2.2.1 Pipe and fittings shall be in good condition and free of mechanical damage, leakage, and corrosion.	The word "misalignment" was removed from this paragraph because the committee believes that checking misalignment from the floor level is impossible. Verifying misalignment is not consistent with the scope of the document.
5.3.1.1.1.2 Sprinklers manufactured using fast-response elements that have been in service for 20 years shall be replaced, or representative samples shall be tested. They shall be retested at 10-year intervals.	Editorial change adding the words "representative samples" to make this section consistent with 5.3.1.1.3.
5.3.1.1.1.5* Dry sprinklers that have been in service for 10 years shall be replaced, or representative samples shall be tested. They shall be retested at 10-year intervals.	Editorial change removing the words "If maintained and serviced" to be consistent with 5.3.1.1.3.
5.3.1.3 Where one sprinkler within a representative sample fails to meet the test requirement, all sprinklers within the area represented by that sample shall be replaced.	Added the words "within the area" to clarify that this testing is for environmental purposes and that not all sprinklers should be replaced if one fails in a given environment.
5.3.3.1 Mechanical waterflow devices including, but not limited to, water motor gongs, shall be tested quarterly.	The word "mechanical" was added to differentiate from electrical devices.
5.3.3.2* Vane-type and pressure switch–type waterflow devices shall be tested semiannually.	Pressure switches were added to this paragraph because the committee believes that the test data from *NFPA 72* included these devices.
5.3.3.3 Testing the waterflow alarms on wet pipe systems shall be accomplished by opening the inspector's test connection.	Annex A paragraph was deleted because the information is contained elsewhere in the standard.
Table 5.3.4.1(b) Antifreeze Solutions To Be Used If Potable Water Is Connected to Sprinklers	The values for glycerin were updated to match those of NFPA 13.
5.3.4.3 The antifreeze solution shall be tested at its most remote portion and where it interfaces with the wet-pipe system. When antifreeze systems have a capacity larger than 150 gal (568 L), tests at one additional point for every 100 gal (379 L) shall be made. If the test results indicate an incorrect freeze point at any point in the system, the system shall be drained, the solution adjusted, and the systems refilled. For premixed solutions, the manufacturer's instructions shall be permitted to be used with regard to the number of test points and refill procedure.	It is the intent to require a system to be drained and refilled with antifreeze when a test point indicates an incorrect freeze point.
5.4.1.3* Special and quick-response sprinklers as defined by NFPA 13, *Standard for the Installation of Sprinkler Systems*, shall be replaced with sprinklers of the same orifice, size, temperature range and thermal response characteristics, and *K*-factor.	It is not the intent of the standard to require an exact replacement with the same make and model of a sprinkler, provided that the characteristics listed in this paragraph are provided in the replacement sprinkler.

Subject/2008 Edition Text	Notes
5.5.1 Whenever a component in a sprinkler system is adjusted, repaired, reconditioned, or replaced, the actions required in Table 5.5.1 shall be performed.	Tests for component replacement were added to avoid a complete acceptance test for replaced components.
Table 5.5.1 Summary of Component Replacement Action Requirements	Tests for component replacement were added to clarify the requirements for replaced components.
Chapter 6 Standpipe and Hose Systems **Table 6.1 Summary of Standpipe and Hose Systems Inspection, Testing, and Maintenance**	All quarterly inspections and tests were changed to annual because the committee felt that the performance of this equipment does not warrant quarterly inspection and testing.
6.3.1.1* A flow test shall be conducted every 5 years at the hydraulically most remote hose connections of each zone of an automatic standpipe system to verify the water supply still provides the design pressure at the required flow.	This paragraph was revised to indicate that this test is required to be performed at multiple standpipes (if present) and not at a single outlet.
6.3.1.3 All systems shall be flow tested and pressure tested at the requirements for the design criteria in effect at the time of the installation.	It is the intent of this standard to test at the appropriate "design criteria" for the standpipe in accordance with NFPA 14.
6.3.2.1 Hydrostatic tests of not less than 200 psi (13.8 bar) pressure for 2 hours, or at 50 psi (3.4 bar) in excess of the maximum pressure, where maximum pressure is in excess of 150 psi (10.3 bar), shall be conducted every 5 years on manual standpipe systems and automatic-dry standpipe systems, including piping in the fire department connection.	It is the intent of the standard to require a hydrostatic test on all manual and automatic-dry standpipe systems.
6.3.3 Alarm Devices. Where provided, waterflow alarm and supervisory devices shall be tested in accordance with 13.2.6 and 13.3.3.5.	Correlates with *NFPA 72* by subdividing "alarm devices" into three categories: waterflow devices, valve supervisory devices, and supervisory signal devices.
Table 6.5.1 Summary of Component Replacement Action Requirements	Tests for component replacement were added to clarify the requirements for replaced components.
Chapter 7 Private Fire Service Mains **Table 7.2.2.4 Dry Barrel and Wall Hydrants**	Editorial change
Table 7.2.2.5 Wet Barrel Hydrants	Editorial change
Table 7.5.1 Summary of Component Replacement Action Requirements	Tests for component replacement were added to clarify the requirements for replaced components.
Chapter 8 Fire Pumps **8.2.2*** The pertinent visual observations specified in the following checklists shall be performed weekly:	Editorial change
8.3.3.8* Where engines utilize electronic fuel management control systems, the backup electronic control module (ECM), and the primary and redundant sensors for the ECM, shall be tested annually.	Due to clean air laws and their impact on engine exhaust emission requirements, engines with electronic fuel management systems are starting to be used in lieu of engines with mechanically controlled fuel systems. As engines with the newer type of fuel management systems make their way

Subject/2008 Edition Text	Notes
	into fire pump driver applications, NFPA 25 now addresses the minimum testing requirements of the electronic fuel management systems in order to ensure the overall reliability of the fire pump installation.
8.3.5.2.1 Theoretical factors for correction to the rated speed shall be applied where determining the compliance of the pump per the test	It is difficult to compare any engine-driven pump with the factory curve. All factory curves are performed at a single speed. If data were not corrected for speed on an annual fire pump test, then the majority of pumps would fail due to the 5 percent degrade required in 8.3.5.4.
8.3.5.3 The fire pump assembly shall be considered acceptable if either of the following conditions is shown during the test: (1)* The test is no less than 95 percent of the pressure at rated flow and rated speed of the initial unadjusted field acceptance test curve, provided that the original acceptance test curve matches the original certified pump curve by using theoretical factors. (2) The fire pump is no less than 95 percent of the performance characteristics as indicated on the pump nameplate.	Changes in items 1 & 2 address the 5 percent degradation of the test results.
8.3.5.5 Current and voltage readings whose product does not exceed the product of the rated voltage and rated full-load current multiplied by the permitted motor service factor shall be considered acceptable.	Editorial change
8.3.5.6 Voltage readings at the motor within 5 percent below or 10 percent above the rated (i.e., nameplate) voltage shall be considered acceptable.	Editorial change
Table 8.5.3 Summary of Fire Pump Inspection, Testing, and Maintenance	In 8.5.3 D.5(g) a requirement to clean terminals annually was added; in 8.5.3 D.5(c) the requirement to remove corrosion was eliminated.
Table 8.6.1 Summary of Component Replacement Testing Requirements	Tests for component replacement were added to clarify the requirements for replaced components.
Chapter 9 Water Storage Tanks **9.1* General.** This chapter shall provide the minimum requirements for the routine inspection, testing, and maintenance of water storage tanks dedicated to fire protection use. Table 9.1 shall be used to determine the minimum required frequencies for inspection, testing, and maintenance.	The committee felt that this paragraph was being inappropriately applied to municipal water storage tanks and private tanks that supply both domestic and fire protection water and therefore added the words "dedicated to fire protection."
Table 9.1 Summary of Water Storage Tank Inspection, Testing, and Maintenance	The test frequency for temperature alarms was changed from monthly to quarterly to coordinate with *NFPA 72*.
9.2.4.2 The temperature of water in tanks with low temperature alarms connected to a constantly attended location shall be inspected and recorded monthly during the heating season when the mean temperature is less than 40°F (4.4°C).	*Heating season* should be more defined to include a mean temperature to verify temperature conditions. Therefore, the 40°F has been added. Monitoring the temperature during heating season is necessary.

Subject/2008 Edition Text	Notes
9.2.4.3 The temperature of water in tanks without low temperature alarms connected to a constantly attended location shall be inspected and recorded weekly during the heating season when the mean temperature is less than 40°F (4.4°C).	*Heating season* should be more defined to include a mean temperature to verify temperature conditions. Therefore, the 40°F has been added. Monitoring the temperature during heating season is necessary.
9.4 Maintenance.	
9.4.1 Voids discovered beneath the floors of tanks shall be filled by pumping in grout or accessing the sand and replenishing.	Editorial change
9.4.6.1 The maintenance of ESCF tanks shall be completed in accordance with this section and the tank manufacturer's instructions.	Editorial change
Table 9.5.1.1 Summary of Automatic Tank Fill Valve Inspection and Testing	The automatic fill valves on water tanks that serve as a suction supply to a fire pump are an integral part of the fire protection system and are now covered by NFPA 25.
Table 9.6.1 Summary of Component Replacement Action Requirements	Tests for component replacement were added to clarify the requirements for replaced components.
Chapter 10 Spray Fixed Systems	
Table 10.1 Summary of Water Spray Fixed System Inspection, Testing, and Maintenance	The inspection frequency for hangers, nozzles, and pipe was changed from quarterly to annual. This change correlates with Table 5.1 for sprinkler systems.
10.2.3 Automatic Detection Equipment	
10.2.3.2 Automatic fire detection equipment not covered by *NFPA 72, National Fire Alarm Code*, shall be inspected, tested, and maintained to ensure that the detectors are in place, securely fastened, and protected from corrosion, weather, and mechanical damage and that the communication wiring, control panels, or tubing system is functional.	This paragraph is intended to apply to detection equipment such as wet and dry pilot sprinklers.
10.3.4.1.2 Under test conditions, the flammable gas detection system, where exposed to a standard test gas concentration, shall operate within the time frame specified in the system design.	This language was added to match that of NFPA 15.
10.3.4.3.1* The water discharge patterns from all of the open spray nozzles shall be observed to ensure that patterns are not impeded by plugged nozzles, to ensure that nozzles are correctly positioned, and to ensure that obstructions do not prevent discharge patterns from wetting surfaces to be protected. **10.3.4.3.1.1** Where the nature of the protected property is such that water cannot be discharged, the nozzles shall be inspected for proper orientation and the system tested with air to ensure that the nozzles are not obstructed.	Provides alternate methods of testing where water cannot be discharged.
Table 10.5.1 Summary of Component Replacement Action Requirements	Tests for component replacement were added to clarify the requirements for replaced components.
Chapter 11 Foam-Water Sprinkler Systems	
Table 11.1 Summary of Foam-Water Sprinkler System Inspection, Testing, and Maintenance	The inspection frequency for pipe corrosion, pipe damage, fitting corrosion, fitting damage, and hangers/supports was changed from quarterly to annual.

Subject/2008 Edition Text	Notes
11.2.4 Hangers and Supports. Hangers and supports shall be inspected for the following and repaired <u>or replaced</u> as necessary: (1) Condition (e.g., missing or damaged paint or coating, rust, and corrosion) (2) Secure attachment to structural supports and piping (3) Damaged or missing hangers	Language added for consistency.
Table 11.5.1 Summary of Component Replacement Action Requirements	Tests for component replacement were added to clarify the requirements for replaced components.
Chapter 12 Water Mist Systems (new chapter)	Chapter 13 of NFPA 750 was extracted since it is the intent of this standard to address the inspection, testing, and maintenance requirements for all types of water-based systems.
Chapter 13 Valves, Valve Components, and Trim **13.2.5.2** When there is a 10 percent reduction in full flow pressure when compared to the original acceptance test or previously performed tests, the cause of the reduction shall be identified and corrected if necessary.	The standard previously required that a "large" drop normally is indicative of a dangerously reduced water supply. "Large" needed to be quantified. In addition, the owner needs to be directed to find and repair the problem. The revised language provides for correction of deficiencies.
13.3.1.1 Systems that have more than one control valve that must be closed to work on a system shall have a sign on each affected valve referring to the existence and location of other valves.	It is the intent to provide for identification of each valve.
13.3.3.4 A main drain test shall be conducted any time the control valve is closed and reopened at a system riser.	Provides for further clarification as to when a main drain test is required.
13.4.1.1* Alarm valves and system riser check valves shall be externally inspected monthly and shall verify the following: (1) The gauges indicate normal supply water pressure is being maintained. (2) The valve is free of physical damage. (3) All valves are in the appropriate open or closed position. (4) The retarding chamber or alarm drains are not leaking.	Added reference to system riser check valves.
13.4.3.1 Inspection	Editorial change
13.4.3.1.1 Valve enclosure heating equipment for preaction and deluge valves subject to freezing shall be inspected daily during cold weather for its ability to maintain a minimum temperature of at least 40°F (4.4°C).	Editorial change
13.4.3.1.1.1 Valve enclosures equipped with low temperature alarms shall be inspected weekly.	Editorial change
13.4.3.2.2* Each deluge valve shall be trip tested annually at full flow in warm weather and in accordance with the manufacturer's instructions. Protection shall be provided for any devices or equipment subject to damage by system discharge during tests. **13.4.3.2.2.1*** Where the nature of the protected property is such that water cannot be discharged for test purposes, the trip test shall be conducted in a manner that does not necessitate discharge in the protected area.	Reference to preaction valve has been removed because filling a preaction system annually is thought to accelerate corrosion.
13.4.3.2.3 Except for preaction systems covered by 13.4.3.2.5, every 3 years the preaction valve shall be trip tested with the control valve fully open.	Provides for 3-year trip test similar to that of dry systems and prevents accelerated corrosion.

Subject/2008 Edition Text	Notes
13.4.3.2.4 During those years when full flow testing in accordance with 13.4.3.2.3 is required, the preaction valve shall be trip tested with the control valve partially open. **13.4.3.3.3*** Auxiliary drains in preaction or deluge systems shall be operated after each system operation and before the onset of freezing conditions.	The term *auxiliary* replaced *low point* for clarity.
13.4.4.1.3 Systems with auxiliary drains shall require a sign at the dry or preaction valve indicating the number of auxiliary drains and location of each individual drain.	This language was added because the committee felt that the property owner or contractor maintaining the system needs to know if there are any auxiliary drains, how many there are, and where they are located. There have been many cases of frozen systems when all drains are not known and utilized.
13.4.4.2.9 Dry pipe systems shall be tested once every three years for air leakage, using one of the following test methods: (1) A pressure test at 40 psi for two hours. The system shall be permitted to lose up to 3 psi (0.2 bar) during the duration of the test. Air leaks shall be addressed if the system loses more than 3 psi (0.2 bar) during this test. (2) With the system at normal system pressure, shut off the air source (compressor or shop air) for 4 hours. If the low air pressure alarm goes off within this period, the air leaks shall be addressed.	The previous 10 psi per week requirement of 12.4.4.3.1 was difficult to measure and enforce. Without a specific air test, it was impossible to determine if there was a significant problem. The dry-pipe system needs to be capable of staying closed (without any false trips of the dry-pipe valve) during a short power outage without needing the compressor to maintain pressure. The user is given two different methods to accomplish this test, which are practical and easy to perform.
13.4.4.3.2* Auxiliary drains in dry pipe sprinkler systems shall be drained after each operation of the system, before the onset of freezing weather conditions, and thereafter as needed.	Previously, this paragraph imposed requirements not in line with NFPA 13. Deluge and preaction systems have no requirements for drainage beyond that of wet systems, except for preaction systems that are subject to freezing. The term *low point* is arbitrary when used alone; therefore, the term *auxiliary* is used. All trapped sections of piping in preaction systems that are subject to freezing are required by NFPA 13 to have an auxiliary drain.
13.5.1 Inspection and Testing of Sprinkler Pressure Reducing Valves. Sprinkler pressure reducing valves shall be inspected and tested as described in 13.5.1.1 and 13.5.1.2.	Editorial – the word "control" was removed.
13.5.4 Master Pressure Reducing Valves. **13.5.4.1*** Valves shall be inspected weekly to verify that the valves are in the following condition: (1)* The downstream pressures are maintained in accordance with the design criteria. (2) The supply pressure is in accordance with the design criteria. (3) The valves are not leaking. (4) The valve and trim are in good condition. **13.5.4.2*** A partial flow test adequate to move the valve from its seat shall be conducted quarterly. **13.5.4.3*** A full flow test shall be conducted on each valve annually and shall be compared to previous test results.	Master pressure reducing valves are commonly used in fire sprinkler systems and standpipe systems for a variety of reasons. NFPA fire protection system installation standards have been hesitant to directly address requirements for master pressure reducing valves until there is guidance in NFPA 25 specifically for inspection, testing, and maintenance of such valves, as maintenance is a major concern. As implied by the previous requirements of 12.2.6.1, master pressure reducing valves require more frequent inspection, testing, and maintenance,

Subject/2008 Edition Text	Notes
13.5.4.4 When valve adjustments are necessary, they shall be made in accordance with the manufacturer's instructions. **13.5.5 Pressure Reducing Valves.** **13.5.5.1** All pressure reducing valves installed on fire protection systems not covered by 13.5.1, 13.5.2, 13.5.3, or 13.5.4 shall be inspected in accordance with 13.5.1.1. **13.5.5.2** All pressure reducing valves installed on fire protection systems not covered by 13.5.1, 13.5.2, 13.5.3, or 13.5.4 shall be tested in accordance with 13.5.1.2. **13.6.2.1*** (1) A forward flow test shall be conducted at the designed flow rate, including hose stream demand, of the system, where hydrants or inside hose stations are located downstream of the backflow preventer.	because of their critical placement. Since the common placement for a master pressure reducing valve is in or immediately downstream of a fire pump discharge, the inspection and testing frequency is associated with the weekly and annual fire pump tests. The partial flow test frequency is that previously required in 12.2.6.1 and is necessary to "exercise" the valves.
Chapter 14 Obstruction Investigation **14.2.1** An inspection of piping and branch line conditions shall be conducted every 5 years by opening a flushing connection at the end of one main and by removing a sprinkler toward the end of one branch line for the purpose of inspecting for the presence of foreign organic and inorganic material.	The word "inspection" was inserted in place of "investigation" because the committee intends to segregate the requirements for obstruction inspection and investigation.
14.3 Ice Obstruction. Dry pipe or preaction sprinkler system piping that protects or passes through freezers or cold storage rooms shall be inspected internally on an annual basis for ice obstructions at the point where the piping enters the refrigerated area.	The word "prevention" was deleted to clarify that this requirement is intended to detect, not prevent, ice obstruction.
Chapter 15 Impairments **15.5.2** (3) Recommendations have been submitted to management or the property owner/manager. Where a required fire protection system is out of service for more than 10 hours in a 24-hour period, the impairment coordinator shall arrange for one of the following: (a) Evacuation of the building or portion of the building affected by the system out of service (b)* An approved fire watch (c)* Establishment of a temporary water supply (d)* Establishment and implementation of an approved program to eliminate potential ignition sources and limit the amount of fuel available to the fire	The time allowance for a system impairment was increased to 10 hours because the committee felt that this amount of time is a more reasonable requirement as it extends beyond the normal work shift.
Annex A Explanatory Material **A.1.1** Generally accepted NFPA installation practices for water-based fire protection systems relevant to this standard are found in the following: NFPA 13, *Standard for the Installation of Sprinkler Systems.* NFPA 13R, *Standard for the Installation of Sprinkler Systems in Residential Occupancies up to and Including Four Stories in Height.* NFPA 14, *Standard for the Installation of Standpipe and Hose Systems.* NFPA 15, *Standard for Water Spray Fixed Systems for Fire Protection.* NFPA 16, *Standard for the Installation of Foam-Water Sprinkler and Foam-Water Spray Systems.* NFPA 20, *Standard for the Installation of Stationary Pumps for Fire Protection.* NFPA 22, *Standard for Water Tanks for Private Fire Protection.* NFPA 24, *Standard for the Installation of Private Fire Service Mains and Their Appurtenances.* NFPA 750, *Standard on Water Mist Fire Protection Systems.*	A reference to the appropriate installation standard is needed when repair, replacement, alteration, or extension of a system is contemplated.

Subject/2008 Edition Text	Notes
For systems originally installed in accordance with one of these standards, the repair, replacement, alteration or extension of such systems should also be performed in accordance with that same standard. When original installations are based on other applicable codes or standards, repair, replacement, alteration, or extension practices should be conducted in accordance with those other applicable codes or standards.	
A.4.1.4.1 System deficiencies not explained by normal wear and tear, such as hydraulic shock, can often be indicators of system problems and should be investigated and evaluated by a qualified person or engineer. Failure to address these issues could lead to catastrophic failure. Examples of deficiencies that can be caused by issues beyond normal wear and tear are as follows: **Pressure Gauge** (1) Gauge not returning to zero (2) Gauge off scale (3) Gauge with bent needle **Support Devices** (1) Bent hangers and/or rods (2) Hangers pulled out/off structure (3) Indication of pipe or hanger movement such as the following: (a) Hanger scrape marks on pipe, exposed pipe surface where pipe and hangers are painted (b) Fire stop material damaged at pipe penetration of fire rated assembly **Unexplained System Damage** (1) Unexplained system damage beyond normal wear and tear (2) Bent or broken shafts on valves (3) Bent or broken valve clappers (4) Unexplained leakage at branch lines, cross main, or feed main piping (5) Unexplained leakage at close nipples (6) Loose bolts on flanges and couplings **Fire Pump** (1) Fire pump driver out of alignment (2) Vibration of fire pump and/or driver (3) Unusual sprinkler system piping noises (sharp report, loud bang)	Examples of potential causes for deficiencies were added.
A.4.4.3 See Section B.2 for information regarding sample forms.	The sample report forms previously printed in this standard were removed because these forms are now available electronically via the Internet. See the NFPA 25 Forms CD for a complete collection of inspection, testing, and maintenance forms.
A.5.2.1.1.1 Sprinkler orientation includes the position of the deflector in relation to the ceiling slope. The deflector is generally required to be parallel to the slope of the ceiling. The inspection should identify any corrections made where deficiencies are noted, for example, pipe with welded outlets and flexible grooved couplings that have "rolled" out of position.	Guidance was added regarding deflector position and ceiling slope.
A.5.2.4.1 It is normal, though, that the pressure above the alarm or system check valve is typically higher than that of the water supply as a result of trapped pressure surges.	Provides additional information to the existing annex material.

Subject/2008 Edition Text	Notes
A.5.3.1 Where documentation of the installation date is not available, the start date for the in-service interval should be based upon the sprinkler's manufacture date.	On older systems it is difficult to accurately define how long a system has been in service. Using the manufacture date of the sprinkler is a reasonable approach.
A.5.3.4 Many refractometers are calibrated for a single type of antifreeze solution and will not provide accurate readings for the other types of solutions.	The change provides additional information for users.
FIGURE A.5.3.4.1 Isothermal Lines — Lowest One-Day Mean Temperature (°F). [24:Figure A.10.5.1]	There are currently no guidelines for the property owner on what level of protection should be maintained. The isothermic map provides some guidelines for freeze protection.
A.5.4.1.1 To help in the replacement of like sprinklers, unique sprinkler identification numbers (SINs) are provided on all sprinklers manufactured after January 1, 2001. The SIN accounts for differences in orifice size, deflector characteristics, pressure rating, and thermal sensitivity.	This change provides useful information for those restoring sprinkler systems.
A.5.4.2 Conversion of dry pipe systems to wet pipe systems on a seasonal basis causes corrosion and accumulation of foreign matter in the pipe system and loss of alarm service.	Converting a dry system to a wet system does not cause corrosion. A dry system experiences an extremely accelerated rate of corrosion compared to wet systems. This is evident in the fact that NFPA 13 assigns a Hazen-Williams C-factor of 100 for dry systems, while a wet system is assigned a C-factor of 120. However, it is the committee's intent to discourage seasonal conversions, not temporary due to repair or other issues.
A.5.4.3 For repairs affecting the installation of less than 20 sprinklers, a test for leakage should be made at normal system working pressure.	This is similar guidance to that offered in NFPA 13.
A.13.4.4.2.2.2 For dry pipe systems that were designed and installed using either a manual demonstration or a computer calculation to simulate multiple openings to predict water delivery time, a full flow trip test from a single inspector's test connection should have been conducted during the original system acceptance and a full flow trip test from the single inspector's test should continue to be conducted every 3 years. The system is not required to achieve water delivery to the inspector's test connection in 60 seconds, but comparison to the water delivery time during the original acceptance will determine if there is a problem with the system.	Clarifies the intent regarding water delivery time and recognizes that NFPA 13 permits a manual demonstration for dry-pipe systems.
A.14.2.2 For obstruction investigation procedures, see Section D.3. The type of obstruction investigation should be appropriately selected based on the observed condition. For instance, ordering an internal obstruction investigation would be inappropriate where the observed condition was broken public mains in the vicinity. On the other hand, such an investigation would be appropriate where foreign materials are observed in the dry pipe valve.	It is the committee's intent to clarify that an internal piping investigation prescribed by 14.2.3.2 is not necessary in all 14 listed conditions in 14.2.2.
Annex B Forms for Inspection, Testing, and Maintenance	
B.2 Sample forms are available for downloading at www.nfpa.org, www.nfsa.org, and www.sprinklernet.org.	It is not necessary to publish several versions of sample report forms. These forms are available via the Internet or on CD.

Subject/2008 Edition Text	Notes
Annex D Obstruction Investigation **D.2.6 Forms of Corrosion.** Corrosion is defined as the deterioration of a material, usually a metal, resulting from a chemical or electrochemical reaction. The eight main forms of corrosion include: (1) uniform corrosion, (2) pitting, (3) galvanic corrosion, (4) crevice corrosion, (5) selective leaching (parting), (6) erosion corrosion, (7) environmental cracking, and (8) intergranular corrosion. Microbiologically influenced corrosion (MIC) is included herein as a ninth form of corrosion, although it is usually a secondary factor that accelerates or exacerbates the rate of another form of corrosion. Definitions of the different forms of corrosion are discussed next. (1) *Uniform (or general) corrosion:* A regular loss of a small quantity of metal over the entire area or over a large section of the total area, which is evenly distributed within a pipe(s). (2) *Pitting:* A localized form of corrosion that results in holes or cavities in the metal. Pitting is considered to be one of the more destructive forms of corrosion and is often difficult to detect. Pits may be covered or open and normally grow in the direction of gravity — for example, at the bottom of a horizontal surface. (3) *Galvanic corrosion:* An electric potential exists between dissimilar metals in a conductive (corrosive) solution. The contact between the two materials allows electrons to transfer from one metal to the other. One metal acts as a cathode and the other as an anode. Corrosion usually occurs at anodic metal only. (4) *Crevice corrosion:* A localized form of corrosion that occurs within crevices and other shielded areas on metal surfaces exposed to a stagnant corrosive solution. This form of corrosion usually occurs beneath gaskets, in holes, surface deposits, in thread and groove joints. Crevice corrosion is also referred to as gasket corrosion, deposit corrosion, and under-deposit corrosion. (5) *Selective leaching:* The selective removal of one element from an alloy by corrosion. A common example is dezincification (selective removal of zinc) of unstabilized brass, resulting in a porous copper structure. (6) *Erosion corrosion:* Corrosion resulting from the cumulative damage of electrochemical reactions and mechanical effects. Erosion corrosion is the acceleration or increase in the rate of corrosion created by the relative movement of a corrosive fluid and a metal surface. Erosion corrosion is observed as grooves, gullies, waves, rounded holes, or valleys in a metal surface. (7) *Environmental cracking:* An acute form of localized corrosion caused by mechanical stresses, embrittlement, or fatigue. (8) *Intergranular corrosion:* Corrosion caused by impurities at grain boundaries, enrichment of one alloying element, or depletion of one of the elements in the grain boundary areas. (9) *Microbiologically influenced corrosion (MIC):* Corrosion initiated or accelerated by the presence and activities of microorganisms, including bacteria and fungi. Colonies (also called bio-films and slimes) are formed in the surface of pipes among a variety of types of microbes. Microbes deposit iron, manganese, and various salts into the pipe surfaces, forming nodules, tubercles, and carbuncles. The formation of these deposits can cause obstruction to flow and dislodge, causing blockage (plugging) of system piping, valves, and sprinklers.	Microbiologically influenced corrosion (MIC) is only one form of corrosion associated with a fire sprinkler system. More common in a fire sprinkler system are uniform corrosion, pitting, and crevice corrosion. Other types of corrosion that may be occurring in a fire sprinkler system should also be identified.

Answer Key

CHAPTER 1

1. NFPA 25 specifically states that it does not apply to design or installation flaws. Subsection 1.1.2 states that "where a system has not been installed in accordance with generally accepted practices, the corrective action is beyond the scope of this standard."

2. Yes. Section 1.3 permits the use of alternative programs, provided that a level of integrity and performance equivalent to that prescribed by NFPA 25 is maintained.

3. The leading reasons for unsatisfactory sprinkler performance include a shut valve, inadequate maintenance, obstruction to water distribution, and frozen systems.

4. Low-, medium-, and high-expansion foam systems are not required to be inspected, tested, and maintained under the requirements of NFPA 25. Requirements for inspection, testing, and maintenance of such systems can be found in NFPA 11, *Standard for Low-, Medium-, and High-Expansion Foam.*

CHAPTER 4

1. The property owner; however, if the owner is not the occupant, he or she may pass on the authority for inspection, testing, and maintenance to an occupant, management firm, or managing individual through a lease, written use agreement, or management contract, per 4.1.2.3.

2. Original records, such as a contractor's material and test certificate, must be kept for the life of the system. Subsequent records must be retained for a period of one year after the next inspection, test, or maintenance required by the standard.

3. The four factors that must be considered when evaluating the fire protection systems in a building that has changed occupancy, use, or processes are the following:

    (1) Occupancy changes such as converting office or production space to storage
    (2) Process or material changes
    (3) Building modifications such as new or moved walls, ceilings, or mezzanines
    (4) Modifications to heating systems in spaces where pipes are subject to freezing

4. As a minimum, the inspector should ask the property owner about any required safety equipment such as safety glasses, hearing protection, and fall protection. The inspector should also ask the property owner to identify any confined spaces or hazardous materials that may be present.

CHAPTER 5

1. According to 5.4.1.5, there must be six spare sprinklers. Since at least two of each type must be provided, an equal number of 165°F (74°C) and 286°F (141°C) sprinklers should be provided due to the quantity of each type installed, as well as a sprinkler wrench for each type of sprinkler.

2. Extra high-temperature sprinklers should be tested every five years, per 5.3.1.1.1.3.

3. Alarms such as water motor alarm gongs must be tested quarterly; vane-type water flow devices and pressure switch devices must be tested semi-annually.

4. A complete obstruction investigation, in accordance with the requirements of Chapter 13 of NFPA 25, must be performed.

5. A plastic or paper bag not thicker than 0.003 in. (0.076 mm) can be placed over the sprinkler. It is important to remember to replace the bag when overspray accumulates.

6. Only sprinklers that have a light and loose coating of dust that can be easily removed should be cleaned. Any other type of material that is adhering to the sprinkler should cause the replacement of that sprinkler. Sprinklers should never be cleaned with soap or solvents.

7. NFPA 25 does not require licensing of individuals in order to perform the work specified herein. The

standard does require that individuals be qualified, however. The state or local jurisdiction may require licensing, and the state or local fire marshal should be consulted for any such requirements.

8. NFPA 25 does not require licensing as stated in the answer to Question 7. Most licensing laws in most jurisdictions only require licensing for those individuals or companies that are engaged in the business of providing these services for a fee. If the property owner is performing this work for his or her own equipment, licensing may not be required. Again, the state or local fire marshal should be consulted for any such requirements. NFPA 25 merely requires that anyone performing this type of work be qualified to do so.

CHAPTER 6

1. A hydrostatic test is required when a system has been modified or repaired or every five years for a manual wet standpipe system, manual dry system, or the dry portion of a wet system.

2. The flow and design pressure for systems installed after 1993 must be 250 gpm (946 L/min) for each of the top two hose valves at 100 psi (6.9 bar) for the hydraulically most demanding riser and 500 gpm (1893 L/min) at 65 psi (4.5 bar) for systems installed prior to 1993.

3. The air test is to be conducted at 25 psi (1.7 bar) prior to introducing water into the system.

CHAPTER 7

1. From flow tables, in inch/pound units,

$$951 \text{ gpm} \times 0.8 = 775 \text{ gpm}$$

and in SI units,

$$3665 \text{ L/min} \times 0.8 = 2932 \text{ L/min}$$

2. In inch/pound units,

$$Q = 29.83 \, (c)(d^2) \times \sqrt{p}$$
$$Q = 29.83 \, (0.97) \, (1.125)^2 \text{ in.} \times \sqrt{48} \text{ psi}$$
$$Q = 253.7 \text{ gpm}$$

and in SI units,

$$Q = 0.0666(c)(d^2) \times \sqrt{p}$$
$$Q = 0.0666(0.97)(29)^2 \text{ mm} \times \sqrt{331} \text{ kPa}$$
$$Q = 989 \text{ L/min}$$

3. The property owner, alarm company, insurance company, and fire department should be notified.

4. Repairs should be made prior to the onset of freezing weather.

CHAPTER 8

1. During weekly testing, an electric pump must run for a minimum of 10 minutes. A diesel pump must run for at least 30 minutes.

2. The pump must be tested at minimum, rated, and peak flows, commonly referred to as churn, rated capacity, and overload.

3. Approximately one drop per second is necessary to keep packing lubricated (see Table A.8.2.2).

4. The three methods of flow testing a fire pump during the annual test are use of pump discharge via hose streams, pump discharge via the bypass flowmeter to a drain or suction reservoir, and pump discharge via the bypass flowmeter to the pump suction (see 8.3.3.1.2.1, 8.3.3.1.2.2, and 8.3.3.1.2.3).

5. The test must be conducted every three years (see 8.3.3.1.3).

CHAPTER 9

1. An interior inspection should include a check for corrosion, pitting, debris, and aquatic growth. Corrosion or pitting can result in failure of the tank. Debris or aquatic growth could cause a piping obstruction.

2. Paragraph 9.2.5.2(1) requires the area to be free of brush or other materials that could present a fire exposure hazard.

3. Wood storage tanks have been replaced by elevated steel tanks.

4. The water temperature in a storage tank must be maintained at 40°F (4°C).

5. When an interior inspection of a tank is performed, it is critical to comply with proper safety precautions, including confined space entry requirements.

CHAPTER 10

1. The test is intended to allow observation of discharge patterns of fixed directional water spray nozzles, measurement of nozzles and deluge valve pressures for hydraulic calculation evaluation, and confirmation that systems function properly and drain correctly.

2. These systems are used to protect equipment such as cable trays, belt conveyors, pumps, compressors, vessels, transformers, structures, and miscellaneous equipment. Water spray systems are also used on flammable and combustible liquid pool fires and are frequently used for exposure protection.

3. Detection systems used to actuate water spray fixed systems include electronic detection systems (using smoke, heat, or infrared detectors) and wet and dry pilot detection systems.

4. An automatic directional water spray nozzle is equipped with a fusible element designed to operate at a specific temperature. A non-automatic water spray nozzle is open and is actuated by means of a detection system.

CHAPTER 11

1. Two methods are permitted — refractometric and conductivity. Refractometric testing measures the difference between the refractometric index of the test sample and that of the prism, using a refractometer. The conductivity method evaluates the changes in electrical conductivity as foam concentrate is added to a solution.

2. The quantity is calculated as follows:

    750 gpm $\times$ 10 min = 7500 gal

    7500 gal $\times$ 0.03 = 225 gal of foam concentrate

    2,839 L/min $\times$ 10 min = 28,390 L

    28,390 L $\times$ 0.03 = 851.7 L of foam concentrate

    Thus, 225 gal (851.7 L) of foam concentrate is needed.

3. No. Paragraph 11.3.5.3 requires that the test results be within 10 percent of the acceptance test.

CHAPTER 12

1. The responsibility for properly maintaining a water mist system is that of the property owner.

2. The corrective action should be documented in the inspection or test report and brought to the attention of the property owner or occupant immediately. The property owner or occupant is directly responsible for correcting or repairing deficiencies founds during inspection, testing, or maintenance.

3. The design and installation of a water mist system should be re-evaluated when any of the following conditions occur: occupancy change, process or material change, structural modifications (relocated walls, horizontal or vertical obstructions added, or ventilation modifications), removal of heating systems in areas with water filled piping.

4. Component replacement or system repairs should be inspected and tested in accordance with Chapter 12 of NFPA 750.

5. In addition to the impairment procedures outlined in Chapter 15, other ancillary actions integrated in a cause and effect protocol such as automatic door closers, emergency power shutdown, stopping or starting of ventilation fans, opening or closing of automatic fire dampers, and stopping of production lines should be known, evaluated, and verified prior to impairing the system.

CHAPTER 13

1. A main drain test must be conducted annually, unless the sole water supply is through a backflow preventer or pressure reducing valve, in which case it must be performed quarterly. In addition, a main drain test must be performed any time a water supply control valve is closed (including the annual test).

2. The interior of a preaction or deluge valve must be inspected annually unless the valve can be reset externally, in which case every five years is acceptable.

3. Water must be delivered to the inspector's test connection within 60 seconds, starting at the normal air pressure on the system and at the time of a fully opened inspectors test connection.

4. The test is required annually unless water restrictions prohibit the test or the annual fire pump test produces flow through the device, in which case it need not be conducted. For a sprinkler system installed under the requirements of the 1999 edition of NFPA 13, a test connection of sufficient capacity should be in place to perform the test. For all other systems, a bypass around the FDC check valve can be installed, the FDC check valve can be temporarily reversed for the test, or the FDC clappers can be temporarily held open to accommodate the test.

5. According to the requirements of 13.5.1.2 and 13.5.2.2, these valves must be flow tested every five years.

CHAPTER 14

1. An obstruction investigation must be conducted whenever any of the conditions listed in 14.2.2 is found. If the cause of the obstruction is not corrected, an obstruction investigation is required to be performed every five years.

2. Investigations must examine the interior of the system valve, riser, cross main, and branch line.

3. A visual examination may be conducted, or alternative nondestructive examination methods such as ultrasound or the use of remote video cameras may be used.

4. A sample of the tubercule or slime should be tested for indications of MIC.

5. An inspection is required every five years and can usually be accomplished when the system is impaired for

the internal system control valve inspection required by Chapter 3. This inspection is very different from the obstruction investigation required by 14.2.2.

CHAPTER 15

1. The procedures that should be used for an impairment that will last for more than one working shift are the following:

    (1) Evacuation of the building or portion of the building affected by the impairment
    (2) Establishment of an approved fire watch
    (3) Establishment of a temporary water supply
    (4) Establishment and implementation of an approved fire prevention program to eliminate potential ignition sources and fuel

2. The basic elements of an impairment program are the following:

    (1) Notification of all interested parties prior to system impairment
    (2) Procurement of tools, equipment, and replacement parts prior to system impairment
    (3) Implementation of an impairment tag procedure
    (4) Verification of systems testing and start-up
    (5) Notification of system start-up
    (6) Closeout of impairment tag procedure
    (7) Impairment procedure documentation

3. An emergency impairment may occur when a system responds to a fire or in the event of failure of a major component, such as a fire pump or water control valve.

4. The tag provides notification that the valve is closed for a specific reason and should not be opened (for safety reasons or to prevent water damage). The tag also indicates that the valve was not closed accidentally or maliciously.

Supplement 3

1. Generally, the weekly and monthly inspection, test, and maintenance requirements of NFPA 25 can be accomplished by the property owner with minimal training.

2. As a minimum, the property owner is required to advise anyone working on the premises of any hazards present. This notification can be in the form of a material safety data sheet (MSDS) or requirements for confined space entry or hazardous processes.

3. Ladder, flashlight, binoculars, tape measure, 12-in. pipe wrench, crescent wrench, calibrated test gauge, voltmeter, ammeter, tachometer, Pitot tube, stopwatch, and heat lamp are some basic tools needed by an inspector.

4. Hard hat, safety shoes, eye protection, hearing protection, and safety harness should all be used by an inspector. It may be necessary for inspectors to wear special clothing when measuring high voltage. Inspectors should avoid wearing loose clothing.

NFPA 25 Index

A

Accessibility, 4.1.1, A.4.1.1
Adjustments *see* Repairs, reconditioning, replacements, or adjustments
Alarm devices; *see also* Waterflow alarms
  False alarms *see* Supervisory service, notification to
  Fire pumps, 8.1.2, A.8.1.2
  Foam-water sprinkler systems, Table 11.5.1
  Inspection, Table 5.1
  Private fire service mains, Table 7.5.1
  Sprinkler systems, 5.2.6, 5.3.3, Table 5.5.1, A.5.3.3.2, A.5.3.3.5
  Standpipe and hose systems, 6.3.3, Table 6.5.1
  Valves, 13.2.6
  Water mist systems, 12.3.6
  Water spray fixed systems, Table 10.5.1
  Water tanks *see* Water tanks
Alarm receiving facility (definition), 3.3.1
Alarm valves, Table 13.1, 13.4.1, A.13.1, A.13.4.1.1, A.13.4.1.2
Antifreeze sprinkler systems, 5.3.4, A.5.3.4
  Antifreeze solution, Table 5.1, 5.3.4, Table 5.5.1, A.5.3.4
  Control valve information sign *see* Signs, information
  Definition, 3.6.4.1
Application of standard, 1.3, A.1.3
Approved (definition), 3.2.1, A.3.2.1
Asian clams, obstruction by, D.2.4, D.4.4
Authority having jurisdiction (definition), 3.2.2, A.3.2.2
Automatic detection equipment, Table 10.1, 10.2.3, 10.3.4.1, 10.3.4.2, 10.4.2, Table 11.1, 11.2.2, 11.3.2.4, 12.3.6, A.10.3.4.1
  Component replacement action requirements, Table 5.5.1, Table 10.5.1, Table 11.5.1
  Definition, 3.3.2, A.3.3.2
Automatic operation (definition), 3.3.3

B

Backflow prevention assemblies, Table 13.1, 13.6, A.13.1, A.13.6.1.2, A.13.6.2.1; *see also* Reduced-pressure principle backflow prevention assemblies (RPBA)
Ball valves, Table 11.5.1, A.13.1
Biological growth, obstruction caused by, D.2.4; *see also* Microbiologically influenced corrosion (MIC)
Bladder tank proportioners, Table 11.1, 11.2.9.5.2, 11.4.4, Table 11.5.1, A.11.2.9.5.2, A.11.4.4.2
  Definition, 3.3.27.1, Fig. A.3.3.27.1
Buildings, inspection of, Table 5.1, 5.2.5

C

Calcium carbonate, obstruction by, D.2.5, D.4.5
Changes
  In hazard, 4.1.6, 12.1.5
  In occupancy, use, process, or materials, 4.1.5, 12.1.5, A.4.1.5
Check valves, Table 9.1, Table 13.1, 13.4.2, A.13.1; *see also* Double check valve assembly (DCVA)
Combined standpipe and sprinkler systems (definition), 3.6.1
Commercial-type cooking equipment, sprinklers and nozzles for, 5.4.1.9
Concealed spaces, sprinklers in, 5.2.1.1.4, 5.2.2.3, A.5.2.1.1.4, A.5.2.2.3
Confined spaces, entry to, 4.8.1
Connections
  Fire department *see* Fire department connections
  Flushing, D.5.1
  Hose *see* Hose connections
Controllers, pump, 8.1.6, A.8.1.6, C.3
Control valves, 13.3, A.13.3.1 to A.13.3.3.5
  Component action requirements, 5.5.1.2, 6.5.1.2, 7.5.1.2, 9.6.1.2, 10.5.1.2, 11.5.3
  Definition, 3.5.1, A.3.5.1
  Impairments, signs indicating, 10.1.4.2
  Information signs *see* Signs, information
  Inspection, Table 5.1, Table 6.1, Table 9.1, Table 13.1, 13.3.2, 13.5.2.1, A.13.3.2.2
  Maintenance, Table 9.1, 12.3.3, 13.3.4
  Obstruction investigations, closure during, D.3
  Testing, 13.3.3, A.13.3.3.2, A.13.3.3.5
Conventional pin rack (definition), 3.3.16.1, Fig. A.3.3.16.1
Corrective action, 4.1.4, 4.3, A.4.1.4
Corrosion, 5.2.2.1, 6.2.3, 10.2.4.2, 10.2.41, 10.3.7.2.1, Table 11.1; *see also* Microbiologically influenced corrosion (MIC)
  Corrosion-resistant sprinklers, A.5.4.1.8
    Definition, 3.3.30.1
  Corrosive atmospheres or water supplies, 5.3.1.1.2, A.5.3.1.1.2
  Obstruction due to corrosion products, D.1, D.2.1, D.2.6, D.5.2

D

Deficiency (definition), 3.3.4
Definitions, Chap. 3
Deluge sprinkler systems, Table 5.1, 5.2.4.2

Definition, 3.6.4.2
Foam-water sprinkler and foam-water spray systems (definitions), 3.4; *see also* Foam-water sprinkler systems
Deluge valves, Table 13.1, 13.4.3, A.13.1, A.13.4.3.2.1 to A.13.4.3.3.3
Definition, 3.5.2
Foam-water sprinkler systems, 11.2.1
Water spray fixed systems, Table 10.1, 10.2.1.5, 10.2.2
Detection equipment, automatic *see* Automatic detection equipment
Discharge devices
Definition, 3.3.5
Foam-water, Table 11.1, 11.2.5, Table 11.5.1, A.11.2.5
Discharge patterns
Deluge and preaction systems, 13.4.3.2.2.3
Foam-water sprinkler systems, 11.3.2.6
Water spray fixed systems, 10.3.4.3, A.10.3.4.3
Discharge time
Foam-water sprinkler systems, 11.3.2.5
Water spray fixed systems, 10.3.4.2
Double check detector assembly (DCDA), 13.6.1.1, A.13.1
Double check valve assembly (DCVA), 13.6.1.1, A.13.1
Definition, 3.3.6
Drainage
Foam-water sprinkler systems, Table 11.1, 11.2.8
Water spray fixed systems, 10.2.8, 10.3.7.2
Drains
Component replacement action requirements, Table 5.5.1, Table 6.5.1, Table 10.5.1, Table 11.5.1
Low point, Table 5.1, 10.3.7.2
Main *see* Main drains
Sectional (definition), 3.3.7.2
Drip valves, Table 11.5.1, A.13.1
Driver, pump, 8.1.5
Dry barrel hydrants, Table 7.1, 7.2.2.4, 7.3.2.3, 7.3.2.6
Definition, 3.3.9.1, Fig. A.3.3.9.1
Dry pipe sprinkler systems
Control valve information sign *see* Signs, information
Definition, 3.6.4.3
Gauges, Table 5.1, 5.2.4.2, 5.2.4.4, A.5.2.4.4
Inspection, Table 5.1
Maintenance, 5.4.2, A.5.4.2
Obstructions, 14.2.2, A.14.2.2, D.3.2, D.3.3, D.4.1, D.5.2, D.5.4
Dry pipe valves, Table 13.1, 13.4.4, A.13.1, A.13.4.4.1.2.3 to A.13.4.4.3.2
Dry sprinklers, 5.3.1.1.1.5, 5.4.1.4.2.1, A.5.3.1.1.1.5
Definition, 3.3.30.2

E

Early suppression fast-response (EFSR) sprinkler (definition), 3.3.30.3
Electrical safety, 4.9, A.4.9
Emergency impairments, 15.6
Definition, 3.3.17.1

Exposed piping, private fire service mains
Inspection, Table 7.1, 7.2.2.1
Testing, Table 7.1, 7.3.1, A.7.3.1
Extended coverage sprinkler (definition), 3.3.30.4

F

Fall protection, 4.8.2
False alarms *see* Supervisory service, notification to
Fire department connections, Table 5.1, 5.1.1, 6.3.2.1, 9.1.1, Table 13.1, 13.7
Component replacement action requirements, Table 10.5.1, Table 11.5.1
Definition, 3.3.8
Flushing outlets, used as, D.5.1
Impairments, signs indicating, 10.1.4.2
Fire hose, maintenance of, 7.1.2; *see also* Standpipe and hose systems
Fire hydrants, Table 7.5.1
Definition, 3.3.10, A.3.3.10
Dry barrel *see* Dry barrel hydrants
Maintenance, Table 7.1, 7.4.2, A.7.4.2.2
Monitor nozzle (definition), 3.3.9.2, Fig. A.3.3.9.2
Testing, Table 7.1, 7.3.2
Wall *see* Wall hydrants
Wet barrel *see* Wet barrel hydrants
Fire pumps, Chap. 8
Auxiliary equipment, 8.1.2, A.8.1.2
Component replacement action requirements, Table 10.5.1
Controllers, 8.1.6, Table 8.1, Table 8.6.1, A.8.1.6, C.3
Definition, 3.6.2
Diesel engine systems, Table 8.1, 8.2.2(4), 8.3.1.3, 8.3.2.2(3), Table 8.5.3, Table 8.6.1
Driver, 8.1.5, Table 8.6.1, Table 10.1
Electrical system, Table 8.1, 8.2.2(3), 8.3.1.2, 8.3.2.2(2), Table 8.5.3, Table 8.6.1, Table 10.1
Energy source, 8.1.4, 8.3.4.1
Impairments, 5.3.3.4, 8.1.7
Inspection, 8.1, 8.2, 8.4.1, Table 8.5.3, Table 10.1, A.8.1, A.8.2.2
Maintenance, 8.1, 8.5, Table 10.1, 12.3.3, 12.3.5, A.8.1, A.8.5.1
Obstruction investigation, 14.2.2, A.14.2.2, D.3
Problems, possible causes of, Annex C
Reports, 8.4, A.8.4.2
Steam systems, 8.2.2(5), 8.3.2.2(4), Table 8.6.1, Table 10.1, A.8.2.2(5)
Supervisory service, notification to, 8.1.8
Testing, 4.6.3, 8.1, 8.3, 8.4, Table 8.5.3, Table 10.1, A.8.1, A.8.3, A.8.4.2
Annual, 8.3.3, A.8.3.3.1 to A.8.3.3.8
Component replacement testing requirements, 8.6
At each flow condition, Table 8.1, 8.3.3.1, 8.3.3.2(2), A.8.3.3.1
At no-flow condition (churn), Table 8.1, 8.3.1, 8.3.3.2(1)
Results and evaluation, 8.3.5, A.8.3.5.1 to A.8.3.5.4
Weekly, 8.3.2, A.8.3.2.2

Fire watch, 15.5.2(3)(b), A.15.5.2(3)(b)
Fittings see Piping
Fixed nozzle systems see Water spray fixed systems
Flow tests, Table 6.1, 6.3.1, Table 7.1, 7.3.1, 7.3.2, 13.5.1.2, 13.5.2.2, 13.5.2.3, 13.5.3.2, 13.5.3.3, 13.5.4.2, A.6.3.1.1, A.7.3.1, A.13.5.1.2, A.13.5.2.2, A.13.5.4.2
Flushing procedures, Table 10.1, 14.2.2, 14.2.3.1, 14.2.4, A.14.2.2, A.14.2.4, D.4.2, D.5
Foam concentrates
   Definition, 3.3.10
   Samples, 11.2.10
   Testing, 11.3.5
Foam discharge devices, 11.2.5, A.11.2.5
   Definition, 3.3.11
Foam-water spray system (definition), 3.4.1; see also Foam-water sprinkler systems
Foam-water sprinkler systems, Chap. 11
   Component action requirements, 11.5
   Definition, 3.4.2
   Impairments, 11.1.4
   Inspection, 11.1, 11.2, A.11.2.5 to A.11.2.9.5.6(2)
     Tanks, 11.2.9.4, A11.2.9.4
   Maintenance, 11.1, 11.4, A.11.4
   Proportioners see Proportioners
   Return to service after testing, 11.3.6
   Supervisory service, notification to, 11.1.5
   Testing, 11.1, 11.3, A.11.3
Freezers, systems protecting see Ice obstruction prevention

G
Gauges
   Component replacement action requirements, Table 5.5.1, Table 6.5.1, Table 7.5.1, Table 10.5.1, Table 11.5.1
   Fire pumps, 8.1.2, 8.2.2(5), 8.3.2.2, A.8.1.2, A.8.2.2(5), A.8.3.2.2
   Sprinkler systems, Table 5.1, 5.2.4, 5.3.2, A.5.2.4.1, A.5.2.4.4, A.5.3.2
   Standpipe and hose systems, 6.3.1.5.2
   Valves, 13.2.7
   Water tanks, Table 9.1, 9.3.6, Table 9.6.1

H
Hangers
   Component replacement action requirements, Table 10.5.1
   Foam-water sprinkler systems, Table 11.1, 11.2.4, Table 11.5.1
   Sprinkler systems, Table 5.1, 5.2.3, A.5.2.3
   Standpipe and hose systems, Table 6.5.1
   Water mist systems, 12.3.5
   Water spray fixed systems, 10.2.4.2, A.10.2.4.2
Hazardous materials, 4.8.4, A.8.4
Hazards, 4.1.6, 4.8.3
Horizontal rack (hose storage) (definition), 3.3.16.2, A.3.3.16.2
Hose see Fire hose; Standpipe and hose systems
Hose connections, 5.1.4
   Definition, 3.3.12

Pressure-reducing valves, Table 13.1, 13.5.2, A.13.5.2.2
Standpipe and hose systems, Table 6.1, Table 6.2.2
Hose houses, Table 6.5.1, Table 7.1, 7.2.2.7, Table 7.5.1
   Definition, 3.3.13, Figs. A.3.3.13(a) to (c)
Hose nozzles, Table 6.1, Table 6.2.2
   Definition, 3.3.15
Hose reel (definition), 3.3.16.3, A.3.3.16.3
Hose station (definition), 3.3.15
Hose storage devices, Table 6.1, Table 6.2.2, Table 6.5.1
   Conventional pin rack (definition), 3.3.16.1
   Horizontal rack (definition), 3.3.16.2
   Hose reel (definition), 3.3.16.3
   Semiautomatic hose rack assembly (definition), 3.3.16.4
Hose valves, 13.5.6, A.13.5.6.2.1
   Definition, 3.5.3
Hydrants, fire see Fire hydrants
Hydraulic nameplate, Table 5.1, 5.2.7, A.5.2.7
Hydrostatic tests, Table 6.1, 6.3.2, 12.2.6, A.6.3.2.2

I
Ice obstruction prevention, 5.2.4.4, 14.3, A.5.2.4.4
Impairments, 4.2, Chap. 15; see also System shutdown
   Coordinator, 15.2
   Definition, 3.3.17, A.3.3.17
   Emergency, 15.6
     Definition, 3.3.17.1
   Equipment involved, 15.4
   Fire pumps, 5.3.3.4, 8.1.7
   Foam-water sprinkler systems, 11.1.4
   Preplanned programs, 15.5, A.15.5
     Definition, 3.3.17.2
   Private fire service mains, 7.1.3
   Restoring systems to service, 12.1.6, 15.7
   Sprinkler systems, 5.1.2, 5.3.3.4
   Standpipe and hose systems, 6.1.2
   Tag system, 15.3, A.15.3.1, A.15.3.2
   Water mist systems, 12.1.3.1
   Water spray fixed systems, 10.1.4, A.10.1.4
   Water tanks, 9.1.2
Information signs see Signs, information
In-line balanced pressure proportioners, Table 11.1, 11.2.9.5.5, 11.4.7, A.11.2.9.5.5(1), A.11.2.9.5.5(2)
   Definition, 3.3.27.2, Fig. A.3.3.27.2
Inspection, testing, and maintenance service (definition), 3.3.19
Inspections, 4.5, A.4.5; see also Impairments; Supervisory service, notification to
   Backflow prevention assemblies, 13.6.1, A.13.6.1.2
   Definition, 3.3.18
   Fire department connections, 13.7
   Fire pumps, Table 8.1, 8.2, A.8.2.2
   Foam-water sprinkler systems, Table 11.1, 11.2, A.11.2.5 to A.11.2.9.5.6(2)
   Forms for, Annex B
   Owner/occupant responsibilities, 4.1.1 to 4.1.4, 4.8.4.2, A.4.1.1 to A.4.1.4

Performance-based program, 4.6.1.1.1, A.4.6.1.1.1
Private fire service mains, 7.2, A.7.2.2
Sprinkler systems, 5.1, 5.2, A.5.2
Standpipe and hose systems, Table 6.1, 6.2
Steam system conditions, 8.2.2(5), A.8.2.2(5)
Valves, Table 5.1, 13.1
   Alarm, 13.4.1, A.13.4.1.1, A.13.4.1.2
   Automatic tank fill valves, 9.5.1
   Check, 13.4.2.1
   Deluge, 13.4.3.1
   Dry pipe/quick opening devices, 13.4.4.1, A.13.4.4.1.2.3
   Hose, 13.5.6.1, A.13.5.6.1
   Preaction, 13.4.3.1
   Pressure-reducing, 13.5.1, 13.5.2.1, 13.5.3.1, 13.5.4.1, 13.5.5.1, A.13.5.1.2, A.13.5.4.1
   Pressure relief, 13.5.7
   Water mist systems, 12.2, A.12.2.4
   Water spray fixed systems, Table 10.1, 10.2, A.10.2.4 to A.10.2.7

L

Large drop sprinklers (definition), 3.3.30.5
Line proportioners, Table 11.1, 11.2.9.5.3, 11.4.5, A.11.2.9.5.3(1), A.11.2.9.5.3(2)
   Definition, 3.3.27.3, Fig. A.3.3.27.3
Listed (definition), 3.2.3, A.3.2.3

M

Main drains
   Component replacement action requirements, Table 5.5.1, Table 6.5.1, Table 10.5.1
   Definition, 3.3.7.1
   Test, Table 5.1, 5.5.1.2, Table 6.1, 6.3.1.5, 6.5.1.2, 7.5.1.3, 9.6.1.2, 10.3.7.1, 10.5.1.2, Table 11.5.1, 11.5.3, Table 13.1, 13.2.5, A.13.2.5
Mainline strainers
   Foam-water sprinkler systems, 11.2.7.1
   Private fire service mains, 7.2.2.3, Table 7.5.1, A.7.2.2.3
   Water spray fixed systems, 10.2.1.7, 10.2.7, A.10.2.7
Mains
   Private fire service *see* Private fire service mains
   Yard, 14.2.2, A.14.2.2, D.3, D.5.1
Maintenance, 4.6.3, 4.7, 6.4, A.4.7; *see also* Impairments
   Backflow prevention assemblies, 13.6.3
   Definition, 3.3.20
   Fire pumps, 8.5, A.8.5.1
   Foam-water sprinkler systems, 11.4, A.11.4
   Forms for, Annex B
   Owner/occupant responsibilities, 4.1.1 to 4.1.4, 4.8.4.2, A.4.1.1 to A.4.1.4
   Performance-based program, 4.6.1.1.1, A.4.6.1.1.1
   Private fire service mains, 7.4, A.7.4.2.2
   Sprinkler systems, 5.1, 5.4, A.5.4.1.1 to A.5.4.4
   Standpipe and hose systems, Table 6.1, 6.2.2, 6.2.3, 6.4
   Valves, Table 5.1, Table 6.1, 12.3.3, 12.3.5, 13.1
      Alarm, 13.4.1.3
      Check, 13.4.2.2
      Control, Table 6.1, 13.3.4
      Deluge, 13.4.3.3, A.13.4.3.3.3
      Dry pipe/quick- opening devices, 13.4.4.3, A.13.4.4.3.2
      Hose, 13.5.6.3
      Preaction, 13.4.3.3, A.13.4.3.3.3
      Pressure relief, 13.5.8
   Water mist systems, Table 12.2.2, 12.3, A.12.3.10
   Water spray fixed systems, Table 10.1, 10.2, A.10.2.4 to A.10.2.7
   Water tanks, Table 9.1, 9.4
Manual operation
   Definition, 3.3.21
   Foam-water sprinkler systems, Table 11.1, 11.3.4, Table 11.5.1
   Preaction and deluge valves, 13.4.3.2.8
   Standpipe and hose systems, 6.3.2.1, 6.3.2.2.1
      Definition, 3.3.31.2
   Water spray fixed systems, 10.3.6
Marine sprinkler systems, 5.4.4, A.5.4.4
Master pressure reducing valve, 13.5.4, A.13.5.4.1 to A.13.5.4.3
   Definition, 3.5.5.1, A.3.5.5.1
Materials, changes in, 4.1.5, 12.1.4.2, A.4.1.5
Measurement, units of, 1.4, A.1.4
Microbiologically influenced corrosion (MIC), D.1, D.2.1, D.2.6, D.2.7
Monitor nozzle hydrant (definition), 3.3.9.2, Fig. A.3.3.9.2
Monitor nozzles
   Component replacement action requirements, Table 7.5.1
   Definition, 3.3.22.1, Fig. A.3.3.22.1(a), Fig. A.3.22.1(b)
   Inspection, Table 7.1, 7.2.2.6
   Maintenance, Table 7.1, 7.4.3
   Testing, Table 7.1, 7.3.3
Multiple systems, testing, 10.3.5, 11.3.3, 13.4.3.2.7

N

Nameplate, hydraulic, 5.2.7, A.5.2.7
Nozzles
   Hose *see* Hose nozzles
   Monitor *see* Monitor nozzles
   Sprinkler (definition), 3.3.30.6
   Water mist, 12.3.5, 12.3.6, 12.3.10, A.12.3.10
   Water spray *see* Water spray nozzles

O

Obstructions
   Ice, 5.2.4.4, 14.3, A.5.2.4.4
   Investigation, Table 5.1, Chap. 14, Annex D
      Procedure, D.3
      Sources of obstruction, D.2
   Prevention, 10.2.6.2, 14.2, 14.3, A.10.2.6.2, A.14.2, D.4
Occupancy, changes in, 4.1.5, 12.1.4.2, 12.1.5, A.4.1.5
Old-style/conventional sprinklers, 5.4.1.1.1, A.5.4.1.1.1
   Definition, 3.3.30.7
Open sprinkler (definition), 3.3.30.8

Operation
Automatic (definition), 3.3.3
Manual *see* Manual operation
Orifice plate proportioning, 11.2.9.5.6, A.11.2.9.5.6(1), A.11.2.9.5.6(2)
Definition, 3.3.23
Ornamental/decorative sprinkler (definition), 3.3.30.9
Owner/occupant responsibilities, 4.1, 4.8.4.2, 12.1, A.4.1.1 to A.4.1.5

P
Pendent sprinklers, D.2.5, D.5.2
Definition, 3.3.30.10
Piers, sprinklers for, 5.4.1.1.2
Piping
Flushing, Table 10.1, 14.2.2, 14.2.3.1, 14.2.4, A.14.2.2, A.14.2.4, D.4.2, D.5
Foam-water sprinkler systems, Table 11.1, 11.2.3, Table 11.5.1
Obstructions *see* Obstructions
Private fire service mains, Table 7.5.1
  Exposed, Table 7.1, 7.2.2.1, 7.3.1, A.7.3.1
  Underground, Table 7.1, 7.2.2.2, 7.3.1, A.7.3.1
Sprinkler systems, Table 5.1, 5.2.2, Table 5.5.1, A.5.2.2, D.3 to D.5
Standpipe and hose systems, Table 6.1, Table 6.2.2, Table 6.5.1
Water spray fixed systems, Table 10.1, 10.2.4, 10.2.6, Table 10.5.1, A.10.2.4, A.10.2.6.2
Preaction sprinkler systems
Control valve information sign *see* Signs, information
Definition, 3.6.4.4
Gauges, Table 5.1, 5.2.4.2, 5.2.4.4, A.5.2.4.4
Inspection, Table 5.1
Obstructions, D.4.1
Preaction valves, Table 13.1, 13.4.3, A.13.4.3.2.1 to A.13.4.3.3.3
Preplanned impairments, 15.5, A.15.5
Definition, 3.3.17.2
Pressure control valves
Definition, 3.5.4
Standpipe and hose systems, Table 6.1
Pressure readings
Deluge and preaction systems, 13.4.3.2.6
Foam-water sprinkler systems, 11.3.2.7, A.11.3.2.7
Water spray fixed systems, 10.3.4.4
Pressure-reducing valves, Table 6.1, Table 13.1, 13.5, 13.5.5, A.13.5.1.2 to A.13.5.6.2.2; *see also* Relief valves (fire pump)
Definition, 3.5.5
Fire pumps, Table 13.1
Hose connection, Table 13.1, 13.5.2, A.13.5.2.2
Hose rack assembly, Table 13.1, 13.5.3
Master pressure reducing valve, 13.5.4, A.13.5.4.1 to A.13.5.4.3
  Definition, 3.5.5.1, A.3.5.5.1
Sprinkler, Table 13.1, 13.5.1, A.13.5.1.2
Standpipe and hose systems, 6.3.1.4

Pressure regulating devices, Table 6.1, 6.3.1.4; *see also* Pressure control valves; Pressure-reducing valves; Relief valves (fire pump)
Definition, 3.3.24, A.3.3.24
Pressure relief valves *see* Relief valves
Pressure restricting devices (definition), 3.3.25
Pressure vacuum vents, Table 11.1, 11.4.8
Definition, 3.3.26, Fig. A.3.3.26
Private fire service mains, Chap. 7
Component action requirements, 7.5
Definition, 3.6.3, A.3.6.3
Impairments, 7.1.3
Inspection, 7.1, 7.2, A.7.2.2
Maintenance, 7.1, 7.4, A.7.4.2.2
Supervisory service, notification to, 7.1.4
Testing, 7.1, 7.3, A.7.3.1
Process, changes in, 4.1.5, 12.1.4.2, A.4.1.5
Proportioners, Table 11.1, 11.2.9, Table 11.5.1, A.11.2.9
Definitions, 3.3.27.1 to 3.3.27.5, Fig. A.3.3.27.1
Pumps *see* Fire pumps
Purpose of standard, 1.2, A.1.2

Q
Qualified (definition), 3.3.28
Quick- opening devices, Table 13.1, 13.4.4, A.13.4.4.1.2.3 to A.13.4.4.3.2
Quick-response early suppression (QRES) sprinklers (definition), 3.3.30.11
Quick-response extended coverage sprinklers (definition), 3.3.30.12
Quick-response (QR) sprinklers, 5.4.1.3, A.5.4.1.3
Definition, 3.3.30.13

R
Recessed sprinklers (definition), 3.3.30.14
Records, 4.4, A.4.4.1, A.4.4.3
Owner/occupant responsibilities, 4.4.3, A.4.4.3
Valves, 13.2.8
Reduced-pressure detector assemblies (RPDA), 13.6.1.2, A.13.1
Reduced-pressure principle backflow prevention assemblies (RPBA), 13.6.1.2, A.13.1
Definition, 3.3.29
References, Chap. 2, Annex E
Relief valves (fire pump)
Circulation relief, 8.3.3.2(1), Table 13.1, 13.5.7.1
Pressure relief, 8.3.3.2(1), 8.3.3.3, Table 13.1, 13.5.7.2, A.8.3.3.3
Repairs, reconditioning, replacements, or adjustments, 4.1.4, A.4.1.4; *See also* Maintenance
Fire department connections, 13.7.3
Fire pumps, 8.6.1
Foam-water sprinkler systems, 11.5.1
Private fire service mains, 7.5.1
Sprinklers, 5.4.1.1, A.5.4.1.1
Sprinkler systems, 5.5.1

Standpipe and hose systems, 6.5.1
Water mist systems, 12.1.3
Water spray fixed systems, 10.5.1
Water storage tanks, 9.6.1
Residential sprinkler (definition), 3.3.30.15
Response time
Foam-water sprinkler systems, 11.3.2.4
Water spray fixed systems, 10.3.4.1, 10.4.5, A.10.3.4.1

S

Safety, 4.8, 4.9, A.4.8.4, A.4.9
Scope of standard, 1.1, A.1.1
Sectional drains (definition), 3.3.7.2
Seismic braces, Table 5.1, 5.2.3, Table 6.5.1, Table 10.5.1, Table 11.5.1, A.5.2.3
Semiautomatic hose rack assembly (definition), 3.3.16.4, Fig. A.3.3.16.4
Shall (definition), 3.2.4
Should (definition), 3.2.5
Shutdown, system see Impairments; System shutdown
Shutoff valves, 4.1.7
Signs, information, 4.1.8, Table 6.5.1, Table 7.5.1, 10.1.4.2, Table 10.5.1, Table 11.5.1
Special sprinklers, 5.4.1.3, A.5.4.1.3
Definition, 3.3.30.16
Spray coating areas, sprinklers protecting, 5.4.1.7
Spray sprinklers, 5.4.1.1.1, A.5.4.1.1.1
Definition, 3.3.30.17
Standard spray sprinkler (definition), 3.3.30.18
Sprinklers; see also Dry sprinklers; Pendent sprinklers; Quick-response (QR) sprinklers
Component replacement action requirements, Table 5.5.1
Concealed spaces, in, 5.2.1.1.4, A.5.2.1.1.4
Corrosion-resistant, A.5.4.1.8
Definition, 3.3.30.1
Definitions, 3.3.30.1 to 3.3.30.19
Inspection, Table 5.1, 5.2.1, A.5.2.1.1, A.5.2.1.2
Maintenance, 5.4.1, A.5.4.1.1 to A.5.4.4
Nozzles (definition), 3.3.30.6
Old-style, 5.4.1.1.1, A.5.4.1.1.1
Definition, 3.3.30.7
Spare, Table 5.1, 5.2.1.3, 5.4.1.4, 5.4.1.5, A.5.4.1.4
Spray, 5.4.1.1.1, A.5.4.1.1.1
Definition, 3.3.30.17
Testing, Table 5.1, 5.3.1, A.5.3.1
Sprinkler systems, Chap. 5; see also Foam-water sprinkler systems
Combined standpipe and sprinkler (definition), 3.6.1
Component action requirements, 5.5
Definition, 3.6.4, A.3.6.4
Flushing see Flushing procedures
Impairments, 5.1.2
Inspection, 5.1, 5.2, A.5.2
Installation, 5.4.3, 5.5.1.1, A.5.4.3
Maintenance, 5.1, 5.4, A.5.4.1.1 to A.5.4.4
Marine systems, 5.4.4, A.5.4.4
Obstructions, 14.2.2, A.14.2.2

Piping see Piping
Supervisory service, notification to, 5.1.3
Testing, 5.1, 5.3, 5.4.3, A.5.3.1 to A.5.3.4.1, A.5.4.3
Upright (definition), 3.3.30.19
Standard balanced pressure proportioners, Table 11.1, 11.2.9.5.4, 11.4.6, A.11.2.9.5.4(1), A.11.2.9.5.4(2)
Definition, 3.3.27.4, Fig. A.3.3.27.4
Standard (definition), 3.2.6
Standard pressure proportioners, 11.2.9.5.1, 11.4.3, A.11.2.9.5.1, A.11.4.3.2
Definition, 3.3.27.5, Fig. A.3.3.27.5
Standard spray sprinkler (definition), 3.3.30.18
Standpipe and hose systems, Chap. 6
Alarm devices, 6.3.3
Combined sprinkler and standpipe system (definition), 3.6.1
Component action requirements, 6.5
Components, 6.2.1, Table 6.2.2
Definition, 3.3.30, 3.3.31, A.3.3.30, A.3.3.31
Dry standpipe system, 6.3.2.1
Definition, 3.3.32.1
Impairments, 6.1.2
Inspection, 6.1, 6.2
Maintenance, 6.1, 6.2.2, 6.2.3, 6.4
Manual standpipe system, 6.3.2.1, 6.3.2.2.1
Definition, 3.3.31.2
System types
Class III system (definition), 3.3.31.1.3
Class II system (definition), 3.3.31.1.2
Class I system (definition), 3.3.30.1.1
Testing, 6.1, 6.2.2, 6.3, A.6.3.1.1, A.6.3.2.2
Wet standpipe system, 6.3.2.2.1
Definition, 3.3.31.3
Storage tanks see Water tanks
Strainers
Definition, 3.3.32, A.3.3.32
Foam concentrate, 11.2.7.2
Foam-water sprinkler systems, 11.2.7
Mainline see Mainline strainers
Nozzle, 10.2.1.6
Valves, Table 13.1, A.13.1
Water mist systems, 12.3.3, Table 12.3.4, 12.3.11
Water spray fixed systems, Table 10.1
Water tank automatic tank fill valves, 9.5.2.3
Suction screens, 8.2.2(2), 8.3.3.7, A.8.3.3.7, C.1.2
Obstructions, D.3, D.4.3
Supervision (definition), 3.3.33
Supervisory service, notification to
Fire pumps, 8.1.8
Foam-water sprinkler systems, 11.1.5
Private fire service mains, 7.1.4
Sprinkler systems, 5.1.3
Water tanks, 9.1.3, A.9.1.3
Supervisory signal devices, Table 5.1, Table 6.1, 6.3.3
Supports, 10.2.4.2, Table 11.1, 11.2.4, A.10.2.4.2
System shutdown; see also Impairments
Notification of system shutdown, 4.1.3
Owner/occupant responsibilities, 4.1.3
Restoring systems to service, 4.1.3.2

2008 Water-Based Fire Protection Systems Handbook

Shutdown procedures, 4.1.7
Shutoff valves, location of, 4.1.7
Sprinkler inspection, 5.2.1.1.5, 5.2.2.4
System valves, 13.4, A.13.4.1.1 to A.13.4.4.3.2

T

Tanks, water *see* Water tanks
Temporary fire protection, 15.5.2(3)(c), A.15.5.2(3)(c)
Testing, 4.6, A.4.6.1 to A.4.6.7; *see also* Flow tests; Hydrostatic tests; Impairments; Supervisory service, notification to
   Acceptance, 5.4.3, A.5.4.3
   Backflow prevention assemblies, 13.6.2, A.13.6.2.1
   Definition, 3.3.34
   Fire pumps *see* Fire pumps
   Foam-water sprinkler systems, Table 11.1, 11.3, A.11.3
   Forms for, Annex B
   Hydrants, Table 7.1, 7.3.2
   Main drain test *see* Main drains
   Owner/occupant responsibilities, 4.1.1 to 4.1.4, 4.8.4.2, A.4.1.1 to A.4.1.4
   Performance- based program, 4.6.1.1.1, A.4.6.1.1.1
   Private fire service mains, Table 7.1, 7.3, A.7.3.1
   Sprinklers, Table 5.1, 5.3.1, A.5.3.1
   Sprinkler systems, 5.1, 5.3, 5.4.3, A.5.3.1 to A.5.3.4.1, A.5.4.3
   Standpipe and hose systems, Table 6.1, 6.2.2, 6.3, A.6.3.1.1, A.6.3.2.2
   Valves, 13.1, 13.2.5, A.13.2.5
      Control, 13.3.3, A.13.3.3.2, A.13.3.3.5
      Deluge, 13.4.3.2, A.13.4.3.2.1 to A.13.4.3.2.11
      Dry pipe/quick- opening devices, 13.4.4.2, A.13.4.4.2.1 to A.13.4.4.2.4
      Hose, 13.5.6.2, A.13.5.6.2
      Preaction, 13.4.3.2, A.13.4.3.2.1 to A.13.4.3.2.11
      Pressure-reducing, 13.5.1, 13.5.2.2, 13.5.2.3, 13.5.3.2, 13.5.3.3, 13.5.5.2, A.13.5.1.2, A.13.5.2.2
      Pressure relief, 13.5.7.2.2
   Water mist systems, 12.2, A.12.2.4
   Water spray fixed systems, Table 10.1, 10.2.1.3, 10.3, 10.4, A.10.3.3 to A.10.3.4.3.1
   Water tanks, 9.1, Table 9.1, 9.3, 9.5.3, Table 10.1, A.9.1, A.9.3.1 to A.9.3.5

U

Ultra-high-speed water spray systems operational tests, 10.4
Underground piping, private fire service mains
   Inspection, 7.2.2.2
   Testing, 7.3.1, A.7.3.1
Units of measurement, 1.4, A.1.4
Upright sprinklers (definition), 3.3.30.19
Use, changes in, 4.1.5, A.4.1.5

V

Valves, Chap. 13; *see also* Alarm valves; Check valves; Control valves; Deluge valves; Dry pipe valves; Hose valves; Pressure control valves; Pressure-reducing valves; Relief valves (fire pump)
   Automatic tank fill valves, 9.5, Table 9.6.1
   Ball, Table 11.5.1, A.13.1
   Check, 13.4.2
   Component action requirements, Table 6.5.1, Table 7.5.1, Table 9.6.1, Table 10.5.1, Table 11.5.1
   Drip, Table 11.5.1, A.13.1
   Flushing, used for, D.5.3
   Gauges, 13.2.7
   Inspection *see* Inspections
   Location, 4.1.7
   Maintenance *see* Maintenance
   Preaction, 13.4.3, A.13.4.3.2.1 to A.13.4.3.3.3
   Protection of, 13.2.3
   Records, 13.2.8
   Return to service after testing, 13.4.3.2.9
   Shutoff, 4.1.7
   System, 13.4, A.13.4.1.1 to A.13.4.4.3.2
   Testing *see* Testing
Valve supervisory devices, Table 5.1, Table 6.1, Table 11.5.1, 13.3.3.5, A.13.3.3.5
   Component replacement action requirements, Table 5.5.1, Table 6.5.1, Table 7.5.1
   Water tanks, Table 9.6.1
Ventilating systems, sprinklers for, 5.4.1.9
Vents, pressure vacuum, 11.4.8

W

Wall hydrants, Table 7.1, 7.2.2.4, 7.3.2.3
   Definition, 3.3.9.3, Fig. A.3.3.9.3
Waterflow alarms, Table 13.1, 13.2.6, A.13.1
   Foam-water sprinkler systems, Table 11.1, 11.3.1.1, Table 11.5.1
   Sprinkler systems, Table 5.1, 5.3.3, Table 5.5.1, A.5.3.3.2, A.5.3.3.5
   Standpipe and hose systems, Table 6.1, 6.3.3, Table 6.5.1
   Water mist systems, 12.3.1.3
   Water spray fixed systems, Table 10.5.1
Water mist systems, Chap. 12
   High pressure cylinders, 12.2.6
   Inspection, maintenance, and testing, 12.2, 12.3, A.12.2.4, A.12.3.10
   Responsibility of owner or occupant, 12.1
   Training, 12.4
Water spray (definition), 3.3.35, A.3.3.35
Water spray fixed systems, Chap. 10; *see also* Fire pumps; Water tanks
   Component action requirements, 10.5
   Definition, 3.6.5
   Foam-water spray system (definition), 3.4.1; *see also* Foam-water sprinkler systems
   Impairments, 10.1.4, A.10.1.4
   Inspection and maintenance procedures, 10.1, 10.2, A.10.2.4 to A.10.2.7
      Automatic detection equipment, 10.2.3, 10.4.2
      Deluge valves, 10.2.1.5, 10.2.2
      Drainage, 10.2.8
      Piping, 10.2.4, 10.2.6, A.10.2.4, A.10.2.6.2

Strainers, 10.2.7, A.10.2.7
Valves, 10.4.4, Chap. 13
Water supply, 10.2.6, A.10.2.6.2
Manual operations, 10.3.6
Return to service after testing, 10.3.7
Supervisory service, notification to, 10.3.2.1
Testing, 10.1, 10.2.1.3, 10.3, 10.4, A.10.3.3 to A.10.3.4.3.1

Water spray nozzles, 5.4.1.9, Table 10.1, 10.2.5, 10.3.4.3, A.10.2.5, A.10.3.4.3
Component replacement action requirements, Table 10.5.1
Definition, 3.3.22.2, A.3.3.22.2

Water supply
Definition, 3.3.36
Foam-water sprinkler systems, 11.2.6, A.11.2.6.2
Obstructions, Annex D
Water spray fixed systems, 10.2.6, 10.3.5, 10.3.7.1, A.10.2.6.2

Water tanks, Chap. 9
Alarm devices, Table 9.1, 9.2.1, 9.2.3, 9.2.4.2, 9.3.3, 9.3.5, Table 9.6.1, A.9.2.1.1, A.9.3.5
Component action requirements, 9.6
Definition, 3.6.6
ESCF, maintenance of, Table 9.1, 9.4.6
Heating systems, Table 9.1, 9.2.3, Table 9.6.1
Impairments, 9.1.2

Inspection, 9.1, 9.2, 9.5.1, Table 10.1, A.9.1, A.9.2.1.1 to A.9.2.6.5
Maintenance, 9.1, 9.4, 9.5.2, Table 12.3.4, A.9.1
Obstruction investigations, D.3
Pressure tanks, Table 9.1, 9.2.2, Table 10.1
Supervisory service, notification to, 9.1.3, A.9.1.3
Testing, 9.1, 9.3, 9.5.3, Table 10.1, A.9.1, A.9.3.1 to A.9.3.5

Wet barrel hydrants, Table 7.1, 7.2.2.5
Definition, 3.3.9.4, Fig. A.3.3.9.4

Wet pipe sprinkler systems
Building inspection, 5.2.5
Definition, 3.6.4.5, A.3.6.4.5
Gauges, 5.2.4.1, A.5.2.4.1
Inspection, Table 5.1
Obstructions, 14.2.2, A.14.2.2, D.2.5, D.3.4, D.5.4

Wharves, sprinklers for, 5.4.1.1.2

Y
Yard mains, 14.2.2, A.14.2.2, D.3, D.5.1

Z
Zebra mussels, obstruction by, D.4.6

Commentary Index

A

Accessibility to equipment, A.4.1.1
Acid-producing bacteria (APB), D.2.6
Additive storage cylinders
 For water mist systems, Table 12.2.2
AFFF. *See* aqueous film-forming foam (AFFF) concentrate
Aircraft hangers
 Monitor nozzles used in, 10.1.1
Air pressure tests
 Dry pipe sprinkler systems, 13.4.4.2.9
Air relief valves
 Fire pumps, 8.1.2(1)(b), Exhibit 8.3
Air supply
 For dry pipe systems, A.5.2.4.4
Air testing
 For confined spaces, 4.8.1
Alarm companies
 Impairment program notification, A.15.5.2(3)(d)
Alarm sensors
 Fire pumps, 8.1.2(4), 8.3.3.3.5
Alarm valves, A.3.6.4.5, A.13.1, 13.2.2, 13.2.6.1, Exhibit 13.9
 Inspection, A.13.4.1.1, A.13.4.1.2, Exhibit 13.14
 Maintenance of, 13.4.1.3.1
 Valve repair kits, 13.4.1.3.1
Alcohol-resistant foam concentrate, 11.1.2.1
Algae, Table 12.2.2
Altitude float, 8.2.2(2)(e)
ANSI Z49.1, *Safety in Welding, Cutting, and Allied Processes*, A.15.5.2(3)(d)
Antifreeze solutions
 Concentrations, A.5.3.4
 For early-suppression fast-response (ESFR) sprinklers, A.5.3.4
 Freezing point of, 5.3.4
Antifreeze sprinkler systems
 Defined, 3.6.4.1
 Draining and remixing solution, A.5.3.4
 Testing, 5.3.4, Exhibit 5.13
Anti-vortex plates
 Water storage tanks, 9.2.6.7, Exhibit 9.11
Approved
 Defined, A.3.2.1, A.3.2.3
Aqueous film-forming foam (AFFF) concentrate, 11.1.2.1, Table 12.2.2
 Additive, 12.2.1.2

Asian clams, D.2.4
ASME B40.1, *Gauges—Pressure Indicating Dial Type—Elastic Element, Grade AA*, A.7.3.1
Authority having jurisdiction (AHJ)
 Defined, A.3.2.2
 Follow-up after impairment, 4.1.3.1
 Impairment program notification, A.15.5.2(3)(d)
 Impairment program review by, Chapter 15
 Responsibilities of, 1.2
Automatic detection equipment
 Defined, A.3.3.2
Automatic pressure recording
 Fire pumps, A.8.3.3.1.2
Auxiliary drains, Exhibit 3.2
 Defined, 3.3.7.2
 In high-rise buildings, 13.2.4
 Information signs for, 4.1.8, Exhibit 4.1

B

Backflow preventers, A.13.1
 Annual fire pump testing, 13.6.2.1.4
 Exercising, 13.2.5.1
 Inspection, 13.6.1
 Spare parts kit, 13.6.2.3.2
Backflow prevention valves, 13.1
 Backflow tests, A.13.6.2.1
 Flow tests, A.13.6.2.1
Backflow tests, A.13.6.2.1
Balanced pressure proportioning, 11.2.9
 Defined, A.3.3.27.2
Ball valves, A.13.1
Binoculars
 Hanger and seismic brace inspection with, 5.2.3
 Pipe inspection with, 5.2.2
 Sprinkler inspection with, 5.2.1.1
Bladder tank proportioner, Exhibit 3.6
 Defined, A.3.3.27.1
Break tanks, Chaper 12.2.2, Chapter 9
Buildings
 Exposure of wet pipe systems to freezing, 5.2.5
 High-rise, 13.2.4
 Revisions, sprinkler system evaluation and, 4.1.5.1
Butterfly valves, A.13.1

C

Calcium carbonate (CaCO3) deposits, D.2.5, Exhibit D.7
Check valves, A.13.1
Chemical scouring, A.14.2.4
Churn
  Fire pump tests at, 8.3.3.1, Exhibit 8.22
Circulation relief valves
  Fire pumps, 8.1.2(1)(d), 8.3.3.2(1)(a), Exhibit 8.5
Classified
  Defined, A.3.2.3
Cold flow, 5.4.1.4.2
Combined sprinkler/standpipe systems, Exhibit 3.14
  Defined, 3.3.31, 3.6.1, Exhibit 3.15
Commercial divers, 9.2.6
Component replacement tables, 5.5
  Fire pumps, 8.6
  Foam-water sprinkler systems, 11.5
  Private fire service mains, 7.5
  Standpipe and hose systems, 6.5
  Water spray fixed sprinkler systems, 10.5
  Water storage tanks, 9.6
Components
  Accessibility of, A.4.1.1
Compressed gas cylinders
  For water mist systems, Exhibit 12.1, Exhibit 12.2, Table 12.2.2
Concealed spaces
  Sprinklers in, A.5.2.1.1.4
Condensate nipple (low point drain), 13.4.4.3, Exhibit 13.20
Conductivity method
  For determining foam concentration, 11.3.5.2
Confined spaces
  Air testing, 4.8.1
  Entry permits, 4.8.1, Exhibit 4.3
  Safety precautions in, 4.8.1
  Warning signs for, 4.8.1, Exhibit 4.2
Contaminants
  On sprinklers, 5.2.1.1.2
Control equipment
  For water mist systems, Table 12.2.2
Control valves, 13.1. *See also* valves
  Electrically supervised, 13.3.1.4
  Shut, sprinkler failure and, A.3.5.1
  Standpipe and hose system, 6.1
  Testing, A.13.3.1.2, 13.3.3.1
  For water mist systems, Table 12.2.2
Corrosion. *See also* microbiologically influenced corrosion (MIC)
  On dry pipe sprinkler systems, 3.6.4.3, 5.3.1.1.1.5, A.5.4.2, Exhibit 5.15
  Galvanized piping and, A.10.2.4
  Maintenance and, 14.2.2
  Minimizing, A.5.4.2
  Sprinkler failure caused by, 14.2.2 Case Study
  On sprinklers, 5.2.1.1.2
  Standpipe and hose systems, 6.2.1

Corrosion-resistant sprinklers, 3.3.30.1, Exhibit 3.8
Coupling alignment
  Fire pumps, 8.3.4.4, Exhibit 8.25
Cycling flow
  Water mist systems, Table 12.2.2

D

Deficiency
  Defined, 3.3.4
Definitions, Chapter 3
Deluge sprinkler systems, Exhibit 3.19
  Defined, 3.4.2, 3.6.4.2
  Foam concentrate pumps for, 11.2.9
  Hydraulic calculations for, 13.4.3.2.6.1
Deluge valves, 3.6.5, Exhibit 13.10, Exhibit 13.16
  External inspection, 13.4.3.1.6
  Externally reset, 13.4.3.1.7.1, Exhibit 13.17
  Features of, 13.4.1
  Low-temperature supervisory signals, 13.4.3.1.2
Design, installation, operation, and maintenance (DIOM) manual
  Water mist systems, 12.1.1.3
Design pressure
  Standpipe and hose systems, A.6.3.1.1, 6.3.1.3
Detection systems
  Water spray fixed sprinkler systems, 10.2.3, Exhibit 10.7
Detector check valves, Exhibit 13.3
Diesel fire pumps
  Annual tests, A.8.3.3.1.2
  Battery electrolyte levels, 8.2.2(4)(k)
  Battery supplies, 8.3.2.2(3)(a)
  Cooling water pressure, 8.3.2.2(1)(a)
  Fire pump controller, 8.2.2(4)(k)
  Fuel contaminants, 8.2.2(4)
  Fuel tanks, Exhibit 8.15
  Heat exchanger, 8.3.2.2(3)(e)
  Overspeed governor, 8.3.3.2(1)(b)
  Tachometers, 8.3.2.2(3)(b), Exhibit 8.20
  Temperature and, 8.2.2(1)(a)
  Testing, 8.3.1.3
Differential-sensing valve relief port, A.13.6.1.2
Directional non-automatic water spray nozzles, A.3.3.22.2, Exhibit 3.4
Discharge outlets
  Obstructions, Annex D
Double check valve assembly, 3.3.6, Exhibit 3.1
Double interlock preaction systems, 3.6.4.4, Exhibit 3.22
  Waterflow alarm testing, 13.2.6.1
Dry barrel (frostproof) hydrants
  Frozen, 7.3.2.3 Case Study
  Inspection, 7.2.2.4
Dry pilot systems
  Water spray fixed sprinkler systems, 10.2.3, Exhibit 10.8
Dry pipe sprinkler systems, Exhibit 3.20
  Air leakage tests, 13.4.4.2.9
  Air supply for, A.5.2.4.4

Corrosion of, 3.6.4.3, 5.3.1.1.1.5, A.5.4.2, Exhibit 5.15
Defined, 3.6.4.3
Draining, A.5.4.2
Failure rates of, 5.3.1.1.1.5
In freezing environments, 5.2.5 Case Study
Impairment programs and, A.15.5.2(3)(c) Case Study
Low point drain (condensate nipple), 13.4.4.3, Exhibit 13.20
Sprinkler types, 3.3.30.2
Water delivery time, 13.2.6.1, 13.4.4.2.2.2, Commentary Table 13.1
Dry pipe system valves, 13.4.4, Exhibit 13.19
Failure to operate, 13.4.4.2.2.1 Case Study
Trip time for, 13.4.4.2.2.2, A.13.4.4.2.2.2(9)
Dry pipe valves, Exhibit 3.21
Defined, 3.6.4.3
Dry standpipe systems
Limitations of, 3.3.31.1

E

Early-suppression fast-response (ESFR) sprinklers, 3.3.30.3, Exhibit 3.10
Antifreeze solutions for, A.5.3.4
Electrical components
Inspection, testing, and maintenance requirements, 1.1.1
Overloaded switches, 4.1.5 Case Study
Electrical safety
Hazard/risk category classification, Commentary Table 4.1
Work practices and procedures, A.4.9
Electric motor-driven fire pumps
Annual tests, A.8.3.3.1.2
Pump speed, 8.3.3.2(2)(b)
Qualifications of persons recording data, 8.3.3.2(2)(a)
Safety precautions, 8.3.3.6
Tachometers, 8.3.3.2(2)(b)
Testing, 8.3.1.2
Electronic water level indicators
Water storage tanks, 9.3(7), Exhibit 9.12
Electronic water temperature sensors
Water storage tanks, 9.3.3, Exhibit 9.13
Elevated tanks, 3.6.6
Embankment-supported fabric suction water tanks, A.9.1, Exhibit 9.8
Maintenance, 9.4.6
Emergency generators
Testing, 8.3.1
Emergency impairments, 3.3.17.2, Chapter 15
Defined, 3.3.17.1, 15.6.1
Private fire service mains, 7.1.3
Enforcement, A.4.1.2, 4.1.3.1
Entry permits
Confined spaces, 4.8.1, Exhibit 4.3
Evaluation
Building revision and, 4.1.5.1
Defined, 3.3.18

By registered design professionals, 3.3.18
Of water-based fire protection systems, 1.1.2
Existing buildings, 1.1.2
Extra hazard areas
Foam-water sprinkler systems for, Chapter 11

F

Fall protection, 4.8.2
False alarms, 4.1.3
Film-forming fluoroprotein (FFFP) foam concentrate, 11.1.2.1
Filters
For water mist systems, Table 12.2.2
Fire department connections, 13.1
With clear access to caps, Exhibit 13.31
Inspection, 13.7.1, Exhibit 13.28
Inspector qualifications, 13.7.2
Missing caps, 13.7.1, Exhibit 13.30
Obstructed, 13.7.1, Exhibit 13.29
Fire departments
Impairment program notification, A.15.5.2(3)(d)
Fire prevention programs, A.15.5.2(3)(c)
Fire protection
Temporary water supply for, A.15.5.2(3)(c)
Fire Protection Handbook, 7.3.1.1
Fire pump controllers, 8.1.6
Diesel fire pumps, 8.2.2(4)(k)
Full-service, 8.1.6
Hazard/risk categories, A.4.9
Limited-service, 8.1.6
As motor control center (MCC), 8.3.3.6
Fire pump pressure relief valves, 13.5.7.2.2, Exhibit 13.27
Fire pumps, Chapter 8
Alarm sensors, 8.1.2(4), 8.3.3.3.5
Annual tests, 8.3.1, 8.3.3.1, A.8.3.3.1.2
Assembly, Exhibit 8.1
Automatic air relief valves, 8.1.2(1)(b), Exhibit 8.3
Automatic pressure recording, A.8.3.3.1.2
Automatic starting, 8.3.1.1
Batter, 8.3.2.2(1)(c)
Circulation relief valves, 8.1.2(1)(d), 8.3.3.2(1)(a), Exhibit 8.5
Component replacement tables, 8.6
Costs, Chapter 8
Coupling alignment, 8.3.4.4, Exhibit 8.25
Defined, 3.6.2
Diesel, 8.2.2(1)(a), 8.2.2(4), 8.3.1.3, 8.3.2.2, A.8.3.3.1.2, 8.3.3.2(1)(b), Exhibit 8.15
Discharge pressure gauge readings, 8.3.2.2(1)(a)
Electrical components, 8.2.2(3)(e)
Electric motor-driven, 8.3.1.2, A.8.3.3.1.2, 8.3.3.2(2)(a), 8.3.3.2(2)(b), 8.3.3.6
Pump speed, 8.3.3.2(2)(b)
Energy sources, 8.1.4, A.8.1.6
Failure of, 3.6.2, 8.3 Case Study
Flowmeters, 8.1.2(2), Exhibit 8.7

Flow tests, A.8.1, A.8.3.3.1.2
Inspection, 8.2.1
Internal components, Exhibit 8.2
Isolating switch, 8.2.2(3)(c)
Low water conditions, A.8.3.3.7
Main relief valve, 8.3.1.4
Maintenance, Chapter 8
Manufacturer's maintenance recommendations, 8.5.3
Measurement procedures, A.8.3.3.1.2
Minimum motor run time, 8.3.2.2(1)(c)
Net fire pump performance, A.8.3.5.3(1)
Operation, A.8.3.3.1.2
Packing glands, A.8.2.2, Exhibit 8.17
Pressure gauges, 8.1.2(1)(c), Exhibit 8.4
Pressure maintenance (jockey) pumps, 8.1.2(6), Exhibit 8.8
Pressure relief valves, 8.3.3.2(1)(b), A.8.3.3.3, 8.3.3.3.2
Pressure sensing line, 8.1.2(2)(6), 8.3.1.1, Exhibit 8.12, Exhibit 8.18
Pump house inspection, 8.2.2
Pump system inspection, 8.2.2(2)(a)
Pump test devices, 8.1.2(2)
Relief valves, 8.1.2(3)
Reverse phase alarm pilot, 8.2.2(3)(d)
Right-angle gear drives, 8.1.2(5)
Running tests, A.8.1
Safety precautions, 8.3.3.6
Settings, A.8.3.3.1.2
Steam-driven, A.8.3.3.1.2
Suction gauges, 8.1.2(1)(c)
Suction line pressure, 8.2.2(2)(c)
Suction reservoir, 8.2.2(2)(3)
Suction supply, 8.1.2
System line pressure, 8.2.2(2)(d)
System suction, 8.3.2.2(1)(a)
Test headers, 8.1.2(2), Exhibit 8.6
Testing frequency, A.4.6.1.1.1, 8.3.1
Testing with pumps on, 5.3.3.4
Transfer switch, 8.2.2(3)(b), Exhibit 8.14
Weekly observations, A.8.3.2.1, Table A.8.3.2.2
Weekly testing, 8.3.1, 8.3.1.1, 8.3.2

Fire watch, A.15.5.2(3)(b)
Fire watch personnel
Training, A.15.5.2(3)(b)
Flanged fittings
Water spray fixed sprinkler systems, A.10.2.4.1
Flow-bypass ("unloader") valves
For water mist systems, Table 12.2.2
Flowmeters
Accuracy of, 8.3.3.1.3
Fire pumps, 8.1.2(2), Exhibit 8.7
Flow tests
Backflow prevention valves, A.13.6.2.1
Fire pumps, A.8.1, A.8.3.3.1.2
Obstruction and, 14.2.3.2
Obstructions and, D.2.5
Private fire service mains, Chapter 7
Standpipe and hose systems, A.6.3.1.1, Exhibit 6.4

Fluoroprotein-foam concentrate, 11.1.2.1
Foam concentrate pumps, 11.2.9
Foam concentrates
Annual testing, A.11.3
Conductivity graphs, Exhibit 11.6
Conductivity method, 11.3.5.2
Factors affecting efficiency, A.11.3
Proportioning, 3.3.27.3
Quantity calculations, A.11.2.9.4
Refractive index method for, 11.3.5.2
Reserve supply of, 3.3.17.1
Shelf-life of, A.11.3
Substitution of, 11.1.2.1
Test connections, A.11.3
Test samples, A.11.3
Types of, 11.1.2.1, Chapter 11
Verifying concentration of, 11.3.5.2
Foam discharge device, 3.3.11, Exhibit 3.3
Foam-water discharge devices, 11.2.5, Exhibit 11.1
Foam-water spray systems, Chapter 11
Discharge devices, 11.2.5
Features of, 11.1.2.1
Foam-water sprinkler systems, Chapter 11
Alterations to, 11.1.3.1
Automatic detection equipment, 11.2.2
Component replacement tables, 11.5
Defined, 3.4.2
Discharge devices, 11.2.5, Exhibit 11.1
Features of, 11.1.2.1
Foam-water test ports, A.11.3.1
Full flow conditions, 11.3.2.3
Hydraulically operated concentrate valves, 11.2.9, Exhibit 11.2
Maintenance tests, A.11.3.1
Notification to supervisory service, 11.1.5
Operational tests, A.11.3, 11.3.2.1, 11.3.2.2
Pressure removal, A.11.2.9.5.1
Proportioning systems, 11.2.9, A.11.2.9.5.6(2)
Test precautions, 11.3.1
Uses of, Chapter 11
Forms
For inspections, testing, and maintenance, 4.4.2
Frostproof hydrants, 7.2.2.4
Fuel tanks
For diesel fire pumps, 8.2.2(4), Exhibit 8.15

G

Galvanized piping, A.10.2.4
Gauges, Exhibit 5.7
Accuracy of, 5.3.2
Inspection, 5.2.4.1, 13.2.7.1
Testing, 5.3.2
Test-quality, A.7.3.1
Glass bulb sprinklers, 5.2.1.1.3, 5.3.1.1.1.2
Grooved fittings, A.10.2.4.1

H

Hangers
　Replacing or refastening, 5.2.3.1
　Visual inspection, 5.2.3
Hazard Communication Standard (OSHA), 4.8.4.2
Hazardous materials, 4.8.4.2
Hazard/risk categories
　Classifications, Commentary Table 4.1
　Fire pump controllers, A.4.9
　Protected by water spray fixed sprinkler systems, 10.1
Header control valves, 8.2.2(2)(a)
Heat exchanger, 8.3.2.2(3)(e)
Heating systems
　Equipment inspection, 5.2.5
　For water storage tanks, 9.2.2
HI 3.6, *Rotary Pump Tests*, A.8.3.3.1.2
High pressure cylinders
　For water mist systems, 12.2.6, Exhibit 12.5
High-rise buildings
　Auxiliary drain lines, 13.2.4
　Sprinkler risers in, 13.2.4
High-temperature zones, 5.2.5, Exhibit 5.8
Hose. *See also* standpipe and hose systems
　Inspection, 6.1
　Inspection frequency, 6.2.1
　Pressure, 6.3.1.4 Case Study
　Repair and replacement, 6.4.1
Hose connections
　Connection cap, 13.5.2.1
　Pressure reducing valves for, 13.5.2
Hose houses, 7.2.2.7
Hose rack assembly, Chapter 6, Exhibit 6.1
　Pressure reducing valves, 13.5.3
Hose valves
　Exercising, A.13.5.6.2.1
　Fire pumps, 8.2.2(2)(a)
　Testing and maintenance qualifications, 13.5.6.2
Host employers
　Defined, 4.8.1
Hot work, A.15.5.2(3)(d)
Hydrant cap and gauge, A.7.3.1, Exhibit 7.7
Hydrant flow tests, A.7.3.1, Exhibit 7.5
　Flow test arrangements, Exhibit 7.8
　Procedure, A.7.3.1
　Verifying outlet coefficient, Exhibit 7.9
Hydrants
　Drainage, 7.3.2.3
　Dry barrel (frostproof), 7.2.2.4, 7.3.2.3 Case Study
　Indicating poles, A.7.4.2.2, Exhibit 7.11
　Outlet coefficients, Exhibit 7.9, Exhibit 7.10
　Snow burial, A.7.4.2.2
　Testing, 7.3.2
　Wet barrel ("California"), 7.2.2.5, Exhibit 7.3
Hydraulically operated concentrate valves, 11.2.9, Exhibit 11.2

Hydraulic nameplates
　Inspection, A.5.2.7, Exhibit 5.9
Hydrostatic tests
　Following system repair, 15.7
　Standpipe and hose systems, 6.3.2.1

I

IBV (indicating butterfly valve) valves, 13.1
Ice obstruction, 14.3, 14.3.1
Impairment coordinators, 4.1.7
　Defined, 15.2.1, 15.6.3
　Tools and equipment, 15.2.1
Impairment notices, 15.7, Exhibit 15.2
　Preplanned impairments, A.145.5.2(3)(d)
Impairment programs. *See also* preplanned impairment programs
　AHJ review of, Chapter 15
　Private fire service mains, 7.1.3
　Restoring systems to service following, 15.7
Impairments, Chapter 15. *See also* emergency impairments; preplanned impairment programs
　Defined, 15.1
　Emergency, 3.3.17.1, 3.3.17.2, Chapter 15
　Fire exposure and, Chapter 15
　Formal notification procedures for, 4.1.3
　Minimizing, 5.4.1.4
　Preplanned, 3.3.17.2, Chapter 15
　Private fire service mains, 7.1.3
　Procedures following, 4.1.3.1, 4.6.3
　Return to service following, 4.2.2
Impairment tags, A.15.3.1, 15.7, Exhibit 15.1
Information signs, 4.1.8, Exhibit 4.1
In-line balanced pressure proportioner
　Defined, A.3.3.27.2
Inspections
　Alarm valves, A.13.4.1.1, A.13.4.1.2, Exhibit 13.14
　Backflow preventers, 13.6.1
　Binoculars for, 5.2.1.1, 5.2.2, 5.2.3
　Deluge valves, 13.4.3.1.6
　Dry barrel (frostproof) hydrants, 7.2.2.4
　Electrical components, 1.1.1
　Exemptions, pipes, A.5.2.2.3
　Fire department connections, 13.7.1, Exhibit 13.28
　Fire pumps, 8.2.1
　Forms for, 4.4.2
　Gauges, 5.2.4.1, 13.2.7.1
　Hangers, 5.2.3
　Heating systems, 5.2.5
　Hose, 6.1, 6.2.1
　Hydraulic nameplates, A.5.2.7, Exhibit 5.9
　Inspection intervals, A.4.5
　Intent of, 1.1.2
　Monitor nozzles, 7.2.2.6, Exhibit 7.4
　For "normal wear and tear," 1.1.2
　Obstructions, 14.2.3.2

OS&Y (outside screw and yoke) valves, 13.6.1.1.1
Performance-based options, A.1.3
Pipe fittings, 5.2.2
Piping, 5.2.2, A.5.2.2.3, 5.2.3
Preaction valves, 13.4.3.1.6
Pressure reducing valves, 13.5.1.1
Pressure-relief valves, 13.5.1.1
Private fire service mains, 7.1, 7.2.1
Property maintenance personnel responsibilities, Chapter 5
Property owners' responsibilities, Chapter 5
Pump house, 8.2.2
Pump system, 8.2.2(2)(a)
Purpose of, 3.3.18
Qualifications for, 4.1.2.2, Chapter 5
Qualified contractors' responsibilities, Chapter 5
Records, A.4.4.1, 4.4.5
Responsibility for, A.4.1.2, Chapter 5
Scope of, 1.1.2
Seismic braces, 5.2.3
Sprinklers, 5.2.1.1, 5.4.1.7.2
Sprinkler systems, Chapter 5
Standpipe and hose systems, 6.1, 6.2.1
System control valves, 13.4.1
Valves, 13.2.2
Water-based fire protection systems, 1.1, A.1.3, A.4.1.2, 4.1.2.1
Water mist systems, 12.1.1.2
Water quality, 13.4.3.1.8
Water spray fixed sprinkler systems, 10.1.2, 10.2.1.1, 10.2.3, Chapter 10
Water storage tanks, 9.2.5.1, 9.2.5.2, 9.2.6, 9.2.6.1.2, 9.2.6.2, 9.27
Wet barrel ("California") hydrants, 7.2.2.5

Insurance carriers
Impairment program notification, A.15.5.2(3)(d)
Intermediate-temperature zones, 5.2.5, Exhibit 5.8
Introduction to Employee Fire and Life Safety, A.15.5.2(3)(d)
Isolating switches, 8.2.2(3)(c)
Isolation valves, 13.3.1.4

J
Jockey (pressure maintenance) pumps, 8.1.2(6)

L
Lever-handled ball valves, Table 12.2.2
Licensing, 3.3.28
Listed
Defined, A.3.2.3
Lock-out/tag-out procedures, A.4.9
Low point drains (condensate nipples), 13.4.4.3, Exhibit 13.20
Information signs for, 4.1.8, Exhibit 4.1
Low-temperature supervisory signals, 13.4.3.1.2

Low water conditions
Fire pumps, A.8.3.3.7
Low water level alarms, 9.3.5

M
Main drain tests, A.3.5.1, Exhibit 3.2, Exhibit 13.6
Backflow preventers and, 13.2.5.1
Following system repair, 15.7
Obstructions and, D.2.5
Pressure readings, 13.2.5.2
Purpose of, 3.3.7.1, A.13.2.5, 13.3.3.4
Maintenance
Alarm valves, 13.4.1.3.1
Corrosion and, 14.2.2
Defined, 3.3.20
Electrical components, 1.1.1
Embankment-supported fabric suction water tanks, 9.4.6
Fire pumps, 8.1.2(6), 8.5.3, Chapter 8, Exhibit 8.8
Foam-water sprinkler systems, A.11.3.1
Forms for, 4.4.2
Hose valves, 13.5.6.2
Nozzles, 10.2.5.1, 10.2.5.2
Performance-based options, A.1.3
Preventive, water mist systems, 12.3.2, Table 12.2.2
Property maintenance personnel responsibilities, Chapter 5
Property owners' responsibilities, Chapter 5
Qualifications for, 4.1.2.2, Chapter 5
Qualified contractors' responsibilities, Chapter 5
Records, A.4.4.1, 4.4.5
Responsibility for, A.4.1.2
Spare sprinklers, 5.4.1.4.2
Sprinklers, 5.4.1.1
Sprinkler systems, Chapter 5
Standpipe and hose systems, 6.4.1
Valves, 13.2.1
Water-based fire protection systems, 1.1, A.1.3, A.4.1.2, 4.1.2.1
Water mist systems, 12.1.1.3, 12.3, 12.3.7
Maintenance logs, A.13.4.3.2.11
Maintenance reports, A.13.4.3.2.11
Master pressure reducing valves, 3.5.5.1, 13.5.4
Multiple, 13.5.4
Material Safety Data Sheets (MSDS), 4.8.4.2
Measurement procedures, A.8.3.3.1.2
Measurement units, 1.4.2
Mercury gauges, 9.3(7)
Microbiologically influenced corrosion (MIC), 14.2.1, 14.2.1.2, Exhibit 14.5, Exhibit 14.6
Bacteria responsible for, D.2.6
Chemical scouring for, A.14.2.4
Pinhole leaks and, 14.2.2
Monitor nozzles, 7.3.3.2
Inspection, 7.2.2.6, Exhibit 7.4
Used in aircraft hangers, 10.1.1
Motor control center (MCC), 8.3.3.6

Multiple system manifold, 13.4.3.2.7, Exhibit 13.18
Multiple systems
Simultaneous testing of, 13.4.3.2.7

N

Net fire pump performance, A.8.3.5.3(1)
Net positive suction head (NPSH), Table 12.2.2
NFPA 1, *Uniform Fire Code,* 1.1.2
NFPA 10, *Standard for Portable Fire Extinguishers,* A.15.5.2(3)(c)
NFPA 13, *Standard for the Installation of Sprinkler Systems,* 3.3.7.2, 4.2.2, 5.2.1.2, A.5.2.7, A.5.3.4, A.10.2.4, 13.2.6.1, 13.4.4.2.2.2, A.13.6.2.1, A.15.5.2(3)(d), Chapter 11, D.2.2
NFPA 14, *Standard for the Installation of Standpipe and Hose Systems,* 3.3.31.1.3, A.6.3.1.1, A.13.5.2.2
NFPA 15, *Standard for Water Spray Fixed Systems for Fire Protection,* A.10.1, 10.1.2, A.10.1.2, A.10.2.4.1, 11.1.2.1, Chapter 10
NFPA 16, *Standard for the Installation of Foam-Water Sprinkler and Foam-Water Spray Systems,* 11.1.2.1, 11.2.9
NFPA 20, *Standard for the Installation of Stationary Pumps for Fire Protection,* 5.3.3.4, 8.1.2, A.8.1.2, 8.1.2(3), Chapter 9, Table 12.2.2
NFPA 22, *Standard for Water Tanks for Private Fire Protection,* 3.6.6, A.9.1, Chapter 9
NFPA 24, *Standard for the Installation of Private Fire Service Mains and Their Appurtenances,* 7.1
NFPA 25, *Standard for the Inspection, Testing, and Maintenance of Water-Based Fire Protection Systems*
Administrative requirements, Chapter 1
Application, A.1.3
Enforcement of, A.4.1.2, 4.1.3.1
Purpose, 1.2
Referenced publications, Chapter 2
Requirements, Chapter 4
NFPA 72, *National Fire Alarm Code,* 1.1.1, A.5.3.3.2, 10.1, 10.2.3, 12.1.1.2, 13.2.6.1
Terminology, 3.3.1
NFPA 110, *Standard for Emergency and Standby Power Systems,* 8.3.1, 11.1.2.2
NFPA 291, *Recommended Practice for Fire Flow Testing and Marking of Hydrants,* A.7.3.1
NFPA 409, *Standard on Aircraft Hangars,* 10.1.1
NFPA 750, *Standard on Water Mist Fire Protection Systems,* 12.1.1.3, 12.3.11, Chapter 12
NFPA 5000, *Building Construction and Safety Code,* 1.1.2
NFPA 51B, *Standard for Fire Prevention During Welding, Cutting, and Other Hot Work,* A.15.5.2(3)(b)
NFPA 13D, *Standard for the Installation of Sprinkler Systems in One- and Two-Family Dwellings and Manufactured Homes,* 1.1.2, A.9.1
NFPA 70E, *Standard for Electrical Safety in the Workplace,* A.4.9, 8.3.3.2(2)(a), 8.3.3.6

Nondestructive obstruction examination methods, 14.2.1.1, 14.2.3.3, 14.3.1
Non-interlock preaction systems, 3.6.4.4
Notification procedures
For system impairment, 4.1.3, A.15.5.2(3)(d), 15.7
Nozzles
Automatic spray, A.10.2.5, Exhibit 10.9
With blow-off cap, Exhibit 10.11
Corrosion of, 10.2.5.1
Effective cross-sectional area, A.10.2.6.2
Maintenance, 10.2.5.1, 10.2.5.2
Misaligned, 10.2.5.1, 10.2.5.3
Monitor, 7.2.2.6, 7.3.3.2, Exhibit 7.4
Non-automatic (open) spray, A.10.2.5, Exhibit 10.10
Orifice size, A.7.2.2.3
Water spray fixed sprinkler systems, 10.1.1, 10.1.2, A.10.2.5, 10.2.5.1, Exhibit 10.9, Exhibit 10.10, Exhibit 10.11
Nozzle strainers
Water spray fixed sprinkler systems, 10.2.1.6

O

Obstructions, Annex D, Chapter 14
Flow tests and, 14.2.3.2
Flushing procedures, A.14.2.4
Ice, 14.3, 14.3.1
Investigation frequency, 14.2.3.1
Nondestructive examination methods, 14.2.1.1, 14.2.3.3, 14.3.1
Sediment blockage, 14.2.1 Case Study
Types of, D.2.2, Exhibit 14.2
Visual inspection, 14.2.3.2
Occupational Safety and Health Administration (OSHA), 4.8.1
Hazard Communication Standard, 4.8.4.2
Orifice plate proportioning, 3.3.23
OSHA 29 CFR 1910.252 "Welding, Cutting, and Brazing, General," A.15.5.2(3)(d)
OS&Y (outside screw and yoke) valves, 13.1, A.13.1, Exhibit 13.4
Inoperable, 13.6.1.1.1 Case Study
Inspection, 13.6.1.1.1
Overload
Fire pump tests at, 8.3.3.1, Exhibit 8.24
Overspeed governor
Diesel fire pumps, 8.3.3.2(1)(b)

P

Packing glands
Fire pumps, A.8.2.2, Exhibit 8.17
Pendent sprinklers, 3.3.30.10, Exhibit 3.11
"Permit-Required Confined Spaces," 4.8.1
Personal protective equipment, A.4.9, Exhibit 4.5
Petcock blow-off, A.7.3.1
Pilot sprinklers, 13.4.3.1.5

Pinhole leaks, Exhibit D.11
  MIC and, 14.2.2
Pipe fittings
  Inspection, 5.2.2
Piping
  Corrosion, 6.2.1, Exhibit D.9, Exhibit D.10
  Discharge of water into, 13.4.3.2.2.3(B)
  External loads, 5.2.2.2
  Inspection, 5.2.2, 5.2.3
  Inspection exemptions, A.5.2.2.3
  Leaks, in wet standpipe systems, 6.3.2.1, 6.3.2.2.1
  MIC in, D.2.6, Exhibit D.8
  Obstructions, Annex D, Chapter 14
  Pinhole leaks, Exhibit D.11
  Pitting, Exhibit D.12, Exhibit D.13
  Standpipe and hose systems, 6.2.1, Chapter 6
  Water spray fixed sprinkler systems, 10.1, A.10.2.4, 10.2.4.1
  Welding, A.15.5.2(3)(d)
Pitot-tube-and-gauge assembly, A.7.3.1, Exhibit 7.6
Plant air
  For water mist systems, Table 12.2.2
Plunge test, 5.3.1.1
Plunge test apparatus, 5.3.1.1, Exhibit 5.10
Pneumatic valves, Table 12.2.2
Positive displacement pumps, Exhibit 12.3, Table 12.2.2
Positive pressure proportioning, 11.2.9
Preaction sprinkler systems, Exhibit 3.22
  Supervisory alarm testing, 13.2.6.1
  Supervisory pressure, 13.4.3.1.4
  Types of, 3.6.4.4
Preaction valves, Exhibit 13.16
  External inspection, 13.4.3.1.6
  Features of, 13.4.1
  Low-temperature supervisory signals, 13.4.3.1.2
Preplanned impairment programs, 3.3.17.2, Chapter 15
  Defined, 15.1
  Fire risk and, A.15.5.2(3)(c) Case Study
  Fire watch personnel, A.15.5.2(3)(b)
  Lack of, A.15.5.2(3)(c) Case Study
  Notifications, A.15.5.2(3)(d)
Pressure-control valves
  Application of, 3.5.5
Pressure dampeners
  Fire pump tests, 8.3.3.2(2)(c)
Pressure gauges
  Fire pumps, 8.1.2(1)(c), Exhibit 8.4
Pressure maintenance (jockey) pumps, 8.1.2(6), Exhibit 8.8, Exhibit 8.10
Pressure proportioner, A.11.2.9.4
Pressure reducing valves, 13.1, Exhibit 13.22
  Adjustable, 13.5.1.2.1
  Application of, 3.5.5
  Hose connections, 13.5.2
  Hose rack assembly, 13.5.3
  Inspection, 13.5.1.1
  Master, 13.5.4, Exhibit 13.23
  Partial flow tests, 13.5.1.3
  Test assembly, Exhibit 13.25
  Typical installation, Exhibit 13.26
Pressure regulating valves, 13.1
Pressure relief valves, 13.1, Exhibit 13.24
  Fire pumps, 8.3.3.2(1)(b), A.8.3.3.3, 8.3.3.3.2
  Inspection, 13.5.1.1
  With sight glass, Exhibit 13.27
Pressure restricting devices, Exhibit 3.5
  Defined, 3.3.25
Pressure sensing line
  Fire pumps, 8.1.2(2)(6), 8.3.1.1, Exhibit 8.12, Exhibit 8.18
Pressure switches, 13.2.6.1, Exhibit 13.8, Exhibit 13.11
  Pressure maintenance (jockey) pumps, 8.1.2(6), Exhibit 8.9, Exhibit 8.11
Pressure tanks, 3.6.6
Pressure water storage tanks, Chapter 9
Preventive maintenance
  Water mist systems, 12.3.2, Table 12.2.2
Private
  Defined, Chapter 7
Private fire service mains, 3.6.3, Chapter 7
  Component replacement tables, 7.5
  Engineering analysis, 7.3.1.1
  Flow tests, Chapter 7
  Hydrant flow test equipment, A.7.3.1
  Hydrant flow tests, A.7.3.1, Exhibit 7.5
  Hydraulic gradient analysis, 7.3.1.1
  Impairment programs, 7.1.3
  Inspection, 7.1, 7.2.1
  Notification to supervisory service, 7.1.4
  Public/private equipment boundary, Chapter 7, Exhibit 7.1
  Strainers, A.7.2.2.3, Exhibit 7.2
  Underground and exposed piping flow tests, 7.3.1
  Water main break, 7.2.2.2 Case Study
Property maintenance personnel
  Inspection, testing, and maintenance responsibilities, Chapter 5
Property owners/managers
  Impairment program notification, A.15.5.2(3)(d)
  Inspection, testing, and maintenance responsibilities, Chapter 5
  Responsibilities of, A.4.1.1
  Valve maintenance and, 13.2.1
Protein-foam concentrate, 11.1.2.1
Public fire service mains
  Public/private equipment boundary, Chapter 7, Exhibit 7.1
Pumped water supply, Table 12.2.2
Pump house
  Inspection, 8.2.2
  Valves, 8.2.2(2)
  Ventilating louvers, 8.2.2(1)(b), Exhibit 8.13

Q

Quagga mussels, 14.2.1
  Distribution of (map), Exhibit 14.2

Qualifications
　Of contractors, Chapter 5
　For inspection of fire department connections, 13.7.2
　For inspections, 4.1.2.2, Chapter 5
　For maintenance, 4.1.2.2, Chapter 5
　For maintenance of hose valves, 13.5.6.2
　For recording data for electric motor-driven fire pumps, 8.3.3.2(2)(a)
　For repairing water mist systems, 12.1.3.2
　For servicing water mist systems, 12.1.1.4
　For testing, 4.1.2.2, Chapter 5
　For testing hose valves, 13.5.6.2
　For testing waterflow alarms, A.5.3.3.2
Qualified
　Defined, 3.3.28
Quick-response (QR) sprinklers, 3.3.30.13, Exhibit 3.12
　Testing frequency, 5.3.1.1.1.2

R
Rated capacity
　Fire pump tests at, 8.3.3.1, Exhibit 8.23
Rate-of-rise thermal detection systems, 13.4.3.1.5
Recessed sprinklers, 3.3.30.14, Exhibit 3.13
Records
　For inspections, testing, and maintenance, A.4.4.1, 4.4.5
　Obstructions and, D.2.5
Reduced-pressure backflow prevention devices, A.13.1
Reduced-pressure principle backflow prevention assembly (RPBA), 3.3.29, Exhibit 3.7
Referenced publications, Annex E
　Mandatory, Chapter 2
　Nonmandatory, Chapter 2
Refractive index method
　For determining foam concentration, 11.3.5.2
Registered design professionals, 3.3.18
Relief valves, 8.1.2(3)
Requirements, Chapter 4
Response time index (RTI), 5.3.1.1.1
Reverse phase alarm pilot, 8.2.2(3)(d)
Right-angle gear drives, 8.1.2(5)
Rubber-gasketed fittings, A.10.2.4.1, 10.2.4(5)
Running tests
　Fire pumps, A.8.1

S
Safety
　In confined spaces, 4.8.1
"Safety and Health Regulations for Construction," A.4.9
Sectional control valves, Table 12.2.2
Sectional drains
　Defined, 3.3.7.2
Sectional drain valves, A.13.5.1.2
Sediment, 14.2.1 Case Study
Seismic braces
　Replacing or refastening, 5.2.3.1
　Visual inspection, 5.2.3

Shall
　Defined, 3.2.4
Shop welding, A.15.5.2(3)(d)
Should
　Defined, 3.2.5
Single interlock preaction systems, 3.6.4.4
"Slave" cylinders, Table 12.2.2
Spare sprinkler cabinets, 5.2.1.3, 5.3.1.2, 5.4.1.4, Exhibit 5.6
Spare sprinklers, 5.2.1.3, 5.4.1.4
　Maintenance of, 5.4.1.4.2
Sprinkler risers, 13.2.4
Sprinklers
　Cleaning, 5.2.1.1.2
　Corroded, 5.2.1.1.2, 14.2.2 Case Study
　Corrosion-resistant, 5.2.1.1.2
　Design performance, 5.4.1.1
　Determining age of, 5.3.1.1.1.2
　Distribution pattern obstructions, 5.2.1.2, 5.2.1.2 Case Study, Exhibit 5.4
　Failure of, 5.3.1.2, 5.3.1.3
　Glass bulb, 5.2.1.1.3, 5.3.1.1.1.2
　Harsh environments, A.5.3.1.1.2
　Impairment program and, A.15.5.2(3)(c)
　Incorrectly installed, 5.2.1.1.2
　Inspection, 5.2.1.1, 5.4.1.7.2
　Leaking, 5.2.1.1.2, Exhibit 5.5
　Loaded with contaminants, 5.2.1.1.2
　Maintenance, 5.4.1.1
　Minimum number to test, 5.3.1.2
　Painted, 5.2.1.1.2
　Protecting, 5.4.1.6, 5.4.1.7.2, A.15.5.2(3)(c), Exhibit 5.14
　Relocating, 5.4.1.2
　Replacing, 5.2.1.1.2, 5.3.1.1, 5.3.1.2, 5.4.1.1, 5.4.1.3, Exhibit 5.1
　Spare, 5.2.1.3, 5.4.1.4, 5.4.1.4.2
　Testing, 5.3.1.1, 5.3.1.1.1
　Visibly damaged, 5.3.1.1
Sprinkler systems, Chapter 5
　Building revision and, 4.1.5.1
　Defined, 3.6.4
　Effectiveness of, A.1.2
　Failure of, A.3.5.1
　Frozen burst pipes, A.1.2 Case Study
　Inspection, testing, and maintenance requirements, Chapter 5
　Out-of-service, 4.1.2 Case Study
　Shut valves and, Chapter 13
　Turned off, due to concerns for freezing, 5.2.5 Case Study
　In unheated spaces, 4.1.5.1
　Unsatisfactory performance of, Commentary Table 1.1
Sprinkler wrenches, 5.4.1.6
Standard pressure proportioner
　Defined, A.3.3.27.5
Standby pumps, Table 12.2.2
Standpipe and hose systems, Chapter 6. *See also* hose
　Component replacement tables, 6.5
　Corrosion, 6.2.1

Defined, 3.3.31
Design pressure, A.6.3.1.1, 6.3.1.3
Flow tests, A.6.3.1.1, Exhibit 6.4
Hose and nozzle, Exhibit 6.1
Hose installation, 3.3.31.1.3
Hose pressure, 6.3.1.4 Case Study
Hydrostatic tests, 6.3.2.1
Impairment program and, A.15.5.2(3)(c)
Inspection, 6.1
Inspection frequency, 6.2.1
Installation date, 6.3.1.3
Maintenance, 6.4.1
Manual, A.6.3.1.1
Responsibility for, 6.1
Testing, 6.1, 6.3, A.13.5.2.2
Steam-driven fire pumps, A.8.3.3.1.2
Strainers
For obstructions, D.2.4
For private fire service mains, A.7.2.2.3, Exhibit 7.2
For water mist systems, Table 12.2.2
For water spray fixed sprinkler systems, 10.2.7.1
Suction gauges, 8.1.2(1)(c)
Suction line pressure, 8.2.2(2)(c)
Suction reservoir, 8.2.2(2)(e)
Suction water storage tanks, Chapter 9
Sulfate-reducing bacteria (SRB), D.2.6
Supervisory pressure, 13.4.3.1.4
Supervisory service
Notification of, 7.1.4
Supervisory switches, 13.3.3.5.2, Exhibit 13.13
Supply valves, 4.1.7
System design information sign, 12.1.4.2(4)
System line pressure, 8.2.2(2)(d)
System valves
Inspection, 13.4.1
For water mist systems, Table 12.2.2

T
Tachometers
Diesel fire pumps, 8.3.2.2(3)(b), Exhibit 8.20
Electric motor-driven fire pumps, 8.3.3.2(2)(b)
Temperature
Lowest one-day mean, 9.2.3, Exhibit 9.9
Test headers
Fire pumps, 8.1.2(2), Exhibit 8.6
Testing
Antifreeze sprinkler systems, 5.3.4, Exhibit 5.13
Backflow preventers, 13.6.2.1.4
Confined air atmosphere, 4.8.1
Control valves, A.13.3.1.2, 13.3.3.1
Diesel fire pumps, 8.3.1.3
Electrical components, 1.1.1
Electric motor-driven fire pumps, 8.3.1.2
Emergency generators, 8.3.1
Fire pumps, A.4.6.1.1.1, 5.3.3.4, 8.3.1, A.8.3.3.1.2

Foam concentrates, A.11.3
Forms for, 4.4.2
Frequency requirements, 4.6.1.1, A.4.6.1.1.1
Gauges, 5.3.2
Hose valves, 13.5.6.2
Hydrants, 7.3.2
Multiple systems, 13.4.3.2.7
Performance-based options, A.1.3, A.4.6.1.1.1
Preaction sprinkler systems, 13.2.6.1
Qualifications for, 4.1.2.2, Chapter 5
Quick-response (QR) sprinklers, 5.3.1.1.1.2
Records, A.4.4.1, 4.4.5
Responsibility for, A.4.1.2
Sprinklers, 5.3.1.1, 5.3.1.1.1
Sprinkler systems, Chapter 5
Standpipe and hose systems, 6.1, 6.3, A.13.5.2.2
Valves, 13.2.2
Water-based fire protection systems, 1.1, A.1.3, A.4.1.2, 4.1.2.1
Waterflow alarms, A.5.3.3.2, 13.2.6.1, Commentary Table 5.1
Water mist systems, 12.1.1.2, 12.1.6.2, A.12.2.4
Water quality, Table 12.2.2
Water spray fixed sprinkler systems, A.10.3.3, Chapter 10
Transfer switches, 8.2.2(3)(b), Exhibit 8.14
Trim
Defined, 13.1
Trip tests, 13.4.3.2.2.1, 13.4.4.2.2.2, A.13.4.4.2.2.2(9)

U
Ultra-high-speed water spray systems (UHSWSS), 10.4
Unheated spaces, 4.1.5.1
"Unloader" valves, Table 12.2.2
Upright sprinklers, 3.3.30.19

V
Vacuum induction, 11.2.9
Valve components
Defined, 13.1
Valves, Chapter 13. *See also specific types of valves*
Alarm devices, 13.2.2, 13.2.6.1, Exhibit 13.9
Clearance dimensions, A.13.2.3, Exhibit 13.5
Defined, 13.1
Inspection, 13.2.2
Main drain test and, 13.3.3.4
Operation and maintenance manuals, 13.2.1
Owner responsibilities, 13.2.1
Sprinkler system failure and, Chapter 13
Supervisory switches, 13.3.3.5.2, Exhibit 13.13
Testing, notifications required for, 13.2.2
Waterflow alarm tests, 13.2.6.1
Vane-type waterflow devices, 13.2.6.1, Exhibit 13.7
Ventilating louvers
Pump house, 8.2.2(1)(b), Exhibit 8.13

W

Wall post indicator valve (WPIV)
  Inoperable, 13.3.2, Exhibit 13.12
Warning signs
  Confined spaces, 4.8.1, Exhibit 4.2
Water-based fire protection systems
  Administrative requirements, Chapter 1
  Complete evaluation of, 1.1.2
  Inspection, testing, and maintenance requirements, 1.1, 4.1.2.1
  "Normal wear and tear" of, 1.1.2
  Performance-based inspection, testing, and maintenance, A.1.3
  Responsibility for inspection, testing, and maintenance, A.4.1.2
  Scope, 1.1
Waterflow alarms, 13.2.6.1, Exhibit 13.7
  Alarm valves and, A.13.1
  Supervisory service notification and, 7.1.4
  Testing, 13.2.6.1
  Testing frequency, A.5.3.3.2, Commentary Table 5.1
  Water storage tanks, 9.1.3
"Water Mist System Questionnaire," 12.3.11
Water mist systems, Chapter 12
  Additive storage cylinders, Table 12.2.2
  Additive systems, 12.2.1.2
  Algae and, Table 12.2.2
  As-built drawings and acceptance test records, 12.3.2
  Auxiliary functions, 12.1.6.2
  Components, 12.2.2
  Composite pump curve, Exhibit 12.4
  Compressed gas cylinders, Exhibit 12.1, Exhibit 12.2, Table 12.2.2
  Control concepts, Chapter 12
  Control equipment for, Table 12.2.2
  Cycling flow, Table 12.2.2
  Design, installation, operation, and maintenance (DIOM) manual, 12.1.1.3
  Flow-bypass ("unloader") valves for, Table 12.2.2
  Functionality testing, 12.1.6.2
  Hardware, Chapter 12
  High pressure cylinders, 12.2.6, Exhibit 12.5
  Impairment notification, 12.1.3.1
  Inspection, 12.1.1.2
  Maintenance, 12.3, 12.3.7
  Net positive suction head (NPSH), Table 12.2.2
  Owner/occupant responsibilities, 12.1.1.2
  Plant air for, Table 12.2.2
  Pneumatic valves for, Table 12.2.2
  Positive displacement pumps, Exhibit 12.3, Table 12.2.2
  Preventive maintenance, 12.3.2, Table 12.2.2
  Pumps and drivers for, Table 12.2.2
  Qualifications for repairing, 12.1.3.2
  Qualifications for servicing, 12.1.1.4
  Restoration after testing, A.12.2.4
  Shutdown notification, 12.1.2.1, 12.1.2.3
  Spare components cabinet, 12.3.9
  Specialized equipment for, 12.2.5
  Specialized training for, 12.4.1, 12.4.2
  Standby pumps for, Table 12.2.2
  System control valves, Table 12.2.2
  System design information sign, 12.1.4.2(4)
  Test connections, Table 12.2.2
  Testing, 12.1.1.2
  Water flow tests, Table 12.2.2
  Water quality testing, Table 12.2.2
  Water recirculation (break) tanks, Chaper 12.2.2
  Water storage cylinders, Table 12.2.2
  Water storage tanks, Table 12.2.2
  Water supplies, Table 12.2.2
Water motor gongs, A.5.3.3.2, 13.2.6.1, Exhibit 5.12
Water quality
  Inspection frequency and, 13.4.3.1.8
  Testing, for water mist systems, Table 12.2.2
Water recirculation (break) tanks, 12.2.2
Water spray fixed sprinkler systems, 3.6.5, Chapter 10
  Component replacement tables, 10.5
  Components, 10.1
  Conveyor belt protection, Exhibit 10.4
  Defined, Chapter 10
  Detection systems, 10.2.3, Chapter 10, Exhibit 10.7
  Discharge systems, Exhibit 10.1
  Discharge tests, 10.2.1.3, 10.3.4.1
  Dry pilot systems, 10.2.3, Exhibit 10.8
  Exposure to freezing, 5.2.5
  Flanged fittings, A.10.2.4.1
  Galvanized piping and fittings, A.10.2.4
  Grooved fittings, A.10.2.4.1
  Hazards protected by, 10.1, 10.1.4.1
  Horizontal chemical tank nozzle arrangements, Exhibit 10.2
  Inspection, 10.1.2, 10.2.1.1, 10.2.3, Chapter 10
  Mechanical damage, 10.2.4(2)
  Misalignment or trapped sections, 10.2.4(3)
  Monitor nozzles, 10.1.1
  Nonfunctioning, 10.1.4.1 Case Study
  Notification of shutdowns, 10.2.1.1
  Nozzles, A.10.2.5, 10.2.5.1, Exhibit 10.9, Exhibit 10.10, Exhibit 10.11
  Nozzle strainers, 10.2.1.6
  Pipe rack nozzle arrangement, Exhibit 10.3
  Piping and fittings, 10.1, A.10.2.4, 10.2.4.1
  Pressure readings, 10.3.4.4.1, 10.3.4.4.3
  Rubber-gasketed fittings, A.10.2.4.1, 10.2.4(5)
  Strainers, 10.2.7.1, Exhibit 10.6
  System components, 10.2.4
  System design, 10.1
  Testing, A.10.3.3, Chapter 10
  Transformer, Exhibit 10.5
  Ultra-high speed, 10.4
  Wet pilot systems, 10.2.3
Water storage cylinders, Table 12.2.2
Water storage tanks, Chapter 9

Alternate sources of water, A.9.1.3
Anti-vortex plates, 9.2.6.7, Exhibit 9.11
Applicability, 9.1
Automatic tank fill valves, 9.5, Exhibit 9.15
Break tanks, Chapter 9
Capacity, A.9.2.1.1
Component replacement tables, 9.6
Electronic water level indicators, 9.3(7), Exhibit 9.12
Electronic water temperature sensors, 9.3.3, Exhibit 9.13
Elevated tanks, A.9.1, Exhibit 9.2
Embankment-supported fabric suction tanks, A.9.1, Exhibit 9.8
Exterior inspection, 9.2.5.1, 9.2.5.2
Fiberglass-reinforces plastic tanks, Exhibit 9.5
Ground-level suction tanks, A.9.1, A.9.2.6.5, Exhibit 9.6
Heating systems for, 9.2.2
Interior inspection, 9.2.6, 9.2.6.1.2, 9.2.6.2, 9.27
Leaks, A.9.2.6.5
Low water level alarms, 9.3.5
Mercury gauges, 9.3(7)
Pressure tanks, 9.2.2.1, Chapter 9
Silt removal frequency, 9.4.5
Sizing, A.9.1
Steel ground-level tanks, A.9.1, Exhibit 9.1
Steel pressure tanks, A.9.1, Exhibit 9.7
Suction tanks, Chapter 9
Underwater obstruction, 9.2.6.2, Exhibit 9.10
Voids, 9.4.1
Waste materials in, 9.4.4, Exhibit 9.14
Water levels in, A.9.2.1.1, A.9.2.1.1 Case Study, 9.4.2, Table 12.2.2
For water mist systems, Table 12.2.2
Wooden, with space below roof, 9.4.4, Exhibit 9.4
Wooden gravity tanks, A.9.1, Exhibit 9.3, Exhibit 9.4
Water supplies
Acceptable options, 3.3.36

Impairment procedures following shutoff, 4.6.3
Problem identification, Exhibit 5.7
Temporary fire protection, A.15.5.2(3)(c)
For water mist systems, Table 12.2.2
Water tanks, 3.6.6
Weld coupons
Obstruction by, Exhibit 14.3, Exhibit D.2
Welding, A.15.5.2(3)(d)
Wet barrel (" California") hydrants, Exhibit 7.3
Inspection, 7.2.2.5
Wet pilot systems, 10.2.3
Wet-pipe sprinkler systems, Exhibit 3.23
Defined, A.3.6.4.5
Effectiveness of, A.1.2 Case Study
Exposure to freezing, 5.2.5
Inoperable OS&Y (outside screw and yoke) valve, 13.6.1.1.1 Case Study
Response to fire, 4.1.5 Case Study
Risers, Exhibit 13.1
Wet standpipe systems
Automatic, 3.3.31.3
Manual, 3.3.31.3
Pipe leaks, 6.3.2.1, 6.3.2.2.1
Water main break, 7.2.2.2 Case Study
Work gloves
Obstruction by, D.2.2, Exhibit 14.3, Exhibit D.3

Z
Zebra mussels, 14.2.1, 14.2.2, D.2.4, Exhibit 14.1, Exhibit D.4, Exhibit D.5
Distribution of (map), Exhibit 14.2, Exhibit D.6
Investigation frequency, 14.2.3.1
Treatment of, D.2.4
Zone control valves, Table 12.2.2

IMPORTANT NOTICES AND DISCLAIMERS CONCERNING NFPA® DOCUMENTS

NOTICE AND DISCLAIMERS OF LIABILITY CONCERNING THE USE OF NFPA DOCUMENTS

NFPA codes, standards, recommended practices, and guides, including the documents contained herein, are developed through a consensus standards development process approved by the American National Standards Institute. This process brings together volunteers representing varied viewpoints and interests to achieve consensus on fire and other safety issues. While the NFPA administers the process and establishes rules to promote fairness in the development of consensus, it does not independently test, evaluate, or verify the accuracy of any information or the soundness of any judgments contained in its codes and standards.

The NFPA disclaims liability for any personal injury, property or other damages of any nature whatsoever, whether special, indirect, consequential or compensatory, directly or indirectly resulting from the publication, use of, or reliance on these documents. The NFPA also makes no guaranty or warranty as to the accuracy or completeness of any information published herein.

In issuing and making these documents available, the NFPA is not undertaking to render professional or other services for or on behalf of any person or entity. Nor is the NFPA undertaking to perform any duty owed by any person or entity to someone else. Anyone using these documents should rely on his or her own independent judgment or, as appropriate, seek the advice of a competent professional in determining the exercise of reasonable care in any given circumstances.

The NFPA has no power, nor does it undertake, to police or enforce compliance with the contents of these documents. Nor does the NFPA list, certify, test or inspect products, designs, or installations for compliance with these documents. Any certification or other statement of compliance with the requirements of these documents shall not be attributable to the NFPA and is solely the responsibility of the certifier or maker of the statement.

ADDITIONAL NOTICES AND DISCLAIMERS

Updating of NFPA Documents

Users of NFPA codes, standards, recommended practices, and guides should be aware that these documents may be superseded at any time by the issuance of new editions or may be amended from time to time through the issuance of Tentative Interim Amendments. An official NFPA document at any point in time consists of the current edition of the document together with any Tentative Interim Amendments and any Errata then in effect. In order to determine whether a given document is the current edition and whether it has been amended through the issuance of Tentative Interim Amendments or corrected through the issuance of Errata, consult appropriate NFPA publications such as the National Fire Codes® Subscription Service, visit the NFPA website at www.nfpa.org, or contact the NFPA at the address listed below.

Interpretations of NFPA Documents

A statement, written or oral, that is not processed in accordance with Section 6 of the Regulations Governing Committee Projects shall not be considered the official position of NFPA or any of its Committees and shall not be considered to be, nor be relied upon as, a Formal Interpretation.

Patents

The NFPA does not take any position with respect to the validity of any patent rights asserted in connection with any items which are mentioned in or are the subject of NFPA codes, standards, recommended practices, and guides, and the NFPA disclaims liability for the infringement of any patent resulting from the use of or reliance on these documents. Users of these documents are expressly advised that determination of the validity of any such patent rights, and the risk of infringement of such rights, is entirely their own responsibility.

NFPA adheres to applicable policies of the American National Standards Institute with respect to patents. For further information, contact the NFPA at the address listed below.

Law and Regulations

Users of these documents should consult applicable federal, state, and local laws and regulations. NFPA does not, by the publication of its codes, standards, recommended practices, and guides, intend to urge action that is not in compliance with applicable laws, and these documents may not be construed as doing so.

Copyrights

The documents contained in this volume are copyrighted by the NFPA. They are made available for a wide variety of both public and private uses. These include both use, by reference, in laws and regulations, and use in private self-regulation, standardization, and the promotion of safe practices and methods. By making these documents available for use and adoption by public authorities and private users, NFPA does not waive any rights in copyright to these documents.

Use of NFPA documents for regulatory purposes should be accomplished through adoption by reference. The term "adoption by reference" means the citing of title, edition, and publishing information only. Any deletions, additions, and changes desired by the adopting authority should be noted separately in the adopting instrument. In order to assist NFPA in following the uses made of its documents, adopting authorities are requested to notify the NFPA (Attention: Secretary, Standards Council) in writing of such use. For technical assistance and questions concerning adoption of NFPA documents, contact NFPA at the address below.

For Further Information

All questions or other communications relating to NFPA codes, standards, recommended practices, and guides and all requests for information on NFPA procedures governing its codes and standards development process, including information on the procedures for requesting Formal Interpretations, for proposing Tentative Interim Amendments, and for proposing revisions to NFPA documents during regular revision cycles, should be sent to NFPA headquarters, addressed to the attention of the Secretary, Standards Council, NFPA, 1 Batterymarch Park, Quincy, MA 02169-9101.

For more information about NFPA, visit the NFPA website at www.nfpa.org.

A Guide to Using the Water-Based Fire Protection Systems Handbook

This second edition of the *Water-Based Fire Protection Systems Handbook* contains the complete text of the 2008 edition of NFPA 25, *Standard for the Inspection, Testing, and Maintenance of Water-Based Fire Protection Systems*.

The commentary text contains Frequently Asked Questions shown in the margin as FAQs. The FAQs answer the most commonly asked questions throughout the handbook.

78 Chapter 5 • Sprinkler Systems

of failure for these sprinklers are possible, such as corrosion around the seat, contamination, and loading.

5.3.1.1.1.3* Representative samples of solder-type sprinklers with a temperature classification of extra high 325°F (163°C) or greater that are exposed to semicontinuous to continuous maximum allowable ambient temperature conditions shall be tested at 5-year intervals.

A.5.3.1.1.1.3 Due to solder migration caused by the high temperatures to which these devices are exposed, it is important to test them every 5 years. Because of this phenomenon, the operating temperature can vary over a wide range.

5.3.1.1.1.4 Where sprinklers have been in service for 75 years, they shall be replaced or representative samples from one or more sample areas shall be submitted to a recognized testing laboratory acceptable to the authority having jurisdiction for field service testing. Test procedures shall be repeated at 5-year intervals.

5.3.1.1.1.5* Dry sprinklers that have been in service for 10 years shall be replaced, or representative samples shall be tested. They shall be retested at 10-year intervals.

FAQ ▶
Do dry sprinklers have to be tested or can they be replaced?

Dry sprinklers have experienced a much higher failure rate than standard sprinklers and are more susceptible to corrosion both internally (when moisture condenses inside the device) and externally. Corrosion at the water seal and at the weep hole at the bottom of the sprinkler has been reported in sprinklers older than 10 years. These sprinklers are usually installed in harsh environments, further compounding this problem.

The failure rate of dry-type sprinklers installed for 10 years is approximately 50 percent. Because of this high failure rate, it is very important that these sprinklers be identified and tested as required. Dry-type sprinklers are custom made in exact lengths, and therefore NFPA 13 does not require spare dry-type sprinklers in the spare sprinkler cabinet. The inspector has to look for typical applications for dry-type sprinklers and perform a close inspection of a sampling of the sprinklers themselves to determine the date on the sprinklers.

The 10-year threshold requirement noted in 5.3.1.1.1.5 was added to NFPA 25 in the 2002 edition because all of the conditions that cause failure are not well understood, and the frequency of failure is higher for sprinklers that have been in service for more than 10 years. In cases where only a few dry sprinklers are installed or where corrosion is noted, it may be more cost-effective to replace the sprinklers rather than test them.

Paragraph 5.3.1.1.1.5 is referring to the listed dry-type sprinkler. This paragraph does not apply to standard spray sprinklers installed on a dry pipe system.

Does 5.3.1.1.1.5 apply to listed dry sprinklers, or does it apply to all sprinklers installed in a dry pipe system?
FAQ ▶

A.5.3.1.1.1.5 See 3.3.30.3.

5.3.1.1.2* Where sprinklers are subjected to harsh environments, including corrosive atmospheres and corrosive water supplies, on a 5-year basis, sprinklers shall either be replaced or representative sprinkler samples shall be tested.

A.5.3.1.1.2 Examples of these environments are paper mills, packing houses, tanneries, alkali plants, organic fertilizer plants, foundries, forge shops, fumigation areas, pickle and vinegar works, stables, storage battery rooms, electroplating rooms, galvanizing rooms, steam rooms of all descriptions including moist vapor dry kilns, salt storage rooms, locomotive sheds or houses, driveways, areas exposed to outside weather, around bleaching equipment in flour mills, all portions of cold storage areas, and portions of any area where corrosive vapors prevail. Harsh water environments include water supplies that are chemically reactive.

Are walk-in refrigerators and walk-in freezers considered a "harsh environment" and therefore subject to this standard?
FAQ ▶

Paragraph A.5.3.1.1.2 provides examples of areas that are considered to be harsh environments and includes "cold storage area." While the annex does not further define the description of a cold storage area, a walk-in refrigerator or walk-in freezer could fall into this category.

2008 Water-Based Fire Protection Systems Handbook

6.4 Maintenance

Maintenance and repairs shall be in accordance with 6.2.3 and Table 6.2.2.

6.4.1 Equipment that does not pass the inspection or testing requirements shall be repaired and tested again or replaced.

This requirement relates to hose only. Following operation, hose must be cleaned, dried, and inspected, as specified in 6.4.1, prior to being returned to service. If not dried, wet hose can deteriorate (especially unlined hose) and fail the next time it is used. Hose that is cut or otherwise damaged during use can also fail during subsequent use. Hose that cannot be repaired should be replaced.

6.5 Component Action Requirements

Component replacement tables were added in the 2008 edition to offer guidance to the user of the standard when system components are adjusted, repaired, rebuilt, or replaced. It is not necessary in each case to require a complete acceptance test for each component when maintenance is performed.

6.5.1 Whenever a component in a standpipe and hose system is adjusted, repaired, reconditioned or replaced, the action required in Table 6.5.1, Summary of Component Replacement Action Requirements, shall be performed.

6.5.1.1 Where the original installation standard is different from the cited standard, the use of the appropriate installing standard shall be permitted.

6.5.1.2 A main drain test shall be required if the control valve or other upstream valve was operated in accordance with 13.3.3.4.

6.5.1.3 These actions shall not require a design review, which is outside the scope of this standard.

SUMMARY

Chapter 6 of NFPA 25 covers the inspection, testing, and maintenance of all types of standpipe systems covered in NFPA 14, including Class I standpipes (for fire department use), Class II standpipes (for building occupant use), and Class III standpipes (intended for use by both fire departments and building occupants). Failure of this equipment can create an unsafe condition for the user; therefore, it is extremely important to verify the operation of a standpipe system.

REVIEW QUESTIONS

1. When are hydrostatic tests required on a standpipe system?
2. What design pressure and required flow must be measured during the five-year flow test?
3. When conducting an air test of a standpipe system, how much air pressure should be used?

REFERENCE CITED IN COMMENTARY

National Fire Protection Association, 1 Batterymarch Park, Quincy, MA 02169-7471.
NFPA 14, *Standard for the Installation of Standpipe and Hose Systems*, 2007 edition.

2008 Water-Based Fire Protection Systems Handbook

This handbook contains a Summary and Review Questions located at the end of each chapter. The Summary provides a synopsis of the chapter content and the Review Questions reinforce the material covered in the chapter.